Microbial Products for Health and Nutrition

Vijay Kothari • Subhasree Ray • Prasun Kumar
Editors

Microbial Products for Health and Nutrition

Springer

Editors
Vijay Kothari
Institute of Science
Nirma University
Ahmedabad, Gujarat, India

Subhasree Ray
Department of Life Sciences
Sharda University
Greater Noida, Uttar Pradesh, India

Prasun Kumar
Department of Biotechnology
Sharda University
Greater Noida, Uttar Pradesh, India

ISBN 978-981-97-4234-9 ISBN 978-981-97-4235-6 (eBook)
https://doi.org/10.1007/978-981-97-4235-6

This Springer imprint is published by the registered company Springer Nature Singapore Pte Ltd.
The registered company address is: 152 Beach Road, #21-01/04 Gateway East, Singapore 189721, Singapore

Dedicated to our families and teachers

Preface

Microorganisms, the invisible heroes of our ecosystem, have long been recognized for their diverse roles in human health and nutrition. From the fermenting microbes that transform simple ingredients into flavorful foods to the production of life-saving pharmaceuticals, the potential of these tiny organisms is vast and ever-expanding. In the realm of health and nutrition, the utilization of microbial products has emerged as a promising frontier, offering innovative solutions to address various challenges.

This book, *Microbial Products for Health and Nutrition*, delves into the multifaceted world of microbial applications in promoting well-being and enhancing nutritional outcomes. Through a comprehensive exploration of diverse topics, ranging from biopharmaceutical production to the use of microbial pigments in food, this volume aims to provide a holistic understanding of the potential and significance of microbial products in shaping the future of health and nutrition.

The journey begins with an in-depth exploration of phage display technology, offering insights into its pivotal role in recombinant monoclonal antibody production. From there, the focus shifts to the utilization of algae and cyanobacteria as food supplements, highlighting their nutritional benefits and potential as sustainable sources of essential nutrients. Moving forward, the book delves into the realm of oral healthcare, examining the emerging trends in probiotic bacteriotherapeutic approaches. It then explores the diagnostic potential of microbial enzymes, paving the way for innovative strategies in disease detection and management. As the chapters unfold, readers are introduced to a diverse array of topics, including the production of organic acids, microbial enzymes, and fat-soluble vitamins. The role of microorganisms in the production of non-alcoholic beverages, bakery products, and dairy industry staples is also explored, shedding light on their contributions to the food and beverage sector.

One of the highlights of this volume is the exploration of microbial pigments as potential substitutes for synthetic colorants, offering safer and more sustainable alternatives for the food and healthcare industries. Additionally, the book delves into the production of amino acids, water-soluble vitamins, and antibiotics, showcasing the versatility of microbial products in meeting diverse health and nutritional needs.

Throughout these pages, readers will gain valuable insights into the historical perspectives, contemporary impacts, and future prospects of microbial applications in health and nutrition. From food preservation to biopharmaceutical production, the potential of microbial products to revolutionize our approach to health and wellness is undeniable. As editors of this volume, it is our hope that this book will serve as a valuable resource for researchers, educators, and practitioners alike, inspiring further exploration and innovation in the field of microbial products for health and nutrition. By harnessing the power of microorganisms, we can unlock new possibilities for promoting human health and enhancing nutritional well-being.

Together, let us embark on a journey into the fascinating world of microbial products and discover the endless possibilities they hold for a healthier, more sustainable future.

Ahmedabad, Gujarat, India — Vijay Kothari
Greater Noida, Uttar Pradesh, India — Subhasree Ray
Greater Noida, Uttar Pradesh, India — Prasun Kumar

Contents

About the Editors

Vijay Kothari, Ph.D. is a microbiologist working in India. He holds two decades of experience in teaching and research. He has >100 publications and >20 conference talks to his credit. His publications have enjoyed wide readership across the globe (>277,400 reads from ResearchGate alone). Many of his articles have been ranked in the top 25% among all those ranked by Altmetric. His research interests include Antimicrobial Resistance (AMR), Traditional Medicine (antipathogenic, neuroprotective, prebiotic properties), and Microbial Response to Sonic Stimulation. His lab has identified multiple potential targets in pathogenic bacteria, and currently working on their validation. Vijay is a recipient of AIMS award (Artificial Intelligence Molecular Screen-2019) from Atomwise Inc., USA; Sentinel of Science (2016) from Publons; Future Leaders Mentoring Fellowship from the American Society for Microbiology (ASM—2023–24). He is a member of 25 different professional bodies. Vijay derives great satisfaction from the M.Sc. Dissertation projects guided by him, which hitherto has enabled 73 M.Sc. students to author one or more papers in indexed journals.

Subhasree Ray, Ph.D. is currently working as an Assistant professor at Sharda University, Greater Noida, Uttar Pradesh, India. She earned her Ph.D. degree from CSIR-IGIB, Delhi in 2018. She received the prestigious CSIR-SRF fellowship. Her main research work was focused on the production of biopolymers from waste biomass. After Ph.D., she joined as a postdoctoral researcher at Ewha University, and the University of Seoul, South Korea. Here, her main focus was anaerobic digestion of food wastes for methane production. She also studied methanogenesis at 4000 L pilot-scale plant. After successful completion of 1 year, she joined another project at Yeungnam University, South Korea. During that period, she worked on several fungal toxins and their inhibition from fermented food. She also worked on biofilm inhibition of pathogenic organisms by natural bioactive compounds. To her credit, she has 24 research papers published in peer-reviewed journals, 8 book chapters, and 2 books. In addition, she is a life member of various scientific societies and also a member of various committees at the Sharda University for graduate and undergraduate programs.

Prasun Kumar holds a Ph.D. in Biotechnology from CSIR-Institute of Genomics and Integrative Biology, Delhi, India. He is presently working as an Assistant Professor at Department of Biotechnology, Sharda University, Greater Noida, Uttar Pradesh, India. Earlier, he was working as an Assistant Professor at the Department of Chemical Engineering, Yeungnam University, Republic of Korea. He has over 7 years of experience in Applied Microbiological Research including about 2 years of experience in industrial R&D. His main areas of research are bioinformatics, biopolymers, microbial biodiversity, bioenergy, microbial biofilms, quorum sensing, quorum quenching, and genomics. His present research is oriented toward valorizing lignocellulosic biowastes into value-added products such as biopolymer, 2G ethanol, bioenergy, and antibiofilm compounds. To his credit, there are over 34 articles in SCI journals, 5 books, and 14 chapters with international publishers. He has been serving the scientific society by reviewing articles for several SCI journals and delivering guest lectures. Publons awarded him the peer-reviewed award in the year 2018. He also serves as the editorial board member of a few international journals.

Chapter 1
Phage Display Technology: A Way Forward for Production of Recombinant Monoclonal Antibodies

Ajay Patel, Manisha Parmar, Payal Patel, Aditee Pandya, and Haren Gosai

Abstract Monoclonal antibodies (mAbs) have become an indispensable part of the healthcare sector having various applications in diagnosis, therapeutic and clinical research. For the purpose of sustaining a continuous supply of highly specific and fully human mAbs, scientists have developed several cutting-edge techniques in the last two decades. Among them, phage display technology is one of the techniques that has been on the rise and may become the go-to method for the production of human mAbs in the near future. Phage display is an effective tool to engineer desirable human mAbs in vitro by genetically manipulating bacteriophages to synthesize mAbs on their surface. This chapter delves into the history, breakthroughs and shortcomings of mAbs' production and explores both past and current scenarios in depth. Previous mAbs of chimeric or humanized origins have also been discussed along with the need for fully human mAbs. The chapter will shed light on the exact mechanism behind the synthesis of mAbs using phage display technology. Applications of human mAbs in the healthcare sector have been summarized to offer a better understanding. In the interest of providing a well-rounded opinion on phage display technology, the authors have compared it with other methods of mAbs' production, its drawbacks and its potential impact in future.

Keywords Monoclonal antibody · Phage display technology · Antibody library · Bacteriophage

---

A. Patel · M. Parmar · H. Gosai (✉)
Department of Biosciences, School of Science, Indrashil University, Kadi, Gujarat, India
e-mail: harengiri.gosai@indrashiluniversity.edu.in

P. Patel
Department of Biosciences, School of Science, Indrashil University, Kadi, Gujarat, India

Department of Biotechnology, Bhagwan Mahavir College of Basic and Applied Science, Bhagwan Mahavir University, Surat, Gujarat, India

A. Pandya
Department of Microbiology, School of Sciences, P P Savani University, Kosamba, Gujarat, India

V. Kothari et al. (eds.), *Microbial Products for Health and Nutrition*,
https://doi.org/10.1007/978-981-97-4235-6_1

1.1 Introduction

Monoclonal antibodies (mAbs) have transformed the healthcare industry and modernized the way we treat diseases. They have proven to be indispensable in numerous aspects of medicine, including diagnostics, therapeutics and clinical research. The attractiveness of antibodies lies in their capability to precisely attach to targets and bind to a diverse range of antigens. mAbs have accelerated the development of innovative treatments, enhanced diagnostics and cutting-edge research methods since their discovery (Singh et al. 2018). However, immunogenicity and side effects related to non-human components have been a perpetual challenge. Moreover, the global scientific community and pharmaceutical industries are regularly confronted by the ever-increasing demand for highly specific and entirely human mAbs (Hansel et al. 2010). To overcome this hurdle, researchers have pursued innovative techniques over the past two decades in an attempt to elevate mAbs' production both qualitatively and quantitatively (Nelson et al. 2010). Among these innovative methods, phage display technology stands out as a promising approach, potentially becoming the cornerstone for producing human mAbs in the near future (Hammers and Stanley 2014; Kumar et al. 2019; Alfaleh et al. 2020).

The journey of mAbs as therapeutics is marked by remarkable progress and continued evolution. In 1975, Köhler and Milstein used the hybridoma technique to generate mAbs. Those early mAbs commonly came from murine origins (Köhler and Milstein 1975). This led to the production of mAbs with a significant proportion of non-human sequences. This limitation raised concerns about immunogenicity and potential adverse effects in humans (Dillman et al. 1986; Sgro 1995). Later on, chimeric mAbs were developed, which fused human constant sections with mice variable areas, greatly decreasing immunogenicity without sacrificing antigen specificity (Stasi et al. 2001; Bleumer et al. 2004; Galizia et al. 2007). Building upon this foundation, humanized mAbs were introduced. This further reduced the number of non-human components by exclusively using human frames for incorporating the complementarity-determining regions (CDRs) of mice antibodies (Oliphant et al. 2005; Ionescu et al. 2008). Despite these advancements, a certain degree of adverse effects was still followed in a considerable number of patients due to murine fractions still present in the mAbs (Aarden et al. 2008; West et al. 2008; Knezevic et al. 2015; Doevendans and Schellekens 2019). Hence, a growing consensus within the scientific and medical communities emphasized the need for fully human mAbs, without any foreign elements, to maximize efficacy and minimize potential side effects. The entire timeline of advancements made in mAbs production has been graphically visualized in Fig. 1.1.

In 1985, Smith made a breakthrough when they invented phage display technology, which allowed scientists to avoid the immunogenicity problem and create entirely human mAbs (Smith 1985). This discovery has since then revolutionized the production of mAbs. It utilizes bacteriophages to showcase numerous antibody fragments on their surfaces, which can be selected and developed into full mAbs. Combining this technology with other methods, such as humanizing antibodies,

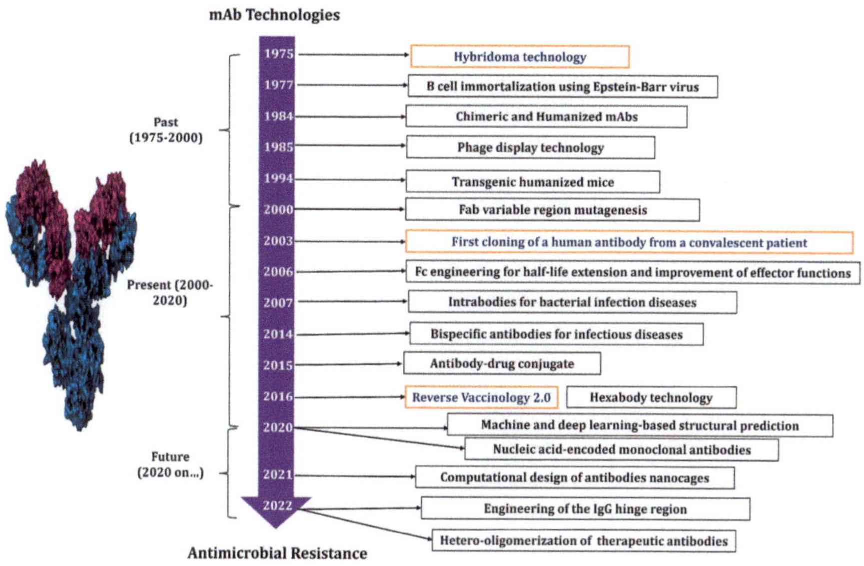

Fig. 1.1 Timeline of mAb technologies from their inception to their current application in the treatment of infectious diseases. Exhibit some promising new technologies that may soon be utilized in the battle against antimicrobial resistance (AMR). Orange boxes show the goals that have been reached in the past, present and in future, respectively

transgenic immunized mice and individual B-cell expression by genetic engineering, allows for the creation of therapeutic antibodies for humans (Aghebati-Maleki et al. 2016; Frenzel et al. 2016; Hess and Jewell 2020). As a result of its many advantages, phage display technology is rapidly becoming the gold standard in antibody engineering and development. This has enabled scientists to efficiently generate highly specific and fully human mAbs on a large scale while overcoming limitations from earlier methods (Frei and Lai 2016; Alfaleh et al. 2020). Additionally, this technology allows for the selection and isolation of antibody fragments from a library of phages that bind to specific targets. It also allows the generation of natural, artificial and immunized phage antibody libraries (Chen and Sidhu 2014; Kügler et al. 2015; Nilvebrant and Sidhu 2018). mAbs can be easily produced in vitro using this technology, making it a comparatively straightforward method for industrial-scale production. At present, over 70 mAbs produced from phages have been subjected to clinical research, resulting in 14 approvals as of May 2020 (Alfaleh et al. 2020).

The utilization of phage display technology presents an attractive and promising alternative to conventional approaches employed in the synthesis of mAbs. Its primary advantage lies in the ability to create phage display libraries of recombinant antibodies from non-mammalian animals (Nishibori et al. 2006). Specifically, immunized chicken-derived phage present an attractive substitute for costly and unreliable polyreactive antibodies. The commercial chicken, *Gallus gallus* serves as a cost-effective and technologically feasible source of excellent mAbs, making it well suited for phage display libraries (Pitaksajjakul et al. 2010; Hu et al. 2012; Li et al. 2016). Key advantages of phage display technology in mAbs' production encompass obtaining high-quality antibodies from non-mammalian animals, targeting diverse epitopes with ease and acquiring antibodies against a broad spectrum of antigens. Furthermore, phage display technology has the capability to extract antibodies specifically targeting self-antigens, hazardous substances, unstable compounds and non-immunogenic antigens by carefully selecting from a library of antibodies (Ledsgaard et al. 2018; Kumar et al. 2019; Sioud 2019).

Phage display technology offers numerous advantages in mAbs' production over the traditional methods that were previously used (Frenzel et al. 2017; Nagano and Tsutsumi 2021; Ledsgaard et al. 2022). Some of these advantages are as follows:

- Extensive antibody libraries: By using this approach, large and varied antibody libraries with billions of variants can be generated, which greatly aids in the search for highly selective and affinity-based mAbs.
- Fully human antibodies: Phage display generates completely human mAbs, reducing immunogenicity and enhancing safety for therapeutic use.
- Swift antibody discovery: The procedure accelerates the process of antibody discovery, enabling swift identification of antibodies that specifically target a certain substance. This capability is crucial for effectively combating newly emerging diseases.
- High affinity and specificity: Multiple rounds of affinity selection yield mAbs with superior binding properties to target antigens.
- Engineering flexibility: Phage display provides versatility in antibody engineering to optimize antibodies for specific clinical requirements.
- Reduced immunogenicity: By generating fully human antibodies, the technology minimizes immunogenicity risks and potential side effects, prioritizing safety in therapy development.
- Diagnostic and research tools: Antibodies derived from phage display are invaluable in disease diagnostics and research, assisting in biomolecule detection and the understanding of biological processes.
- Cost-efficiency: Although initial set-up costs may be high, phage display can result in long-term savings by streamlining antibody development and research.
- Complementary technology: Phage display can be used alongside traditional methods, combining strengths to achieve the best outcomes in mAbs production.

Here, the authors discuss the origins, recent breakthroughs and prospective future applications of mAbs made with this unique technique. This chapter examines mAb manufacturing's advancements, and drawbacks, both past and present. Discussed include chimeric or humanized mAbs and the necessity for completely human ones. The phage display-based mAb manufacturing mechanism will be discussed in this chapter. Medical uses of human mAbs have been simplified for clarity. To give a complete picture of phage display technology, the author compared it to alternative mAb production methods, their drawbacks and potential future effects.

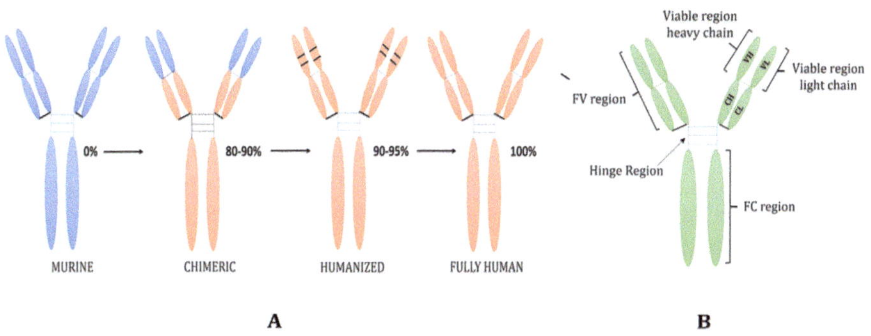

Fig. 1.2 Structure of monoclonal antibody (**a**) monoclonal antibody humanization and (**b**) overall structure of the monoclonal antibody

1.2 Chimeric, Humanized and Fully Human mAbs

Before fully human mAbs became the norm, mAbs with varying degrees of non-human components were used, namely chimeric and humanized mAbs. Significant progress made in protein sequencing has enabled the creation of antibodies using both murine and human sources. Researchers in the field of antibody studies have devoted considerable efforts to enhancing the therapeutic applications of these antibodies in humans, leading to the development of chimeric, humanized and fully human antibodies (Kalayu Yirga et al. 2019). Figure 1.2 shows the different types of mAbs, including murine, chimeric and humanized antibodies.

1.2.1 Chimeric mAbs

Chimera refers to a hybrid creation composed of different animal parts. Here also, chimeric antibodies are created by merging the variable and constant regions of two different types of antibodies. Such antibodies can be either murine or human. The human antibody section possesses well-known sequences for N-linked glycosylation, and the initial amino acid at the N-terminus undergoes cyclization to become pyroglutamic acid (Si et al. 2021). Genetic engineering techniques are employed to synthesize chimeric antibodies, consisting of approximately 65% human regions and 35% murine origin. mAbs of chimeric origin are identified by the suffix "-ximab" in their nomenclature, such as Infliximab, Dinutuximab, Rituximab and Cetuximab (Mayrhofer and Kunert 2018; Damián-blanco et al. 2023). The first chimeric mAb, Rituximab or Rituxan, was authorized for use as a cancer treatment by the FDA in 1997. Rituximab specifically targets CD20, a surface marker present on both normal B cells and the majority of B-cell neoplasms. CD20 expression is detected across all stages of B-cell development, ranging through the pre-B-cell phase to their transformation into plasma cells (Ribatti 2014). Following Rituximab's approval by the FDA, two more chimeric antibodies have been launched for tumour treatment: Dinutuximab and Cetuximab.

The production of cetuximab involved infecting mice with a purified form of the epidermal growth factor receptor (EGFR), collecting their antibodies and then exchanging the constant domain of the receptor for an IgG1 domain derived from humans. The resulting chimeric molecule, C225, showed substantially greater tumour growth inhibition and approximately fivefold higher affinity than the parental mouse antibody (Goldstein et al. 1995; Almagro et al. 2018). Dinutuximab was developed utilizing the GD2 antibody initially discovered in mice (Gillies et al. 1989). While testing with human effector cells, the resultant chimeric molecule, ch14.18, which is an IgG1 antibody produced by humans, showed a significantly higher level of antibody-dependent cell-mediated cytotoxicity (ADCC) compared to the original antibody produced by mice, IgG2a. In fact, the ADCC activity of ch14.18 was found to be 50–100 times more potent than that of the IgG2a antibody (Almagro et al. 2018). While early mouse mAbs did have significant problems, chimerization solved some of those problems. The process led to the development of medicinal compounds with enhanced effector capabilities and binding affinities that were comparable to or even higher than those of the original mouse antibodies. This development was crucial in elevating monoclonal antibody treatments' general safety and efficacy.

1.2.2 Humanized mAbs

There was a need to create humanized mAbs to mitigate the negative impacts of chimeric mAbs containing non-human residues. These antibodies comprise around 95% human-origin components and 5% mouse-origin components (specifically, the complementarity determining region) (Kalayu Yirga et al. 2019). Humanization of antibodies involves the fusion of protein sequences to convert antibodies originally developed in mice into humanized forms (Hu and Nagata 2016). Due to their altered molecular structure, humanized antibodies are particularly valuable in therapy applications that require minimal immunogenicity (Damián-blanco et al. 2023). These humanized mAbs are typically designated by the suffix "-zumab". Some common examples of which include Bevacizumab, Tanibirumab, Alemtuzumab, Pembrolizumab and Necitumumab (Garnock-Jones 2016; Mayrhofer and Kunert 2018; Pelletier and Mukhtar 2020). The initial significant milestone in this field was achieved in 1998 when the FDA granted permission for Trastuzumab (Herceptin) to be used in the treatment of metastatic breast cancer that overexpresses the Her-2 protein (Wilson et al. 2017). The prognosis for breast cancer patients is worse when Her-2 is overexpressed. Trastuzumab consists of a humanized IgG1 antibody that specifically attaches to the extracellular region that contains the human Her-2/neu receptor, which is a part of the EGFR class (Ribatti 2014; Maximiano et al. 2016). Trastuzumab stimulates the synthesis of anti-angiogenic factors, inhibits the expression of pro-angiogenic factors and mediates antibody-dependent cytotoxicity

(Ribatti 2014). Further down the line, numerous humanized antibodies have been synthesized and integrated into the healthcare system. Pembrolizumab, Atezolizumab, Obinutuzumab and Elotuzumab are some of the humanized mAbs that have been used in cancer treatment (Bajaj et al. 2019). These antibodies were developed by inserting non-human complementarity-determining regions into human framework regions with strategically planned backmutations that substitute human remains with the mice counterparts. By using this method, the original non-human antibodies' affinities can be preserved or even improved (Almagro et al. 2018).

1.2.3 Fully Human Antibodies

In order to minimize the immune response triggered by chimeric and humanized mAbs, scientists have been working to create a method for manufacturing human-derived mAbs. A variety of techniques, including phage display and transchromosome mice technology, have been developed to express human Ig genes in mice in order to generate human mAbs. Monoclonal antibodies are designated with names that conclude with the suffix -umab, such as daratumumab, ipilimumab, ofatumumab, tremelimumab and denosumab (Mayrhofer and Kunert 2018; Ashraf 2019; Gonzales Carazas et al. 2021).

1.3 Production of mAbs Using Phage Display Technology

The use of phage display technology to produce mAbs is an incredibly effective and advanced technique that has completely transformed the way antibodies are produced. Identification, selection and creation of antibodies with remarkable affinity and specificity for a desired antigen are all outcomes of this procedure, which comprise multiple clearly defined steps. In this section, we will explore the complexities of each stage, emphasizing the scientific concepts and key discoveries that support this innovative technique.

1.3.1 Antibody Library Creation

The process of phage display-based mAb production begins with the construction of a diverse antibody library. This library serves as the basis for identifying potential mAb candidates. The term "library" in this context refers to a collection of bacteriophages, typically M13, each with a unique antibody variant on its surface. These antibody variants originate from a wide range of organisms, including human donors and other species such as mice (Almagro et al. 2019). The genetic material encoding these antibody variants is cleverly integrated into the bacteriophage genome. This allows displaying a vast repertoire of antibody sequences on the phage surface, creating a diverse and representative pool of potential binders. Phage display libraries generated in this way are an invaluable resource for the screening and selection of antibodies with specific binding properties. Phage display libraries can be designed to encompass an astonishing number of antibody variants, numbering in the billions (Sioud 2019). This diversity is a fundamental advantage when searching for mAbs with the highest specificity and affinity for a particular target antigen. The library's diversity ensures that even rare and highly specific antibodies can be discovered, making it an essential tool in antibody research (Zhang 2023).

1.3.2 Selection of Phage Display Antibodies (Panning)

The next significant phase of phage display technology was the selection of phages (often referred to as "panning") that display antibodies that attach specifically to the antigen of interest. Panning is a process that mimics the natural selection of antibodies in an in vitro environment (Frenzel et al. 2017). Panning is the process of exposing a library of phage displays to a specific antigen. In the context of therapeutic mAb development, this antigen could be a specific protein associated with the disease, a peptide epitope or even the surface of diseased cells. Phage displaying antibodies with the desired specificity bind with the chosen antigen. Phages that exhibit binding affinity are retained, whereas phages that lack binding capability are eliminated through washing. This iterative process is typically repeated with multiple pivots to improve selection accuracy (Shim 2017). As panning progresses, the pool of retained phages becomes enriched with phages that have antibodies exhibiting greater affinity and recognition towards the desired antigen. The process effectively isolates and enriches potential mAb candidates with desirable binding properties (Alfaleh et al. 2017). Now, the entire process of screening, selection and identification can be completed in less than 2 weeks (Steinke et al. 2023).

1.3.3 Amplification and Analysis

After the panning process, the selected phages are amplified using a bacterial host. This amplification step generates sufficient quantities of phages for further analysis and characterization (Aghebati-Maleki et al. 2016). The process begins once enough phages have been produced, the next key step is to extract and sequence the genetic material that encodes the specific antibody fragments (often called antigen-binding fragments, or Fabs) that appear on the phage's surface (Frei and Lai 2016; Shim 2017). Sequence information is important to identify the exact antibody sequence that showed binding to the target antigen. Modern sequencing technologies, like next-generation sequencing (NGS), have greatly facilitated this step, allowing rapid, high-throughput identification of antibody sequences (Yang et al. 2017). The influence of NGS depends on whether it can sequence thousands or even millions of antibody fragments, providing researchers with a comprehensive dataset to search for potential mAbs.

1.3.4 Conversion of Full-Length Antibodies

Once the specific antibody sequences are complete, the next step is to convert these antigen-binding fragments into full-length antibodies (Ledsgaard et al. 2018). Although full-length antibodies (Fabs) displayed on the phage surface are useful for screening and selection, therapeutic mAbs typically need to contain a constant region (Fc region) to confer stability and effector function. This conversion process is achieved through genetic engineering techniques. Researchers clone identified antibody sequences into expression vectors and ensure that they are properly fused to their corresponding constant regions. These expression vectors are then introduced into a suitable host system for antibody production (Hentrich et al. 2018; Dal Ferro et al. 2019).

1.3.5 Expression and Purification

Engineered host systems, often mammalian cells (e.g. Chinese hamster ovary cells or CHO cells) or bacteria (e.g. *Escherichia coli*), serve as factories for antibody production (Jaroszewicz et al. 2022). The introduced antibody genes are transcribed and translated, resulting in the synthesis of full length antibodies. After expression, the antibody is purified to yield a highly concentrated and purified mAb product. Purification processes typically use chromatographic techniques, such as protein A or protein G affinity chromatography, to take advantage of the specific interactions of antibodies with these proteins (Oślizło et al. 2011). The aim is to obtain high-purity and high-quality mAbs for further characterization and downstream applications such as cancer treatment (Sales-Dias et al. 2021).

1.3.6 Additional Technical Measures

In some cases, additional technical measures may be taken to optimize mAb for a specific clinical use. These engineering strategies include affinity maturation to increase binding affinity for target antigens, glycoengineering to alter glycosylation patterns to improve therapeutic efficacy or to tailor-make effector functions to desired outcomes (Almagro et al. 2019). The availability of advanced molecular biology techniques and detailed structural information about antibodies have made such optimization increasingly possible. These customized modifications significantly improve the therapeutic potential of mAbs and expand their range of applications.

In summary, the synthesis of mAbs utilizing the phage display method is a meticulously coordinated process that integrates various antibody libraries, rigorous selection methods, genetic engineering and optional optimization steps. This method has become a cornerstone of antibody research, driving the synthesis of highly specific, fully human mAbs that possess significant therapeutic capabilities. Success in accelerating the production of antibodies has paved the way for further advances in targeted therapies, diagnostics and biomedical research.

1.4 Therapeutic Applications of mAbs

Numerous mAbs have received regulatory approval for the therapeutic treatment of different diseases, demonstrating the immense versatility of mAbs in the medical industry diseases. Table 1.1 catalogues the applications of mAbs in the medical sector.

1.4.1 Cetuximab

Cetuximab is a monoclonal antibody of the IgG1 chimeric class that selectively binds to the EGFR. This mAb is authorized for the management of metastatic colorectal cancer in patients with EGFR expression and wild-type RAS. It is also approved for the therapy of squamous cell carcinoma in the neck and head region (Paci et al. 2020). Cetuximab blocks ligand-induced EGFR phosphorylation because it binds with great affinity to the human EGFR extracellular domain (Baselga 2001; Goldberg 2005). It has the ability to inhibit ligand attachment, halt cell proliferation or repair of DNA, reverse angiogenesis, boost receptor internalization and stimulate antibody-dependent cellular cytotoxicity (ADCC) (Harding and Burtness 2005; Kim et al. 2023). Cetuximab was recommended for treatment at an initial dose of 400 mg/m^2 and weekly doses of 250 mg/m^2 (Trullas et al. 2021). After years of research and development, Becher et al. (2017) released a plasma cetuximab testing that used a combination of liquid chromatography (LC) and mass spectrometry techniques performed in a randomized, controlled trial on 25 individuals' neck and head cancer. On day 14, responders had higher cetuximab concentrations than non-responders. The authors suggested a cut-off of 33.8 mg/mL as the 87% sensitive and specific C_{min} of response accuracy (Becher et al. 2017). Additionally, nowadays, many studies have reported that using a combination of cetuximab for threptic application, e.g. Wu et al. (2019) reported higher response rates with combination therapy compared to monotherapy. Cetuximab and Canerpaturev (C-REV) are two anti-EGFR monoclonal antibodies that have been studied for their potential

Table 1.1 A summary of commercialized anti-cancer therapies based on monoclonal antibodies

Antibody	Brand name	Target	Company	Category	Cancers	References
Blinatumomab	Blincyto	CD19, CD3	Amgen	Murine mAb	Acute lymphoblastic leukaemia	Frenzel et al. (2016)
Tositumomab	Bexxar	CD20	Glaxo	Murine mAb	Non-Hodgkin's lymphoma	Brinkmann and Kontermann (2017)
Rituximab	Rituxan	CD20	IDEC/Genentech	Chimeric mAb	Non-Hodgkin's lymphoma, chronic lymphocytic leukaemia	Mohapatra and Kaur (2021)
Dinutuximab	Unituxin	GD2	United Therapeutics	Chimeric mAb	Paediatric high-risk neuroblastoma	Almagro et al. (2018)
Brentuximab	Vedotin	CD30	Seattle Genetics	Chimeric mAb-drug conjugate	Hodgkin lymphoma	Zaroff and Tan (2019)
Alemtuzumab	Campath	CD52	Genzyme	Humanized mAb	Chronic lymphocytic leukaemia	Ernst et al. (2019)
Atezolizumab	Tecentriq	PD-L1	Genentech	Humanized mAb	Urothelial carcinoma, non-small cell lung cancer	Winkler et al. (2017)
Pembrolizumab	Ketruda	PD-1	Merck	Humanized mAb	Melanoma, non-small cell lung, head and neck squamous cell, gastric, cervical, hepatocellular cancers, Hodgkin lymphoma, urothelial carcinoma	Almagro et al. (2018)
Certolizumab pegol	Cimzia	TNF-α	Union Chimique Belge	Humanized mAb	Crohn's disease, rheumatoid arthritis	Mohapatra and Kaur (2021)
Eculizumab	Soliris	CD59	Alexion	Humanized mAb	Paroxysmal nocturnal haemoglobinuria	Mohapatra and Kaur (2021)
Elotuzumab	Empliciti	SLAMF7	Bristol-Myers Squibb	Humanized mAb	Multiple myeloma	Almagro et al. (2018)

(continued)

Table 1.1 (continued)

Antibody	Brand name	Target	Company	Category	Cancers	References
Avelumab	Bavencio	PD-L1	EMD Serono	Human mAb	Metastatic Merkel cell carcinoma	Almagro et al. (2018)
Golimumab	Simponi	TNF-α	Janssen Biotech	Human mAb	Rheumatoid and psoriatic arthritis, active a kylosing spondylitis	Delegan et al. (2022), Mohapatra and Kaur (2021)
Necitumumab	Portrazza	EGFR	Eli Eilly	Human mAb	Squamous non-small cell lung cancer	Alfaleh et al. (2020)
Durvalumab	Imfinz	PD-L1	AstraZeneca	Human mAb	Urothelial carcinoma	Almagro et al. (2018)
Nivolumab	Opdivo	PD-1	Bristol-Myers Squibb	Human mAb	Melanoma, non-small cell lung cancer, Hodgkin lymphoma, renal cell carcinoma	Deeks (2014)
Olaratumab	Lartruvo	PDGFRA	Eli Lilly	Human mAb	Soft tissue sarcoma	Almagro et al. (2018)
Daratumumab	Darzalex	CD38	Janssen Biotech	Human mAb	Multiple myeloma	Cohen et al. (2021)
Ramucirumab	Cyramza	VEGFR2	Eli Lilly	Human mAb	Gastric cancer	Syed (2020)

anticancer effects. The researchers (Wu et al. 2019) employed an in vitro and in vivo human colorectal xenograft model to examine the effects of each drug alone and in combination. Different human colorectal cancer cell lines (HT-29, WiDr, and CW2) exhibited varying degrees of EGFR expression, and C-REV's ability to kill these cells was altered with time and dosage. Cetuximab also failed to prevent viral replication in in vitro investigations. In HT-29 tumour xenograft models, the combination of cetuximab with C-REV had a synergistic anti-tumour effect by increasing C-REV distribution across the tumour while inhibiting angiogenesis. The tumours regressed more effectively when cetuximab was applied before C-REV than when it was applied after the virus (Wu et al. 2019).

1.4.2 Pertuzumab

Pertuzumab, a recombinant humanized mAb, inhibits the formation of heterodimers between HER2 and along with other HER family receptors by specifically targeting the extracellular dimerization domain of the HER2 receptor (Amiri-Kordestani et al. 2014). It also works well in tandem with trastuzumab. It has been authorized to be used in the therapy of HER2-positive breast cancer in both the United States and the European Union since 2012 and 2013, respectively. The recommended dosage is 420 mg every 3 weeks in addition to trastuzumab plus chemotherapy along with a loading dose of 840 mg for 4 weeks (Paci et al. 2020). There are many reports on the utilization of pertuzumab and trastuzumab. An investigation carried out by Bergen et al. in 2021 found that the combination of pertuzumab and trastuzumab (referred to as TP) was highly beneficial in curing brain metastases in individuals positive for HER2 breast cancer. When used as the initial systemic treatment for brain metastases, TP resulted in a significantly extended overall survival of 44 months in comparison to additional HER2-targeted treatment of 17 months or there was no HER2-targeted treatment at all ($p < 0.001$). Additionally, among patients who received TP as their initial systemic therapy after brain metastasis diagnosis, 35.7% achieved complete intracranial remission, 57.1% achieved partial intracranial remission and 7.1% had stable intracranial disease, while none experienced progressive intracranial disease. Because of this, the intracranial clinical benefit rate (iCBR) was 100% and the intracranial objective response rate (iORR) was 92.9% (Bergen et al. 2021). In an investigation conducted by Ma et al. 2022, it was found that neoadjuvant chemotherapy with trastuzumab (H) and pertuzumab (P) significantly increased the rate of pathological complete response (pCR) in patients with HER2 positive breast cancer. In this retrospective analysis, patients positive for HER2 breast cancer who underwent neoadjuvant therapy using HP and chemotherapy between March 2019 and February 2022, (total pathological complete response) tpCR rates were analysed. (Taxane + carboplatin H + P) TcbHP (73.1%), (Taxane +H + P) THP (52.1%) and (anthracycline and cyclophosphamide+Taxane +H + P) AC-THP (65.1%) exceeded a 64.9% general rate, with higher tpCR in (hormone receptor) HR-negative patients (80.3%). Both (breast pathologic complete response) bpCR (73.5%) and (axillary pathologic complete response) apCR (74.8%) were notable. Independent predictors of tpCR were HR, HER2 status and therapy regimen. TcbHP had higher toxicity levels (3 and 4) compared to THP and AC-THP, but the pCR rate remained high, especially with TcbHP, suggesting manageable adverse effects (Ma et al. 2022).

1.4.3 Adalimumab (Humira)

The first human antibody, Humira, was approved for medicinal purposes by the US Food and Drug Administration (FDA) in 2002. It was developed for the therapy of rheumatoid arthritis. This antibody was classified as a human monoclonal antibody. Adalimumab represents the initial mAb generated from phage display, which has undergone humanization through directed selection (Frenzel et al. 2016). The mAb exhibits a strong affinity for tumour necrosis factor (TNF) and effectively inhibits its interaction with the receptor. The binding of Adalimumab to tumour necrosis factor (TNF) results in the inhibition of various inflammatory responses (Kaymakcalan et al. 2009). Consequently, this medication is employed in the management of several medical conditions, including psoriatic arthritis, ankylosing spondylitis, chronic psoriasis, ulcerative colitis, Crohn's disease, juvenile idiopathic arthritis, rheumatoid arthritis and hidradenitis suppurativa (Kumar et al. 2019).

1.4.4 Belimumab

Belimumab was a completely human recombinant mAb of the IgG1λ family it specifically binds with the cytokine B lymphocyte stimulator (BlyS), which is also known as a B-cell activation factor (BAFF) (Kumar et al. 2019). Belimumab received its initial approval from the FDA in 2011 and was subsequently approved by European Medicines Agency in the month of July of that same year (Vana 2018). Its involvement has a significant impact on B-cell viability as well as function. B-cell activating factor (BAFF) helps B cells, even those that attack themselves, stay alive. Blocking BAFF, on the other hand, causes autoreactive B cells to die by a process called apoptosis. This represents the initial authorized biological intervention for systemic lupus erythematosus. Belimumab functions as an antagonist by inhibiting B-cell activation (Kumar et al. 2019; Wang et al. 2022).

1.4.5 Ranibizumab

Ranibizumab is an antibody fragment (Fab) synthesized through genetic engineering of a murine mAb to the human IgG1-chain (Papadopoulos et al. 2012). The vascular endothelial growth factor A (VEGF-A) intended for the drugs bevacizumab and ranibizumab shares a common mouse antibody parent because they both bind well to and neutralize VEGF-A isoforms (Papadopoulos et al. 2012). The promotion of angiogenesis by VEFG-A and its receptors VEGFR-1 and VEGFR-2 is a key mechanism in degenerative ocular diseases, such as age-related macular degeneration (AMD), retinal vein occlusions and diabetic retinopathy. Wet AMD, in particular, is a leading cause of vision loss among elderly people (Shibuya 2011; Ma et al. 2015). Ranibizumab is more easily absorbed by the retina following intravitreal injection due to its smaller size compared to bevacizumab. Because of its shorter serum half-life and quicker system clearance, ranibizumab is anticipated to be less toxic to normally functioning, healthy cells that produce VEGF-A (Alfaleh et al. 2020). Ranibizumab was FDA approved for wet AMD in 2006 (Chen and Kaiser 2020).

1.4.6 Raxibacumab

The bacterium *Bacillus anthracis (B. anthracis)* which causes anthrax is a spore-forming, aerobic Gram-positive rod shaped that can be found in soils all over the world (Singh et al. 2015; Frenzel et al. 2016). Anthrax has been classified as a Category A bioweapon by the Centres of Disease Control and Prevention, indicating its high potential for use in bioterrorism. (Froude et al. 2011). *Bacillus anthracis* produces two toxins: oedema toxin (ET) and lethal toxin (LT) each consisting of two subunits. Lethal toxin (LT) consists of a lethal component (LF) and protecting antigen (PA), whereas the oedema toxin (ET) is composed of the oedema factor along with PA. Antigens for these toxins plays a crucial role in their cellular uptake hence, development of Raxibacumab can effectively neutralizes those antigens. Liu et al. (2014) used Raxibacumab for treatment of anthrax and obtained an IC_{50} value of 0.21 nM in their experiment (Liu et al. 2014). The binding kinetics (K_d) for Raxibacumab to protect antigens reveal a K_d with 2.78 nM. Raxibacumab had been developed by Human Genome Science (now part of Glaxo Smith Kline) utilizing the human scFv phage display library from CAT (Frenzel et al. 2016). It is now available in the IgG1λ mAb format and is FDA approved for treating inhalational anthrax since 2012 (Tsai and Morris 2015). Combining raxibacumab with antibiotics is recommended (Kummerfeldt 2014). An alternative approach is inhibiting lethal factor complex formation (Frenzel et al. 2016).

1.4.7 *Nectiumumab*

Nectinumab was first authorized by the United States FDA in 2015 for therapy of metastatic squamous lung cancer with non-small cell carcinoma when used in combination with gemcitabine as well as cisplatin (Fala 2016). Necitumumab exclusively binds the EGFR and completely the human IgG1-k monoclonal antibody. Necitumumab hinders the transmission of signals from the EGFR, which plays a crucial role in cell development and angiogenesis. These processes are essential for promoting the growth and dissemination of cancerous cells. As a result, necitumumab restrains the growth of different cancer cell lines (Li et al. 2008; Alfaleh et al. 2020). In particular, it prevents cancer cells from dividing, specializing, adhering, migrating and even surviving by blocking downstream signalling pathways like phosphatidylinositol-4,5-bisphosphate 3-kinase (PI3k) and mitogen-activated protein kinase (MAPK) (Halder et al. 2023).

1.4.8 *Avelumab*

The Food and Drug Administration approved avelumab, a fully human monoclonal antibody of the IgG1 isotype, in 2017 for the therapy of Merkel cell carcinoma (MCC) for individuals aged 12 and older (Kim 2017). MCC is a very deadly form of skin cancer. This immunotherapy is the first of its kind to obtain authorization through the FDA. Bavencio is being created by Merck KGaA, Pfizer and Eli Lilly (Kumar et al. 2019). This monoclonal antibody (mAb) functions by attaching to PD-L1, preventing it from activating its receptor, PD-1, on target cells. This mAb suppresses immunological response, particularly from CD8+ T cells, by interfering with the formation of receptor–ligand complexes. The initial system was created using the Dyax naïve human Fab phage display antibody (Kumar et al. 2019).

1.5 Use of Phage Display-Derived mAbs Against Infectious Diseases

Using phage display, a large number of antibodies have been produced against various pathogens. Different types of pathogens have been targeted by these recombinant antibodies including malaria and influenza. One of the most common diseases spread by mosquitoes is malaria, which is caused by one of five species of *Plasmodium*: *P. falciparum*, *P. vivax*, *P. malariae*, *P. ovale* and *P. knowlesi*. While *P. vivax* is the more prevalent malaria parasite, *P. falciparum* is the most dangerous variant for humans. Studies on malaria and other significant human pathogens have benefited greatly from phage display (Lanzillotti and Coetzer 2008). There was a 12-mer peptide discovered in 2001 (PCQRAIFQSICN) that prevented *Plasmodium* from invading the midgut and salivary glands (Ghosh et al. 2001). Female mosquitoes have their hemocoels injected with a phage library to conduct affinity selection for binding to vector salivary glands. The salivary glands were dissected 30 min later to elute and collect the bound phages. Mosquitoes were fed the phage library to find the ones that bound to the midgut epithelium's luminal side. It is worth noting that the same peptide was chosen using the oral ingestion and hemocoel infection selection procedures.

Epitope mapping, diagnostics, treatments and vaccine development have all benefited greatly from phage display in the field of malaria research (Coley et al. 2001; Casey et al. 2004; Huang et al. 2012). The first monoclonal antibody targeting the apical membrane antigen, AMA-1, was introduced by Greven et al., 2021. Given the high incidence of anti-AMA-1 antibodies in immunological individuals, it presents an intriguing therapeutic target and is one of the leading choices for the blood stage of the vaccine. Antibodies were selected against three different AMA-1 variants using a human malaria Fab immune collection. We lightly shuffled the chains to increase the affinity and functional activity. Surface plasmon resonance (SPR) spectroscopy was used to conduct affinity and competition experiments following subcloning to IgG1/k. Both an in vitro growth assay and an immunofluorescence assay were used to test the antibodies. Here, all antibodies inhibited the growth of four different strains of *Plasmodium falciparum*, one of which was 3D7. Hence, it is possible that such antibodies could serve as the basis for creating an antibody cocktail that could be used for therapeutic or preventive purposes in the fight against malaria (Seidel-Greven et al. 2021).

One of the most contagious diseases in the world, seasonal flu, is caused by viruses called influenza (Keilman 2019). The "swine flu" (a variation of H1N1) and "bird flu" (H5N1) have taken centre stage in recent years, garnering both public and academic interest due to their worldwide transmission (Trombetta et al. 2015). The selection of antibodies from a naïve scFv library that recognizes the H5 haemagglutinin ectodomain in vaccine and diagnostic formulation may be hindered by the ongoing occurrence of novel influenza variants in the future, which can be attributed to various genetic events such as antigenic drift and shift (Roth et al. 2021). Antibodies that target and neutralize the H5N1 influenza haemagglutinin were discovered by Lim et al. (2008). Not long after that, Khurana and colleagues built phage display libraries using fragments of the H5N1 gene and used them to examine monoclonal antibodies and blood samples taken from people infected with H5N1. Results from this study's epitope mapping could help in H5N1 virus diagnostics and vaccine development (Khurana et al. 2009).

Phage display technology has also been used to work against antimicrobial resistance (AMR) to a certain extent. Besides antibodies, phage display technology can be used to produce a wide range of peptides, with some of them being used to inhibit AMR. β-Lactamase enzymes have been found to induce resistance against β-lactam-containing antibodies. β-Lactamase inhibiting peptides have been developed that could be potentially used to fight against resistant bacterial strains (Muteeb et al. 2017). A similar application of peptide derived from phage display has been exhibited against multidrug-resistant *Listeria monocytogenes.* Antimicrobial peptide was synthesized using the combinatorial phage display method and the resulting peptides L2 and L3 were then used to combat *L. monocytogenes.* These two peptides were successful in inhibiting the in vitro growth of *L. monocytogenes* without any haemolytic or toxic side-effects on eukaryotic cells (Flachbartova et al. 2016). Drug-resistant strains of *Staphylococcus aureus* are also being studied with the help of phage display technology. Phages can be used to display virulent factor, cell surface proteins or other protein factors that are synthesized by resistant *S. aureus* strains. With the help of such phage libraries, screening of these bacterial strains can be simplified. Moreover, interaction between these factors and their ligands can be used to develop new drugs that could actually inhibit or kill these strains (Tian et al. 2022).

1.6 Other Methods for Production of mAbs

1.6.1 Transgenic Animals

The foundation of transgenic animal technology mostly revolves around the concept of enhancing mice to exhibit characteristics more closely like those of humans. The process involves the introduction of human antibody loci into various animal species such as mice, chickens and cows (Jiachen et al. 2022). Subsequently, the rearrangement and reactivation of human antibody V region genes occur in the lymphocytes of these transgenic animals. Consequently, these animals are capable of generating B lymphocytes that exhibit complete expression of human antibodies. Summarizing, this procedure can be described as a partial humanization of animals (Laffleur et al. 2012; Brüggemann et al. 2015). Upon exposure to antigens, lymphocytes have the ability to undergo continual cloning and differentiation, resulting in the formation of plasma cells. These plasma cells possess the capacity to produce human antibodies with high affinity (Chen and Murawsky 2018). Furthermore, it is worth noting that transgenic animals that harbour snippets of human DNA exhibit full functionality, including the ability to undergo efficient homologous conversion and affinity maturation. As a result, these animals are capable of generating human antibodies with a high affinity following immunization with any desired target antigen (Elgundi et al. 2017).

1.6.2 Recombinant Antibodies

Production of recombinant antibodies (rAb) is achieved by employing technologies based on recombinant DNA. By utilizing bacterial cells, synthesis of recombinant antibody fragments can be achieved (Charlton 2004). Recombinant antibodies are known for their cost-effectiveness and high level of repeatability in manufacturing.

This is because they are made from a DNA sequence that has been well established and carefully described. In addition, it is possible to enhance the engineering of recombinant antibodies by incorporating various affinity tags. These tags serve to streamline the purification process and enable the immobilization of the antibodies for subsequent applications (Sharma et al. 2016). In an alternative approach, the use of recombinant binders might be employed in a multimerized form to enhance their avidity. Recombinant antibodies can derive from several origins, indicating that they may not consistently adhere to the criterion of being completely devoid of animal components. A common approach is to directly clone gene fragments derived from hybridoma cells and use them in conjunction with a well-established monoclonal antibody sequence to create recombinant antibody fragments (Peltomaa et al. 2022). While it is feasible to produce antibody fragments, such as in *E.coli*, with no animal involvement, it is crucial to acknowledge that the synthesis of the initial monoclonal antibody continues to necessitate animal immunization. Consequently, these recombinant antibodies are classified as animal derived (Peltomaa et al. 2022). In certain cases, the process of isolating functional recombinant fragments of antibodies from hybridoma cell lines can be difficult, which may lead to fragments having lower affinity compared to the original monoclonal antibody counterparts (Li et al. 2013; Hu et al. 2015).

1.7 Limitation and Future Challenges of Phage Display Technology

Avoiding some of these constraints can be achieved by the use of phage-displayed antibody repertoires, which can be administered to animals as either naive or modified libraries. An in vitro selection strategy called phage display allows for the automated or high-throughput generation of antibodies against almost any target (Miersch et al. 2015). In vitro selection allows for a more precise control when selecting antibodies against potentially dangerous targets or proteins, or when strengthening their affinity (Barderas et al. 2008, 2012). However, phage display technology has its own disadvantages and other limitations. A major disadvantage is the probable difficulty in obtaining antibodies that accurately imitate the local form of the target antigen. Phage display depends on the presentation of peptides or proteins on the surface of phages, which may not always reflect the origin structure of the antigen. Consequently, antibodies generated by phage display may have a reduced affinity or affinity for binding compared to antibodies derived from natural immune responses (Kumar et al. 2019). The in vitro nature of the phage display may overlook certain aspects of the in vivo maturation process of antibodies, such as post-translational modifications or glycosylation. These changes can have a major impact on the function and bioactivity of antibodies (Alfaleh et al. 2020; Ledsgaard et al. 2022). Consequently, antibodies produced through phage display may lack important functions that are essential to the therapeutic efficacy or diagnostic application.

1.8 Conclusion

Phage display technology has become a critical and versatile tool for the production of mAbs. With the help of this ground-breaking method, the process of mAbs has been streamlined and made easier. This has greatly enhanced the entire procedure and led to commercialization of 14 FDA-approved mAbs with many more in the pipeline. Phage display technology has made it possible to take a microorganism with a limited value to the human and transform it into a useful tool. Bacteriophages have been previously investigated to counteract bacterial infections in animals and humans. However, it is phage display technology that enabled the unleash of the full potential of bacteriophages. Technological advancements have made it possible to create highly specific or personalized mAbs in the future. mAb libraries could be developed consisting of a variety of mABs with high degree of specificity and affinity. This versatility shown by phage display technology gives clear advantages over conventional methods for mAb production. Innovative therapies for previously difficult targets such as membrane proteins and highly conserved epitopes have been successfully attempted. By isolating mAbs that identify a variety of antigens, these targets could be extensively studied. These features are valuable when designing new drugs for new infectious diseases or rapidly evolving pathogens. Discovery of novel drugs is being explored by observing interactions between phage-derived peptides/proteins and therapeutic compounds. Rapid screening of antibody libraries to identify peptides that could function as antimicrobial agents against drug-resistant bacteria could further improve the process. In addition, phage display is a viable option for large-scale manufacture of mAbs due to its affordability, compared to conventional hybridoma-based techniques. Phage display technology's main drawbacks are the probable false positives and the need for labour-intensive trials involving numerous variables. However, as the field develops, these limitations are continually being addressed by designing increasingly sophisticated display systems.

References

Aarden L, Ruuls SR, Wolbink G (2008) Immunogenicity of anti-tumor necrosis factor antibodies—toward improved methods of anti-antibody measurement. Curr Opin Immunol 20:431–435. https://doi.org/10.1016/J.COI.2008.06.011

Aghebati-Maleki L, Bakhshinejad B, Baradaran B, Motallebnezhad M, Aghebati-Maleki A, Nickho H, Yousefi M, Majidi J (2016) Phage display as a promising approach for vaccine development. J Biomed Sci 23:1–18. https://doi.org/10.1186/s12929-016-0285-9

Alfaleh MA, Jones ML, Howard CB, Mahler SM (2017) Strategies for selecting membrane protein-specific antibodies using phage display with cell-based panning. Antibodies 6:10. https://doi.org/10.3390/ANTIB6030010

Alfaleh MA, Alsaab HO, Mahmoud AB, Alkayyal AA, Jones ML, Mahler SM, Hashem AM (2020) Phage display derived monoclonal antibodies: from bench to bedside. Front Immunol 11. https://doi.org/10.3389/fimmu.2020.01986

Almagro JC, Daniels-Wells TR, Perez-Tapia SM, Penichet ML (2018) Progress and challenges in the design and clinical development of antibodies for cancer therapy. Front Immunol 8. https://doi.org/10.3389/fimmu.2017.01751

Almagro JC, Pedraza-Escalona M, Arrieta HI, Pérez-Tapia SM (2019) Phage display libraries for antibody therapeutic discovery and development. Antibodies 8:1–21. https://doi.org/10.3390/antib8030044

Amiri-Kordestani L, Wedam S, Zhang L, Tang S, Tilley A, Ibrahim A, Justice R, Pazdur R, Cortazar P (2014) First FDA approval of neoadjuvant therapy for breast cancer: pertuzumab for the treatment of patients with HER2-positive breast cancer. Clin Cancer Res 20:5359–5364. https://doi.org/10.1158/1078-0432.CCR-14-1268

Ashraf MA (2019) POVERTY ALLEVIATION AND CONVENTIONAL MFIS CHALLENGES AND PROSPECTS. Nova Science Publishers Inc.

Bajaj G, Suryawanshi S, Roy A, Gupta M (2019) Evaluation of covariate effects on pharmacokinetics of monoclonal antibodies in oncology. Br J Clin Pharmacol 85:2045–2058. https://doi.org/10.1111/bcp.13996

Barderas R, Desmet J, Timmerman P, Meloen R, Casal JI (2008) Affinity maturation of antibodies assisted by in silico modeling. Proc Natl Acad Sci USA 105:9029–9034. https://doi.org/10.1073/pnas.0801221105

Barderas R, Desmet J, Alard P, Casal JI (2012) Chapter 27. Affinity maturation by semirational approaches. p 907. https://doi.org/10.1007/978-1-61779-974-7

Baselga J (2001) The EGFR as a target for anticancer therapy—focus on cetuximab. Eur J Cancer 37:16–22. https://doi.org/10.1016/s0959-8049(01)00233-7

Becher F, Ciccolini J, Imbs DC, Marin C, Fournel C, Dupuis C, Fakhry N, Pourroy B, Ghettas A, Pruvost A, Junot C, Duffaud F, Lacarelle B, Salas S (2017) A simple and rapid LC-MS/MS method for therapeutic drug monitoring of cetuximab: a GPCO-UNICANCER proof of concept study in head-and-neck cancer patients. Sci Rep 7:1–11. https://doi.org/10.1038/s41598-017-02821-x

Bergen ES, Binter A, Starzer AM, Heller G, Kiesel B, Tendl-Schulz K, Bago-Horvath Z, Furtner J, Leitner J, Exner R, Fitzal F, Dieckmann K, Widhalm G, Preusser M, Berghoff AS, Bartsch R (2021) Favourable outcome of patients with breast cancer brain metastases treated with dual HER2 blockade of trastuzumab and pertuzumab. Ther Adv Med Oncol 13:1–12. https://doi.org/10.1177/17588359211009002

Bleumer I, Knuth A, Oosterwijk E, Hofmann R, Varga Z, Lamers C, Kruit W, Melchior S, Mala C, Ullrich S, De Mulder P, Mulders PFA, Beck J (2004) A phase II trial of chimeric monoclonal antibody G250 for advanced renal cell carcinoma patients. Br J Cancer 905(90):985–990. https://doi.org/10.1038/sj.bjc.6601617

Brinkmann U, Kontermann RE (2017) The making of bispecific antibodies. MAbs 9:182–212. https://doi.org/10.1080/19420862.2016.1268307

Brüggemann M, Osborn MJ, Ma B, Hayre J, Avis S, Lundstrom B, Buelow R (2015) Human antibody production in transgenic animals. Arch Immunol Ther Exp 63:101–108. https://doi.org/10.1007/s00005-014-0322-x

Casey JL, Coley AM, Anders RF, Murphy VJ, Humberstone KS, Thomas AW, Foley M (2004) Antibodies to malaria peptide mimics inhibit plasmodium falciparum invasion of erythrocytes. Infect Immun 72:1126–1134. https://doi.org/10.1128/IAI.72.2.1126-1134.2004

Charlton KA (2004) Expression and isolation of recombinant antibody fragments in E. Coli. Methods Mol Biol 248:245–254. https://doi.org/10.1385/1-59259-666-5:245

Chen ER, Kaiser PK (2020) Therapeutic potential of the Ranibizumab port delivery system in the treatment of AMD: evidence to date. Clin Ophthalmol 14:1349–1355. https://doi.org/10.2147/OPTH.S194234
Chen WC, Murawsky CM (2018) Strategies for generating diverse antibody repertoires using transgenic animals expressing human antibodies. Front Immunol 9:1–7. https://doi.org/10.3389/fimmu.2018.00460
Chen G, Sidhu SS (2014) Design and generation of synthetic antibody libraries for phage display. Methods Mol Biol 1131:113–131. https://doi.org/10.1007/978-1-62703-992-5_8/COVER
Cohen YC, Oriol A, Wu KL, Lavi N, Vlummens P, Jackson C, Garvin W, Carson R, Crist W, Fu J, Feng H, Xie H, Schecter J, San-Miguel J, Lonial S (2021) Daratumumab with cetrelimab, an anti–PD-1 monoclonal antibody, in relapsed/refractory multiple myeloma. Clin Lymphoma Myeloma Leuk 21:46–54.e4. https://doi.org/10.1016/j.clml.2020.08.008
Coley AM, Campanale NV, Casey JL, Hodder AN, Crewther PE, Anders RF, Tilley LM, Foley M (2001) Rapid and precise epitope mapping of monoclonal antibodies against plasmodium falciparum AMA1 by combined phage display of fragments and random peptides. Protein Eng 14:691–698. https://doi.org/10.1093/protein/14.9.691
Dal Ferro M, Rizzo S, Rizzo E, Marano F, Luisi I, Tarasiuk O, Sblattero D (2019) Phage display technology for human monoclonal antibodies. Methods Mol Biol 1904:319–338. https://doi.org/10.1007/978-1-4939-8958-4_15/COVER
Damián-blanco P, Ahuexoteco-sánchez S, Carbajal-gallardo AA (2023) Use of monoclonal antibodies in cancer immunotherapy: types and mechanisms of action. Bol Med Hosp Infant Mex 80:153–164. https://doi.org/10.24875/BMHIM.22000123
Deeks ED (2014) Nivolumab: a review of its use in patients with malignant melanoma. Drugs 74:1233–1239. https://doi.org/10.1007/s40265-014-0234-4
Delegan Y, Sushkova S, Minkina T, Filonov A, Kocharovskaya Y, Demin K, Gorovtsov A, Rajput VD, Zamulina I, Grigoryeva T, Dudnikova T, Barbashev A, Maksimov A (2022) Diversity and metabolic potential of a PAH-degrading bacterial consortium in Technogenically contaminated haplic chernozem. Southern Russia Process 10:2555. https://doi.org/10.3390/PR10122555/S1
Dillman RO, Beauregard JC, Halpern SE, Clutter M (1986) Toxicities and side effects associated with intravenous infusions of murine monoclonal antibodies. J Immunother 5
Doevendans E, Schellekens H (2019) Immunogenicity of innovative and biosimilar monoclonal antibodies. Antibodies 8. https://doi.org/10.3390/antib8010021
Elgundi Z, Reslan M, Cruz E, Sifniotis V, Kayser V (2017) The state-of-play and future of antibody therapeutics. Adv Drug Deliv Rev 122:2–19. https://doi.org/10.1016/j.addr.2016.11.004
Ernst D, Williams BA, Wang XH, Yoon N, Kim KP, Chiu J, Luo ZJ, Hermans KG, Krueger J, Keating A (2019) Humanized anti-CD123 antibody facilitates NK cell antibody-dependent cell-mediated cytotoxicity (ADCC) of Hodgkin lymphoma targets via ARF6/PLD-1. Blood Cancer J 9:6. https://doi.org/10.1038/s41408-018-0168-2
Fala L (2016) Portrazza (Necitumumab), an IgG1 monoclonal antibody, FDA approved for advanced squamous non-small-cell lung cancer. Am Health Drug Benefits 9:119–122
Flachbartova Z, Pulzova L, Bencurova E, Potocnakova L, Comor L, Bednarikova Z (2016) Inhibition of multidrug resistant listeria monocytogenes by peptides isolated from combinatorial phage display libraries. Microbiol Res 188–189:34–41. https://doi.org/10.1016/j.micres.2016.04.010
Frei JC, Lai JR (2016) Protein and antibody engineering by phage display, 1st edn. Elsevier Inc.
Frenzel A, Schirrmann T, Hust M (2016) Phage display-derived human antibodies in clinical development and therapy. MAbs 8:1177–1194. https://doi.org/10.1080/19420862.2016.1212149
Frenzel A, Kügler J, Helmsing S, Meier D, Schirrmann T, Hust M, Dübel S (2017) Designing human antibodies by phage display. Transfus Med Hemotherapy 44:312–318. https://doi.org/10.1159/000479633
Froude JW, Stiles B, Pelat T, Thullier P (2011) Antibodies for biodefense. MAbs 3(6):517–527. https://doi.org/10.4161/mabs.3.6.17621

Galizia G, Lieto E, De Vita F, Orditura M, Castellano P, Troiani T, Imperatore V, Ciardiello F (2007) Cetuximab, a chimeric human mouse anti-epidermal growth factor receptor monoclonal antibody, in the treatment of human colorectal cancer. Oncogene 2625(26):3654–3660. https://doi.org/10.1038/sj.onc.1210381

Garnock-Jones KP (2016) Necitumumab: first global approval topical collection on lung cancer. Drugs 76:283–289. https://doi.org/10.1007/s40265-015-0537-0

Ghosh AK, Ribolla PEM, Jacobs-Lorena M (2001) Targeting plasmodium ligands on mosquito salivary glands and midgut with a phage display peptide library. Proc Natl Acad Sci USA 98:13278–13281. https://doi.org/10.1073/pnas.241491198

Gillies SD, Lo KM, Wesolowski J (1989) High-level expression of chimeric antibodies using adapted cDNA variable region cassettes. J Immunol Methods 125:191–202. https://doi.org/10.1016/0022-1759(89)90093-8

Goldberg RM (2005) Cetuximab. Nat Rev Drug Discov Suppl S10-1:S10. https://doi.org/10.1038/nrd1728

Goldstein NI, Prewett M, Zuklys K, Rockwell P, Mendelsohn J (1995) Biological efficacy of a chimeric antibody to the epidermal growth factor receptor in a human tumor xenograft model. Clin Cancer Res 1:1311–1318

Gonzales Carazas MM, Pinto JA, Casado FL (2021) Biological bases of cancer immunotherapy. Expert Rev Mol Med 23:e3. https://doi.org/10.1017/erm.2021.5

Halder S, Basu S, Lall SP, Ganti AK, Batra SK, Seshacharyulu P (2023) Targeting the EGFR signaling pathway in cancer therapy: what's new in 2023? Expert Opin Ther Targets 27:305–324. https://doi.org/10.1080/14728222.2023.2218613

Hammers CM, Stanley JR (2014) Antibody phage display: technique and applications. Early Hum Dev 83:255–262. https://doi.org/10.1038/jid.2013.521.Antibody

Hansel TT, Kropshofer H, Singer T, Mitchell JA, George AJT (2010) The safety and side effects of monoclonal antibodies. Nat Rev Drug Discov 9:325–338. https://doi.org/10.1038/nrd3003

Harding J, Burtness B (2005) Cetuximab: an epidermal growth factor receptor chimeric human-murine monoclonal antibody. Drugs of Today 41:107–127. https://doi.org/10.1358/dot.2005.41.2.882662

Hentrich C, Ylera F, Frisch C, Ten Haaf A, Knappik A (2018) Monoclonal antibody generation by phage display: history, state-of-the-art, and future. Elsevier Inc.

Hess KL, Jewell CM (2020) Phage display as a tool for vaccine and immunotherapy development. Bioeng Transl Med 5:1–15. https://doi.org/10.1002/btm2.10142

Hu G, Nagata LP (2016) Opportunities and challenges of therapeutic monoclonal antibodies as medical countermeasures for biodefense. J Bioterror Biodef 7. https://doi.org/10.4172/2157-2526.1000149

Hu ZQ, Liu JL, Li HP, Xing S, Xue S, Zhang JB, Wang JH, Nölke G, Liao YC (2012) Generation of a highly reactive chicken-derived single-chain variable fragment against fusarium verticillioides by phage display. Int. J. Mol. Sci. 13:7038–7056. https://doi.org/10.3390/IJMS13067038

Hu ZQ, Li HP, Wu P, Li YB, Zhou ZQ, Zhang JB, Liu JL, Liao YC (2015) An affinity improved single-chain antibody from phage display of a library derived from monoclonal antibodies detects fumonisins by immunoassay. Anal Chim Acta 867:74–82. https://doi.org/10.1016/j.aca.2015.02.014

Huang JX, Bishop-Hurley SL, Cooper MA (2012) Development of anti-infectives using phage display: biological agents against bacteria, viruses, and parasites. Antimicrob Agents Chemother 56:4569–4582. https://doi.org/10.1128/AAC.00567-12

Ionescu RM, Vlasak J, Price C, Kirchmeier M (2008) Contribution of variable domains to the stability of humanized IgG1 monoclonal antibodies. J Pharm Sci 97:1414–1426. https://doi.org/10.1002/JPS.21104

Jaroszewicz W, Morcinek-Orłowska J, Pierzynowska K, Gaffke L, Węgrzyn G (2022) Phage display and other peptide display technologies. FEMS Microbiol Rev 46:1–25. https://doi.org/10.1093/FEMSRE/FUAB052

Jiachen LU, Ding J, Zhaoxia LIU, Chen T (2022) Retrospective analysis of the preparation and application of immunotherapy in cancer treatment (review). Int J Oncol 60:12. https://doi.org/10.3892/ijo.2022.5302

Kalayu Yirga S, Lin M, Huang Z, Hu J (2019) Development of therapeutic antibody technology and recent perspectives for leukemia-lymphoma treatment. J Leuk 07. https://doi.org/10.35248/2329-6917.19.7.259

Kaymakcalan Z, Sakorafas P, Bose S, Scesney S, Xiong L, Hanzatian DK, Salfeld J, Sasso EH (2009) Comparisons of affinities, avidities, and complement activation of adalimumab, infliximab, and etanercept in binding to soluble and membrane tumor necrosis factor. Clin Immunol 131:308–316. https://doi.org/10.1016/j.clim.2009.01.002

Keilman LJ (2019) Seasonal Influenza (Flu). Nurs Clin North Am 54:227–243. https://doi.org/10.1016/j.cnur.2019.02.009

Khurana S, Suguitan AL, Rivera Y, Simmons CP, Lanzavecchia A, Sallusto F, Manischewitz J, King LR, Subbarao K, Golding H (2009) Antigenic fingerprinting of H5N1 avian influenza using convalescent sera and monoclonal antibodies reveals potential vaccine and diagnostic targets. PLoS Med 6:1–14. https://doi.org/10.1371/journal.pmed.1000049

Kim ES (2017) Avelumab: first global approval. Drugs 77:929–937. https://doi.org/10.1007/s40265-017-0749-6

Kim DM, Lee SY, Lim JC, Cho EH, Park UJ (2023) RUNX3 regulates the susceptibility against EGFR-targeted non-small cell lung cancer therapy using 47Sc-conjugated cetuximab. BMC Cancer 23:652. https://doi.org/10.1186/s12885-023-11161-1

Knezevic I, Kang HN, Thorpe R (2015) Immunogenicity assessment of monoclonal antibody products: a simulated case study correlating antibody induction with clinical outcomes. Biologicals 43:307–317. https://doi.org/10.1016/J.BIOLOGICALS.2015.06.009

Köhler G, Milstein C (1975) Continuous cultures of fused cells secreting antibody of predefined specificity. Nature 256(5517):495–497. https://doi.org/10.1038/256495a0

Kügler J, Wilke S, Meier D, Tomszak F, Frenzel A, Schirrmann T, Dübel S, Garritsen H, Hock B, Toleikis L, Schütte M, Hust M (2015) Generation and analysis of the improved human HAL9/10 antibody phage display libraries. BMC Biotechnol 15:1–15. https://doi.org/10.1186/s12896-015-0125-0

Kumar R, Parray HA, Shrivastava T, Sinha S, Luthra K (2019) Phage display antibody libraries: a robust approach for generation of recombinant human monoclonal antibodies. Int J Biol Macromol 135:907–918. https://doi.org/10.1016/j.ijbiomac.2019.06.006

Kummerfeldt CE (2014) Raxibacumab: potential role in the treatment of inhalational anthrax. Infect Drug Resist 7:101–109. https://doi.org/10.2147/IDR.S47305

Laffleur B, Pascal V, Sirac C, Cogné M (2012) Production of human or humanized antibodies in mice. Methods Mol Biol 901:149–159. https://doi.org/10.1007/978-1-61779-931-0_9

Lanzillotti R, Coetzer TL (2008) Phage display: a useful tool for malaria research? Trends Parasitol 24:18–23. https://doi.org/10.1016/j.pt.2007.09.007

Ledsgaard L, Kilstrup M, Karatt-Vellatt A, McCafferty J, Laustsen AH (2018) Basics of antibody phage display technology. Toxins (Basel) 10. https://doi.org/10.3390/toxins10060236

Ledsgaard L, Ljungars A, Rimbault C, Sørensen CV, Tulika T, Wade J, Wouters Y, McCafferty J, Laustsen AH (2022) Advances in antibody phage display technology. Drug Discov Today 27:2151–2169. https://doi.org/10.1016/J.DRUDIS.2022.05.002

Li S, Kussie P, Ferguson KM (2008) Structural basis for EGF receptor inhibition by the therapeutic antibody IMC-11F8. Structure 16:216–227. https://doi.org/10.1016/j.str.2007.11.009

Li X, Li P, Lei J, Zhang Q, Zhang W, Li C (2013) A simple strategy to obtain ultra-sensitive single-chain fragment variable antibodies for aflatoxin detection. RSC Adv 3:22367–22372. https://doi.org/10.1039/c3ra42706d

Li J, Xu Y, Wang X, Li Y, Wang L, Li X (2016) Construction and characterization of a highly reactive chicken-derived single-chain variable fragment (scFv) antibody against Staphylococcus aureus developed with the T7 phage display system. Int Immunopharmacol 35:149–154. https://doi.org/10.1016/J.INTIMP.2016.02.024

Lim APC, Chan CEZ, Wong SKK, Chan AHY, Ooi EE, Hanson BJ (2008) Neutralizing human monoclonal antibody against H5N1 influenza HA selected from a Fab-phage display library. Virol J 5:1–10. https://doi.org/10.1186/1743-422X-5-130

Liu S, Moayeri M, Leppla SH (2014) Anthrax lethal and edema toxins in anthrax pathogenesis. Trends Microbiol 22:317–325. https://doi.org/10.1016/j.tim.2014.02.012

Ma IT, McConaghy S, Namachivayam K, Halloran BA, Kurundkar AR, Mohankumar K, Maheshwari A, Ohls RK (2015) VEGF mRNA and protein concentrations in the developing human eye. Pediatr Res 77:500–505. https://doi.org/10.1038/pr.2015.15

Ma X, Zhang X, Zhou X, Ren X, Zhang W, Liu Y (2022) Real-world research of Trastuzumab and Pertuzumab combined with chemotherapy in neoadjuvant treatment of HER2-positive breast cancer. Cancer Res Prev Treat 49:46–52. https://doi.org/10.3971/j.issn.1000-8578.2022.21.0802

Maximiano S, Magalhães P, Guerreiro MP, Morgado M (2016) Trastuzumab in the treatment of breast cancer. BioDrugs 30:75–86. https://doi.org/10.1007/s40259-016-0162-9

Mayrhofer P, Kunert R (2018) Nomenclature of humanized mAbs: early concepts, current challenges and future perspectives. Hum Antibodies 27:37–51. https://doi.org/10.3233/HAB-180347

Miersch S, Li Z, Hanna R, McLaughlin ME, Hornsby M, Matsuguchi T, Paduch M, Sääf A, Wells J, Koide S, Kossiakoff A, Sidhu SS (2015) Scalable high throughput selection from phage-displayed synthetic antibody libraries. J Vis Exp 1–15. https://doi.org/10.3791/51492

Mohapatra C, Kaur G (2021) An overview of monoclonal antibodies and their therapeutic applications. Nat Volatiles Essent oils 8:4121–4130

Muteeb G, Tabish Rehman M, Zafar Ali S, Al-Shahrani AM, Amjad Kamal M, Md Ashraf G (2017) Phage display technique: a novel medicinal approach to overcome an tibiotic resistance by using peptide-based inhibitors against β-lactamases. Curr Drug Metab 18(2):90–95

Nagano K, Tsutsumi Y (2021) Phage display technology as a powerful platform for antibody drug discovery. Viruses 13:178. https://doi.org/10.3390/V13020178

Nelson AL, Dhimolea E, Reichert JM (2010) Development trends for human monoclonal antibody therapeutics. Nat Rev Drug Discov 910(9):767–774. https://doi.org/10.1038/nrd3229

Nilvebrant J, Sidhu SS (2018) Construction of synthetic antibody phage-display libraries. Methods Mol Biol 1701:45–60. https://doi.org/10.1007/978-1-4939-7447-4_3/COVER

Nishibori N, Horiuchi H, Furusawa S, Matsuda H (2006) Humanization of chicken monoclonal antibody using phage-display system. Mol Immunol 43:634–642. https://doi.org/10.1016/J.MOLIMM.2005.04.002

Oliphant T, Engle M, Nybakken GE, Doane C, Johnson S, Huang L, Gorlatov S, Mehlhop E, Marri A, Chung KM, Ebel GD, Kramer LD, Fremont DH, Diamond MS (2005) Development of a humanized monoclonal antibody with therapeutic potential against West Nile virus. Nat Med 115(11):522–530. https://doi.org/10.1038/nm1240

Oślizło A, Miernikiewicz P, Piotrowicz A, Owczarek B, Kopciuch A, Figura G, Dabrowska K (2011) Purification of phage display-modified bacteriophage T4 by affinity chromatography. BMC Biotechnol 11:1–9. https://doi.org/10.1186/1472-6750-11-59/TABLES/1

Paci A, Desnoyer A, Delahousse J, Blondel L, Maritaz C, Chaput N, Mir O, Broutin S (2020) Pharmacokinetic/pharmacodynamic relationship of therapeutic monoclonal antibodies used in oncology: part 1, monoclonal antibodies, antibody-drug conjugates and bispecific T-cell engagers. Eur J Cancer 128:107–118. https://doi.org/10.1016/j.ejca.2020.01.005

Papadopoulos N, Martin J, Ruan Q, Rafique A, Rosconi MP, Shi E, Pyles EA, Yancopoulos GD, Stahl N, Wiegand SJ (2012) Binding and neutralization of vascular endothelial growth factor (VEGF) and related ligands by VEGF trap, ranibizumab and bevacizumab. Angiogenesis 15:171–185. https://doi.org/10.1007/s10456-011-9249-6

Pelletier JPR, Mukhtar F (2020) Passive monoclonal and polyclonal antibody therapies. Immun Concepts Transf Med 2020:251–348

Peltomaa R, Barderas R, Benito-Peña E, Moreno-Bondi MC (2022) Recombinant antibodies and their use for food immunoanalysis. Anal Bioanal Chem 414:193–217. https://doi.org/10.1007/s00216-021-03619-7

Pitaksajjakul P, Lekcharoensuk P, Upragarin N, Barbas CF, Ibrahim MS, Ikuta K, Ramasoota P (2010) Fab MAbs specific to HA of influenza virus with H5N1 neutralizing activity selected from immunized chicken phage library. Biochem Biophys Res Commun 395:496–501. https://doi.org/10.1016/J.BBRC.2010.04.040

Ribatti D (2014) From the discovery of monoclonal antibodies to their therapeutic application: an historical reappraisal. Immunol Lett 161:96–99. https://doi.org/10.1016/j.imlet.2014.05.010

Roth KDR, Wenzel EV, Ruschig M, Steinke S, Langreder N, Heine PA, Schneider KT, Ballmann R, Führer V, Kuhn P, Schirrmann T, Frenzel A, Dübel S, Schubert M, Moreira GMSG, Bertoglio F, Russo G, Hust M (2021) Developing recombinant antibodies by phage display against infectious diseases and toxins for diagnostics and therapy. Front Cell Infect Microbiol 11:1–38. https://doi.org/10.3389/fcimb.2021.697876

Sales-Dias J, Ferreira A, Lamy M, Domenici G, Monteiro SMS, Pires A, Lemos AR, Kucheryava K, Nobre LS, Sousa PMF, Bandeiras TM, Silva G, Barbas A (2021) Development of antibodies against the notch ligand Delta-Like-1 by phage display with activity against breast cancer cells. New Biotechnol 64:17–26. https://doi.org/10.1016/J.NBT.2021.05.003

Seidel-Greven M, Addai-Mensah O, Spiegel H, Chiegoua Dipah GN, Schmitz S, Breuer G, Frempong M, Reimann A, Klockenbring T, Fischer R, Barth S, Fendel R (2021) Isolation and light chain shuffling of a Plasmodium falciparum AMA1-specific human monoclonal antibody with growth inhibitory activity. Malar J 20:1–20. https://doi.org/10.1186/s12936-020-03548-3

Sgro C (1995) Side-effects of a monoclonal antibody, muromonab CD3/orthoclone OKT3: bibliographic review. Toxicology 105:23–29. https://doi.org/10.1016/0300-483X(95)03123-W

Sharma S, Byrne H, O'Kennedy RJ (2016) Antibodies and antibody-derived analytical biosensors. Essays Biochem 60:9–18. https://doi.org/10.1042/EBC20150002

Shibuya M (2011) Vascular endothelial growth factor (VEGF) and its receptor (VEGFR) signaling in angiogenesis: a crucial target for anti- and pro-angiogenic therapies. Genes Cancer 2:1097–1105. https://doi.org/10.1177/1947601911423031

Shim H (2017) Antibody phage display. Adv Exp Med Biol 1053:21–34. https://doi.org/10.1007/978-3-319-72077-7_2/COVER

Si Y, Melkonian AL, Curry KC, Xu Y, Tidwell M, Liu M, Zaky AF, Liu XM (2021) Monoclonal antibody-based cancer therapies. Chinese J Chem Eng 30:301–307

Singh LK, Dhasmana N, Sajid A, Kumar P, Bhaduri A, Bharadwaj M, Gandotra S, Kalia VC, Das TK, Goel AK, Pomerantsev AP (2015) *clpC* operon regulates cell architecture and sporulation in *Bacillus anthracis*. Environ Microbiol 17(3):855–865. https://doi.org/10.1111/1462-2920.12548

Singh S, Kumar NK, Dwiwedi P, Charan J, Kaur R, Sidhu P, Chugh VK (2018) Monoclonal antibodies: a review. Curr Clin Pharmacol 13:85–99. https://doi.org/10.2174/1574884712666170809124728

Sioud M (2019) Phage display libraries: from binders to targeted drug delivery and human therapeutics. Mol Biotechnol 614(61):286–303. https://doi.org/10.1007/S12033-019-00156-8

Smith GP (1985) Filamentous fusion phage: novel expression vectors that display cloned antigens on the virion surface. Science 228(4705):1315–1317. https://doi.org/10.1126/science.4001944

Stasi R, Pagano A, Stipa E, Amadori S (2001) Rituximab chimeric anti-CD20 monoclonal antibody treatment for adults with chronic idiopathic thrombocytopenic purpura. Blood 98:952–957. https://doi.org/10.1182/BLOOD.V98.4.952

Steinke S, Roth KDR, Ruschig M, Langreder N, Polten S, Schneider KT, Ballmann R, Russo G, Zilkens KJK, Schubert M, Bertoglio F, Hust M (2023) Antibody selection via phage display in microtiter plates. Methods Mol Biol 2702:247–260. https://doi.org/10.1007/978-1-0716-3381-6_12/COVER

Syed YY (2020) Ramucirumab: a review in hepatocellular carcinoma. Drugs 80:315–322. https://doi.org/10.1007/s40265-020-01263-6

Tian L, Jackson K, Chan M, Saif A, He L, Didar TF, Hosseinidoust Z (2022) Phage display for the detection, analysis, disinfection, and prevention of Staphylococcus aureus. Smart Med 1:1–16. https://doi.org/10.1002/SMMD.20220015

Trombetta C, Piccirella S, Perini D, Kistner O, Montomoli E (2015) Emerging influenza strains in the last two decades: a threat of a new pandemic? Vaccine 3:172–185. https://doi.org/10.3390/vaccines3010172

Trullas A, Delgado J, Koenig J, Fuerstenau U, Dedorath J, Hausmann S, Stock T, Enzmann H, Pignatti F (2021) The EMA assessment of encorafenib in combination with cetuximab for the treatment of adult patients with metastatic colorectal carcinoma harbouring the BRAFV600E mutation who have received prior therapy. ESMO Open 6:100031. https://doi.org/10.1016/j.esmoop.2020.100031

Tsai CW, Morris S (2015) Approval of Raxibacumab for the treatment of inhalation anthrax under the US food and drug administration 'animal rule'. Front Microbiol 6:1–5. https://doi.org/10.3389/fmicb.2015.01320

Vana DR (2018) Belimumab: therapeutic mechanism and current status of clinical trials. Biomed Res 29:3233–3238. https://doi.org/10.4066/biomedicalresearch.29-18-841

Wang D, Shan C, Liu J, Zhang R, Zhu G, Gao T, Chang H, Gao S, Bai C, Nie N, Zhang Q, Lin Y (2022) Efficacy and safety of belimumab for the treatment of refractory childhood-onset systemic lupus erythematosus: a single-center, real-world, retrospective study. Front Immunol 13:1–10. https://doi.org/10.3389/fimmu.2022.1067721

West RL, Zelinkova Z, Wolbink GJ, Kuipers EJ, Stokkers PCF, Van Der Woude CJ (2008) Immunogenicity negatively influences the outcome of adalimumab treatment in Crohn's disease. Aliment Pharmacol Ther 28:1122–1126. https://doi.org/10.1111/J.1365-2036.2008.03828.X

Wilson FR, Coombes ME, Wylie Q, Yurchenko M, Brezden-Masley C, Hutton B, Skidmore B, Cameron C (2017) Herceptin® (trastuzumab) in HER2-positive early breast cancer: protocol for a systematic review and cumulative network meta-analysis. Syst Rev 6:1–8. https://doi.org/10.1186/s13643-017-0588-2

Winkler MT, Bushey RT, Gottlin EB, Campa MJ, Guadalupe ES, Volkheimer AD, Weinberg JB, Patz EF (2017) Enhanced CDC of B cell chronic lymphocytic leukemia cells mediated by rituximab combined with a novel anti-complement factor H antibody. PLoS One 12:1–12. https://doi.org/10.1371/journal.pone.0179841

Wu Z, Ichinose T, Naoe Y, Matsumura S, Villalobos IB, Eissa IR, Yamada S, Miyajima N, Morimoto D, Mukoyama N, Nishikawa Y, Koide Y, Kodera Y, Tanaka M, Kasuya H (2019) Combination of Cetuximab and oncolytic virus Canerpaturev synergistically inhibits human colorectal cancer growth. Mol Ther Oncolytics 13:107–115. https://doi.org/10.1016/j.omto.2019.04.004

Yang W, Yoon A, Lee S, Kim S, Han J, Chung J (2017) Next-generation sequencing enables the discovery of more diverse positive clones from a phage-displayed antibody library. Exp Mol Med 49. https://doi.org/10.1038/emm.2017.22

Zaroff S, Tan G (2019) Hybridoma technology: the preferred method for monoclonal antibody generation for in vivo applications. BioTechniques 67:90–92. https://doi.org/10.2144/btn-2019-0054

Zhang Y (2023) Evolution of phage display libraries for therapeutic antibody discovery. MAbs 15(1):2213793. https://doi.org/10.1080/19420862.2023.2213793

Chapter 2
Algae and Cyanobacteria as Food Supplements

Abiola M. Asowata-Ayodele, Olumoye Ezekiel Oyetunji, Temitayo Esther Adeyeoluwa, Taiwo Godspower Olawoyin, Babawale Peter Olatunji, and Timothy Bamgbose

Abstract Algae and cyanobacteria have emerged as valuable sources of dietary food supplements due to their rich nutrient composition and potential applications in various fields, including biotechnology. *Spirulina* is a prominent example, widely recognized and utilized as a "health food," available in tablet or dried powder form at supermarkets. Notably, cyanobacteria encompass a spectrum of nutrients, including proteins, amino acids, fructose, fatty acids, pigments, minerals, digestive and restricting enzymes, iron, and vitamin B6. Consumption of these microorganisms as food supplements offers a myriad of health benefits, along with medicinal, nutraceutical, pharmaceutical, and therapeutic advantages. For instance, *Microcystis aeruginosa* and *Thalassiosira weissflogii*, among others, boast significant concentrations of essential amino acids, rendering them particularly valuable in supplement formulations. Moreover, microalgae have attracted increased attention in recent years, primarily due to their pivotal role in dietary supplementation and their potential utility in diverse biotechnological applications. This growing interest underscores the significance of algae and cyanobacteria as vital contributors to human nutrition and health, paving the way for innovative advancements in food science and biomedicine. As research continues to unravel the multifaceted benefits of these microorganisms, their integration into the food and pharmaceutical industries holds promise for enhancing human well-being and fostering sustainable development.

Keywords Algae · Cyanobacteria · Pigments · Protein · Biomass · Supplements

A. M. Asowata-Ayodele (✉) · O. E. Oyetunji · T. E. Adeyeoluwa · T. G. Olawoyin · B. P. Olatunji
Department of Biosciences and Biotechnology, Faculty of Science, University of Medical Sciences, Ondo, Ondo State, Nigeria
e-mail: mayodeleasowata@unimed.edu.ng

T. Bamgbose
Microbiology Unit, Department of Biological Sciences, Kings University, Odeomu, Osun State, Nigeria

V. Kothari et al. (eds.), *Microbial Products for Health and Nutrition*,
https://doi.org/10.1007/978-981-97-4235-6_2

2.1 Introduction

Cyanobacteria, one of the earliest forms of life on earth, dating back about 3 billion years (Olson 2006), have received considerable attention for their multifaceted benefits. They were originally classified as phototrophic organisms with cell walls similar to those of a subgroup of Gram-negative bacteria and were usually grouped with microalgae in the phylum Plantae (Kollmen and Strieth 2022). In addition to their basic ecological role, cyanobacteria offer several advantages in various fields. The consumption of cyanobacteria as a dietary supplement holds enormous promise due to their various health, medicinal, pharmaceutical, and therapeutic benefits. In particular, they act as irreplaceable resources to promote plant growth, produce fertilizers, and produce biofuels, reflecting their versatility and importance in sustainable practices.

In addition, blue-green algae play a key role in human nutrition and health and act as a dietary supplement that increases food intake. Rich in essential nutrients and bioactive compounds, they promote overall well-being and offer potential tools to combat malnutrition and food insecurity. In addition, the biochemical richness of cyanobacteria extends to their utility in various industrial applications. Their natural pigments, including chlorophyll-a and phycobiliproteins, are used as dyes in many fields, from food and cosmetics to textiles and medicine. In addition, cyanobacteria are valuable sources of metabolites with antibiotic, fungicidal, and antiviral properties, providing opportunities for drug discovery and development (Chakdar et al. 2012).

One way to improve health is through the use of dietary supplements, which have attracted the attention of various researchers due to the growing desire for natural solutions that promote general health and well-being (Singh et al. 2017). More interesting emerging competitors are algae and cyanobacteria. These microalgae have emerged as a new and promising source of dietary supplements, gaining attention due to their impressive nutritional composition. Dinu et al. (2016) shed light on their rich biochemical composition and revealed a treasure trove of essential nutrients. From proteins and essential amino acids to fatty acids, pigments, minerals such as iron, vitamins, and even digestive and regulatory enzymes, microalgae offer a compelling nutritional profile, particularly suitable for vegetarians and vegans looking to address potential nutritional deficiencies.

However, the potential of algae extends far beyond simply providing essential nutrients. Extensive research efforts are exploring the exciting possibilities of integrating microalgae into the world of nutraceuticals, a broad category encompassing products that bridge the gap between food and medicine. Nutraceuticals include a diverse range of options, from dietary supplements and herbal extracts to personalized dietary plans and even fortified food products like cereals, sauces, and beverages. All these options share a common thread: they offer functional benefits beyond basic nutrition, aiming to address specific health concerns and promote well-being. The discovery of microalgae's potential as a dietary supplement presents a groundbreaking opportunity to expand the landscape of nutritional interventions and redefine our approach to health maintenance. Recent research, reported by Charles Adetunji et al. (2023), looked at the unique ecological roles of algae and cyanobacteria. These marine-dominated organisms, encompassing a vast array of species, occupy a significant portion of our planet; nearly 70% of the earth's surface is dominated by marine environments, with algae playing a leading role in these ecosystems.

Considering that each species contains a different biochemical composition, this offers a unique advantage for the targeted selection of new functional food ingredients with specific health benefits. A good example of this targeted approach is the well-documented research on *Spirulina*, a type of cyanobacteria (commonly called blue-green algae) that has received considerable attention for its potential nutritional applications. However, it is critical to recognize the potential challenges associated with consuming cyanobacteria. Studies such as Teresa et al. (2016) emphasized the importance of addressing concerns about potential contaminants such as arsenic. Their research shows that processing algae products causes changes in the digestive system that can release a significant part of the arsenic they contain. This emphasizes the importance of strict safety standards and strict regulations. Many governmental organizations have recognized this potential risk and established strict rules for the introduction of new products, including algae and cyanobacteria (Gantar and Zorica 2008). These regulations ensure consumer safety and promote responsible manufacturing practices in the food industry.

While these challenges require careful consideration, ongoing research into the health benefits of algae and cyanobacteria continues to yield promising findings. Studies point toward potential applications in areas like cardiovascular health, immune function, and metabolic regulation. Their impressive array of functional properties includes antioxidant capabilities, anti-inflammatory effects, and the potential to influence the gut microbiota, all of which contribute to promoting overall well-being through a holistic approach. With the growing demand for algae-based supplements, the development of sustainable cultivation methods becomes paramount. These methods not only ensure a reliable supply chain but also minimize environmental impact. Advancements in biotechnology and aquaculture hold the key to scalable and eco-friendly production processes, ensuring the availability of high-quality algae supplements for consumers worldwide.

The exciting health benefits of algae and cyanobacteria as dietary supplements will be elucidated here, looking into their health benefits and the risks of consumption. By unlocking their full potential through scientific research and innovation, cyanobacteria stand to revolutionize various sectors and contribute to a more resilient and sustainable healthy future, which is part of the sustainable development goal.

2.2 Cultivation of Algae for Commercial Purposes

Cultivation of algae for commercial purposes involves a variety of methods, with the two main approaches being natural and laboratory cultivation. Industrial production systems generally use shallow passages where blade mixing facilitates the integration of *Spirulina* cultures. Although many industrial operations depend on these controlled environments, *Spirulina* is still harvested accidentally from wild populations. For example, in the Valley of Mexico, the Sosa Texcoco Company collected *Spirulina maxima* from a part of Texcoco located at 2200 m above sea level. In this semi-tropical environment, the average annual temperature remains at 18 °C. The extraction of *Spirulina* from Lake Texcoco began in 1968 after research in collaboration with the French Petroleum Institute. In particular, this facility is the largest *Spirulina* biomass production unit. The farming method is called "semi-natural," which involves both day and night harvesting to optimize yields.

Spirulina cultivation encompasses a range of mediums, particularly those containing inorganic and partially degraded organic nutrients. To assess growth and biochemical characteristics under controlled conditions, various *Spirulina* species have been cultivated. In a study conducted by Bhattacharya and Shivaprakash (2005), three species of *Spirulina* were evaluated: *Spirulina platensis*, *Spirulina laxissima*, and *Spirulina lonar*. Among these species, *S. platensis* exhibited the highest growth rate, biomass production, pigment concentration, and lowest intracellular phenolics. Notably, *S. platensis* demonstrated the quickest growth rate, emphasizing the importance of laboratory cultivation methods for maximizing *Spirulina* production. The advancements in cultivation techniques have led to the development of innovative approaches to optimize *Spirulina* yield and quality. Photobioreactors, for example, offer controlled environments for algae cultivation, allowing precise regulation of environmental parameters such as light intensity, temperature, and nutrient availability. These closed systems minimize contamination risks and enable continuous monitoring and optimization of growth conditions to maximize productivity.

Moreover, efforts are underway to explore sustainable cultivation practices that minimize environmental impact and resource consumption. Integrated algae cultivation systems, which utilize wastewater or carbon dioxide emissions as nutrient sources, demonstrate the potential for algae cultivation to contribute to environmental remediation and carbon sequestration. By harnessing waste streams as nutrient sources, these systems mitigate pollution and reduce reliance on conventional fertilizers, thereby promoting sustainability in algae cultivation. The cultivation of algae, particularly *Spirulina*, for commercial purposes, encompasses a range of methods aimed at maximizing yield, quality, and sustainability. While traditional raceway systems remain prevalent in industrial operations, laboratory cultivation techniques offer precise control over growth conditions to optimize productivity. Continued research and innovation in cultivation methods hold the key to unlocking the full potential of algae as a sustainable source of nutrition, bioactive compounds, and environmental solutions (Fig. 2.1).

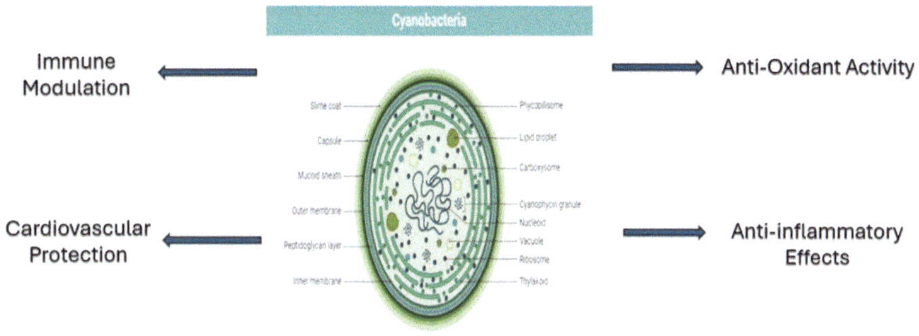

Fig. 2.1 Showing a basic overview of the cyanobacteria cell structure, the specific components and potential health benefits. Their relative abundance of the components can vary depending on the species of cyanobacteria. (Source: created in Biorender app and Microsoft PowerPoint)

2.3 Medical and Health Benefits of Using Cyanobacteria as Food Supplements

Because cyanobacteria can be antibacterial, antifungal, or insecticidal, they are a potential therapeutic agent, especially for diseases caused by other species. Shaieb et al. (2014) investigated the effectiveness of 15 cyanobacteria against seven bacteria and one fungus. All cyanobacterial extracts exhibited antibacterial or antifungal activity, although not all strains are universally effective against bacteria and fungi. In addition, the activity was affected by the solvent (ethanol or water). It is used to develop bioactive chemicals on a commercial basis for use in the pharmaceutical and nutritional industries (Chai et al. 2013).

The development of bioactive substances derived from cyanobacteria presents exciting opportunities for the pharmaceutical and nutraceutical industries, particularly when considering current and predicted market trends (as discussed by Chai et al. 2013). Research has revealed a wealth of health benefits associated with cyanobacteria, including antiviral, anticancer, antioxidant, anti-inflammatory, antiallergenic, and antibacterial properties. Additionally, they have been shown to lower cholesterol levels. These multifaceted properties make cyanobacteria a promising resource for combating various health concerns. Their ability to suppress chronic inflammation and oxidative stress suggests potential applications in preventing cardiovascular diseases (CVD) and nonalcoholic fatty liver disease (NAFLD).

Cyanobacteria create a wide range of unique biologically active chemicals, many of which are expected to be used in drug development. Some of the potential anticancer, antibacterial, antiviral, anti-inflammatory, and other qualities of some of the active ingredients from cyanobacteria may have been used at a commercial level. There is information available on the applications of cyanobacteria and microalgae within the human diet, as well as information on the regulations that control their use.

2.3.1 Antimicrobial Properties

Due to *Spirulina's* high amino acids and ease of production in ponds, it has historically been valued as a food source. The Aztecs often eat it, and it might be an important part of nutrition in the tropics. In addition, it is used in several Asian dishes. In the USA, *Spirulina* is extensively used as "health food" and is frequently purchased in supermarkets as tablets or dried powder. Both microalgae and cyanobacteria (blue-green algae) are eaten by people. They have drawn considerable interest recently because of their potential applications in biotechnology.

Microalgae and cyanobacteria are powerhouses of bioactive chemicals, boasting antiviral, antibacterial, antifungal, and even anticancer properties. These potent compounds can be produced by the algae when grown under specific conditions and stresses. Their potential in drug development and nutraceuticals is undeniable. Understanding how to discover, extract, screen, and utilize these bioactive treasures unlocks exciting possibilities for pharmaceutical and other industries, as highlighted by Hassan et al. (2022). Cyanobacteria have traditionally been included in human diets in non-Western civilizations. Currently, microalgae and cyanobacteria are gathered from their native habitats, grown under careful monitoring, and then dispersed.

2.3.2 Biomass Components

Cyanobacteria, renowned for their rich nutritional profile, offer a wealth of biomolecules within their biomass, making them invaluable as dietary supplements. Carotenoids, vitamins, and a plethora of other essential biomolecules are encapsulated within the microalgal biomass, serving as vital constituents of dietary formulations. Moreover, the biomass of cyanobacteria contains substantial quantities of protein and various other essential elements crucial for human health. For instance, Arthrospira, a genus of cyanobacteria, not only synthesizes lipids (5–6%) and fatty polyunsaturated acids (1.5–2%), but also produces a remarkable protein composition ranging from 50 to 70%. This protein content, coupled with the presence of diverse vitamins, underscores the nutritional significance of cyanobacterial biomass, positioning it as a promising resource for enhancing dietary intake and promoting overall well-being. (Kollmen and Strieth 2022).

2.3.3 Antioxidants and Anti-inflammatory Potential

Both the gradual fat accumulation in artery walls and atherosclerosis are inflammatory diseases. According to a study conducted by Rasool et al. (2006), oral treatment of 800 mg of *Spirulina* once a day for 8 days dramatically reduced adjuvant-induced arthritis in mice. This implies that *Spirulina* may be useful in avoiding inflammatory bowel disease (Coskun et al. 2011). Cyanobacteria's gamma-linolenic acid (GLA) and phycocyanin (PC) are two components that have been associated with their anti-inflammatory capabilities. SP contains 1.3% GLA. GLA may be transformed into dihomo-GLA, which is subsequently exposed to oxidative damage by cyclo-oxygenases and lipoxygenases to produce anti-inflammatory eicosanoids (Chai et al. 2013). GLA is often used to treat rheumatoid arthritis, atopic eczema, acute respiratory distress syndrome, lung cancer, and asthma, and to improve serum lipid levels (Chai et al. 2013). The anti-inflammatory effects of *Spirulina* are usually linked with the antioxidant and antiviral properties of Cyanobacteria-Phycocyanin (C-PC) and Rhodophyceae-Phycocyanin (R-PC) (Chai et al. 2013). C-PC, a naturally occurring blue pigmentation, takes up around 14% of the dry matter of *Spirulina*, according to Deng and Chow (2010). The paw edema index was reduced in mice treated with C-PC when inflammation was induced by glucose oxidase, confirming the efficient use of C-PC in *Spirulina* as an anti-inflammatory (Chai et al. 2013).

Also, research indicates that cyanobacteria are rich in essential amino acids. For instance, a study by Schmidtlr and Jonasdottir (1997) found that *Microcystis aeruginosa* contains 44% more essential amino acids compared to *Thalassiosira weissflogii*. While both species boast significant protein content (approximately 40% dry weight for *T. weissflogii* and 30% for *M. aeruginosa*), diatoms like *T. weissflogii* tend to have higher concentrations of fructose and rhamnose, making them a good source of energy, along with lower glucose levels compared to other autotrophs. Freshly prepared meals containing cyanobacteria offer a compelling nutritional profile. These meals provide essential amino acids, gamma-linolenic acid (GLA), dietary fiber, B vitamins, minerals like calcium, phosphorus, and iron, and a range of bioactive pigments such as beta-carotene, xanthophyll, and chlorophyll.

Microalgae hold a wealth of bioactive compounds, and research is increasingly focusing on the health benefits associated with specific peptides released through enzymatic or other hydrolytic treatments. These peptides exhibit promising antioxidant and free radical-scavenging properties.

For example, studies have identified peptides like Leu-Asn-Gly-Asp-Val-Trp from *Chlorella ellipsoidea* and Val-Glu-Cys-Tyr-Gly-ProAsn-Arg-Pro-Gln-Phe (chlorella-11) from *Chlorella vulgaris*. Notably, *chlorella*-11 appears to demonstrate anti-inflammatory effects. In animal models, it has been observed to suppress blood levels of TNF-α, a key inflammatory marker, and reduce the production of prostaglandin E2 (PGE2) after exposure to lipopolysaccharide (LPS), a known inflammatory trigger. Furthermore, the anti-inflammatory properties of peptides and other components in *Chlorella* products might contribute to their potential antidiabetic effects. Research suggests a connection between low-grade inflammation and diabetes, and *chlorella*'s anti-inflammatory properties could play a role in managing the disease. Similarly, studies exploring the antihypertensive effects of *Chlorella vulgaris* point towards specific oligopeptides within the algae. When administered to spontaneously hypertensive rats at a dose of 200 mg/kg body weight, these oligopeptides, such as Ile-Val-Val-Glu, Ala-Phe-Leu, Phe-Ala-Leu, Ala-Glu-Leu, and

Val-Val-Pro-Pro-Ala (Ramos-Romero et al. 2021), have demonstrated the ability to inhibit angiotensin-converting enzyme (ACE), a key player in blood pressure regulation.

2.3.4 Effect of the Cyanobacteria on Cholesterol

Nostoc commune var. *sphaeroides* controlled the total plasma cholesterol; triglyceride levels were also reduced by 20% with the *N. commune* supplement (Heather et al. 2009). When *N. commune* was given to mice, fecal neutral sterol synthesis skyrocketed while intestinal cholesterol absorption sharply reduced. The activation of genes such as 3-hydroxy-3-methylglutaryl coenzyme and sterol regulatory factor interacting protein-2 that influence cholesterol levels, A reductase levels were increased when *N. commune* was administered to mice and *N. commune* supplementation exhibited a hypocholesterolemic impact in mice by reducing intestinal cholesterol absorption and promoting fecal free sterol outflow (Heather et al. 2009).

Research in mice fed a high-fat diet by Ramos-Romero et al. (2021) suggests that an ethanolic extract of *A. maxima* (at doses of 150 or 450 mg/kg/day) may offer various metabolic advantages, such as:

(a) Reduced blood sugar and lipid levels: The extract appears to decrease fasting glucose and blood lipid concentrations.
(b) Weight management: *A. maxima* may contribute to weight management by lowering body weight and reducing both subcutaneous and visceral adipose tissue (fat accumulation).
(c) Regulation of fat metabolism: The extract might influence gene expression related to the browning of adipose tissue (a process promoting energy expenditure) and potentially suppress genes associated with fat cell formation.

Similarly, studies in rats fed a high-fat diet have shown that dried *A. maxima powder* (at doses of 62.5, 125, or 250 mg/kg) may help with:

(a) Weight gain prevention: The powder appears to reduce weight gain and mitigate an increase in the white adipose tissue (WAT) index caused by the high-fat diet.
(b) Metabolic improvements: *A. maxima* may help lessen changes in metabolic markers like lipid profile, serum levels of adiponectin (regulates fat storage), leptin (regulates appetite), TNF-α (inflammatory marker), glucose, and insulin. These findings suggest that *A. maxima*'s potential benefits for metabolism might be linked to its ability to activate the AMPK pathway, a cellular signaling system that plays a crucial role in regulating energy balance.

2.3.5 Cyanobacteria's Atheroprotective Properties

Atheroprotective properties are observed in lipid-lowering cyanobacteria. For example, Riss et al. (2007) found that hamsters fed an atherogenic diet supplemented with phycocyanin (PC) rich in *Spirulina* showed significantly reduced fatty lesion formation and superoxide anion production in the heart. According to studies, rats fed neutral detergent fibers and oxalic acid solubilizer had much lower blood cholesterol levels than control animals. Thus, fiber has significant benefits for lowering blood cholesterol levels (Chai et al. 2013).

For example, dyslipidemia is characterized by elevated plasma levels of low-density lipoprotein cholesterol (LDL-C) and reduced levels of high-density lipoprotein cholesterol (HDL-C) and stands as a prominent risk factor for cardiovascular diseases (CVD) (Betteridge 2011). While statins have conventionally served as a primary intervention to lower LDL-C and total cholesterol levels in individuals with hypercholesterolemia, their use often accompanies adverse effects such as myalgia and hepatic impairment (McKenney et al. 2006). Consequently, the pursuit of alternative hypocholesterolemic interventions has gained traction, with various natural products and dietary components emerging as promising substitutes for statins in managing dyslipidemia.

Among these alternatives, soy, plant sterols, water-soluble fibres, omega-3 fatty acids, and soy products have garnered attention for their potential lipid-lowering effects (Theuwissen and Mensink 2008). These dietary interventions offer multifaceted benefits beyond mere cholesterol reduction, including anti-inflammatory properties, antioxidant effects, and modulation of lipid metabolism. Incorporating these components into the diet presents a holistic approach to managing dyslipidemia, addressing not only cholesterol levels but also mitigating associated cardiovascular risks. Moreover, unlike pharmacological interventions like statins, these natural products typically exhibit favorable safety profiles, minimizing the likelihood of adverse reactions and enhancing overall cardiovascular health. Hence, due to its many advantages, cyanobacteria is used as a major dietary and therapeutic component in various parts of the world.

2.3.6 Lipid-Lowering Effect of Cyanobacteria

Spirulina stands out as a nutritional powerhouse. Unlike many other foods, it boasts a remarkably low cholesterol content. At the same time, it delivers a wealth of essential nutrients, including protein, vitamins, minerals, and a balanced blend of omega-3 and omega-6 fatty acids. In comparison to other food sources, these algae have a significantly lower percentage of total fat and cholesterol. This remains true even though these algae also contain a much smaller percentage of total fat. Furthermore, phycocyanin pigment, polysaccharides, and phenols are among the many bioactive substances found in high concentrations in *Spirulina*. These chemicals participate in several biological processes, including anti-inflammatory and antioxidant ones. This is true even though these algae also contain a significant amount of total fat. Consequently, *Spirulina* has been incorporated into a wide range of functional foods, therapeutic foods, and dietary supplements (Bashandy et al. 2011).

A blend of unsaturated fatty acids identified in the ethanolic fraction of *Anthospira platensis* biomass has been observed to induce hypolipidemia in Wistar rats fed with a high-fat (HF) diet. Notably, this effect is accompanied by an upregulation of the AMPK-α pathway in the liver, while pathways involving SREBP-1c, 3-hydroxy-3-methyl glutaryl coenzyme A reductase, and acetyl CoA (HMG-CoA) are downregulated. Furthermore, the extract appears to modulate the gut microbiota composition, leading to a decrease in the population of *Turicibacter*, *Romboutsia*, *Phascolarctobacterium*, *Olsenella*, and *Clostridium*, alongside an increase in putatively beneficial bacteria such as *Prevotella*, *Alloprevotella*, *Porphyromonadaceae*, *Barnesiella*, and *Paraprevotella*. Interestingly, these alterations correlate positively with serum levels of triglycerides (TAG), total cholesterol (TC), and low-density lipoprotein cholesterol (LDL-C), while showing a negative association with high-density lipoprotein cholesterol (HDL-C) levels. It has been shown that Cyanobacteria *Spirulina* species' ability to reduce cholesterol levels helps the body burn fat more efficiently. When *Spirulina* species were added with up to 15% by weight of food, the serum lipid levels of hyperlipidemic rats and mice fed with a high-fat/high-cholesterol diet or a high-fructose diet significantly improved (Bashandy et al. 2011).

In human clinical trials, supplementation with several species of *Spirulina* resulted in lipid reduction. Subjects with type 2 diabetes who consumed 2 g of *Spirulina* (SP) daily for 2 months had significantly lower plasma triglyceride levels and total cholesterol, including HDL:HDL and LDL:HDL ratios (Chai et al. 2013). Eight grams of SP supplementation per day for 12 weeks significantly reduced plasma triglyceride levels and blood pressure in patients with elevated baseline type 2 diabetes, while patients with elevated baseline total cholesterol and LDL cholesterol significantly reduced plasma lipid levels (Lee et al. 2008). In patients with hyperlipidemic nephrotic syndrome, 1 g/day of *Spirulina* successfully lowered plasma total cholesterol, LDL cholesterol, and triglycerides for 2 months (Samuels et al. 2002) (Table 2.1).

2.3.7 General Food Supplements Obtained from Algae

Cyanobacteria are abundant sources of minerals and vitamins, boasting a rich nutritional profile that includes significant concentrations of essential nutrients. Vitamin B12, in particular, is found in abundance in cyanobacteria, along with vitamins A, E, H, folacin, inositol, and pantothenic acid. These vitamins play crucial roles in various physiological processes, supporting overall health and well-being. In addition to vitamins, cyanobacteria are also rich in carbohydrates (polysaccharides) and proteins, providing essential macronutrients necessary for optimal bodily function

Table 2.1 Cyanobacteria used as supplements with nutrients and bioactive compounds

S/No	Cyanobacteria	Content	Comments	Commercial product
1	*Spirulina* (*Arthrospira plantensis*)	>60% proteins, carbohydrates, PFAs, sterols, some level of magnesium, manganese, iron, zinc, calcium, and selenium magnesium, manganese, iron, selenium, zinc, and calcium, among the best sources of vitamin B in nature, vitamins E, C, tocopherols, and a wide range of xanthophyll and mixed carotene phytopigments found in nature. Rich in gamma linolenic acid (GLA), which is transformed into prostaglandin E2 after first being changed to arachidonic acid in the body	Eaten as food in some parts of South America and the Philippines. Considered a "wonder food" and has been proposed by the United Nations to be the potential "food of the future." Grown on a large scale and reaped by modest procedures. Processed and marketed into different forms such as powder, tablet, flakes, and syrups. Easily digestible. Prostaglandin E2 reduces blood pressure and is crucial for the metabolism of fats	Powders, extracts, tablets, and beverages produced by Earthrise Nutritionals (California, USA), Hainan Simai Pharmacy Co. (China), Cyanotech Corp. (Hawaii, USA), and Myanmar Spirulina factory (Myanmar) Production yield: 0.06–0.09 g/L/day
2	*Anabaena* spp. (Fadl et al. 2020)	Contains several nutrients, including proteins, lipids, vitamins and minerals	Eaten as food in some parts of South America and the Philippines. Enhances growth and immunity of fish	Production yield: 0.06–1.27 g/L/day
3	*Nostoc commune* (Sies 1996; Kelman et al. 2009)	Rich in fibre and protein	Eaten as food in some parts of South America and the Philippines. Plays an important physiological and nutritional role	Production yield: 0.06–0.5 g/L/day
4	*Nostoc sphaeroides* (Zhu et al. 2022)	Includes minerals (Fe, K, Ca, Na, and P), vitamins (B1, B2, B5, and B8), fatty acids (ω-3 and ω-6), and protein. Also contains bioactive fats, phycobilin, polysaccharides, and phycobiliproteins	Potential ingredient for utilization in medicinal goods, dietary supplements, and cosmetics	Production yield: 16.69% weight

(continued)

Table 2.1 (continued)

S/ No	Cyanobacteria	Content	Comments	Commercial product
5	*Microalgae Spirulina* (Bishop and Zubeck 2012)	Contains about 70% proteins, vitamins B and E, phycocyanin, minerals, and omega-6 fatty acids	Documented as efficient in managing high blood pressure, diabetes, and in losing weight	Spirulina in powder form -(i. Spiruvita -C, Spirulina with amla, ii. Diavita-C, Spirulina with gymnema)
6	*Microalgae chlorella* (Nicoletti 2016; Naguib 2000)	Contains >50% protein, about 5% chlorophyll, carbohydrates, about 20% dietary fibre, several vitamins, and minerals	Immune stimulant and detoxicant. Effective in managing hypertension, ulcerative colitis, and in relieving fibromyalgia	Sun Chlorella, Lucky Vitamin™, Puritan's Pride, Prime Chlorella™ Distribution Inc., HerbMark. Production yield: 0.22 g/dL
7	*Aphanizomenon flos aquae* (AFA) (Carmichael et al. 2000; Bishop and Zubeck 2012; Amato et al. 2021)	Contains proteins, polyunsaturated fatty acids, sterols, minerals, carotenoids, polycyanins, and a significant amount of -phenylethylamine (PEA)	Considered a "superfood"	Supplements produced by Life Enthusiast Co-op, Cell Tech International Inc., AquaSource and Klamath Valley Botanicals, Inc. in North America. Production yield: 1106 kg/annum

Additional cyanobacteria used as food supplements are *Scytonema bohneri, Lyngbya limnetica, Oscillatoria foreaui, Oscillatoria acuminate, Oscillatoria calcuttensis, Spirulina platensis, and Spirulina pacifica* (Mišurcova et al. 2010), *Calothrix fusca, Gloeocapsa livida* (Rajeshwari and Rajashekhar 2011), *Phormidium abronema, Lyngbya cryptovaginata, Oscillatoria limnetica, Tolypothrix tenuis*, and *Oscillatoria tenuis* (Subramanian et al. 2014)

(Mišurcova et al. 2010; Hii et al. 2016). Furthermore, cyanobacteria contain fatty acids, which are essential for cellular structure and function, contributing to overall health and vitality (Amato et al. 2021).

Pigments found in cyanobacteria contribute not only to their color but also to their nutritional value. These pigments, such as chlorophyll and carotenoids, possess antioxidant properties and play a role in protecting cells from oxidative damage (Zhu et al. 2022; Nicoletti 2016). Additionally, cyanobacteria contain a variety of phytochemicals, including phenolic compounds, phenylpropanoids, and flavonoids, which have been associated with various health benefits, including anti-inflammatory and antioxidant effects (Singh et al. 2017).

2.4 Health Risks of Consuming Cyanobacteria

The utilization of algae-derived supplements has experienced a surge in popularity owing to their inherent nutritional composition and the possible positive impacts they may have on human health. Nevertheless, it is crucial to take into account the existence of arsenic in these supplements and its potential for harm. There are multiple types of arsenic species, which can be categorized into inorganic species (i-As) like arsenite [As (III)] and arsenate [As (V)] and organic compounds (o-As) that contain an As-C link. Organic arsenic compounds that have been identified include arsenosugars (As-sug), arsenocholine (AC), arsenobetaine (AB), arsenolipids (AsL), monomethylarsonic acid (MMA), and dimethylarsinic acid (DMA), among other chemicals.

2.4.1 *Chronic Arsenic Toxicity*

In a study by Yokoi and Konomi (2012), rats fed a diet containing 3% powder exhibited a significant buildup of arsenic in their blood and tissues. The study observed a correlation between elevated arsenic accumulation and a series of physiological disruptions in the subjects. These disruptions included elevated body temperatures observed in mice with higher arsenic levels. Also, abnormal blood chemistry was reported from the blood tests carried out, which revealed abnormalities in blood chemistry, specifically a rise in plasma alkaline phosphatase activity and inorganic phosphorus levels. These findings are strongly suggestive of arsenic poisoning, as disruptions are characteristic symptoms associated with arsenic exposure. In a report by Cheyns et al. (2021), the BMDL-0.5 value for inorganic arsenic, namely 3 μg kg bw^{-1} day^{-1}, has been designated for the "general" population. Additionally, a BMDL-01 value of 0.34 μg kg bw^{-1} day^{-1} has been determined for "sensitive" population groups. The phrase "sensitive population groups" pertains to individuals who have an elevated susceptibility to cancer due to other factors, such as tobacco users and other related factors. The rationale behind employing distinct benchmark dose lower limit (BMDL) values for the two demographic cohorts stems from the observed tendency of smokers to have a heightened susceptibility to lung cancer subsequent to exposure to As in comparison to individuals who have never engaged in smoking.

Further, Cheyns et al. (2021) identified that the inorganic arsenic species were the most dangerous forms of arsenic, and these species have been extensively studied in terms of their toxicological qualities. The association between prolonged exposure to inorganic arsenic and various detrimental effects on human health is evident in the development of skin lesions, cancer (classified as Group 1 carcinogenic by the IARC in 2012), toxicity during development, impairment of neurological function, cardiovascular diseases, as well as abnormal glucose metabolism, and the onset of diabetes.

2.4.2 Enhance Bioavailability and Bioaccessibility of Arsenic Exposure

Numerous research endeavors have been conducted to examine the levels of arsenic present in supplements derived from algae. The significance of considering both the concentration of arsenic, particularly inorganic forms and its bioavailability was discussed. It has been documented that roughly 80% of arsenic is released from algal food items as a result of the transformations that take place in the gastrointestinal tract (GIT). Consequently, this released arsenic has the potential to become available for absorption. The term "bioavailability" pertains to the quantity of a contaminant found in food that is absorbed by the intestines, potentially entering the circulatory system and subsequently causing harmful effects.

The gastrointestinal tract (GIT) allows for the retention of substances, making them accessible for absorption within the intestines. According to Desideri et al. (2018), the bioaccessibility of arsenic is influenced by its chemical form and behavior. Therefore, it is important to consider the potential risk of harmful consequences associated with arsenic exposure through the eating of seaweed, especially one's found in contaminated water bodies. The bioaccessibility of inorganic arsenic exhibited a notable increase after the cooking process. Specifically, the bioaccessibility reached 73% in *Porphyra* sp. and 88% in *H. fusiforme*, suggesting that the potentially harmful inorganic arsenic can be absorbed by the intestinal mucosa. In a separate investigation, the bioaccessible percentage of inorganic arsenic in cooked *H. fusiforme* was found to be 26 μg/g of seaweed. This concentration raises concerns about the potential toxicological risk associated with the consumption of this product. The observed values exhibited a notable increase in comparison to the measurements obtained from unprocessed seaweed samples. Research has demonstrated that the process of cooking, specifically boiling, has a noteworthy effect on enhancing the bioaccessibility of inorganic arsenic (iAs) in edible seaweeds as compared to uncooked samples (Desideri et al. 2018).

2.4.3 Carcinogenicity

Cholpraipimolrat et al. (2017) conducted a study that examined the potential risks connected with the eating of Hijiki (a type of seaweed known to have high levels of inorganic arsenic). The findings indicate that the existence of arsenic in seaweed may partially contribute to the development of colorectal cancer cells. The growth stimulation by Hijiki may involve EGFR (Epidermal growth factor receptor) dependent and EGFR-independent ERK1/2 signaling pathways, with the potential involvement of mAChR.

In this study, Molin et al. (2014) examined how the poisonous and cancer-causing form of arsenic, known as iAs, is changed in the human body and eliminated through urine as two carcinogens called dimethylarsinate (DMA) and methylarsonate (MA).

This transformation process generates reactive substances. Limited information is available regarding the biotransformation of organoarsenicals. However, recent findings suggest that the conversion of arsenosugars and arsenolipids in seafood, specifically shellfish and seaweed, leads to the excretion of DMA in urine, potentially generating reactive trivalent arsenic compounds as well. According to the Agency for Toxic Substances and Disease Registry (ATSDR 2007), the administration of multiple doses of macro algae (MA) has been found to elicit various effects on the gastrointestinal tract, kidney, thyroid, and reproductive systems. There was a positive correlation between an increased proportion of urine MA and an elevated susceptibility to arsenic-related illnesses. It was determined that the presence of sulfur-reducing bacteria in the human gut is essential for the conversion of MA (V) into the thiolated monomethyl monothioarsonate compound. DMA has also been identified in marine macroalgae, such as kelp. According to the categorization of carcinogenicity by the International Agency for Research on Cancer (IARC Working Group on the Evaluation of Carcinogenic Risks to Humans 2012), the arsenic species known as MMA and DMA are classified as Group 2B, indicating their potential to be carcinogenic.

The utilization of $CaCo^{-2}$ cells as a representative model for the intestinal barrier has revealed that iAs exhibits significantly greater permeability (62%) across the intestinal wall compared to arsenosugars (Leffers et al. 2013a). However, it is worth noting that when human subjects consumed a pure arsenosugar compound, up to 98% of it was detected in urine (Raml et al. 2009). The bioavailability of arsenosugars to bladder cells is comparatively low compared to inorganic arsenic (iAs). However, the uptake of thio-DMAV, which is a cytotoxic metabolite of arsenosugars, is higher than that of As(III) in these cells (Leffers et al. 2013b). Additional research is needed to fully understand the metabolic processes, distribution within the body, and toxicological characteristics of organic arsenic compounds. However, existing evidence indicates that arsenosugars undergo metabolism and are thus believed to be bioavailable to humans. An instance of potential toxicity can be observed in As-sugar and As-lipids due to their metabolic conversion into other arsenicals, including DMAV. DMAV is a metabolite of inorganic arsenic that has been detected in urine and has been established as a tumor promoter and proven carcinogen in experimental animals (Luvonga et al. 2020).

Toxicity can manifest even at minimal concentrations ranging from 0.05 to 1 mg kg^{-1}. This includes the presence of arsenic (As) and other heavy metals, including mercury (Hg), cadmium (Cd), and lead (Pb), which have the ability to induce carcinogenic consequences when accumulated in the human body (Chijioke et al. 2020). It is noteworthy to emphasize that heavy metal ions have the potential to induce cancerous effects. This is attributed to their interactions with DNA and nuclear proteins, leading to subsequent damage (Briffa et al. 2020). In a study by Khandaker et al. (2021), it was found that the carcinogen potency factors for arsenic (As), nickel (Ni), and hexavalent chromium (Cr-VI) exhibited increased values. Notably, exposure to these heavy metals was identified as the primary contributor to the overall cancer risk associated with the lifelong use of seaweeds.

According to a study conducted by Yang et al. (2010), the consumption of *Porphyra* spp. showed a negative correlation with the incidence of breast cancer in premenopausal women. However, this association was not observed in postmenopausal women. Michikawa et al. (2012) conducted a study that revealed a positive correlation between the consumption of seaweed and the likelihood of developing thyroid cancer, particularly papillary carcinoma, in women who had reached menopause.

2.4.4 Cytotoxicity

In an investigation by Luvonga et al. (2020), it was reported that methylated arsenic metabolites are capable of crossing the placental barrier through in vivo tests. There is a hypothesis that MMAIII might present a higher potential for toxicity towards liver, skin, and lung cells in comparison to inorganic arsenic. Through a distinct inquiry, it has been recorded that DMAIII demonstrates greater toxicity in comparison to DMAV and inorganic arsenic. This is attributed to its neutral charge, which enables effective penetration into cells (up to 16%).

The toxicity of the methylation trivalent arsenicals, specifically MMAIII and DMAIII, is higher compared to that of As(III) and As(V). Furthermore, the toxicity of As(III) and As(V) is greater than that of the methylated pentavalent arsenicals, MMAV, and DMAV. The harmful effects of arsenic are closely connected to its metabolic processes and are significantly affected by the level of methylation and the valence state of its metabolites.

2.5 Conclusion

Microalgae and cyanobacteria have emerged as exciting contenders in the world of dietary supplements, offering a natural source of essential nutrients and potentially boasting a range of health benefits. While some species have been explored for their medicinal properties, the vast potential of this aquatic world remains largely untapped. The promise lies in the immense diversity of these organisms. Each species possesses a unique biochemical profile, potentially harboring specific bioactive compounds with targeted health applications. Research suggests potential benefits ranging from anti-inflammatory effects and cholesterol reduction to enhanced fat metabolism. However, translating this potential into reality necessitates a multi-prolonged approach. Firstly, large-scale cultivation practices need to be developed and optimized. The surge in demand for specific microalgal strains with proven health benefits highlights the need for sustainable and efficient production methods. This ensures a reliable supply chain while minimizing environmental impact. Secondly, rigorous scientific research is crucial. While preliminary studies are encouraging, further investigation is needed to validate the efficacy of specific algae strains for various health conditions. Additionally, research on the optimal extraction methods for bioactive compounds and determining safe and effective doses is essential. Therefore, establishing clear regulations and safety standards for algae-based supplements is paramount. This will ensure consumer protection and foster confidence in the industry. By addressing these challenges and harnessing the incredible potential of microalgae and cyanobacteria, it will be possible to unlock a new frontier in health and well-being.

References

Adetunji CO, Oloke JK, Dwivedi N, Ummalyma SB, Dwivedi S, Hefft DI, Adetunji JB (eds) (2023) Next-generation algae: Volume II: Applications in medicine and the pharmaceutical industry. Scrivener Publishing LLC, pp 177–196

Agency for Toxic Substances and Disease Registry (ATSDR) (2007) Toxicological profile for Arsenic. U.S. Department of Health and Human Services, Public Health Service, Atlanta, GA

Amato A, Terzo S, Marchesa P, Maffongelli A, Martorana M, Scoglio S, Mulè F (2021) Spasmolytic effects of *Aphanizomenon Flos Aquae* (AFA) extract on the human colon contractility. Nutrients 13(10):3445

Bashandy SA, Alhazza IM, El-Desoky GE, Al-Othman ZA (2011) Hepatoprotective and hypolipidemic effects of *Spirulina platensis* in rats administered Mercuric chloride. Afr J Pharm Pharmacol 5:175–182

Betteridge DJ (2011) Lipid control in patients with diabetes mellitus. Nat Rev Cardiol 8:278–290

Bhattacharya S, Shivaprakash MK (2005) Evaluation of three *spirulina* species grown under similar conditions for their growth and biochemicals. J Sci Food Agric 85:333–336

Bishop M, Zubeck M (2012) Evaluation of microalgae for use as nutraceuticals and nutritional supplements. J Nutr Food Sci 2(5):1000147

Briffa J, Sinagra E, Blundell R (2020) Heavy metal pollution in the environment and their toxicological effects on humans. Heliyon 6(9):e04691. https://doi.org/10.1016/j.heliyon.2020.e04691

Carmichael W, Drapeau C, Anderson D (2000) Harvesting of Aphanizomenon flos-aquae Ralfs ex Born. & Flah. Var. flos-aquae (cyanobacteria) from Klamath Lake for human dietary use. J Appl Phycol 12:585–595. https://doi.org/10.1023/A:1026506713560

Chai SK, Yang Y, Park Y, Lee J (2013) Health benefits of blue-green algae: prevention of cardiovascular disease and nonalcoholic fatty liver disease. J Med Food 16(2):103–111. https://doi.org/10.1089/jmf.2012.2468

Chakdar H, Jadhav SD, Dhar DW, Pabbi S (2012) Potential applications of blue green algae. J Sci Ind Res 71:13–20

Cheyns K, Demaegdt H, Waegeneers N, Ruttens A (2021) Intake of food supplements based on algae or cyanobacteria may pose a health risk due to elevated concentrations of arsenic species. Food Addit Contam Part A Chem Anal Control Expo Risk Assess 38(4):609–621. https://doi.org/10.1080/19440049.2021.1877834

Chijioke NO, Khandaker MU, Tikpangi KM, Bradley DA (2020) Metal uptake in chicken giblets and human health implications. J Food Compos Anal 85:103332

Cholpraipimolrat W, Suriyo T, Rangkadilok N, Nookabkaew S, Satayavivad J (2017) Hijiki and sodium arsenite stimulate growth of human colorectal adenocarcinoma cells through ERK1/2 activation. Food Chem Toxicol 110:33–41. https://doi.org/10.1016/j.fct.2017.09.052

Coskun ZK, Kerem M, Gurbuz N, Omeroglu S, Pasaoglu H, Demirtas C, Lortlar N, Salman B, Pasaoglu OT, Turgut HB (2011) The study of biochemical and histopathological effects of Spirulina in rats with TNBS-induced colitis. Bratisl Lek Listy 2011(112):235–243

Deng R, Chow TJ (2010) Hypolipidemic, antioxidant, and antiinflammatory activities of microalgae *Spirulina*. Cardiovasc Ther 28:33–45

Desideri D, Roselli C, Feduzi L, Ugolini L, Meli MA (2018) Applicability of an in vitro gastrointestinal digestion method to evaluation of toxic elements bioaccessibility from algae for human consumption. J Toxic Environ Health A 81(8):212–217

Dinu M, Abbate R, Gensini GF, Casini A, Sofi F (2016) Vegetarian, vegan diets and multiple health outcomes: a systematic review with meta-analysis of observational studies. Crit Rev Food Sci Nutr 57(17):3640–3649

Fadl S, Elsadany A, el Shenawy M, Sakr O, El Gammal G, Gad D, Abonorag M, Eissa I (2020) Efficacy of cyanobacterium *Anabaena* sp. as a feed supplement on productive performance and immune status in cultured *Nile tilapia*. Aquac Rep 17:100406

Gantar M, Zorica S (2008) Microalgae and cyanobacteria: food for thought. J Phycol 44(2):260–268. https://doi.org/10.1111/j.1529-8817.2008.00469.x

Hassan S, Meenatchi R, Pachillu K, Bansal S, Brindangnanam P, Arockiaraj J, KiranG S, Selvin J (2022) Identification and characterization of the novel bioactive compounds from microalgae and cyanobacteria for pharmaceutical and nutraceutical applications. J Basic Microbiol 62(9):999–1029. https://doi.org/10.1002/jobm.202100477

Heather ER, Kara RB, Elliot DJ, Chai SK, Park Y, Lu F, Carr TP, Lee J (2009) Hypocholesterolemic effect of Nostoc commune var. sphaeroides Ku¨tzing, an edible blue-green alga Eur. J Nutr 48:387–394. https://doi.org/10.1007/s00394-009-0025-y

Hii SL, Lim J, Ong WT, Wong CL (2016) Agar from Malaysian red seaweed as potential material for synthesis of bioplastic film. J Eng Sci Technol 7:1–15

IARC Working Group on the Evaluation of Carcinogenic Risks to Humans (2012) Arsenic, metals, fibres, and dusts. IARC Monogr Eval Carcinog Risks Hum 100(Pt C):11–465

Kelman D, Ben-Amotz A, Berman-Frank I (2009) Carotenoids provide the major antioxidant defence in the globally significant N2-fixing marine cyanobacterium Trichodesmium. Environ Microbiol 11:1897–1908

Khandaker MU, Chijioke NO, Heffny NA, Bradley DA, Alsubaie A, Sulieman A, Faruque MRI, Sayyed MI, Al-Mugren KS (2021) Elevated concentrations of metalloids in seaweed and the concomitant exposure to humans. Foods (Basel, Switzerland) 10(2):381. https://doi.org/10.3390/foods10020381

Kollmen J, Strieth D (2022) The beneficial effects of cyanobacterial co-culture on plant growth. Life 12:223. https://doi.org/10.3390/life12020223

Lee EH, Park JE, Choi YJ, Huh KB, Kim WY (2008) A randomized study to establish the effects of spirulina in type 2 diabetes mellitus patients. Nutr Res Pract 2:295–300

Leffers L, Ebert F, Taleshi MS, Francesconi KA, Schwerdtle T (2013a) In vitro toxicological characterization of two arsenosugars and their metabolites. Mol Nutr Food Res 57(7):1270–1282. https://doi.org/10.1002/mnfr.201200821

Leffers L, Unterberg M, Bartel M, Hoppe C, Pieper I, Stertmann J, Ebert F, Humpf HU, Schwerdtle T (2013b) In vitro toxicological characterisation of the S-containing arsenic metabolites thiodimethylarsinic acid and dimethylarsinic glutathione. Toxicology 305:109–119. https://doi.org/10.1016/j.tox.2013.01.007

Luvonga C, Rimmer CA, Yu LL, Lee SB (2020) Organoarsenicals in seafood: occurrence, dietary exposure, toxicity, and risk assessment considerations—a review. J Agric Food Chem 68(4):943–960. https://doi.org/10.1021/acs.jafc.9b07532

McKenney JM, Davidson MH, Jacobson TA, Guyton JR (2006) Final conclusions and recommendations of the National Lipid Association Statin Safety Assessment Task Force. Amer J Cardiol 97:89–94

Michikawa T, Inoue M, Shimazu T, Sawada N, Iwasaki M, Sasazuki S, Yamaji T, Tsugane S, Japan Public Health Center-based Prospective Study Group (2012) Seaweed consumption and the risk of thyroid cancer in women: the Japan public health center-based prospective study. Eur J Cancer Prev 21(3):254–260. https://doi.org/10.1097/CEJ.0b013e32834a8042

Mišurcova L, Kračmar S, Klejdus B, Vacek J (2010) Nitrogen content, dietary fiber, and digestibility in algal food products. Czech J Food Sci 28(1):27–35

Molin M, Ulven SM, Dahl L, Goessler W, Fliegel D, Holck M, Sloth JJ, Oshaug A, Alexander J, Meltzer HM, Ydersbond TA (2014) Urinary excretion of arsenicals following daily intake of various seafoods during a two weeks intervention. Food Chem Toxicol 66:76–88. https://doi.org/10.1016/j.fct.2014.01.030

Naguib YMA (2000) Antioxidant activities of astaxanthin and related carotenoids. J Agric Food Chem 48(4):1150–1154

Nicoletti M (2016) Microalgae nutraceuticals Foods 5(3):1–13

Olson JM (2006) Photosynthesis in the Archean era. Photosynth. Res. 88:109–117

Rajeshwari KR, Rajashekhar M (2011) Biochemical composition of seven species of cyanobacteria isolated from different aquatic habitats of western ghats, southern India. Braz Arch Biol Technol 54(5):849–857

Raml R, Raber G, Rumpler A, Bauernhofer T, Goessler W, Francesconi KA (2009) Individual variability in the human metabolism of an arsenic-containing carbohydrate, 2′,3′-dihydroxypropyl 5-deoxy-5-dimethylarsinoyl-beta-D-riboside, a naturally occurring arsenical in seafood. Chem Res Toxicol 22(9):1534–1540. https://doi.org/10.1021/tx900158h

Ramos-Romero S, Torrella JR, Pagès T, Viscor G, Torres JL (2021) Edible microalgae and their bioactive compounds in the prevention and treatment of metabolic alterations. Nutrients 13(2):563. https://doi.org/10.3390/nu13020563

Rasool M, Sabina EP, Lavanya B (2006) Anti-inflammatory effect of *Spirulina fusiformis* on adjuvant-induced arthritis in mice. Biol Pharm Bull 29:2483–2487

Riss J, Decorde K, Sutra T, Delage M, Baccou JC, Jouy N, Brune JP, Oreal H, Cristol JP, Rouanet JM (2007) Phycobiliprotein C-phycocyanin from Spirulina platensis is powerfully responsible for reducing oxidative stress. J Agric Food Chem 55:7962–7967

Samuels R, Mani UV, Iyer UM, Nayak US (2002) Hypocholesterolemic effect of spirulina in patients with hyperlipidemic nephrotic syndrome. J Med Food 5:91–96

Schmidtlr K, Jonasdottir SH (1997) Nutritional quality of two cyanobacteria: how rich is 'poor' food? Mar Ecol Prog Ser 151:1–10

Shaieb FA, Issa AA, Meragaa A (2014) Antimicrobial activity of crude extracts of cyanobacteria Nostoc commune and Spirulina platensis. Acta Biomed Sci 2:34–41

Sies H (1996) Antioxidants in disease, mechanisms and therapy. Academic Press, New York

Singh DP, Prabha R, Verma S, Meena KK, Yandigeri M (2017) Antioxidant properties and polyphenolic content in terrestrial cyanobacteria. 3 Biotech 7(2):134

Subramanian G, Anusha MB, Bai ALG, Latha R, Sasikala J, Chandrasakaran R (2014) Carbohydrate and protein content of five species of marine cyanobacteria from Pamban and Vadakadu (Rameswaram) coastal regions, Tamil Nadu, India. Int J Adv Interdiscipl Res 1(6):18–20

Teresa C-C, Beshai M, Maher W, Kelly T, Foster S (2016) Bioaccessibility and degradation of naturally occurring arsenic species from food in the human gastrointestinal tract. Food Chem 212:189–197

Theuwissen E, Mensink RP (2008) Water-soluble dietary fibers and cardiovascular disease. Physiol Behav 2008(94):285–292

Yang YJ, Nam SJ, Kong G, Kim MK (2010) A case-control study on seaweed consumption and the risk of breast cancer. Br J Nutr 103(9):1345–1353. https://doi.org/10.1017/S0007114509993242

Yokoi K, Konomi A (2012) Toxicity of so-called edible seaweed (*Sargassum fusiforme*) containing inorganic arsenic. Regul Toxicol Pharmacol 63(2):291–297

Zhu S, Xu J, Adhikari B, Lv W, Chen H (2022) *Nostoc sphaeroides* Cyanobacteria: a review of its nutritional characteristics and processing technologies. Crit Rev Food Sci Nutr 13:1–17

Chapter 3
Probiotic Bacteriotherapeutic Approaches in Oral Healthcare

Vedam Venkata Kanthi Vaishnavi, Ganapathy Sivadas, and Subramani Parasuraman

Abstract "Oral microbiome" denotes to a colony of microorganisms that typically reside inside the mouth and is thought to be the second most densely populated collection of bacteria in our body. The group of oral microbiota has a substantial impact on mouth and systemic health. The oral microbial biofilm is made up of over 700 different kinds of microorganisms, and each habitat supports a unique, species-specific microbial community. Numerous host and environmental variables impact the formation, growth, and perseverance of the oral microbiota. The longevity of various microbial communities primarily depends on the symbiotic relationship between them. In the mouth, saliva is a fluid medium to support this microbial community. Probiotics are live microbial species of bacteria that are shown to have positive effects when taken on systemic health. Most oral disorders that occur today, including tooth caries and periodontal diseases, occur mainly due to the perturbation of specific pathogenic microorganisms due to environmental and ecological changes. Ecological pressures include factors like diet-rich carbohydrates, decreased salivary flow, and acidic pH. Oral diseases are thus controlled first by bacteriotherapy against the pathogenic microorganisms directly or second by modulation of ecological changes in the oral environment. Probiotics in dentistry are used for the evasion of common pathological diseases. Healthy bacteria having a positive effect on the host to oral commensal status thereby decreasing the pathogenic effects on

V. V. K. Vaishnavi
Department of Oral Pathology, Faculty of Dentistry, AIMST University, Bedong, Malaysia
e-mail: vaishnavi@aimst.edu.my

G. Sivadas
Department of Pedodontics, Faculty of Dentistry, AIMST University, Bedong, Malaysia
e-mail: sivadas@aimst.edu.my

S. Parasuraman (✉)
Department of Pharmacology, Toxicology and Basic Health Sciences, Faculty of Pharmacy, AIMST University, Bedong, Malaysia

Centre for Global Health Research, Saveetha Medical College and Hospital, Chennai, India
e-mail: parasuraman@aimst.edu.my

V. Kothari et al. (eds.), *Microbial Products for Health and Nutrition*, https://doi.org/10.1007/978-981-97-4235-6_3

the host in oral diseases. In this chapter, the effect of probiotic bacteriotherapeutic approaches in oral healthcare is discussed.

Keywords Probiotics · *Lactobacillus* · Microflora · Oral microbiome · *Streptococcus*

3.1 Oral Microbiome

"Oral microbiome" denotes the group of microorganisms that normally found in the oral cavity (Cho et al. 2021). This term was invented by Nobel Laureate *Joshua Lederberg* to showcase the natural community of normal and pathogens that are elements of health and disease. In specific, the "oral microbiota" was documented by *Antony Van Leeuwenhoek* on plaque biofilm as living animalcules (Morrison et al. 2023).

Most often *oral microbiome* and *oral microbiota* are used synonymously; these terms are often interchangeable. But both terminologies represent different but related concepts. *Microbiota* points to the composition and taxonomy of useful or harmful microorganisms that are human-dependent living in a given environment, while *microbiome* refers to the joint genome of microbes in a particular atmosphere. They exist in the oral mucosal tissues of the host as "harmless" bacteria. This bacterium arrests the aggregation of pathogenic species, provides safeguard to human by a robust immunity, and alters inflammatory cytokines and growth factors production, thereby assisting in upholding the balance with human tissues (Jain and Sharma 2012).

Although several groups of microorganisms exist in the human body, oral microbiota is considered to be the second most densely populated and exhibits diverse types of microbial communities. Over 700 species of microbes constitute the oral microbial biofilm; each habitat provides a distinct species-specific microbial community (Faran Ali and Tanwir 2012). A microbial ecosystem comprises bacteria, viruses, fungi, protozoa, and archaea residing in various sites (microbial niches) of the mouth (Table 3.1) (Sharma et al. 2018; Chugh et al. 2020; Giordano-Kelhoffer et al. 2022).

Molecular studies revealed the presence of distinct microbial communities on dental plaque biofilm and other soft tissue mucosal sites. They are usually seen on the tongue, cheeks, gums, palate, tonsils, and floor of the mouth and throat (Yamashita and Takeshita 2017). Of this, the tongue microbiota is chosen for estimation as it is elevated in biomass and could influence the composition of microbial communities at other mucosal sites. There are six major phyla (96% taxonomy) that establish the oral bacteria including *Firmicutes (Staphylococcus, Streptococcus, Lactobacillus, Peptostrptococcus, Gemella Veillonella), Bacteroidetes (Prevotella, Capnocytophaga, Porphyromonas, Bacteriodes), Proteobacteria (Neisseria,*

Table 3.1 Normal microflora of the oral cavity

Dental plaque	Keratinized mucosa	Non-keratinized mucosa	Specialized mucosa (Dorsum of the tongue)	Saliva
Supragingival plaque: *Actinomyces, Veillonella, Corynebacterium, Neisseriaceae, Fusobacterium, Streptococcus, Capnocytophaga* **Subgingival plaque:** *Streptococcus, Prevetollaceae, Fusobacterium, Gemella, Neisseriaceae, Campylobacter, Actinomycetales, Corynebacterium*	**Hard palate:** *Streptococcus, Prevotellaceae, Firmicutes, Pasteurellaceae, Veillonella, Neisseria, Gemella, Mogibacterium, Catonella* **Gingiva:** *Streptococcus, Veillonella, Pasteurellaceae, Proteobacteria*	*Atopobium, Streptococcus, Bacilli, Gemella, Pasteurellaceae, Prevotellaceae, Acidobacteriaceae, Neisseriaceae, Catonella, Bacteriodetes, Actinobacteria Firmicutes, Proteobacteria*	*Staphylococcus, Clostridiales, Streptococcus, Firmicutes, Actinomycetales, Proteobacteria, Haemophilus, Bacteroidetes, Prevotellaceae, Neisseriaceae, Synergistes, Gemella, Peptostreptococcus, Fusobacterium, Veillonella, Bacilli, Lactobacillales, Treponema, Pasteurellaceae*	*Proteobacteria, Firmicutes, Veillonella, Bacteriodes, Streptococcus, Neisseria, Prevotella, Actinomycetales, Pasteurellaceae, Fusobacterium, Neisseria, Tannerella*
Common Pathological lesions of the oral cavity				
Dental caries	*Actinomyces, Lactobacillus, Bifidobacterium, Propionibacterium, Streptococcus mutans, Candida albicans*			
Periodontitis	*Prevotella, Actinomyces, Fusobacterium, Treponema denticola, Tannerella forsythia, Actinobacillus actinomycetemcomitans, Eubacterium saphenum, Filifactor alocis, Anaeroglobus germinates*			
Gingivitis	*Treponema, Fusobacterium, Streptococcus, Actinomyces, Leptotrichia buccalis, Porphyromonas gingivalis*			

Haemophilus, *Eikenella*, and *campylobacter)*, *Actinobacteria (Actinomyces*, *Corynebacterium*, *Propionibacterium)*, *Spirochaetes (Treponema)*, and *Fusobacteriota (Fusobacterium)*. The residual 4% of the taxa are comprised of the phyla *Euryarchaeota*, *Chloroflexi*, *Chlamydia*, *Tenericutes*, *Synergistetes*, and *Saccharibacteria* (Barboza-Solís and Acuña-Amador 2019).

"Human Microbiome Project (HMP)" was established by the National Institute of Health (NIH) to characterize the healthy microbiota of humans in five sites of the body, namely nasal passages, skin, mouth, gastrointestinal mucosa, and urogenital region, and evaluate the role in health and pathology (Pujia et al. 2017). It was observed that the mouth is one of the sites with extensive microbial colonization and diversity. Microbial analysis included samples collected from supragingival dental plaque, subgingival dental plaque, nonkeratinized mucosa, keratinized mucosa (palate and gingiva), specialized areas (dorsum of the tongue and tonsils), and saliva (McBain et al. 2009). These microorganisms live in symbiotic relations with other microbes in a stable environment. This project was later extended to "Human Microbiome Project 2 (HMP2)," which included host microbial communication, molecular mechanisms, and immunity (Cho et al. 2021).

3.1.1 Oral Microbiome in Children

The microorganisms of the oral cavity increase progressively in children. The literature states that as the womb of the mother is sterile, and at first all babies were considered to be microbial-free (Escherich 1989). However, recent studies showed the presence of intrauterine environmental colonization of microbes in 70% of pregnant women (Deo and Deshmukh 2019). The similarity of the oral microorganisms with the placenta of pregnant women indicates an early connection with the microorganisms of the mother through the umbilical cord (Nuriel-Ohayon et al. 2016).

Communities of oral microflora in children are greatly subject to change by the mode of delivery and the feeding habits at later stages. During the delivery, the infants develop the microorganisms from the vagina of the women through normal delivery or the skin following the cesarian delivery. *Fusobacterium nucleatum* is the most cultivable microorganism found in infants. Breast-fed babies have lactic acid bacteria which is absent in formula milk-fed infants (Van Den Elsen et al. 2019).

Initially, microbial colonization begins soon after birth in babies with *Streptococcus salivarius*, and thereafter continues with the accumulation of more *Streptococcus*, *Lactobacillus*, *Actinomyces*, *Neisseria*, and *Veillonella*. As teeth emerge (approx. 6–8 months), more teeth surfaces and gum regions for microbial colonization develop. Also, this acquisition of microflora continues in infants for several years. With time, microbial diversity occurs due to changes in the environment, habitat, and vaccination status in the children. The oral environment can alter the microbial biology of the oral cavity. Finally, the transition of deciduous-permanent-no teeth, and microflora changes and returns to the same state as the children in the geriatric conditions (Kilian et al. 2016).

3.1.2 Factors Influencing the Growth of "Oral Microbiome"

The formation, growth, and persistence of the oral microbiome vary based on a multitude of human and environment-related factors. These factors are broadly divided into intrinsic and extrinsic factors (Tables 3.2 and 3.3) (Cornejo Ulloa et al. 2019; Giordano-Kelhoffer et al. 2022);

3.1.3 Oral Biofilm/Dental Plaque Biofilm

An oral biofilm is a group of microorganisms involved in a hard or soft tissue border within an extracellular matrix. These extracellular polysaccharides allow microbes to combat antibiotics and human immune systems (Tahmourespour 2012; Jhajharia et al. 2015).

Table 3.2 Intrinsic factors associated with oral microbiome

Host-related factors	Oral environment
• Genetics • Ethnicity • Geography • Circadian rhythm • Immune response	• Saliva • Gingival crevicular fluid (GCF): Albumin, glycoproteins, iron, mucin, lactate, bicarbonate, and nitrate alter the microbial load • Nutrition, • Attachment surfaces • Temperature

Table 3.3 Extrinsic factors associated with oral microbiome

Host modulation	Unmodified	Miscellaneous
• Lifestyle modification • Habits-smoking, alcohol • Diet (carbohydrate-rich food and micronutrient deficiencies) • Oral hygiene • Antibiotics • Systemic diseases	• Socioeconomic status • Environment	• Surface ultrastructure • Oxygen (higher on the tooth surface and lesser in the gums) • Nutrition • Host–microbial interaction and biofilm maturation

3.1.3.1 Formation of Dental Plaque Biofilm

3.1.3.1.1 Adhesion, Colonization, and Maturation

Oral microflora can bind to various hard tissue and soft tissue surfaces of the mouth. The ability of this microbial flora to persevere in the mouth depends on the inherent adherence and colonization potential in the mouth. In addition, this microbial community can adhere to the teeth and gums by effective microbe–microbe adherence mechanisms. This includes syntrophy, co-aggregation, quorum sensing, and adaptation to the host. (Kalia et al. 2014; Barboza-Solís and Acuña-Amador 2019).

The colonization potential of the microbes on specific tooth surfaces and gums is also influenced by nutrients, water, oxygen partial pressure, redox potential, temperature, and pH factors. Over a period, there is a slow transition of the microbial status from a microbial-free infant state to the mixture of aerobes–anaerobic microbial community. Among them, microbial talking plays a major part in the maintenance of microflora organized as a plaque biofilm. These interactions among different microbes are complex, i.e., may predominantly be antagonistic or synergistic (Diaz and Valm 2020).

The survival of different communities of microbes primarily depends on the symbiotic relationship between them. In the mouth, saliva is a fluid medium to support this microbial community. The rate and composition of saliva (particularly "Planktonic" organisms-saliva inhabiting microbes) and quorum sensing mechanism influence the growth of other microbial species. This quorum sensing mechanism is essential to maintain a balance between the bacterial and fungal organisms for biofilm development and maintenance (Kovács and Majoros 2020; Kumar et al. 2020).

In addition to the quorum sensing mechanism, bacterial–fungal species interaction occurs due to yeast–mycelia transformation. Particularly, in dental caries of children (early childhood caries), Candida appears to enhance the growth of other bacterial species. This is mainly attributed to the increase in the extra polysaccharides (EPS), carbohydrate-related genes, proteins, and metabolic pathways. In particular, the binding of glucosyl transferases (GTs) by *S. mutans* with the *C. albicans* mannan layer encourages the development of the amalgamation of microbial species biofilms (Sultan et al. 2018).

3.1.3.1.2 Co-aggregation

The oral microflora plays a major role in oral and general well-being. It is essential to characterize these microbial species relevant to various oral diseases. Broadly, the initial stages of the oral microbial biofilm based on the acquired pellicle are associated with Gram-positive microorganisms (*Firmicutes* and *Actinobacteria* genera predominantly *Streptococcus*, *Veillonella*, and *Actinomyces*), while the later stages are associated with Gram-negative filamentous microorganisms *(Fusobacteria*, *Prevotella*, and *Porphyromonas)*. In the mouth, these microbial communities change in composition between supra-gingival and sub-gingival sites (above and below the gums), respectively. As time progresses, subgingival microbial biofilm builds more and exhibits Gram-negative microbial species in predominance. This strong adhesive ability of *Fusobacterium nucleatum*, *Corynebacterium*, and *Veillonella* plays an important role between early and late bacterial colonizers. Co-aggregation of microbial species is promoted by further binding between adhesins and polysaccharides on the other bacterial species (Li et al. 2022).

In a healthy individual, there exists a symbiotic association between dental plaque biofilm and adjacent tooth tissues. As time progresses, changes in the genetic makeup of microbes with host environmental changes result in the alteration from a "planktonic state" to a "pathogenic state" of dental biofilm formation. This dental plaque occurs as supragingival plaque (microbes sited on the teeth above gums) and subgingival plaques (microbes within the gum sulcus) (Mark Welch et al. 2020). This shift in microbial configuration and persistence takes place due to the buildup of cytotoxic metabolites (endotoxin and lytic enzymes) and alteration in the rate and composition of gum fluid (gingival crevicular fluid). Other minor factors associated with adherence include selective adherence to the teeth or epithelium, cell–cell interactions, and host–cell–environment interactions.

3.1.4 Dysbiosis

Dysbiosis is a slight disruption in this microbial equilibrium/homeostasis. In oral health, a dynamic balance and harmonious coexistence are always required with mutual benefits between the host and microorganisms. This is always maintained by bidirectional communication and regulatory mechanisms between the host and environment. Dysbiosis results in the overgrowth of indigenous microbes of resident microflora, host immune homeostasis, inflammatory host response eventually oral systemic diseases such as tooth decay, periodontitis, and oral cancer (Bizzini et al. 2012).

3.2 Probiotics

Probiotics are referred to *live microbes that exhibit benefits when administered at appropriate amounts and dosages to the human* (FAO/WHO 2002; Fenster et al. 2019). To date, these probiotics have been consumed for the management of gastric diseases, allergic and atopic reactions, and urogenital and respiratory diseases. Recently, these bacteria are an evolving method in defensive management for oral diseases such as dental decay, gum infections, halitosis, fungal infections, and some allergic reactions (Li et al. 2016).

3.2.1 Sources of Probiotics

Lactobacillus and *Bifidobacteria* are available naturally within the host or supplemented from external sources at appropriate dosages. However, to use probiotics in commercial products, they should meet the standard criteria set by the Food and Agriculture Organization (FAO) of the United Nations and the World Health Organization (WHO). This includes the following criteria (Sanders 2008).

- Identification of class and strains of all probiotic bacteria in the marketable product
- Safety assessment based on the microbial strain characterization
- Endorsement of beneficial effects of probiotics in human studies based on the identification and quantity of microorganisms
- Labeling substance and effectiveness claims truthfully and without deception until the end of the shelf life

Additional factors that may affect the choice of the probiotic bacteria include (1) capability to modify the stability of the host microbiota; (2) have a proven favorable outcome on the host; (3) harmless with no opposing side effects; (4) ability to live in the stomach; (5) a sufficient quantity of live cells to provide health support; (6) maintenance of the desired properties throughout the storage time, and labeled correctly; and (7) characterization of strain, and dosage evaluation of the probiotics and (8) miscellaneous factors-concentration, stability, good sensory properties, enteric coating, metabolism, and interference with other ingredients in the final product (Kechagia et al. 2013; Sareen et al. 2012; Baral et al. 2021).

3.2.2 Characteristics of Probiotics

Probiotic microbes can be extracted from the targeted animal, human intestine, breast milk, or other fermented food products using suitable molecular methods. The following is the list of characteristics for the microbe to qualify as a suitable probiotic (Kaur et al. 2021).

- **Host-Associated Tolerance Ability:** Ability to withstand digestion by intestinal and gastric enzymes, acid and bile concentrations, and pH 2–5. Among them, *lactobacillus* strains can withstand low pH.
- **Adhesive Properties:** Microbes must be able to well adhere to the host mucosal and epithelial tissue to maximize the probiotic effect on the host.
- **Antimicrobial Activity:** By competitive mechanisms, the microbial species must thrive the secretion of lacticins, bacteriocin, lactic acid, antibodies, and increased leukocyte phagocytic activity.
- **Enhances the Host Tissue Response:** Ability to produce immunoglobulin, cytokines, IgA, and IgM-secreting cells to alter the host immune response.

- **Host-Related Practical Criteria:** Host-related factors include systemic factors like antihyperlipidemic, antiobesity, antianxiety and antidepression. In addition, inhibition of mast cell activation results in a decrease in glucose concentration. Finally enzymes like superoxide dismutase and glutathione dismutase aid in antioxidant activity.
- **Good Technological Properties:** The ability of the microbial species to live a long duration, efficiency, and viability.

The primary sources of probiotics are fermented foods and dietary supplements. Among them, fermented foods such as dairy and nondairy products have been the most accepted and best source of probiotics. The fermented substrate should be active; the probiotic microbe must contain sufficient colony-forming units (CFU)/ adequate dose to produce maximum health benefit. The list of common probiotic foods that have proven the maximum health benefits is summarized in Tables 3.4 and 3.5 (Fontana et al. 2013; Zhao et al. 2019; Valero-Cases et al. 2020; Dahiya and Nigam 2022; Singh et al. 2022).

Table 3.4 Association of dairy products with probiotic species

Dairy products (Cow milk/sheep milk/buffalo milk)	Probiotics species
Yoghurt "Milk of eternal life" *Fermented milk—"Yoghurt cultures" (Lactobacillus bulgaricus/Streptococcus thermophiles)*	*Lactobacillus acidophilus*, *Lactobacillus johnsonii*, *Enterococcus faecium*, *Lacticaseibacillus rhamnosus*, *Lactobacillus delbrueckii*, *Lacticaseibacillus casei*, *Lactobacillus gasseri*, *Bifidobacterium bifidum*, *Bifidobacterium lactis*, *Bifidobacterium animalis*, and *Saccharomyces boulardii*
Kefir	*Lacticaseibacillus paracasei*, *Lentilactobacillus kefiri*, *Lactobacillus parabuchneri*, *Lactobacillus kefiranofaciens*, *Lactobacillus delbrueckii*, *Lactobacillus bulgaricus*, *Acetobacter aceti*, *Lactococcus lactis*, *Acetobacter lovaniensis*, *Lactiplantibacillus plantarum*, *Leuconostoc mesenteroides*, *Geotrichum candidum*, *Acetobacter syzygii*, *Enterococcus durans*, *Lactobacillus acidophilus*, *Saccharomyces cerevisiae*, *Saccharomyces unisporus*, and *Kluyveromyces marxianus*
Starter-free cheese	*Lactiplantibacillus plantarum*, *Sterptococcus thermophilus*, *Lactococcus cremoris*, *Lactobacillus deLactobacillus rueckii subsp.* bulgaricus, *Lactococcus lactis*, *Lactobacillus helveticus*, molds, and yeasts
Others	Fermented milk products (names vary based on the country of origin)-sour cream, buttermilk

The mechanism of action of these probiotic species from fermented foods occurs via the release of bacteriocins, lactococcin, nisin, plantaricin, garvieacin, and sakacin

Table 3.5 Association of nondairy products with probiotic species

Nondairy products	Probiotics species
Tempeh (Soy-based products)	*Akkermansia muciniphila, Acetobacter indonesiensis, Bifidobacterium* sp., *Flavobacterium* sp., *Bacillus subtilis, Lactobacillus agilis, Lactobacillus fermentum, Enterobacteria cecorum, Brevundimonas* sp., *Pseudomonas putida, Acinetobacter* sp., *Aeromonas oryzae, Streptococcus* sp., *Pediococcus sp., Escherichia faecium, Candida parapsilosis, Geotrichum candidum, Actinomucor elegans*, and *Rhizopus*
Samerkaut [Cabbage]	*Lactobacillus brevis, Lactobacillus plantarum, Leuconostoc mesenteroides, Leuconostoc fallax, Pediococcus pentosaceus, Lactobacillus curvatus, Latilactobacillus sakei*, and *Lactococcus lactis*
Fermented vegetables (Kimchi) and fruits *Leafy vegetables, radish, cucumber, mustard leaves, bamboo shoots, wax gourd, ginger, cabbage, olive*, and *eggplant*	*Lactobacillus, Lactococcus, Pediococcus*, and *Leuconostoc sp.*
Kombucha (black tea)—Beverages	*Acetobacter xylinum, Acetobacter aceti, Acetobacter xylinoides, Acetobacter pasteurianus, Gluconobacter Brettanomyces, Saccharomyces, Pichia, Zygosaccharomyces*, and *Candida species*
Probiotic functional foods (Probiotic-infused juices, yoghurt-based drinks, liquid probiotics)	*Lactobacillus* and *Bifidobacterium* (species variation based on the manufacturer)
Probiotic drug supplements	*Lactobacillus* and *Bifidobacterium* (Species variation based on the pharmaceutical companies)
Breast milk	*Ligilactobacillus salivaricus, Lactobacillus rhamnosus, Lactobacillus gasseri, Limosilactobacillus fermentum, Lactobacillus acidophilus, Streptococcus salivaricus, Streptococcus thermophilus*, Bifidobacterium sp., and *Mirococcus*
Others—Meat, fish, shrimp, pork, Bread, cupcakes, fruit juices, pickles, bakery, cereals—barley, millet, wheat, black gram, maize, beverages embedded with probiotics	

3.2.3 *Delivery of Probiotics*

The oral probiotics are added into dietary supplements in liquid (health drinks), paste, powdered form, sachets, granules, capsules, chewable gum, gelatin tablets, lozenges, mouthwash oral strips, and toothpaste. The following parameters need to be addressed before a probiotic is selected for future use.

(a) **Viability of Probiotics:** Delivery of probiotics with appropriate dosage is significant for successful action of probiotics. Because of decreased acidity in the mouth, they are beneficial when administered in yoghurt or cheese containing casein phosphopeptides (CPP) for nullification of acidic pH, preventing tooth loss, and aiding in buildup of teeth. To provide humans with the most benefit, probiotics should be consumed five to ten billion CFU/day at a minimum concentration of 1×10^6–10^7 live organisms per g/mL of the product (Nanavati et al. 2021; FAO/WHO 2002).

(b) **Bioavailability of the Probiotic Formulation:** The bioavailability of probiotics can be improvised by using milk and milk-derived items with a high content of calcium or calcium lactate directly that helps shield the tooth surfaces and stop the development of oral microorganisms (Homayouni Rad et al. 2023).

(c) **Adhesion of Probiotics:** *Lactobacilli* and *Bifidobacterium* possessing inherent properties of acidogenicity and aciduricity are considered highly cariogenic to oral health (Padhye 2020). But *Lactobacilli* are deliberated as late colonizers and possess poor adhesion properties.

(d) **Delivery Vehicles of Probiotics:** The delivery vehicle is capable of delivering the probiotic to the site of action. The advantages and disadvantages of each of the delivery vehicles of probiotics are summarized in Table 3.6 (Reddy et al. 2010; Sareen et al. 2012; Fenster et al. 2019).

Table 3.6 Probiotics vehicles Advantages and disadvantages of probiotics vehicles

Mode of delivery of probiotics	Advantages	Disadvantages
Lozenges/chewing tablets/slow-dissolving tablets	• High retention time • Reduces oral disease pathogens • Compact, easy to store and transport	• Contraindicated in children, elders and patients with chewing disabilities-risk of choking • May require taste masking • Brittle
Capsules	• Compact, easy to store and transport	• Contraindicated in children, elders and patients with chewing disabilities-risk of choking • Lower retention time • Brittle • Require water to swallow
Oral strip	• Compact, easy to store and transport • Good retention time • Suitable for all age groups • Target-specific therapy	• May require taste masking • Small dosage only • Affected by oral environmental conditions
Mouthwash	• Good oral care product	• Difficult to deliver specific probiotic dose
Hydrogel/ Buccoadhesive gel	• Compact, easy to store and transport • Suitable for all age groups • High degree of adhesion • Target-specific therapy	• Difficult to deliver specific probiotic dose • May require taste masking

3.2.4 Probiotics and Oral Health

The mouth is the natural habitat of a complex community of microbial species. The well-being of oral health chiefly is determined by the balance maintained between normal commensal and pathogenic microorganisms. This bacterial biofilm exists on various teeth and gums of the oral cavity. At present, almost all oral lesions such as dental decay and periodontitis occur mainly just before the perturbation of specific pathogenic microorganisms due to environmental changes. Ecological pressures include factors like diet-rich carbohydrates, decreased salivary flow, and acidic pH. Oral diseases are controlled by bacteriotherapy against the pathogenic microorganisms directly or second by modulation of ecological changes in the oral environment (Vedam et al. 2023).

Probiotics are used in oral care for the circumvention of lesions in the mouth (Kumar et al. 2018; Vedam et al. 2023). Host is given naturally existing beneficial bacteria to help balance the pathogenic and beneficial bacteria again, reducing tissue damage and improving the oral and general status of the patient (Laleman and Teughels 2015).

Probiotics are classified as (1) lactic acid producing bacteria (LAB) *(Lactobacillus, Bifidobacterium, Streptococcus);* (2) nonlactic acid producing bacterial species *(Bacillus, Propionibacterium);* (3) nonpathogenic yeasts *(Saccharomyces);* and (4) nonspore forming and nonflagellated rod or coccobacillus. Figure 3.1 is a representation of the microbial species that mainly play as probiotics in the mouth (Masood

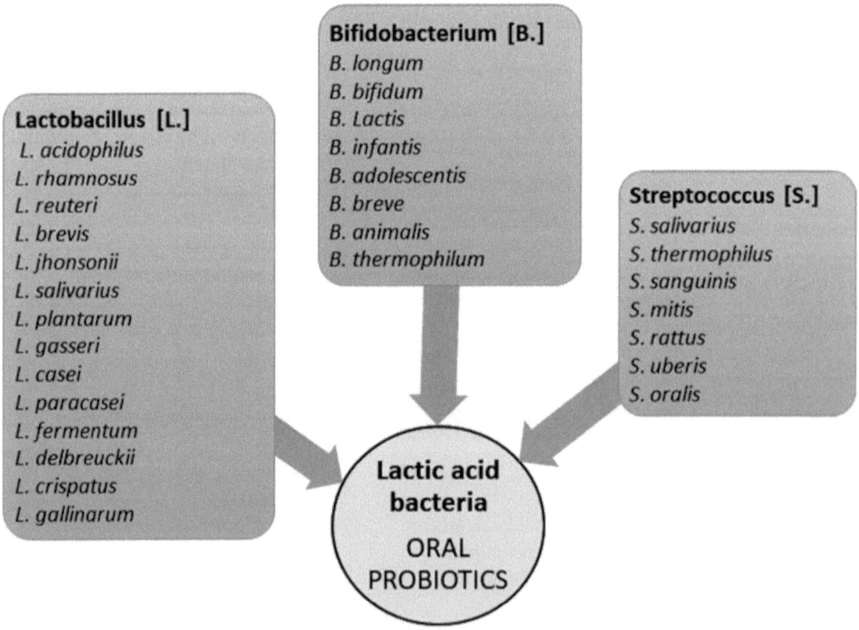

Fig. 3.1 Types of lactic acid bacteria (LAB)

et al. 2011; Jain and Sharma 2012; Tandon et al. 2015; Alok et al. 2017; Nanavati et al. 2021).

Lactobacillus, *Bifidobacterium*, and *Streptococcus* genera species have been termed "Generally recognized as safe (GRAS)" by the FAO/WHO in 2002. Oral biofilm constitutes less than 1% of the cultivable species of *Lactobacillus* (*L. paracasei, L. rhamnosus, L. plantarum, and L. salivarius*) and few *Bifidobacterium* (*B. bifidum, B. longum, and B. dentium*) and they exhibit better colonization in oral mucosal tissues (De Filippis et al. 2020; Reddy et al. 2011).

3.2.5 Mechanism of Action

The information regarding the probiotics in the mouth is essential to use them appropriately to prevent oral diseases. Broadly, probiotic action can be divided into two routes, namely,

Direct action: Probiotics directly act on the oral microflora.

Indirect action: Action of probiotics by moderating the host's humoral and cell-mediated immunity and inflammatory response against the oral pathogen (Fig. 3.2) (Haukioja 2010; Rastogi et al. 2011; Allaker and Stephen 2017; Javanshir et al. 2021; Youssef et al. 2021; Parasuraman et al. 2023).

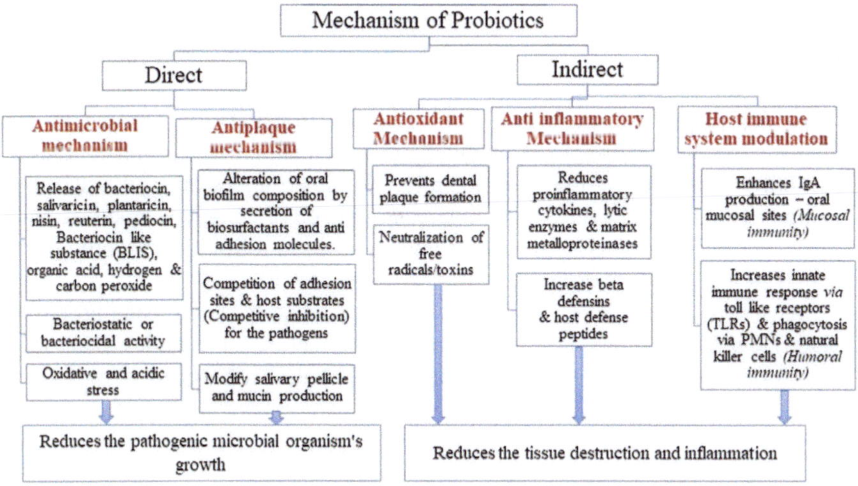

Fig. 3.2 Mechanism of actions of probiotics

3.2.6 Safety and Toxicity Considerations

Probiotics display an advantageous role in the human health that are strain-specific rather than species- or genus-specific. Probiotics are often sold as food or medicinal products. Throughout human history *Lactobacillus*, *Pediococcus*, and *Leuconostoc* species have been used in food technology, and foods containing these live bacteria, dead bacteria, and metabolites of these microorganisms are ingested. Currently, the probiotics *Lactobacilli* strain exhibits health advantages in the human when managed in sufficient amounts. While most of the species and genera seem to be harmless, a few microorganisms may be challenging, particularly *Enterococci* (*E. faecium* and *E. faecalis*) are opportunistic pathogens, and are responsible for nosocomial infections (Merenstein et al. 2023). *Bacillus cereus* is an anaerobic Gram-positive bacterium found in vegetables and processed foods and is widely used as probiotics. *B. cereus* can cause enterotoxin generation, transferable antibiotic resistance, and cytotoxicity against normal cells. *B. cereus* has also a risk of producing enterotoxin; however, some of them have been used as probiotics (Lee et al. 2019). Hence, research on probiotic bacterial strains is necessary to guarantee the protection of their genotype and phenotypic features.

3.2.7 Therapeutic Approaches

Maintaining optimal oral health requires diligent care of the mouth, teeth, and associated structures. Not only is this crucial for basic functions such as eating, breathing, and speaking, but it also has significant psychological implications, affecting factors such as self-esteem, overall well-being, and socialization. Oral health is a dynamic concept that changes throughout one's lifespan, intricately linked to overall health and enabling individuals to fully participate in society and reach their potential. By staying informed and implementing good oral hygiene practices, individuals can maintain healthy teeth and gums, ensuring a lifetime of optimal oral health (FAO/WHO 2002; Baiju et al. 2017).

Diseases of the oral cavity that can be prevented at early stages using probiotics include dental caries, gingival and periodontal infections. Probiotics also, show beneficial effects in patients with halitosis and oral cancer-associated inflammation.

3.2.7.1 Dental Caries: Etiopathogenesis

Dental decay is an active process that affects the enamel and dentin of the teeth in all age groups. The alternating demineralization and remineralization process involves tooth structure, microbial biofilm, sugars, salivary, and genetic factors. The equilibrium between the protective features and the pathogenic features determines the demineralization–remineralization cycle. The whole process of dental caries is not completely understood (Featherstone 2008). Criteria for caries identification as

Table 3.7 ICDAS coding criteria

Code	Description
0	Sound tooth surface: No evidence of caries after 5 s air drying
1	First visual change in enamel: Opacity or discoloration (white or brown) is visible at the entrance to the pit or fissure seen after prolonged air drying
2	Distinct visual change in enamel visible when wet, lesion must be visible when dry
3	Localized enamel breakdown (without clinical visual signs of dentinal involvement) seen when wet and after prolonged drying
4	Underlying dark shadow from dentine
5	Distinct cavity with visible dentine
6	Extensive (more than half the surface) distinct cavity with visible dentine

per the International Caries Detection and Assessment System (ICDAS) are summarized in Table 3.7 (Gugnani et al. 2011).

The earliest theory, a chemo-parasitic theory (Miller in 1881), explains that tooth decalcification is caused by a combination of acids and oral microorganisms. The microorganisms in the oral cavity digest starch and produce organic acid that enables tooth dissolution. The several elements that affect dental health include oral environment and dental plaque (Suddick and Harris 1990).

3.2.7.1.1 Role of the Oral Environment in Dental Caries

The development of tooth biofilm is an ordinary process in dental caries. However, the presence of biofilm alone does not automatically indicate that disease is present. Rather, the development of dental decay is a multifaceted route that contains various host factors, such as tooth, saliva, and an increase in dietary sugars, which interact in intricate ways over time to lead to disease expression. These conditions are influenced by various tooth-related factors, such as tooth composition (including imperfectly formed structures such as hypoplasia), tooth morphology fluoride exposure, as well as factors that impact biofilm thickness and pathogenicity, such as tooth form, arch pattern, tooth-tooth contact, and tooth position (Fejerskov 2004). Developmental defects like enamel hypoplasia, which is described by thin enamel, can increase acid dissolution and lead to tooth loss with plaque stagnation and increased risk of decay in deciduous teeth (Carounanidy and Sathyanarayanan 2009).

Additionally, caries exposure can be influenced by the nearness of teeth to salivary gland openings and salivary thickness at exact tooth sites. Dental devices and faulty restorations can upsurge caries predisposition at particular tooth sites by creating areas of food lodgement that encourage tooth coating formation. Overall, the development of dental decay is a multifactorial development that involves the interplay of various factors, both tooth-related and host-related, and occurs in a highly localized and complex manner at each tooth site (Pitts et al. 2021).

3.2.7.1.2 Role of Dental Plaque in Dental Caries

Dental plaque is a biofilm comprised of organized communities of interacting species that grow on surfaces in the oral cavity, known as the oral microfora. The establishment of dental plaque involves the following stages. Initially, tooth surfaces are coated by a layer of glycoproteins known as the acquired pellicle. This film is derived mostly from saliva, bacteria, gingival crevicular fluid, blood, and food. The pellicle formed later offers sites for the attachment of the first bacterial colonizers on the tooth surface, which leads to the formation of tooth biofilm. Biofilm prevents diffusion by acting as a barrier (Marsh 2006; Kumar et al. 2013).

Bacteria can be held by reversible van der Waal forces to the surface. However, attachment becomes tougher and more fixed if interactions occur between adhesins on the microflora of the mouth and corresponding receptors in the tooth biofilm. Secondary and early bacterial colonizers adhere through co-adhesion, which increases the density of the biofilm (Araujo et al. 2021; Kalia and Kumar 2014).

The biofilm undergoes growth and development, during which numerous for and against bacterial interactions occur. An extracellular matrix is made up of bacterial exopolymers like polysaccharides derived from carbohydrate metabolism. This matrix helps to hold the biofilm on the surface and can influence the movement of molecules within the biofilm. This configuration varies on diverse surfaces of the tooth due to minute differences in the environmental conditions (Pitts et al. 2021; Kalia et al. 2015).

The disturbances occur in two types during the pathogenesis of dental caries. They are listed below.

- Disturbance of host biofilm-related bacterial balance
- Disturbance of mineral equilibrium between the tooth and saliva

Oral fluid includes both saliva and fluid from the gums, both of which play a crucial role in the onset or inhibition of dental decay. Saliva, for instance, neutralizes the acidity formed by the fermentation of carbohydrates by microorganisms, the chief etiology of dental caries. It also contains various enzymes that provide antimicrobial properties by breaking down bacterial cell walls. The gingival crevicular fluid acts as a defense mechanism by flushing out the microbial population and their byproducts from the periodontal pocket (Subbarao et al. 2019).

Disruption of microbial homeostasis in biofilm is a significant cause of the pathogenesis of dental decay. The biofilm is a complex and dynamic ecosystem with various microbial species co-existing in microcosms. Disruption of the delicate balance between these microbial species can lead to the overgrowth of acid-producing bacteria, causing a shift in pH toward acidity and the subsequent demineralization of the tooth's enamel (Marsh 2006).

Disturbance of overall mineral status between the tooth and the saliva is another crucial role in the disease process of dental caries. The oral fluid plays a dynamic role in maintaining a balance between the demineralization and remineralization of the tooth's enamel. A loss of balance can lead to a loss of minerals from the tooth, resulting in the formation of cavities (Carounanidy and Sathyanarayanan 2009).

3.2.7.2 Gingival and Periodontal Diseases: Etiopathogenesis

Good gingival health is usually characterized by a stable equilibrium between the inflammatory response and the immune system. Diagnose gingival health in patients with an intact periodontium by assessing the absence of clinical attachment loss, radiographic bone loss, pocket depth of fewer than 3 mm, and less than 10% bleeding on probing (Lang et al. 1990). Gingivitis is a changeable disease associated with inflammation of the gingiva. It can lead to periodontitis and advanced connective tissue attachment and bone loss if left untreated. Therefore, managing gingival inflammation is critical to preventing periodontitis and its recurrence. However, not all affected areas progress to periodontitis as the relationship between health, gingivitis, and periodontitis is extremely reliant on the human susceptibility and immune-inflammatory-related effects (Cekici et al. 2014).

Dental plaque biofilm-induced gingivitis is a common local irritation caused by microbial accumulation on teeth, but it can be prevented by preserving good oral hygiene practices such as brushing and flossing regularly. Clinicians can help patients understand how to maintain good oral health and prevent plaque-induced inflammation by educating them about the importance of brushing and flossing regularly, maintaining a healthy diet, and avoiding harmful habits such as smoking. Clinical signs of inflammation and gingivitis may cause discomfort and impact the value of life of affected individuals (Takenaka et al. 2019). Clinicians must play a crucial role in managing gingivitis for patients of all ages, with or without a history of periodontal disease, by providing comprehensive care and ensuring that patients have access to the necessary resources and support to maintain good oral health.

3.2.7.3 Halitosis

Halitosis (Bad breath) occurs in an individual due to the putrefaction of sulfur-containing amino compounds like methyl mercaptan, dimethylsulphide and hydrogen sulphide. This has been stated to occur due to the alteration of oral biofilm accumulated on the dorsum of the tongue, interdental spaces of the teeth, appliances or associated with several pathologies. Recently, lactic acid bacteria probiotics have been given in dietary supplements that have shown a positive impact in reducing the bacterial load thereby aiding in a positive impact in reducing halitosis (López-Valverde et al. 2022).

3.2.7.4 Oral Mucositis

It is a common adverse effect following radiotherapy in oral cancer patients. This painful condition occurs due to poor oral hygiene and nutrition. Several over-the-counter medications are prescribed to patients simultaneously to reduce the intensity of oral mucositis. Despite this, the patient's condition remains unchanged. At the molecular level, it has been noticed that these cellular changes are associated with reactive oxygen species. Recently, probiotics with their anti-inflammatory action have shown prominent benefits in the reduction of oral mucositis associated with oral cancer (Liu et al. 2022).

While oral health disorders have been successfully managed with traditional approaches, there has been a hunt for a more recent, safe, and effective solution. Probiotic applications for dental health are the subject of the following section.

3.2.8 Clinical Effects

Probiotics can change the ecology of oral microbes and might be helpful in the treatment of periodontitis. Probiotics suppress periodontal pathogens by competitive inhibition of oral pathogens and reducing oral microflorapathogenicity. Probiotics also alter dynamic gum disease-associated inflammation. Probiotics as an adjuvant to clinical management result in a more significant improvement in the patient's medical state when associated with clinical management alone (Allaker and Stephen 2017). The outcomes of a methodical evaluation of clinical trials on the clinical effect of probiotics showed a significant reduction (65%) in the *S. mutans* number in the oral cavity (Navidifar et al. 2023). The list of probiotic microbiomes used in the management of oral disease is précised in Table 3.8 (Seminario-Amez et al. 2017; Saïz et al. 2021).

3.3 Conclusion

Probiotics are considered to be safe and can be used as an adjuvant or an alternative to conventional therapies. Probiotics have a significant impact on the decrease in tooth decay and oral infections, thereby improving oral health. Although probiotics may have a significant role in the therapeutics of periodontal disorders; the evidence is less strong. The efficiency of the probiotics remains unclear and further clinical studies with diversified and large samples are required. Overall, although probiotics are proven to be nontoxic, additional studies are required to demonstrate their long-term efficacy and safety.

Table 3.8 Probiotic microbiomes used in the treatment of oral disease

Probiotic species	Therapeutic uses
Bifidobacterium animalis, *Bifidobacterium longum*, *Bifidobacterium lactis*, *Bacillus subtilis*, *Bacillus megaterium*, *Bacillus pumulus*, *Bacillus mesentericus*, *Clostridium butyricum*, *Lacticaseibacillus rhamnosus*, *Lactobacillus curvatus*, *Streptococcus faecalis*, *Bifidobacterium bifidum*, *Limosilactobacillus reuteri*, *Lactobacillus brevis*, *Lactiplantibacillus plantarum*, *Lactobacillus acidophilus*, *Lactobacillus sporogenes*, *Ligilactobacillus salivarius*, *Lactobacillus sporogenes*, and *Saccharomyces boulardii*	Gingival health, reduction in plaque formation and improvement of periodontal health
Bifidobacterium animalis, *Bifidobacterium longum*, *Bifidobacterium lactis*, *Lacticaseibacillus rhamnosus*, *L. paracasei*, *Lacticaseibacillus reuteri*, *Lactobacillus brevis*, *Streptococcus oralis*, *Lactobacillus kefiranofaciens*, *Streptococcus dentisani*, *Lactobacillus gasseri*, *Lactobacillus acidophilus*, *Streptococcus uberis*, *Streptococcus rattus*, and *Streptococcus salivarius*	Prevention of dental caries
Lactobacillus reuteri, *Ligilactobacillus salivarius*, *Lactobacillus brevis*, *Lactobacillus delbrueckii*, *Lactobacillus crispatus*, *Streptococcus uberis*, *Streptococcus salivarius*, *Streptococcus oralis*, *Streptococcus salivarius*, *Streptococcus thermophilus*, *Weissella cibaria*, and *Enterococcus faecium*	Management of halitosis
Lactococcus lactis, *Lactobacillus reuteri*, *Lactobacillus brevis*, *Lactococcus lactis*, *Bifidobacterium longum*, and *Enterococcus faecium*	Increase the effectiveness of wound healing and management of oral inflammation related to cancer treatment

References

Allaker RP, Stephen AS (2017) Use of probiotics and oral health. Curr Oral Health Rep 4:309–318. https://doi.org/10.1007/s40496-017-0159-6

Alok A, Singh I, Singh S, Kishore M, Jha P, Iqubal MA (2017) Probiotics: a new era of biotherapy. Adv Biomed Res 6:31. https://doi.org/10.4103/2277-9175.192625

Araujo G, Zheng Z, Oh JJ, Tang JX (2021) Assessment of a weak mode of bacterial adhesion by applying an electric field. Appl Microbiol 1:255–269. https://doi.org/10.3390/applmicrobiol1020019

Baiju RM, Peter E, Varghese NO, Sivaram R (2017) Oral health and quality of life: current concepts. J Clin Diagn Res JCDR 11:ZE21–ZE26. https://doi.org/10.7860/JCDR/2017/25866.10110

Baral KC, Bajracharya R, Lee SH, Han H-K (2021) Advancements in the pharmaceutical applications of probiotics: dosage forms and formulation technology. Int J Nanomedicine 16:7535–7556. https://doi.org/10.2147/IJN.S337427

Barboza-Solís C, Acuña-Amador LA (2019) The oral microbiota: a literature review for updating professionals in dentistry. Part I. Odovtos Int J Dent Sci:143–152. https://doi.org/10.15517/ijds.2020.39178

Bizzini B, Pizzo G, Scapagnini G, Nuzzo D, Vasto S (2012) Probiotics and oral health. Curr Pharm Des 18:5522–5531. https://doi.org/10.2174/138161212803307473

Carounanidy U, Sathyanarayanan R (2009) Dental caries: a complete changeover (part II)-changeover in the diagnosis and prognosis. J Conserv Dent JCD 12:87–100. https://doi.org/10.4103/0972-0707.57631

Cekici A, Kantarci A, Hasturk H, Van Dyke TE (2014) Inflammatory and immune pathways in the pathogenesis of periodontal disease. Periodontol 64:57–80. https://doi.org/10.1111/prd.12002

Cho Y-D, Kim K-H, Lee Y-M, Ku Y, Seol Y-J (2021) Oral microbiome and host health: review on current advances in genome-wide analysis. Appl Sci 11:4050. https://doi.org/10.3390/app11094050

Chugh P, Dutt R, Sharma A, Bhagat N, Dhar MS (2020) A critical appraisal of the effects of probiotics on oral health. J Funct Foods 70:103985. https://doi.org/10.1016/j.jff.2020.103985

Cornejo Ulloa P, Van Der Veen MH, Krom BP (2019) Review: modulation of the oral microbiome by the host to promote ecological balance. Odontology 107:437–448. https://doi.org/10.1007/s10266-019-00413-x

Dahiya D, Nigam PS (2022) Probiotics, prebiotics, Synbiotics, and fermented foods as potential biotics in nutrition improving health via microbiome-gut-brain Axis. Fermentation 8:303. https://doi.org/10.3390/fermentation8070303

De Filippis F, Pasolli E, Ercolini D (2020) The food-gut axis: lactic acid bacteria and their link to food, the gut microbiome and human health. FEMS Microbiol Rev 44:454–489. https://doi.org/10.1093/femsre/fuaa015

Deo P, Deshmukh R (2019) Oral microbiome: unveiling the fundamentals. J Oral Maxillofac Pathol 23:122. https://doi.org/10.4103/jomfp.JOMFP_304_18

Diaz PI, Valm AM (2020) Microbial interactions in oral communities mediate emergent biofilm properties. J Dent Res 99:18–25. https://doi.org/10.1177/0022034519880157

Escherich T (1989) The intestinal bacteria of the neonate and breast-fed infant. Rev Infect Dis 11:352–356. https://doi.org/10.1093/clinids/11.2.352

FAO/WHO (2002) Guidelines for the evaluation of probiotics in food. FAO, Paris, pp 1–11

Faran Ali SM, Tanwir F (2012) Oral microbial habitat a dynamic entity. Journal of oral biology and craniofacial research 2:181–187. https://doi.org/10.1016/j.jobcr.2012.07.001

Featherstone JDB (2008) Dental caries: a dynamic disease process. Aust Dent J 53:286–291. https://doi.org/10.1111/j.1834-7819.2008.00064.x

Fejerskov O (2004) Changing paradigms in concepts on dental caries: consequences for oral health care. Caries Res 38:182–191. https://doi.org/10.1159/000077753

Fenster K, Freeburg B, Hollard C, Wong C, Rønhave Laursen R, Ouwehand AC (2019) The production and delivery of probiotics: a review of a practical approach. Microorganisms 7:83. https://doi.org/10.3390/microorganisms7030083

Fontana L, Bermudez-Brito M, Plaza-Diaz J, Muñoz-Quezada S, Gil A (2013) Sources, isolation, characterisation and evaluation of probiotics. Br J Nutr 109:S35–S50. https://doi.org/10.1017/S0007114512004011

Giordano-Kelhoffer B, Lorca C, March Llanes J, Rábano A, Del Ser T, Serra A, Gallart-Palau X (2022) Oral microbiota, its equilibrium and implications in the pathophysiology of human diseases: a systematic review. Biomedicines 10:1803. https://doi.org/10.3390/biomedicines10081803

Gugnani N, Pandit IK, Srivastava N, Gupta M, Sharma M (2011) International caries detection and assessment system (ICDAS): a new concept. Int J Clin Pediatr Dent 4:93–100. https://doi.org/10.5005/jp-journals-10005-1089

Haukioja A (2010) Probiotics and oral health. Eur J Dent 4:348–355

Homayouni Rad A, Pourjafar H, Mirzakhani E (2023) A comprehensive review of the application of probiotics and postbiotics in oral health. Front Cell Infect Microbiol 13:1120995. https://doi.org/10.3389/fcimb.2023.1120995

Jain P, Sharma P (2012) Probiotics and their efficacy in improving oral health: a review. J Appl Pharm Sci 2:151–163

Javanshir N, Hosseini GNG, Sadeghi M, Esmaeili R, Satarikia F, Ahmadian G, Allahyari N (2021) Evaluation of the function of probiotics, emphasizing the role of their binding to the intestinal epithelium in the stability and their effects on the immune system. Biol Proced Online 23:23. https://doi.org/10.1186/s12575-021-00160-w

Jhajharia K, Parolia A, Shetty KV, Mehta LK (2015) Biofilm in endodontics: a review. J Int Soc Prev Community Dent 5:1–12. https://doi.org/10.4103/2231-0762.151956

Kalia VC, Kumar P (2014) The battle: quorum-sensing inhibitors versus evolution of bacterial resistance. In: Quorum sensing vs quorum quenching: a battle with no end in sight. New Delhi, Springer India, pp 385–391

Kalia VC, Wood TK, Kumar P (2014) Evolution of resistance to quorum-sensing inhibitors. Microb Ecol 68:13–23. https://doi.org/10.1007/s00248-013-0316-y

Kalia VC, Kumar P, Pandian ST, Sharma P (2015) Biofouling control by quorum quenching. In: Springer handbook of marine biotechnology, pp 431–440

Kaur S, Kaur R, Rani N, Sharma S, Joshi M (2021) Sources and selection criteria of probiotics. In: Goel G, Kumar A (eds) Advances in probiotics for sustainable food and medicine, microorganisms for sustainability. Springer Singapore, Singapore, pp 27–43. https://doi.org/10.1007/978-981-15-6795-7_2

Kechagia M, Basoulis D, Konstantopoulou S, Dimitriadi D, Gyftopoulou K, Skarmoutsou N, Fakiri EM (2013) Health benefits of probiotics: a review. ISRN Nutr 2013:481651. https://doi.org/10.5402/2013/481651

Kilian M, Chapple ILC, Hannig M, Marsh PD, Meuric V, Pedersen AML, Tonetti MS, Wade WG, Zaura E (2016) The oral microbiome—an update for oral healthcare professionals. Br Dent J 221:657–666. https://doi.org/10.1038/sj.bdj.2016.865

Kovács R, Majoros L (2020) Fungal quorum-sensing molecules: a review of their antifungal effect against Candida biofilms. J Fungi 6:99. https://doi.org/10.3390/jof6030099

Kumar P, Patel SK, Lee JK, Kalia VC (2013) Extending the limits of *Bacillus* for novel biotechnological applications. Biotechnol Adv 31(8):1543–1561. https://doi.org/10.1016/j.biotechadv.2013.08.007

Kumar VN, Krishnamurthy M, Poorni S, Patil S, Raj AT (2018) Probiotics in caries prevention. J Contemp Dent Pract 19:123–124. https://doi.org/10.5005/jp-journals-10024-2224

Kumar P, Lee JH, Beyenal H, Lee J (2020) Fatty acids as antibiofilm and antivirulence agents. Trends Microbiol 28(9):753–768. https://doi.org/10.1016/j.tim.2020.03.014

Laleman I, Teughels W (2015) Probiotics in the dental practice: a review. Quintessence Int Berl Ger 46:255–264. https://doi.org/10.3290/j.qi.a33182

Lang NP, Adler R, Joss A, Nyman S (1990) Absence of bleeding on probing an indicator of periodontal stability. J Clin Periodontol 17:714–721. https://doi.org/10.1111/j.1600-051X.1990.tb01059.x

Lee N-K, Kim W-S, Paik H-D (2019) Bacillus strains as human probiotics: characterization, safety, microbiome, and probiotic carrier. Food Sci Biotechnol 28:1297–1305. https://doi.org/10.1007/s10068-019-00691-9

Li R, Lo ECM, Liu BY, Wong MCM, Chu CH (2016) Randomized clinical trial on arresting dental root caries through silver diammine fluoride applications in community-dwelling elders. J Dent 51:15–20. https://doi.org/10.1016/j.jdent.2016.05.005

Li X, Liu Y, Yang X, Li C, Song Z (2022) The oral microbiota: community composition, influencing factors, pathogenesis, and interventions. Front Microbiol 13:895537. https://doi.org/10.3389/fmicb.2022.895537

Liu YC, Wu CR, Huang TW (2022) Preventive effect of probiotics on oral mucositis induced by cancer treatment: a systematic review and meta-analysis. Int J Mol Sci 23(21):13268. https://doi.org/10.3390/ijms232113268

López-Valverde N, López-Valverde A, Macedo de Sousa B, Rodríguez C, Suárez A, Aragoneses JM (2022) Role of probiotics in halitosis of oral origin: a systematic review and meta-analysis of randomized clinical studies. Front Nutr 8:787908. https://doi.org/10.3389/fnut.2021.787908

Mark Welch JL, Ramírez-Puebla ST, Borisy GG (2020) Oral microbiome geography: micron-scale habitat and niche. Cell Host Microbe 28:160–168. https://doi.org/10.1016/j.chom.2020.07.009
Marsh PD (2006) Dental plaque as a biofilm and a microbial community—implications for health and disease. BMC Oral Health 6(Suppl 1):S14. https://doi.org/10.1186/1472-6831-6-S1-S14
Masood MI, Qadir MI, Shirazi JH, Khan IU (2011) Beneficial effects of lactic acid bacteria on human beings. Crit Rev Microbiol 37:91–98. https://doi.org/10.3109/1040841X.2010.536522
McBain AJ, Madhwani T, Eatough J, Ledder R (2009) An introduction to probiotics for dental health. Food Sci Technol Bull Funct Foods 6:5–29. https://doi.org/10.1616/1476-2137.15748
Merenstein D, Pot B, Leyer G, Ouwehand AC, Preidis GA, Elkins CA, Hill C, Lewis ZT, Shane AL, Zmora N, Petrova MI, Collado MC, Morelli L, Montoya GA, Szajewska H, Tancredi DJ, Sanders ME (2023) Emerging issues in probiotic safety: 2023 perspectives. Gut Microbes 15:2185034. https://doi.org/10.1080/19490976.2023.2185034
Morrison AG, Sarkar S, Umar S, Lee STM, Thomas SM (2023) The contribution of the human oral microbiome to oral disease: a review. Microorganisms 11:318. https://doi.org/10.3390/microorganisms11020318
Nanavati G, Prasanth T, Kosala M, Bhandari SK, Banotra P (2021) Effect of probiotics and prebiotics on oral health. Dent J Adv Stud 9:01–06. https://doi.org/10.1055/s-0040-1722523
Navidifar T, Mahdizade Ari M, Alipourkermani A, Afifirad R, Asadollahi P, Veisi A, Ghanavati R, Darbandi A (2023) Clinical efficacy of probiotics on oral health: a systematic review of clinical trials. Curr Pharm Biotechnol 24:1916–1927. https://doi.org/10.2174/1389201024666230405135457
Nuriel-Ohayon M, Neuman H, Koren O (2016) Microbial changes during pregnancy, birth, and infancy. Front Microbiol 7:1031. https://doi.org/10.3389/fmicb.2016.01031
Padhye S (2020) Role of oral microbiome in disease predictions—current advances. Biomed J Sci Tech Res 30(4). https://doi.org/10.26717/BJSTR.2020.30.004993
Parasuraman S, Vedam VKV, Sabesan GS (2023) Probiotics: an emerging strategy for oral health care. In: Kothari V, Kumar P, Ray S (eds) Probiotics, prebiotics, synbiotics, and postbiotics. Springer Nature Singapore, Singapore, pp 275–306. https://doi.org/10.1007/978-981-99-1463-0_15
Pitts NB, Twetman S, Fisher J, Marsh PD (2021) Understanding dental caries as a non-communicable disease. Br Dent J 231:749–753. https://doi.org/10.1038/s41415-021-3775-4
Pujia AM, Costacurta M, Fortunato L et al (2017) The probiotics in dentistry: a narrative review. Eur Rev Med Pharmacol Sci 21:1405–1412
Rastogi P, Saini H, Dixit J, Singhal R (2011) Probiotics and oral health. Natl J Maxillofac Surg 2:6. https://doi.org/10.4103/0975-5950.85845
Reddy J, Sampathkumar N, Aradhya S (2010) Probiotics in dentistry: review of the current status. Rev Clín Pesq Odontol Curitiba 6:261–267
Reddy R, Swapna L, Ramesh T, Singh T, Vijayalaxmi N, Lavanya R (2011) Bacteria in oral health-probiotics and prebiotics a review. Int J Biol Med Res 2:1226–1233
Saïz P, Taveira N, Alves R (2021) Probiotics in oral health and disease: a systematic review. Appl Sci 11:8070. https://doi.org/10.3390/app11178070
Sanders ME (2008) Probiotics: definition, sources, selection, and uses. Clin Infect Dis 46 Suppl 2:S58–S61; discussion S144–151. https://doi.org/10.1086/523341
Sareen M, Roy S, Singh S, Gupta A (2012) A review on probiotics and their implications in dentistry. J Dentofac Sci 4:7–10
Seminario-Amez M, López-López J, Estrugo-Devesa A, Ayuso-Montero R, Jané-Salas E (2017) Probiotics and oral health: a systematic review. Med Oral Patol Oral Cirugia Bucal 22:e282–e288. https://doi.org/10.4317/medoral.21494
Sharma N, Bhatia S, Singh Sodhi A, Batra N (2018) Oral microbiome and health. AIMS Microbiol 4:42–66. https://doi.org/10.3934/microbiol.2018.1.42
Singh S, Gupta R, Chawla S, Gauba P, Singh M, Tiwari RK, Upadhyay S, Sharma S, Chanda S, Gaur S (2022) Natural sources and encapsulating materials for probiotics delivery systems: recent applications and challenges in functional food development. Front Nutr 9:971784. https://doi.org/10.3389/fnut.2022.971784

Subbarao KC, Nattuthurai GS, Sundararajan SK, Sujith I, Joseph J, Syedshah YP (2019) Gingival Crevicular fluid: an overview. J Pharm Bioallied Sci 11:S135–S139. https://doi.org/10.4103/JPBS.JPBS_56_19

Suddick RP, Harris NO (1990) Historical perspectives of oral biology: a series. Crit Rev Oral BiolMed 1:135–151. https://doi.org/10.1177/10454411900010020301

Sultan AS, Kong EF, Rizk AM, Jabra-Rizk MA (2018) The oral microbiome: a lesson in coexistence. PLoS Pathog 14:e1006719. https://doi.org/10.1371/journal.ppat.1006719

Tahmourespour A (2012) Probiotics and the reduction of dental caries risk. In: Li M-Y (ed) Contemporary approach to dental caries. InTech. https://doi.org/10.5772/36661

Takenaka S, Ohsumi T, Noiri Y (2019) Evidence-based strategy for dental biofilms: current evidence of mouthwashes on dental biofilm and gingivitis. Jpn Dent Sci Rev 55:33–40. https://doi.org/10.1016/j.jdsr.2018.07.001

Tandon V, Arora V, Yadav V, Singh V, Punia H, Agrawal S, Gupta V (2015) Concept of probiotics in dentistry. Int J Dent Med Res 1:206–209

Valero-Cases E, Cerdá-Bernad D, Pastor J-J, Frutos M-J (2020) Non-dairy fermented beverages as potential carriers to ensure probiotics, prebiotics, and bioactive compounds arrival to the gut and their health benefits. Nutrients 12:1666. https://doi.org/10.3390/nu12061666

Van Den Elsen LWJ, Garssen J, Burcelin R, Verhasselt V (2019) Shaping the gut microbiota by breastfeeding: the gateway to allergy prevention? Front Pediatr 7:47. https://doi.org/10.3389/fped.2019.00047

Vedam V, Sabesan GS, Adhikary AK, Parasuraman S (2023) Biotherapeutic potential of lactobacillus probiotic strains on Streptococcus mutans biofilm in dental caries–pathogenesis revisited. Indian J Pharm Educ Res 57:956–964. https://doi.org/10.5530/ijper.57.4.117

Yamashita Y, Takeshita T (2017) The oral microbiome and human health. J Oral Sci 59:201–206. https://doi.org/10.2334/josnusd.16-0856

Youssef M, Ahmed HY, Zongo A, Korin A, Zhan F, Hady E, Umair M, Shahid Riaz Rajoka M, Xiong Y, Li B (2021) Probiotic supplements: their strategies in the therapeutic and prophylactic of human life-threatening diseases. Int J Mol Sci 22:11290. https://doi.org/10.3390/ijms222011290

Zhao W, Liu Y, Latta M, Ma W, Wu Z, Chen P (2019) Probiotics database: a potential source of fermented foods. Int J Food Prop 22:198–217. https://doi.org/10.1080/10942912.2019.1579737

Chapter 4
Diagnostics Based on Microbial Enzymes

Nowshin Jahan, Ainun Afroz Sristy, Hazrat Ali, Nafisa Azmuda, and Nihad Adnan

Abstract Microbial enzymes have emerged as desirable resources for the diagnostic field in the rapidly expanding healthcare sector. Their biocompatible nature, sustainable and selective quality, and economic viability make them preferable to enzymes derived from plants or animals. This chapter explores the fascinating world of these biocatalytic agents, highlighting their diversity, distinctive capabilities, and versatile applications in the realm of diagnostics. Furthermore, the chapter discusses the associated assay technologies, their principles, novel technology adaptations, and limitations. By highlighting the real-world uses of microbial enzymes as diagnostic tools, this chapter can serve as a valuable resource for academics and healthcare professionals in the ever-evolving field of microbial diagnostics.

Keywords Microbial enzymes · In vitro diagnostics · Biomolecular detection · Point-of-care testing · Immunoassay · Enzymatic assays · Pathogen detection

4.1 Introduction

The modern era of personalized healthcare constantly demands accurate insights into disease diagnostics, prognostics, and treatment. As diagnostic tools become more sensitive, rapid, and cost-effective, the unique properties of microbial enzymes make them crucial players. The high specificity of these enzyme molecules to catalyze biochemical reactions under precise conditions has been utilized in disease diagnosis, environmental monitoring, and the food industry. Commonly, enzymes

N. Jahan
Department of Microbiology, Jahangirnagar University, Dhaka, Bangladesh

Department of Microbiology, Primeasia University, Dhaka, Bangladesh

A. A. Sristy · H. Ali · N. Azmuda · N. Adnan (✉)
Department of Microbiology, Jahangirnagar University, Dhaka, Bangladesh
e-mail: nihad@juniv.edu

V. Kothari et al. (eds.), *Microbial Products for Health and Nutrition*,
https://doi.org/10.1007/978-981-97-4235-6_4

for such purposes can be derived naturally from various sources such as plants, animals, and microorganisms. However, microbial enzymes are favored over traditional plant- and animal-derived enzymes because of their versatility, ease of genetic modification, low processing cost, and consistency in large-scale production (Gurung et al. 2013).

Microbial enzymes have found diverse applications in diagnostics due to their unique properties. They can directly serve as diagnostic markers, signaling abnormal metabolic or physiologic changes linked to infectious diseases, metabolic disorders, cancer, etc. Alternatively, they are incorporated into assay systems to detect toxins or pesticide residues and screen, identify, or quantify biomarker levels in body fluids. From multifunctional alkaline phosphatase to bacterial luciferase, numerous diagnostic microbial enzymes have been identified in bacteria, fungi, and extremophilic organisms (Nigam 2013). Novel microbial enzymes are now being routinely developed to cater to industry demand with the aid of genetic engineering, protein engineering, recombinant DNA technology, high-throughput sequencing, and omics technologies. These enzymes offer a wide range of substrate specificities, environmental stabilities (e.g., extremophilic enzymes), isoforms, and functionalities. As a result, they have been adapted to a wide array of assay technologies (Fig. 4.1) like enzyme immunoassays, nucleic acid amplification tests, diverse biosensors, and methodologies (Dahiya and Nigam 2021).

This chapter will review the prevailing contributions of microbial enzymes in the realm of diagnostics. It will provide a comprehensive look into their diversity, catalyzing functions, and utility in diagnostic applications. Furthermore, the chapter will explore the associated assay technologies, their principles, novel technology

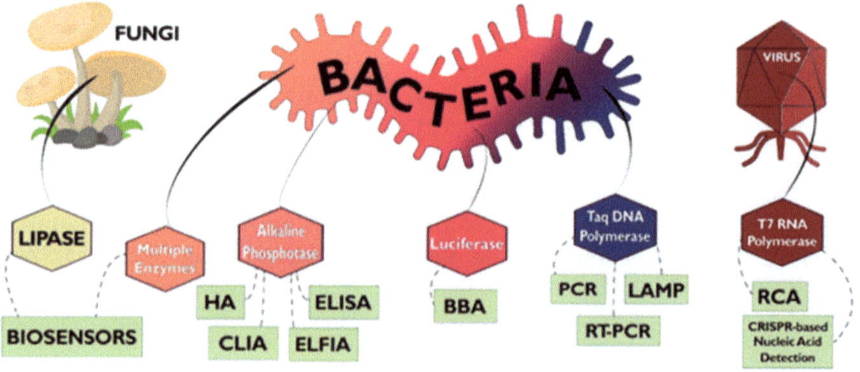

Fig. 4.1 Overview of microbial enzyme-based diagnostic technologies. The represented categories include biosensors, ELISA (Enzyme-Linked Immunosorbent Assay), HA (Hybridization Assay), CLIA (Chemiluminescent Immunoassay), ELFIA (Enzyme-Linked Fluorescent Immunoassay), BBA (Bacterial Bioluminescence Assay), PCR (Polymerase Chain Reaction), LAMP (Loop-Mediated Isothermal Amplification), RT-PCR (Reverse Transcription Polymerase Chain Reaction), RCA (Rolling Circle Amplification), and Biosensors and CRISPR-based Nucleic Acid Detection

adaptations, and limitations while highlighting the real-world uses of microbial enzymes as diagnostic tools.

4.2 Major Microbial Enzymes Used in Diagnostics

Based on their catalytic functions, diagnostic enzymes typically belong to three major classes: hydrolase, oxidoreductase, and transferase. Oxidoreductases oxidize or reduce their substrate and commonly include oxidases and dehydrogenases. Chemical bonds are hydrolyzed by hydrolase enzymes, whereas functional groups are transferred between molecules by transferases. On the other hand, isomerases convert molecules to their isomeric forms, while ligase combines different molecules. Examples of various classes of enzymes derived from microorganisms and their respective applications are outlined in Table 4.1.

4.2.1 Hydrolases

4.2.1.1 Alkaline Phosphatase

Alkaline phosphatase (ALP) is a phosphate metabolizer that hydrolyzes phosphoric acid-containing substrates by catalyzing phosphodiester bonds to release inorganic phosphorus in alkaline conditions (Jackson et al. 2016). It is one of the key enzymes in clinical microbiology (Table 4.1), which can be sourced from animals and microorganisms. Microbial form of the enzyme can be found in different bacteria and filamentous fungi such as *Escherichia coli, Bacillus stearothurmophilus, Bacillus licheniformis, Thermotoga neapolitana, Thermus caldophilus, Humicola* spp., and *Scytalidiym thermophilum* (Nalini et al. 2015).

Commercially, diagnostic reagent-grade alkaline phosphatase (ALP) is available in various forms, either as a lyophilized powder or conjugated to secondary antibodies or streptavidin for binding with target antigens or biotinylated proteins. For example, ALP from *Escherichia coli* is available as a lyophilized powder form from Takara Bio (Cat No. 2120B), ASAHI KASEI ENZYMES (T-08), Sigma Aldrich (P5931), and NZYTech (MB01801). Creative Enzymes provide a recombinant version sourced from *E. coli* (NATE 0806), Roche offers a version from *Pichia pastoris* (ROCHE/03359123001) and Sigma-Aldrich provides ALP of *Streptomyces avidinii* conjugated to streptavidin (S2890).

Alkaline phosphatase serves widely as an enzyme label in bioassays for detecting pathogens and biomolecules. It catalyzes the hydrolysis of a chromogenic or fluorogenic substrate, producing a quantifiable colorimetric or fluorometric absorbance signal indicating the target analyte's presence and concentration in the assays. Traditionally employed in Enzyme-Linked Immunosorbent Assay (ELISA) as documented by Li et al. (2016), this enzyme has more recently been integrated with

Table 4.1 Common microbial enzymes used in diagnostics

Enzyme class	Catalytic function	Enzymes types	Diagnostic applications	Diagnostic platform	Ref.
Hydrolase	Breakdown chemical bonds using water molecules	Alkaline phosphatase	Prostate-specific antigen detection	ELISA	Li et al. (2016)
			Salmonella spp. detection	Hybridization assay	Lucarelli et al. (2006)
			E. coli detection	Biosensor	Bozkurt et al. (2018)
			Bacterial endogenous ALP monitoring	Biosensor	Kang et al. (2018)
			E. coli detection	Phase based assay	Jackson et al. (2016)
			Zearalenone (ZEN) detection	Biosensor	Ma et al. (2023)
		Lipase	Triglyceride concentration determination	Biosensor	Narwal and Pundir (2017), Hasanah et al. (2019)
		Nuclease	Urinary tract infection detection	Hybridization assay	Flenker et al. (2017)
			Staphylococcus aureus infection detection	Biochemical assay	Lagacé-Wiens et al. (2007)
			Serum prostate antigen detection	Biosensor	Fang et al. (2017)
			Cancer exosome	Fluorescence	Wang et al. (2018)
			Cancer biomarker (micro RNA-195)	ELISA	Lei et al. (2022)
			Methicillin-resistant *Staphylococcus aureus* (Mec A gene) detection	PCR	Smith et al. (2014)
			Thrombin protein detection	Aptasensor	Huang et al. (2013)
		Creatinase	Serum/urine creatine detection	Biosensor	Staden and Bokretsion (2006)
			Plasma creatine conc. determination	Enzyme kinetic assay	Jimenez Valencia et al. (2021)
		Cholinesterase	Organophosphate pesticide detection	Biosensor	Liang and Han (2020)
		Cholesterol esterase	Small dense low-density lipoprotein (sd-LDL) conc. determination	Homogenous assay	Ito et al. (2011)
			Total serum cholesterol conc. determination	Biosensor	Román-Pizarro et al. (2023)
		Uracil-DNA glycosylase	Uropathogens detection	RT-LAMP	Zeng et al. (2020)
			Ocular toxoplasmosis diagnosis	RT-LAMP	Rahimi Esboei et al. (2022)
			Influenza virus detection	RT-LAMP	Kim et al. (2016)
		Helicase	*M. tuberculosis*, SARS CoV2 virus detection	HDA/TINY	Shanmugakani et al. (2021)
			SARS- CoV- 2 detection	RT-HDA/Diptick	Shanmugakani and Wu (2022)
			Monkey pox virus detection	HDA/lateral flow	Gulinaizhaer et al. (2023)
			Platelet product	HDA/SYBR green I	Yamket et al. (2023)
		Urease	Serum urea conc. determination	Biosensor	Marchenko et al. (2015)

Oxidoreductase	Redox reaction	Glucose oxidoreductase	Glucose sensor	Biosensor	Santiago et al. (2010)
		Alcohol oxidoreductase	Sweat alcohol conc. determination	Biosensor	Gamella et al. (2014)
			Whole blood alcohol conc. determination	Biosensor	Aymerich et al. 2018
		Cholesterol oxidase	Serum cholesterol determination	Biosensor	Chauhan and Pundir (2011)
		Bilirubin oxidase	Serum bilirubin conc. determination	Fluorometry	Andreu et al. (2000)
			Nucleic acid detection	Hybridization assay	Kim et al. (2004)
			Thrombotic thrombocytopenic purpura (TTP)	FRET assay	Eckmann et al. (2007)
		3-hydroxysteroid dehydrogenase	Serum bile conc. determination	Biosensor	Rani et al. (2004), Zhang et al. (2005)
		Catalase	Protein biomarker	ELISA/volumetric chip	Song et al. (2012)
			SARS CoV2 virus detection	Hybridization-CRISPR/Cas	Silva et al. (2021)
		Lactate dehydrogenase	Alanine transaminase activity determination	Colorimetric assay	Sun et al. (2021)
			Malaria diagnosis	Immunodot enzyme assay	Kaushal and Kaushal (2002)
Transferase	Transfer of functional groups	Kinase	Glucose sensor	Biosensor	Aoki et al. (2005)
			Glucose concentration	Fluorescence competitive assay	D'Auria et al. (2002)
			Serum creatine conc. determination	Enzymatic cycling assay	Ueda and Sakasegawa (2016)
		DNA polymerase	*Streptococcus pneumoniae detection*	RPA	Wang et al. (2022)
			Uropathogen detection	RPA	Chen et al. (2018)
			Cancer detection	Hybridization assay	Li et al. (2022b)
			Micro RNA detection	SDA/biosensor	Chen et al. (2015)
			Bacterial RNA detection	PCR-CRISPR/Cas	Shen et al. (2020)
		Ribozymes	Immunodiagnosis of HCV virus detection	Immunoassay	Kossen et al. (2004)

various other diagnostic technologies such as chemiluminescence assay, hybridization tests, and biosensors. In a hybridization assay designed for *Salmonella* spp., ALP from *Streptomyces avidinii* was used as a signal amplifier. The system precisely detected previously amplified *Salmonella* DNA through a sandwich hybridization process, wherein biotinylated amplicons hybridized with oligonucleotide probes are immobilized on a gold electrode. This hybrid was subsequently bound with streptavidin/ALP-streptavidin-biotin/ALP, creating a layered structure that maximized optical signals via the fluorescence activities of α-naphthol and facilitated the straightforward detection of the amplified amplicons (Lucarelli et al. 2006). A surface-enhanced Raman spectroscopy (SERS) based immunosensor was developed using the sandwich immunoassay with ALP nanoparticles for *E. coli* detection. The system entailed immunomagnetic separation of *E. coli* cells using gold nanoparticles modified with pathogen-specific antibodies through the streptavidin–biotin system. Subsequently, the separated target bacteria were indirectly identified using alkaline phosphatase-labeled *E. coli* antibodies immobilized on gold nanoparticles, achieved through the hydrolysis of 5-bromo-4-chloro-3-indolyl phosphate (BCIP) and the generation of a SERS signal (Bozkurt et al. 2018). Kang et al. developed a solid-phase biosensor utilizing bacterial endogenous alkaline phosphatase (ALP) to detect and kill bacteria directly. The system employs quaternary ammonium with bactericidal activities and a phosphorylated 2-hydroxychalcone (HCAP) probe attached to poly(vinylpyrrolidone) (PVP). If the sensor comes in contact with bacteria, the bacterial alkaline phosphatase (ALP) cleaves the HCAP probe, leading to fluorogenic aggregation and a change in fluorescence emission that indicates the presence of target bacteria while simultaneously killing them (Kang et al. 2018). Jackson et al. have engineered ALP expressing T7 phage, where ALP acts as a reporter protein in a colorimetric assay for *E. coli* detection (Jackson et al. 2016). In a novel technique proposed by Ma et al., mycotoxin zearalenone (ZEN) has been detected in a biosensor utilizing the ALP's dNTP hydrolysis activities. The process involves two steps: first, the ZEN-specific monoclonal antibodies are captured in a competitive indirect ELISA with secondary antibodies conjugated to ALP enzymes. In the second step, when amplified on a PCR sensor, ALP enzymes dephosphorylate dNTP, thus interfering with target amplification that negatively correlates with ALP activities (Ma et al. 2023).

4.2.1.2 Lipase

Lipase is a triacylglycerol ester hydrolases family member, catalyzing the conversion of long-chain triglycerides (TG) into fatty acids and glycerol. Microbial lipases are found in a wide variety of fungi and bacteria, including *Achromobacter* spp., *Alcaligenes* spp., *Arthrobacter* spp., *Bacillus* spp., *Candida* spp. *Penicillium* spp., *Fusarium* spp., *Aspergillus* spp., and *Pseudomonos* spp. for industrial applications. Lipase enzyme is responsible for fat metabolism and is widely distributed throughout the body, playing distinct biosynthetic and metabolic roles in the pancreas, liver, adipocytes, and endothelial cells. Consequently, serum triglyceride levels are an important biomarker for various diseases such as pancreatitis, cardiovascular diseases, and diabetes. The lipase enzyme's unique ability to catalyze reactions at the lipid–water interface and its broad substrate selectivity enable its adaptation into diverse analytical platforms (Chandra et al. 2020). For diagnostic purposes, Sigma Aldrich provides various microbial lipases sourced from *Aspergillus* sp. (62,301, L0777), *Candida rugosa* (L1754), and *Rhizopus oryzae.* (80612). Lipoprotein lipase from *Pseudomonas* is obtainable through Asahi Kasei (T-116), Innovative Enzymes (151622), and Sorachim (LPL-311). Ray Biotech offers a distinct option with human lipase expressed in *E. coli* (228-12238).

Due to their high sensitivity, specificity, and cost-effectiveness, microbial lipases, individually or in combination with other enzymes, are now widely used to construct various categories of triglyceride biosensors. In the study by Narwal et al., lipase nanoparticle from *Candida rugosa* was immobilized along with microbial glycerol kinase (GK) and glycerol-3-phosphate oxidase (GPO) on pencil graphene electrodes for the amperometric determination of serum triglyceride (TG) (Narwal and Pundir 2017). The co-immobilized enzymes facilitate the conversion of triglycerides to fatty acids and further into oxygen gas. The generated electrochemical signal provides a quantitative indication of the triglyceride levels in the sample. In a separate study, *Candida antarctica* lipase b enzyme was entrapped in a pectin hydrogel membrane, catalyzing the hydrolysis of triglycerides into glycerol and fatty acids. The resulting proton release induced a colorimetric pH change, enabling the determination of triglyceride levels through optical signal measurement (Hasanah et al. 2019).

4.2.1.3 Nucleases

Nucleases, a group of enzymes that hydrolyze phosphodiester linkage and primarily function to digest nucleic acids (DNA/RNA), are integral to exo- and endonuclease activities, DNA repair, recombination, and replication along with RNA splicing processing, maturation, and interference activities. These enzymes contribute to the virulence and pathogenicity of many viruses and bacteria, such as *Streptococcus* spp., *Aeromonas hydrophila, Mycobacterium* spp., *and Mycoplasma* (Sharma et al. 2019). In addition to serving as a biomarker, various nucleases have been applied in versatile diagnostic assays for diverse purposes.

Deoxyribonucleases (DNases) enzymes, which break down DNA molecules, are categorized into two major classes based on their biochemical and biological properties: DNAse I and DNAse II. Initially employed in footprinting assays to identify DNA binding sites for regulatory proteins, deoxyribonuclease enzymes (DNA I and II) now find extended applications as diagnostic markers for urinary detection (Flenker et al. 2017), serve as phenotypic markers in clinical microbiology laboratories, aiding in screening, differentiation, and identification of pathogenic bacteria (Lagacé-Wiens et al. 2007), and function as signal amplifiers in bio-assays. In DNase-based assays, if a target analyte is detected, quantum dots or fluorescently labeled aptamers anchored on nanoparticles are released and form analyte–aptamer complexes. DNAse I amplifies the detection signal by cleaving the complex and freeing analyte molecules to bind with more aptamers. The principle has been used to detect serum antigens specific to the prostate (Fang et al. 2017) and cancer exosomes (Wang et al. 2018). S1 nuclease sourced from *Aspergillus oryzae* serves as a nuclease digester within an ELISA platform for detecting Micro RNA-195, a specific biomarker associated with osteosarcoma (Lei et al. 2022).

Restrictions endonuclease (REase), an enzyme that can cleave double strands of DNA at particular locations called restriction sites, is another adaptable nuclease with various uses in molecular diagnostics. The enzyme is predominantly present in most bacteria, occurring in different classes and subclasses, among which RE type II is used in molecular biology. Originally identified as the R endonuclease (now known as HindII) in *Hemophilus influenza*, the use of this enzyme began with DNA mapping (Roberts 2005). Over time, different varieties have been used in a range of DNA detection tests. Bacterial REases are used in different fingerprinting methods, including Restriction Fragment Length Polymorphism (RFLP) (Melo et al. 2021), terminal Restriction Fragment analysis (T-RFLP) (Christensen et al. 2004), Amplified Ribosomal DNA Restriction Analysis (ARDRA) (Stakenborg et al. 2005), Amplified Fragment Length Polymorphism (AFLP) (Wong et al. 2004), Pulsed-field gel electrophoresis (PFGE) (Davis et al. 2003) for typing, identification, screening, and analyzing disease strain variations. In modern assays, REase enzymes play the role of signal amplifiers. In these systems, target analytes are first amplified and hybridized with specific probes labeled with molecular markers, which REase enzymes cleave. The quantity of released markers corresponds to the amount of the target bio-molecule amplified via nucleic acid amplification methods. For instance, methicillin-resistant *Staphylococcus aureus* (MRSA) was detected in a PCR test using the BglII enzyme, which cleaves a molecular marker that colorimetrically indicates the presence of mecA gene amplicons (Smith et al. 2014). A similar principle is applied in another assay system for nucleic acid detection utilizing the mutant version of Restriction Endonuclease (RE) from *Bacillus* spp., which causes single-stranded breaks. The assay system comprises two probes: a closed hairpin DNA probe containing the recognition sequence for the target analyte and a sequence for endonuclease activity, and a GC-rich DNA probe carrying a DNAzyme sequence with peroxidase-like activities. When the analyte is absent, the hairpin probe remains closed, and the DNAzyme activities are blocked by a DNA blocker, maintaining a circular formation that prevents interaction with its substrate. In the presence of the target analyte (thrombin protein) captured on the aptamer sequence, the hairpin probe opens and hybridizes with the GC-rich DNA, initiating the endonuclease-cleaving activities. Repeating the process of hybridization and cleavage generates numerous DNAzyme segments, leading to the oxidation of the substrate and the production of an amplified colored product, which indicates the presence of the target analyte (Huang et al. 2013).

4.2.1.4 Creatinase

Creatinase is an important enzyme for renal function evaluation. It catalyzes creatine to sarcosine and urea. The enzyme is native to *Alcaligenes, Arthrobacter, Pseudomonas, Bacillus, Flavobacterium, Clostridium, Micrococcus,* and

Paracoccus, and it is often expressed through recombinant *E. coli* for more efficient utilization (Afshari et al. 2017). Creatinase enzyme, combined with creatininase and sarcosine oxidase, is routinely used in enzymatic assays to determine serum and urine creatinine, a byproduct of muscle metabolism that serves as a valuable marker for assessing kidney health. In these assays, with these three consecutive enzymes, creatinine is converted from creatine to sarcosine to hydrogen peroxide, which is measured directly or indirectly via the colorimetric method to directly access the original concentration of creatinine and creatine in a biological sample. In one such system, Staden et al. employed creatininase and creatinase enzymes from *Flavobacterium* sp., along with sarcosine oxidase from *Arthrobacterium* to create an electrochemical sensor. The combined tri-enzyme activities in this system offer direct access to serum creatinase concentration (Staden and Bokretsion 2006). A recombinant version of these enzymes from *E. coli* was utilized in a different system to develop an enzymatic kinetic assay for detecting creatine using sarcosine as a limiting enzyme (Jimenez Valencia et al. 2021).

4.2.1.5 Cholinesterase

Cholinesterase or acetylcholine acylhydrolase enzyme aids in neural functions by breaking neurotransmitter acetylcholine into choline and acetic acids.

In clinical diagnostics, cholinesterase enzyme activity is employed to detect nerve agents or pesticides that impede the enzyme's function, causing an accumulation of acetylcholine in the synapse that causes the onset of neurological symptoms. Microbial cells, such as *S. cerevisiae*, are utilized for expressing eukaryotic enzyme versions in biosensor platforms to immobilize and detect anticholinergic compounds (e.g., neurotoxins, pesticides) by assessing the inhibition of cholinesterase enzymes (Liang and Han 2020).

4.2.1.6 Cholesterol Esterase

Cholesterol esterase, an ester hydrolyzing enzyme that breakdown fatty acid esters to free cholesterol, has been reported in different microbial sources such as *Acinetobacter* spp., *Burkholderia* spp., *Pseudomonas* spp., *Streptomyces* spp., *Fusarium oxysporum*, *Saccharomyces cerevisiae*, *Trichoderma* spp., and *Candida* spp. (Vaquero et al. 2016). Diagnostic-grade cholesterol esterase enzymes from microbial sources, including *Pseudomonas* and *Candida*, are available in a lyophilized form from suppliers such as Creative Diagnostics (DAG-WT1275), Creative Enzymes (DIA-134, DIA-135, NATE-1679), Roche (10,634,522,103), Asahi Kasei (T-250), and Sigma Aldrich (C1403). The enzyme is routinely employed in enzymatic assays for blood cholesterol (e.g., total cholesterol, low-density lipoprotein or LDL, high-density lipoprotein or HDL) level determination, an important parameter for monitoring lipid profiles and evaluating cardiovascular health. These assay systems utilizes the analytical platform of the homogeneous assay or biosensor, where cholesterol enzymes with other hydrolases break down cholesterol to hydrogen peroxide, which is detected colorimetrically or flurometrically. In a homogenous assay for small dense LDL (sd-LDL) detection (Ito et al. 2011), *Pseudomonas* sp. cholesterol esterase, microbial cholesterol oxidase, and sphingomyelinase were employed. The assay principle involves the use of two surfactants: one facilitates the dissociation of all blood lipids, including triglyceride-rich lipoproteins and HDL, with the three hydrolyzing enzymes, while the other specifically protects sd-LDL. The sd-LDL, preserved from enzymatic reactions, is subsequently measured colorimetrically using the hydrogen peroxide and peroxidase system. Román-Pizarro et al. co-immobilized cholesterol esterase with cholesterol oxidase on magnetic nanoparticles in a microfluidic biosensor platform to determine total serum cholesterol. The redox activities of the enzymes in this assay convert cholesterol to hydrogen peroxide, which is subsequently detected using the fluorescence activities of oxidized naphthofluorescein (Román-Pizarro et al. 2023).

4.2.1.7 Uracil-DNA Glycosylase (UDG)

Uracil-DNA glycosylase (UDG) is a highly conserved DNA-repairing enzyme, which has been identified in bacteria (e.g., *E. coli, M. tuberculosis, B. subtillis, H. pylori*), fungus (e.g., *S. cerevisiae)*, archaea (e.g., *Methanococcus jannaschii, Pyrococcus furiosus, Sulfolobus solfataricus*), and viruses (e.g., poxvirus, herpes virus) (Schormann et al. 2014). It is a type of glycosylase that hydrolyzes the N-glycosidic link that joins uracil and deoxyribose to carry out DNA repair functions and preserves genome integrity. In in vitro diagnostics, the enzyme's repairing functions have been employed in the real-time loop-mediated isothermal amplification (RT-LAMP) method for the detection of uropathogens (Zeng et al. 2020), ocular toxoplasmosis (Rahimi Esboei et al. 2022), and influenza virus (Kim et al. 2016), where it reduces false-positive results by preventing undesired amplification of contaminated amplicons from previous cycles. For this purpose, recombinant versions of the enzymes are commonly available from *Escherichia coli*, often sourced from extremophilic organisms. The recombinant protein can be accessed from New England BioLabs (M0280S), Transgene Biotech (LU201-01), NZYTech (MB44601), Roche (CAS No 39369-21-7), DiagnoCine (CSB-DEM026), and Meridian Bioscience (BIO-27044).

4.2.1.8 Urease

Urease, a widely utilized hydrolase, is frequently employed in developing urea biosensors for assessing urea levels in biological fluids (e.g., blood and urine). Urea concentration holds significance in evaluating various pathological conditions involving the kidneys, liver, cancer, and gastrointestinal tract, as it is one of the end products of protein metabolism and is excreted in the urine. The extensive applications of this enzyme span diverse categories of biosensor development. In a particular study, a recombinant *E. coli* urease was immobilized on a pH-sensitive field-effect transistor transducer, enabling the determination of serum urea concentration through the potentiometric principle (Marchenko et al. 2015).

4.2.1.9 Helicase

Bacterial helicase enzymes, known for their distinctive ability to disrupt hydrogen bonds and unwind DNA strands, are increasingly being employed as a substitute for the heat denaturation steps traditionally used in PCR (Polymerase Chain Reaction). An array of helicase-dependent amplification (HAD) isothermal assays, integrated with various detection platforms, is currently under development for point-of-care detection of pathogens. Some examples are HDA-TINY (Tiny Iso-thermal Nucleic acid quantification sYstem) for *Mycobacterium tuberculosis* (Shanmugakani et al. 2021), RT-HDA-coupled dipstick for the SARS-CoV-2 virus (Shanmugakani and Wu 2022), an HDA coupled lateral flow assay for Monkeypox virus (Gulinaizhaer et al. 2023), and HDA/SYBR Green I for detecting platelet product contamination (Yamket et al. 2023).

4.2.2 Oxidoreductases

4.2.2.1 Glucose Oxidoreductase

Glucose oxidoreductases fall into two categories: glucose oxidase, which utilizes oxygen as an electron acceptor, and glucose dehydrogenase, which does not. Glucose oxidase (GOx) is a flavoprotein of the glucose/methanol/choline (GMC) family, which catalyzes hydrogen peroxide production from beta-D-glucose. Glucose dehydrogenase (GDH), on the other hand, oxidizes glucose to gluconic acid while reducing co-factors nicotinamide adenine dinucleotide (NAD), flavin adenine dinucleotide (FAD), and pyrroloquinoline quinone (PQQ). The GOx enzyme has been identified in *Aspergillus* spp. and *Penicillium* spp., including a recombinant version in *Saccharomyces cerevisiae* and *Hansenula polymorpha*, while the GDH enzyme has been reported in *Bacillus megaterium*, *Sulfolobus*, and *Aspergillus* (Ferri et al. 2011). Despite the structural and functional differences between the two enzymes, they are routinely utilized for glucose monitoring in clinical settings. The majority of the enzymes from commercial vendors are sourced from the fungus *Aspergillus niger*. Some notable examples are Sigma Aldrich (G7141), SEKISUI Diagnostics (GLOX-70-6457), Central Drug House Ltd. (104835), Goproan Biotech (GPE0008), Roche (11939998103), Krackeler Scientific (SIGMA/G2133), Toyobo Enzymes (GLO-101201), Creative Enzymes (DIA-193), and Lee Biosolutions (277-03).

Glucose sensors have evolved through three generations: the first method relies on the measurement of oxygen or hydrogen peroxide to determine glucose concentration; the second employs mediators as an alternative to oxygen to move electrons between the enzyme and the electrode; and the third uses direct electron transfer between the enzyme and the electrode through either electron relay or nanoparticles (Wong et al. 2008). In addition, glucose oxidase from *Aspergillus niger* immobilized on polypyrrole electropolymerized films has been explored as a promising electrochemical biosensor approach for alkaline phosphatase detection (Santiago et al. 2010).

4.2.2.2 Alcohol Oxidoreductase

Based on electron acceptors, microbial alcohol oxidoreductase enzymes can be of two types: NAD(P) dependent or independent alcohol dehydrogenase and FAD-dependent alcohol oxidase. These enzymes mainly catalyze the transformation of alcohol, aldehyde, and ketones, thus involved in a wide range of biosynthetic, metabolic, and fermentative processes in a variety of microorganisms. For instance, anaerobic bacteria and yeasts use alcohol dehydrogenase to produce alcohol from acetaldehyde for energy generation (Raj et al. 2014). In diagnostics, alcohol oxidase serves as a common biorecognition element in biosensors, facilitating the determination of alcohol intoxication and the evaluation of alcohol concentration in body fluids for clinical, forensic, and law enforcement purposes. Commercial diagnostic-grade alcohol oxidase is primarily sourced from fungi such as *Candida* (Creative Enzymes, DIA-123; SEKISUI Diagnostics, T-38; Albert Bio, E010-1), *Hansenula* (Creative Enzymes, NATE-0046), and *Pichia pastoris* (Creative Enzymes NATE-0047). Yeast alcohol dehydrogenase can be obtained from Roche (11452541103) and Creative Enzymes (NATE-0035). Gamella et al. employed alcohol oxidase from *Pichia pastoris* to quantify transdermal alcohol levels, measuring the amperometric current generated as the enzyme oxidizes ethanol in sweat to hydrogen peroxide. This biosensor featured the immobilization of alcohol oxidase with hydrogen peroxidase enzyme on a composite electrode, where the oxidation reaction was mediated by the reduction of ferrocene to ferricenium (Gamella et al. 2014). Exploring a similar principle, alcohol determination from whole blood samples has been investigated using an alcohol oxidase/hydrogen peroxide system immobilized on alginate hydrogel within a microfluidic platform. This setup allows for alcohol concentration assessment with a digital readout on a smartphone-based platform (Aymerich et al. 2018).

4.2.2.3 Cholesterol Oxidase

Cholesterol oxidase is a bacterial flavoenzyme that oxidizes blood sterols. The enzyme is naturally produced by numerous organisms, both intracellularly and extracellularly. Some examples are *Arthrobacter, Corynebacterium, Nocardia erythropolis, Rhodococcus erythropolis, Mycobacterium, Pseudomonas, Schizopyllum commune, Brevibacterium sterolicum, Streptoverticillium cholesterolicum, Streptomyces violascens,* and *Rhodococcus* spp. However, due to the difficulties associated with bacterial virulence, overexpression, and production costs, the enzyme is now produced in recombinant *E. coli* and *Brevibacterium*. Several vendors manufacture microbial cholesterol oxidases, such as Procurenet (9028-76-6), Asahi Kasei (T-101), Sigma Aldrich (C8649), Creative Enzymes (NATE-0128, DIA-406), and SEKISUI Diagnostics (RECO-70-1221). Cholesterol oxidase is frequently used together with cholesterol esterase and peroxidase enzymes in many diagnostic platforms for detecting cholesterol levels in different serum samples, lipoproteins, cell membranes, and gallstones. In these platforms, cholesterol esterase (ChE) liberates cholesterol molecules from cholesterol esters, and subsequently, cholesterol oxidase (CHOx) converts them into hydrogen peroxide, which is then detected colorimetrically (Pollegioni et al. 2009). Chauhan et al. co-immobilized cholesterol esterase, cholesterol oxidase, and peroxidase on a reusable polyvinyl plastic strip for total serum cholesterol determination, where the sequential reaction of these enzymes resulted in a colored end product, indicating the analyte concentration (Chauhan and Pundir 2011).

4.2.2.4 Bilirubin Oxidase

Bilirubin oxidase (BOx) is one of the multicopper oxidases that catalyze bilirubin to biliverdin. It is primarily utilized in the determination of bilirubin levels in serum, serving as a diagnostic tool for conditions such as jaundice. The enzyme has been identified in *Penicillium janthinellum*, *Bacillus* spp., *Pleurotus ostreatus*, *Trachyderma tsunodae*, *Myrothecium verrucarria*, *Aspergillus sojae*, and *Magnaporthe oryzae* (Mano 2012). Bilirubin oxidase from *Myrothecium verrucaria* is available for use in clinical and diagnostic applications from Asahi Kasei (T-188), Sigma-Aldrich (B0390), SEKISUI Diagnostics (BIOX-70-1021), and Creative Enzymes (NATE-0094). In bilirubin oxidase-based assay systems, the measurement of serum bilirubin levels, encompassing both conjugated and unconjugated forms, is primarily determined by assessing the reduction in bilirubin absorbance, which negatively correlates to bilirubin concentration (Doumas et al. 1987). Andreu et al. developed a direct bilirubin determination method using Bilirubin Oxidase (BOx) from *Myrothecium verrucaria*, detecting bilirubin levels through changes in fluorescence absorbance (Andreu et al. 2000). In addition to its role as a diagnostic agent, the enzyme has other diverse applications in the diagnostic field. Kim et al. have used bilirubin oxidase as an alternative enzyme label in a hybridization assay for DNA detection, utilizing oxygen as a substrate instead of hydrogen peroxide (Kim et al. 2004). Bilirubin oxidase serves as a neutralizer in bioassays to mitigate bilirubin interference, exemplified in diagnostic assays for Thrombotic Thrombocytopenic Purpura (TTP) (Eckmann et al. 2007) and plasma hemoglobin (Wong and Schenkel 1995).

4.2.2.5 3-Hydroxysteroid Dehydrogenase

The enzyme 3-hydroxysteroid dehydrogenase is responsible for the oxidoreduction of the steroid hormone. The enzyme is commonly employed in enzyme-based assays to determine serum bile levels, a crucial parameter in diagnosing liver diseases. The primary sources of the enzyme for such purpose come from *Pseudomonas* sp. from Asahi Kasei (T-58), Sigma Aldrich (H1506), Creative Enzymes (NATE-0007), and SEKISUI Diagnostics (T-58). Rani et al. used 3-hydroxysteroid dehydrogenase from *Pseudomonas testosteroni* in combination with lipoyl dehydrogenase from *Clostridium* sp. to develop an inexpensive colorimetric platform for such a purpose. The platform utilizes arylamine glass beads to immobilize the enzymes to measure the reduction in nicotinamide adenine dinucleotide, which is inversely linked to serum bile concentration (Rani et al. 2004). In another system, a prokaryotic recombinant version of the protein has been used in an enzyme cycling platform for a similar purpose. The determination of serum bile levels is based on the reversible transformation of bile acids to their oxidized form, facilitated by 3a-HSD. In this process, bile acid oxidoreduction occurs in the presence of thio-NAD+ and NADH. The concentration of total bile acids (TBA) is derived from the quantity of thio-NADH generated during these reactions, measuring serum bile levels (Zhang et al. 2005).

4.2.2.6 Microbial Antioxidative Enzymes (Superoxide Dismutase, Catalase, and Peroxidase)

As a byproduct of aerobic respiration, microorganisms produce several reactive oxygen species (ROS), including singlet oxygen, hydroxyl radical, hydrogen peroxide, and superoxide anion, which can kill them directly. In response to such oxygen toxicity, microbes have evolved an antioxidant defense system comprising several oxidative enzymes that break down their reactive substrate into free oxygen. Superoxide dismutase, catalase, glutathione reductase, and glutathione peroxidase enzymes are disease markers in many communicable and noncommunicable diseases (Staerck et al. 2017). Among them, superoxide dismutases are the earliest enzymes associated with ROS detoxification. Rather than being used as a probing agent in diagnostic assays, superoxide dismutase serves as a target analyte in the diagnosis of gastric cancer (Li et al. 2019), EBV infection (Dworzański et al. 2020), and tuberculosis (Chattopadhyay 2021).

Catalase enzyme decomposes hydrogen peroxide into water and free oxygen and is ubiquitous in numerous bacteria, fungi, and extremophiles (Sooch et al. 2014). Catalase from the fungus *Aspergillus* spp., available in both liquid and powder forms, can be obtained for the formulation of diagnostic assays from suppliers such as Creative Enzymes (DIA-131), Sigma Aldrich (219261), and Southern Biological (MC23.05). It plays a significant role in the metabolism, virulence, and pathogenesis of infectious diseases (Mina et al. 2015). In addition, the enzyme has long been used as a signal amplifier in diagnostic assays. Song et al. conjugated catalase to ELISA detection antibodies to break down hydrogen peroxide into measurable oxygen in a microfluidic platform. The microchannels are coupled with a volumetric ink bar chart that responds to pressure fluctuation caused by free oxygen and reports the quantitative level of the target analyte in a multiplexed manner (Song et al. 2012). In a recent cellphone-based, amplification-free system featuring a CRISPR/CAS-dependent enzymatic (CASCADE) assay for the SARS-CoV-2 virus, catalase-hydrogen peroxide activities have been used for target detection. This system involves conjugating catalase enzymes to a single-stranded DNA (ss-DNA) probe synthesized for target viral RNA. Following reverse transcription of viral RNA and hybridization with ss-DNA probe, the Cas enzymes liberate conjugated catalase, producing oxygen gas when hydrogen peroxide is added to the reaction tube. The formation of gas bubbles indicates the presence of SARS-CoV-2 viral RNA, and the results are read out through smartphones (Silva et al. 2021).

Peroxidases are a set of oxidoreductases that act as biocatalysts for the oxidation of organic and inorganic substrates by reducing peroxides. They degrade hydrogen peroxides at the expense of an electron-donating substance in a distinct three-step catalytic mechanism common to all isoforms of the enzymes. This makes peroxidases a useful analytical tool in clinical applications (Regalado et al. 2004). In the clinical field, peroxidases are largely used as a label for enzyme immunoassays. While plant-based horseradish peroxidase enzyme is the more traditional choice, microbial peroxidase has been documented as more efficient in such diagnostic analysis (Akimoto et al. 1990).

4.2.2.7 Lactate Dehydrogenase

Lactate dehydrogenase (LDH) is a carbohydrate metabolism enzyme that catalyzes pyruvate to lactate with the NAD^+/NADH coenzyme system. The enzyme has been documented to originate from various bacterial species, including *Lactobacillus, Bacillus,* and *Pediococcus* spp. LDH has been extensively employed in commercial colorimetric ALT (alanine aminotransferase) assay kits to measure ALT activities in serum. In these systems, LDH catalyzes the conversion of pyruvate to lactate, concomitantly oxidizing NADH to NAD, reducing NADH absorbance. This reduction, inversely correlated with ALT activity, is utilized to report the assay results (Sun et al. 2021). In addition, lactate dehydrogenase of the malaria parasite, which is distinct to the *Plasmodium* spp., serves as a metabolite for the immunodiagnosis of malaria and has been utilized in the development of immunodiagnostics, including immunoblots and dipstick assays (Kaushal and Kaushal 2002).

4.2.3 Transferases

4.2.3.1 Kinase

Carbohydrate kinase enzymes are essential for sugar utilization in microorganisms, thus fundamental for growth and survival. Different bacteria and fungi have five distinct classes of such enzymes: hexokinases, ROK kinases, ribokinases, GHMP kinases, and phosphatidylinositol kinases. Microbial hexokinase primarily falls under the glucokinase category, one of the two main classes of hexokinase, and typically uses substrates like glucose, ribulose, gluconate, xylulose, glycerokinase, fructose, rhamnose, and fucose (Roy et al. 2019). The enzyme is frequently used as a probe, either singly or in combination with other enzymes in various glucose sensors utilizing luminescence, fluorescence, and electrochemical principles, including potentiometry, amperometry, and conductometry (Scognamiglio et al. 2004). Glucokinase of *Bacillus stearothermophilus* has been co-immobilized with hydrolase and glucose-6-phosphate dehydrogenase to construct sensors for determining sugar concentrations (Aoki et al. 2005). In a competitive assay system, bacterial glucokinase labeled with a fluorescence probe binds with glucose, indicating glucose levels through fluorescence activities. Competitive binding with a resonance energy transfer (RET) acceptor molecule, initially bound with glucose to quench fluorescence, is alleviated upon glucokinase binding, providing a fluorescence readout for glucose concentration (D'Auria et al. 2002). Ueda et al. utilized the enzymatic cycling method with *Pyrococcus furiosus* glucokinase to quantify ADP, a product in the cyclic conversion of creatine-to-creatine phosphate, and described its potential for real-time determination of creatine in serum samples (Ueda and Sakasegawa 2016).

4.2.3.2 DNA Polymerases

Family A DNA polymerase, one of seven families of DNA polymerases in charge of nucleic acid metabolism, has extensive applications in molecular diagnostics. Enzymes from this family mostly have prokaryotic origins, such as prokaryotic DNA polymerase I and several bacteriophage polymerases. PCR amplification remains the most impressive diagnostic application of DNA polymerases to date. Following the identification of the *Taq* polymerase of *Thermus aquaticus*, several thermostable DNA polymerases have been identified, boasting fidelity and 3′→5′ exonuclease activities. DNA amplification has been carried out with *KOD* DNA polymerase from *Thermococcus kodakaraensis, pPfu* DNA polymerase from *Pyrococcus furiosus*, *Pwo* from *Pyrococcus woesei, Tfl* from *Thermus flavus*, *Tth* from *Thermus thermophilus*, *Tma* polymerase from *Thermotoga maritima*, *Tne* from *Thermotoga neapolitana*, *Tli* from *Thermococcus litoralis, Tpa* from *Thermococcus pacificus*, and *Tfu* from *Thermococcus fumicolans* (Śpibida et al. 2017). In addition to using recombinant versions of these enzymes for PCR, DNA polymerase is now being modified with mutagenesis and fusion technologies to enhance its thermostability and improve its resistance to inhibitors present in samples. Numerous biotech companies produce different types of DNA polymerases from different microbial sources. Notable examples include QIAGEN, New England Biolabs, Invitrogen, Promega, Tiangen, Sigma Aldrich, Takara Bio, and Ampliqon.

A new range of DNA polymerases has emerged in modern diagnostics because of their functioning properties at constant temperatures and strand-displacement activities. The large fragment of DNA pol I from *Geobacillus stearothermophilus,* known as *Bst* Polymerase, is widely used in several isothermal DNA amplification tests, including loop-mediated isothermal amplification (LAMP), isothermal multimerization and amplification (LIMA), whole genome amplification (WGA), and rolling circle amplification (RCA) (Oscorbin and Filipenko 2023). The Bsu DNA polymerase, derived from *Bacillus subtilis*, has been utilized in the Recombinase Polymerase Amplification (RPA) method, facilitating the detection of microorganisms like *Streptococcus pneumoniae* and uropathogens (Wang et al. 2022; Chen et al. 2018). Additionally, it has been employed in hybridization assays aimed at diagnosing cancer. The assay relies on the enzyme's ability to incorporate Cy5-dATP to 8-oxo-7,8-dihydroguanine (OG) selectively, the oxidation product of the guanosine base found in the telomere DNA of cancer cells. This process involves the hybridization of a biotinylated probe to denatured telomere DNA extracted from cancer cells, where the Bsu enzyme incorporates cytosine opposite to OG, and the amount is quantified fluorometrically (Li et al. 2022b). Another polymerase commonly employed for isothermal amplification is ϕ29 DNA polymerase, classified within the B family of DNA polymerases. This enzyme is valued for its exceptional processivity, capacity for strand displacement, and intramolecular proofreading activity, making it a valuable choice for amplification technologies. In an electrochemiluminescence biosensor system, ϕ29 amplification activity was employed to produce a single-stranded DNA probe from target mRNA via strand-displacement amplification (SDA). The detection of mRNA involved the hybridization of the single-stranded DNA probe with a ferrocene-labeled probe bound to luminophores, resulting in the quenching of an electrochemiluminescence signal, the reduction of which signaled the presence of mRNA (Chen et al. 2015).

The RNA-dependent DNA polymerase of retroviruses, reverse transcriptase, is a highly desirable tool in clinical applications. Due to its distinctive capability to transform RNA into complementary DNA (cDNA) and high fidelity, this enzyme has extensive use in detecting expressed genes, exploring transcript variations, and sequencing. The enzyme is employed in conjunction with PCR, a technique known as reverse transcriptase PCR (RT-PCR), for the quantification and diagnostics of RNA in clinical diagnostics (Martín-Alonso et al. 2021).

T7 DNA polymerase was first identified in a mutant *E. coli* infected with the bacteriophage. Because of the enzyme's high processivity and high fidelity, chemically or genetically engineered enzymes are extensively used in DNA sequencing technologies (Zhu 2014). The enzyme is commonly employed for pathogen detection within CRISPR-coupled amplification systems. A system utilizes a specific allosteric probe (AP) with aptamer, primer binding site, and T7 promoter domains. Upon recognizing the target pathogen, the AP undergoes a conformational change, facilitating primer binding and DNA polymerase-mediated amplification of pathogen DNA. The resulting amplicons are transcribed into single-stranded RNAs (ssRNAs) by T7 RNA polymerase. Detection is achieved by the Cas13a/crRNA system, wherein activated collateral cleavage of RNA reporter probes generates fluorescence signals for recognizing the target pathogenic RNA (Shen et al. 2020).

4.2.3.3 Other Transferases

Several microbial transferase enzymes are used as markers for diagnosing diseases. The detoxifying enzyme aminotransferase or transaminase enzymes transfer amino groups between amino acids and keto acids. Two of these enzymes, aspartate aminotransferase (AST) and alanine transferase (ALT), formerly known as glutamate oxaloacetate transaminase (GOT) and glutamate pyruvate transaminase are frequently used to monitor liver health. Enzymatic activities are assessed using glutamate biosensors, where ALT and AST convert their substrate amino acids to glutamate, subsequently oxidized by glutamate oxidase from *Streptomyces* spp. to produce α-ketoglutarate and hydrogen peroxide, allowing quantification (Song et al. 2009). The glycosyltransferase (Gtf) from *Streptococcus mutans*, participating in glucan polymerization, serves as a diagnostic marker for the development of dental carious lesions, with its activity assessed through the immunoassay system with Gtf-specific monoclonal antibodies (Vacca Smith et al. 2007).

4.2.4 Ribozymes

Ribozymes are RNA molecules with catalytic activity, capable of cleaving and ligating phosphodiester bonds. A recent review has outlined diverse naturally occurring ribozymes such as the hepatitis delta virus (HDV)-like, glucosamine-6-phosphate riboswitch ribozyme (glmS), Neurospora Varkud satellite, hairpin, hammerhead, group I and II of introns, the ribosome, spliceosome, and RNase P, emphasizing their potential use as diagnostic tools (Cardador et al. 2023). The cleaving activity of ribozymes serves as a signal amplifier in diagnostic tools. In a patent (WO 94/13833) for biomolecule detection, the binding of a target molecule with a labeled co-target molecule leads to the liberation of labels by ribozymes, producing a measurable signal that indicates the presence of the target analyte (Shih et al. 1994). Ribozyme's self-cleavage activity was utilized in a colorimetric assay. Ribozyme binding with the target analyte amplifies the reporter sequence, forming single-stranded DNA guanine quadruplexes. These DNA structures possess peroxidase activities, with which they oxidize a colorless substrate (TMB) into a colored product and signal target detection (Liao et al. 2018). Kossen et al. employed a half ribozyme derived from the hepatitis C virus (HCV) in immunodiagnostic platforms, utilizing a fluorometric biotin–streptavidin detection system to identify viral nucleic acids (Kossen et al. 2004).

4.3 Enzyme-Based Diagnostic Techniques

Based on technologies or methods, enzyme-based diagnostics can be broadly classified as immunoassays, nucleic acid-based amplification tests, hybridization assays, biosensors, and conventional biochemical assays.

4.3.1 Serology

4.3.1.1 Enzyme Immunoassay

Enzyme immunoassays (EIA) commonly employ enzymes as labels, which are attached to the desired antigens or antibodies through physical or chemical means, facilitating the detection of immune complexes. These assays come in two forms: homogeneous and heterogeneous assays. Homogeneous enzyme immunoassays involve enzymes losing activity upon antibody binding without the need to separate the antigen from the solution, whereas, in heterogeneous assays, the resulting antigen–antibody complex is immobilized after binding, requiring a crucial washing step to remove the unbound antigen. The latter technique is more widely used in a variety of applications due to its affordability and sensitivity. EIA has employed a wide range of microbial enzymes as labels, including *Leuconostoc mesenteroide*'s glucose 6-phosphate dehydrogenase, *Aspergillus niger*'s glucose oxidase/peroxidase and alkaline phosphatase (ALP), and *E. coli's* and β-galactosidase (Porstmann and Kiessig 1992) in different EIA technologies: enzyme-linked immunosorbent assay (ELISA), enzyme multiplied immunoassay technique (EMIT), fluorescent enzyme immunoassays (FEIAs), and chemiluminescent immunoassays (CLIAs) (Table 4.2).

4.3.1.1.1 ELISA (Enzyme-Linked Immunosorbent Assay)

Enzyme-linked immunosorbent assay (ELISA) is a quantitative tool widely used in clinical laboratories to detect and quantify the presence of a target analyte (e.g., pathogen, antigen, antibody) in biological samples. Its origins date back to 1941 when Albert H. Coons introduced the concept of using a marker to tag analytes. While primarily radioisotopes were used as markers, concerns over radioactivity led scientists to replace them with an enzyme–substrate combination on specific antibodies. Thus, ELISA was first conceptualized in 1971 by two separate European research groups. ELISA techniques are categorized based on their components and experimental setups. Direct ELISA employs an enzyme-labeled antibody to directly attach to the target antigen, with enzyme activity signaling the antigen's presence. In contrast, indirect ELISA involves coating the assay plate with the antigen and using a primary antibody to bind to it and a secondary antibody with an enzyme label to detect the primary antibody, amplifying the signal. Sandwich ELISA captures the antigen between two antibodies: one immobilized on the plate (capture antibody) and the other labeled with an enzyme (detection antibody). Competitive ELISA, on the other hand, quantifies antigen concentration by comparing a labeled antigen binding to a limited number of antibody binding sites in the sample (Aydin 2015). ELISA uses enzyme substrate chromogenic reactions for target detection. In most systems, among the microbial enzymes (Table 4.2), alkaline phosphatase is the most commonly utilized; however, additional enzymes are frequently chosen, such as galactosidase, urease, catalase, laccase, glucose oxidase (GOx), and acetylcholinesterase (AChE) (Li et al. 2008). Several commercially available ready-to-use

Table 4.2 Microbial enzymes in immunoassay technologies

Name of technique	Enzymes used	Target molecule	Microbial source	Ref.
Enzyme-linked immunosorbent assay (ELISA)	Alkaline phosphatase, glucose oxidase, urease, catalase, laccase, acetyl cholinesterase, and galactosidase	Biomolecules	*Aspergillus niger, Escherichia coli*	Porstmann and Kiessig (1992), Li et al. (2008)
Enzyme-labeled fluoroimmunoassay (ELFIA)	Alkaline phosphatase, β-D-galactosidase	4-Methylumbelliferyl phosphate (4MUP), 4-methylumbelliferyigalactoside	*Escherichia coli*	Porstmann and Kiessig (1992), Hicks (1984)
Chemiluminescence immunoassay (CLIA)	Alkaline phosphatase	Alpha-fetoprotein (AFP)	*Streptomyces avidinii*	Chong et al. (2012)
Bacterial bioluminescence assay (BBA)	Luciferase	Biolomecules	*Vibrio harveyi*	Li et al. (2022a)

ELISA kits incorporate alkaline phosphatase (ALP) as the enzyme label or conjugate for efficient detection. Notable examples include the ELISA Flex kits for Human IgA and IgE (Mabtech), CalproLab ELISA (ALP) (Svar Life Science), Human PSA-total ELISA Kit (Sigma Aldrich), AMP'D® HSP70 high-sensitivity ELISA kit (Enzo Life Sciences), Protein A-H ELISA Kit (Human IgG), and Protein A ELISA Kit (Cygnus Technologies).

4.3.1.1.2 ELFIA (Enzyme-Linked Fluorescence Immunoassay)

ELFIA, or enzyme-linked fluorescence immunoassay, is a highly sensitive immunoassay technique that uses fluorogenic substrate (e.g., 4-methylumbelliferyl phosphate) for enzyme labels. This approach uses microbial enzymes like alkaline phosphatase (ALP) and β-D-galactosidase (β-DG) to rapidly generate numerous colored, fluorescent, or luminescent molecules, thereby amplifying the detection process. ELFIA surpasses traditional color-based methods by enhancing sensitivity using fluorogenic substrates emitting fluorescence (Hicks 1984).

4.3.1.1.3 Chemiluminescence Enzyme Immunoassay (CLEIA)

CLEIA quantifies sample concentrations based on the luminescence intensity resulting from the chemical reaction, combining chemiluminescent systems with immunoreactions and utilizing specific reagents as labels to enable quantitative detection upon introducing chemiluminescent substrates. The heterogeneous method is more common in CLIA, which can be direct (using luminophore markers) or indirect (involving enzyme markers), and both can be competitive or noncompetitive. The utilization of microbial alkaline phosphatase (ALP) in chemiluminescence enzyme immunoassay (CLEIA) has been well documented. In a multiplex biomarker detection assay, the chemiluminescent system employs alkaline phosphatase sourced from *Streptomyces avidinii* for highly sensitive tumor marker alpha-fetoprotein (AFP) detection. The innovative approach utilizes fluorescein diphosphate tetraammonium (FDP) as a substrate for streptavidin-tagged ALP bound to the biotinylated AFP immunocomplex. This CLEIA system incorporates 1,1′-oxalyl diimidazole (ODI), which, upon rapid reaction with hydrogen peroxide, imparts high energy to fluorescein dyes, resulting in intense luminescence surpassing typical fluorescence levels (Chong et al. 2012).

4.3.1.2 Bacterial Bioluminescence Assay (BBA)

Bioluminescence assays utilize the emission of light from biological processes to detect and quantify specific molecules, cellular activities, or pathogens. The principle behind these assays is based on the generation of bioluminescent signals through the action of luciferase enzymes, which catalyze the oxidation of luciferin in the presence of oxygen and co-factors. Bacterial bioluminescence has become a valuable tool in bioanalysis and bioimaging, especially as whole-cell biosensors and bio-reporters for studying infectious diseases, cancer therapy, and tracking stem cells (Li et al. 2022a). Some examples are luciferase immunosorbent assay (LISA) for the serodiagnosis of the SARS-CoV-2 virus (Liang et al. 2021), Luciferase-linked Antibody Capture Assay (LACA) for the detection of *Toxoplasma gondii* (Duong et al. 2020), luciferase immunoprecipitation assay (LIPS) for African Swine Fever (Liu et al. 2021), and ATP bio-luminescence-sensing assay for microbial growth recognition (Eed et al. 2016).

4.3.2 NAAT (Nucleic Acid Amplification Test)-Based Techniques

Nucleic acid amplification tests (NAAT), another category of microbial enzyme-based technologies, have become pivotal in modern disease diagnosis, greatly improving sensitivity, accuracy, and speed compared to traditional culture-based methods for detecting infectious diseases (Table 4.3).

4.3.2.1 Polymerase Chain Reaction (PCR)

Polymerase chain reaction, commonly known as PCR, is a widely used molecular diagnostic technique involving duplicating a section of DNA using particular primers and a heat-resistant microbial enzyme called DNA polymerase. The process

Table 4.3 Application of microbial enzymes in Nucleic Acid Amplification Techniques (NAAT)

Name of technique	Enzymes used	Target molecule	Microbial source	Ref.
Polymerase chain reaction (PCR)	Taq DNA polymerase	DNA	*Thermus aquaticus*	Yang and Rothman (2004)
Reverse transcriptase PCR (RT-PCR)	Reverse transcriptase	DNA	Retroviruses (HIV, hepatitis C virus)	Forootan et al. (2017)
Loop-mediated isothermal amplification of DNA (LAMP)	Bst DNA polymerase	DNA	*Bacillus stearothermophilus*	Notomi et al. (2000)
Rolling circle amplification (RCA)	ϕ29 DNA polymerase	Micro RNA	*E. coli*	Liu et al. (2013)
Nucleic acid sequence-based amplification (NASBA)	Reverse transcriptase, RNaseH, T7 RNA polymerase	RNA	T7 bacteriophage	Perbal (2008)
Recombinase polymerase amplification (RPA)	Recombinase (T4 UvsX) Bsu DNA polymerase	DNA	*Bacillus subtilis*, T4 bacteriphage	Wang et al. (2022)
Strand-displacement amplification (SDA)	ϕ29 DNA polymerase	Micro RNA	*Escherichia coli*	Chen et al. (2015)

involves using known target DNA sequences to design specific oligonucleotide primers, initiating cycles of denaturation, primer hybridization, and extension in a thermocycler, resulting in exponential amplification of the target DNA. PCR has been used for pathogen detection, emerging infection surveillance, early bioterrorism threat detection, and assessment of antimicrobial resistance (Yang and Rothman 2004). Various commercial PCR kits and associated reagents have been developed or are currently available to diagnose numerous diseases and detect pathogens. Examples include the FastStart™ High Fidelity PCR System by Sigma Aldrich, the COBAS series by Roche Diagnostics, PrimeSTAR GXL DNA Polymerase by Takara Bio, and the MyTaq™ Blood-PCR Kit by Meridian Bioscience.

4.3.2.2 Quantitative Real-Time PCR (qPCR)

Quantitative Real-Time Polymerase Chain Reaction (qPCR) is an advanced improvement in PCR technology where DNA amplification and simultaneous detection occur within a single reaction vessel. This method eliminates the need for post-amplification steps like gel electrophoresis. It enables real-time measurement of DNA by incorporating dyes or probes that emit fluorescence during DNA synthesis, facilitating accurate and efficient quantification of target DNA. This method offers speed, simplicity, reproducibility, and precise DNA quantification, making it invaluable in clinical and research applications (Smith et al. 2014).

4.3.2.3 Reverse Transcriptase PCR (RT-PCR)

Traditional PCR techniques are inefficient for RNA amplification because they depend on DNA polymerase enzymes. Reverse transcriptase, an RNA-dependent DNA polymerase, is employed to overcome this by transforming RNA templates into complete DNA. The enzyme is naturally found in retroviruses like the human immunodeficiency virus and hepatitis C virus (Table 4.3). Several commonly used reverse transcriptases include avian myeloblastosis virus reverse transcriptase, Moloney murine leukemia virus reverse transcriptase, and engineered enzymes designed to enhance polymerase activity or reduce undesired nuclease activities (Forootan et al. 2017). Some examples of commercially available ready-to-use RT-PCR kits include the ABScript II One Step RT-qPCR Probe Kit by Sigma Aldrich, QIAGEN's One-Step Ahead RT-PCR Kit, Transcriptor's One-Step RT-PCR Kit by Roche, New England Biolabs' OneTaq RT-PCR Kit, LunaScript Multiplex's One-Step RT-PCR Kit, and Promega's AccessQuick RT-PCR System.

4.3.2.4 Isothermal Amplification Method

Isothermal amplification techniques offer a streamlined and exponential way to detect specific nucleic acid sequences without the need for thermal cycling. These methods commonly employ microbial polymerase enzymes (Table 4.3) with strand-displacement activity, facilitating primer binding and amplification initiation without the need for heat denaturation of DNA strands, ensuring an efficient amplification process. Today, molecular diagnostics uses a wide range of amplification methods that take advantage of these enzyme principles. Various isothermal DNA and RNA amplification kits are commercially available, including the WarmStart® LAMP Kit, IsoAmp® II Universal tHDA Kit by New England Biolabs (H0110S), IsoAmp III Universal tHDA kit from Quidel (USA), TwistAmp Basic kit by BioGerm Medical Technology, the NUCLISENS® EASYQ® system by bioMérieux, ALLin™ Isothermal DNA Amplification Kit by highQu, and the NASBA kit for RNA detection offered by AMSBIO.

The most common isothermal amplification technique is known as loop-mediated isothermal amplification (LAMP), a nucleic acid detection technology originally developed for hepatitis B virus (HBV) DNA detection. The amplification was performed with *Bst* DNA polymerase, obtained from *Bacillus stearothermophilus* that operated at a constant temperature with exceptional selectivity. The method employs six primers, including loop primers, along with isothermal polymerase enzymes featuring strand-displacement activity and no 5′–3′ exonuclease activity. DNA replication is achieved through the auto-cycling strand-displacement synthesis principle. Visualization of results is facilitated by turbidity caused by pyrophosphate precipitations and DNA-binding dyes like SYBR green (Notomi et al. 2000).

Rolling Circle Amplification (RCA) stands out among isothermal systems, synthesizing nucleic acid (NA) products with predetermined sequences, mimicking the natural rolling circle replication of circular molecules like plasmids. It employs a circular probe as a template, along with short DNA or RNA primers, to generate long, contiguous, single-stranded DNA or RNA products. The process, adaptable to diverse enzymes and reporter systems, involves five essential components: a unique DNA/RNA polymerase, short primers, a circular padlock probe template, a ligase for probe circularization, and deoxynucleotide triphosphates (dNTPs) or nucleotide triphosphates (NTPs) (Liu et al. 2013).

NASBA (nucleic acid sequence-based amplification) is a specialized isothermal technique tailored for RNA detection. It utilizes a trio of enzymes, including reverse transcriptase, RNaseH, and T7 RNA polymerase, working in harmony to amplify sequences from an initial single-stranded RNA template. Derived from Avian myeloblastosis virus (AMV), an alpha retrovirus causing myeloblastosis in chickens, the reverse transcriptase (RT) predominantly employed in the process comprises α and β subunits. The α subunits contribute to reverse transcriptase and RNaseH activity. The enzymes perform concurrent activities, catalyzing RNA chain reactions, synthesizing cDNA chains, and selectively degrading RNA residues in nucleic acids. Additionally, RNaseH has the capability to eliminate both the RNA probe from previous hybridization and the poly-A tail from the 3′ end of mRNAs. Then, copies of RNA are synthesized by the DNA-dependent 5′–3′ RNA polymerase T7 RNA polymerase from the T7 bacteriophage (Perbal 2008).

RPA, an acronym for recombinase polymerase amplification, was developed by Piepenburg and colleagues in 2006. It is a groundbreaking development in DNA diagnosis. RPA technology relies on three enzymes: recombinase (T4 UvsX), a single-stranded DNA binding protein (SSB) named gp32, and Bsu DNA polymerase. The RPA process begins when a recombinase protein called uvsX, often found in T4-like bacteriophages, attaches to primers in the presence of ATP and a substance that makes the solution denser (a high molecular weight polyethylene glycol). This creates a recombinase–primer complex. This complex then explores double-stranded DNA, looking for a matching sequence, and encourages the primer to enter the DNA at the right spot. Single-stranded binding proteins stabilize the displaced DNA strand to prevent the inserted primer from being pushed out during branch migration. Eventually, the recombinase dissociates, and a DNA polymerase that can displace one of the DNA strands (such as the large fragment of *Bacillus subtilis* Pol 1, Bsu) attaches to the 3′ end of the primer and extends it by adding the necessary building blocks (dNTPs). Executing this procedure in a repetitive sequence results in the exponential amplification of DNA (Wang et al. 2022).

Strand-Displacement Amplification (SDA) is an isothermal technique employing key enzymes, including a restriction endonuclease (e.g., BsmAI) for precise DNA cleavage at specific sites and a DNA polymerase (e.g., ϕ29 DNA polymerase) derived from *Escherichia coli*. The choice of polymerase is based on its capability to replace one DNA strand with another without exhibiting exonuclease activity. The integration of these enzymes in SDA facilitates precise DNA amplification at a constant temperature, offering a versatile and efficient molecular biology technique (Chen et al. 2015).

4.3.2.5 Hybridization Assay (HA)

Hybridization tests involve the specific annealing of labeled nucleic acid probes to their complementary DNA or RNA sequences. This allows for identifying particular genetic sequences, enabling the determination of the presence or absence of specific DNA or RNA targets. When annealed to their matching sequences, the labeled probes are visualized by colorimetric detection. Microbial alkaline phosphatase, bilirubin oxidase, nuclease, and catalase enzymes have been reported for such purposes in multiple studies (Lucarelli et al. 2006; Flenker et al. 2017; Lagacé-Wiens et al. 2007; Kim et al. 2004; Silva et al. 2021).

4.3.3 Microbial Enzyme-Based Biosensors

In biosensor development, microbial polymers and enzymes are favored for their widespread availability, diverse metabolic potential, and ability to thrive in harsh conditions (Ray et al. 2023). In the progress of point-of-care (POC) in vitro diagnostics, biosensor platforms are being employed to miniaturize conventional and sensitive detection techniques for biomolecule analysis. In sensing devices, microbial enzymes immobilized on the transducer surface recognize target analytes, initiating biochemical reactions that generate signals measured through mass-to-frequency conversion (piezoelectric), light absorbance (optical), or electrical signal (electrochemical) principles. Or they are indirectly utilized for biomolecule detection in other biosensors (e.g., nanosensors, immunosensors), where they serve as signal generators by catalyzing substrates to produce colorimetric products (Fig. 4.2). A wide range of microbial oxidoreductases, hydrolases, and transferases have been employed to develop biosensors for diverse biomedical applications. These biosensors utilize various electrode materials, including glass beads, plastic strips, paper, alginate hydrogel, pectin hydrogel, chitosan, polypyrrole electropolymerized films, monocrystalline diamond, and pencil graphene electrodes (Table 4.4). Biosensor-based diagnostic platforms

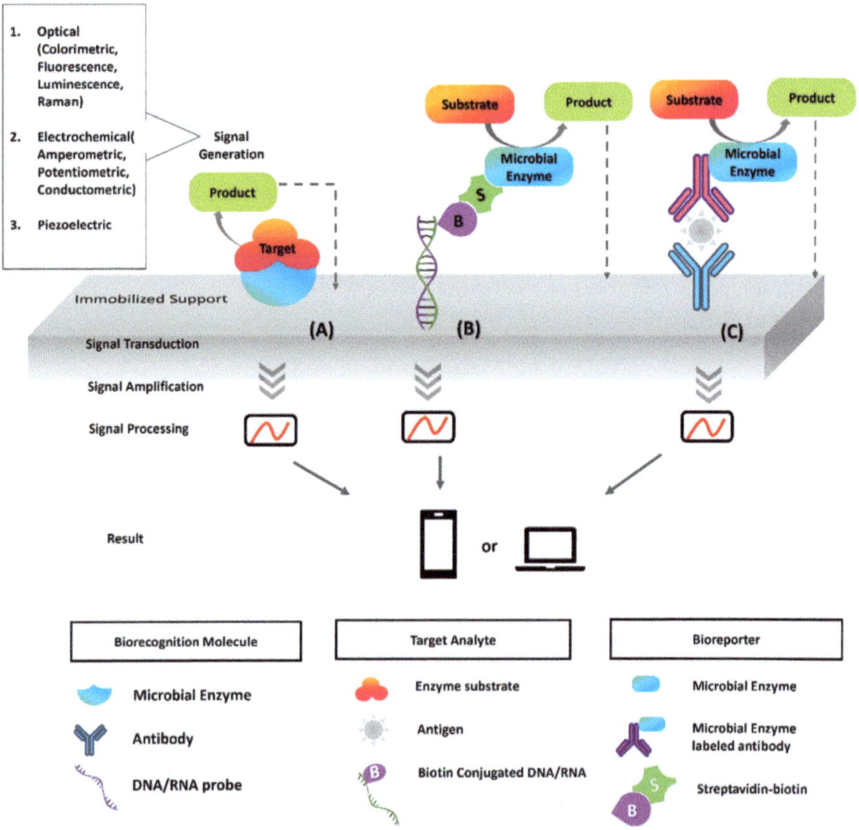

Fig. 4.2 Schematic representation of a microbial enzyme-based biosensor. (**a**) Microbial enzymes immobilized on electrode materials. The enzyme acts as a biorecognition molecule, capturing the target analyte (substrate) and catalyzing it to produce a measurable signal (piezoelectric/optical/electrical). This signal is then absorbed or transduced through the electrode to a signal amplifier and processor, ultimately providing the output result; (**b**) working mechanism of microbial enzyme-based genosensor. DNA/RNA probe immobilized on the electrode capture target nucleic acid, followed by binding a streptavidin/microbial enzyme complex via the biotin molecule to enhance signal; (**c**) the immunosensor mechanism utilizing microbial enzymes. The immunosensor employs immobilized antibodies on the sensor's surface to selectively capture specific antigens. Subsequent binding of a secondary antibody labeled with microbial enzymes enhances the signal through enzyme catalysis

Table 4.4 Microbial enzyme-based biosensors

Biosensor application	Detection technology	Electrode	Enzyme used	Microbial sources	Ref.
Urea sensor	Potentiometric	Polyvinyl alcohol/ styrylpyridinium group (PVA-SbQ) composite	Urease	Recombinant *E. coli*	Marchenko et al. (2015)
Triglyceride sensor	Amperometric	Pencil graphene (PG)	Lipase	*Candida rugosa*	Narwal and Pundir (2017)
Triglyceride sensor	Optical	Pectin hydrogel	Lipase	*Candida antarctica*	Hasanah et al. (2019)
Creatine sensor	Amperometric	Monocrystalline diamond	Creatinase	*Flavobacterium*	Staden and Bokretsion (2006)
Creatine sensor	Amperometric	Monocrystalline diamond	Sarcosine	*Arthrobacterium*	Staden and Bokretsion (2006)
Cholesterol sensor	Optical	Magnetic nanoparticle	Cholesterol esterase	*Pseudomonas* sp.	Román-Pizarro et al. (2023)
Cholesterol sensor	Colorimetric	Paper strip	Cholesterol oxidase	*Streptomyces* sp.	Nguyen et al. (2020)
Glucose sensor	Amperometric	Polypyrrole electropolymerized films	Glucose oxidase	*Aspergillus niger*	Santiago et al. (2010)
Alcohol sensor	Amperometric	Graphite/teflon	Alcohol oxidase	*Pichia pastoris*	Gamella et al. (2014)
Alcohol sensor	Potentiometric	Alginate hydrogel	Alcohol oxidase	*Pichia pastoris*	Aymerich et al. (2018)
Bile detecting sensor	Colorimetric	Glass beads	3-Hydroxysteroid dehydrogenase	*Pseudomonas testosteroni*	Rani et al. (2004)

with immobilized microbial enzymes are routinely utilized for monitoring alcohol (Aymerich et al. 2018; Gamella et al. 2014), creatinine (Staden and Bokretsion 2006; Jimenez Valencia et al. 2021), cholesterol (Ito et al. 2011; Narwal and Pundir 2017; Hasanah et al. 2019), glucose (Santiago et al. 2010), bile (Rani et al. 2004; Zhang et al. 2005), and bilirubin (Andreu et al. 2000) levels in blood, serum, urine, and sweat samples to analyze different health issues. Recently, conventional molecular diagnostics like ELISA and PCR are being miniaturized in biosensor systems to transform these sensitive technologies into cost-effective platforms to facilitate biomolecular analysis (Ma et al. 2023).

Lucarelli et al. developed a genosensor for bacterial DNA detection, where microbial alkaline phosphatase was used to amplify the fluorescence signal by developing multiple layers of the enzyme with the streptavidin–biotin complex in a dendritic pattern to increase enzyme–substrate activities (Lucarelli et al. 2006). A bacterial endonuclease's single-strand cleavage activities were utilized in an aptasensor to release an enzyme with colorimetric substrate activities for protein detection (Huang et al. 2013).

4.3.4 Biochemical Tests

Despite all the advances in diagnostic technologies, biochemical tests remain a fundamental tool in clinical microbiology laboratories for detecting, differentiating, and profiling of pathogens. Instead of immobilizing or tagging any biomolecule with enzymes in assay platforms, these tests directly utilize microbial enzymatic activities (e.g., catalase, oxidase) for the phenotypic detection of pathogens. Traditionally, these assays are conducted using tube methods. However, due to the laborious nature of the work, attempts have been made to miniaturize multiple tests in a single panel or a plate, which is then interpreted manually or automatically. Examples of such successful systems are the BBL Minitek Differential Identification System, Rapid Pos ID Panel, VITEK-2 system, Baxter MicroScan, API series, IDS RapID, STR System, Remel RapID series, AuxoTab, Enterotube, PathoTec, and the CHROMagar series (Hinnebusch et al. 1991; Maloney et al. 2014; Nord et al. 1974; Ohkusu 2000).

4.4 Enzyme-Based Novel Technologies

4.4.1 CRISPER-Cas Technology

CRISPR-Cas (Clustered Regularly Interspaced Short Palindromic Repeats and CRISPR-Associated Proteins) technology is a revolutionary gene-editing system that allows for the precise modification of DNA in a wide range of organisms. It is based on a naturally occurring mechanism in bacteria and archaea that protects

Table 4.5 Application of Cas enzymes and their sources

Enzyme name	Sources	Application	Ref.
CRISPR-Cas3	*Escherichia coli*	SARS CoV 2 and influenza virus detection	Yoshimi et al. (2022)
CRISPR-Cas9	*Streptococcus pyogenes*	Zika virus, MRSA detection	Li et al. (2023)
CRISPR-Cas12	*Francisella novicida*, *Acidaminococcus* spp., *Alicyclobacillus acidiphilus*	Diagnosis of SARS-CoV-2 and HIV	Ding et al. (2020)
CRISPR-Cas13	*Leptotrichia shahii, Leptotrichia wadei, Ruminococcus flavefacien*	RNA virus diagnostics	Freije et al. (2019)
CRISPR-Cas14 (C2c2-III)	*Acidaminococcus* and *Prevotella* bacteria	Viral diagnostics	Wang et al. (2020)

against viral infections. In contemporary applications, this bacterial defense system is harnessed for diagnostics, often integrated with other detection platforms such as ELISA and nucleic acid amplification systems, contributing to point-of-care development. CRISPR/Cas functions primarily as a detection system for target analytes in these systems. It utilizes CRISPR-RNA (crRNA) or guide RNA (gRNA) to guide the Cas enzyme to the target sequence. The binding of the CRISPR/Cas enzyme to the target sequence results in cleavage, releasing the reporter molecule and generating a signal indicating the presence of the target analytes. Several CRISPR-based diagnostic systems have been commercialized or are in development, including FnCas9 Editor-Limited Uniform Detection Assay (FELUDA), Specific High-sensitivity Enzymatic Reporter Unlocking (SHERLOCK), DNA Endonuclease-Targeted CRISPR Trans Reporter (DETECTR), All-in-One Dual CRISPR-Cas12a Assay (AIOD-CRISPR), Allosteric Probe-Initiated Catalysis and CRISPR-Cas13a (APC-Cas) system (Shen et al. 2020), and Cellphone-Based Amplification-Free System with CRISPR/Cas-Dependent Enzymatic (CASCADE) Assay (Silva et al. 2021; Oishee et al. 2021). Different categories of Cas enzymes from different microbial sources have been utilized in diagnostic applications (Table 4.5).

4.4.2 Nanotechnology

The utilization of nanoparticles in conjunction with microbial enzymes is a growing trend, increasingly employed in the development of diverse diagnostic platforms for biomedical applications. Nanostructured materials, including gold nanoparticles, magnetic nanoparticles, graphene, and carbon nanotubes, have been employed as an immobilization medium for enzymes to concentrate analytes from samples at the electrode surface, enhancing the analytical signal in biosensors. Enzyme molecules are undergoing transformation into nanoparticles through aggregation and cross-linking, enhancing their surface area and leveraging their catalytic, optical, electrical, and thermal properties to improve detection signals. Microbial lipase, glycerol kinase, and glycerol-3-phosphate oxidase have been converted into nanoparticles to develop a biosensor device designed to measure serum cholesterol (Narwal and Pundir 2017). Gold nanoparticles have been used to modify microbial ALP enzymes to develop plasmonic ELISA (Li et al. 2016) and SER-active immunosensor for *E. coli* detection (Bozkurt et al. 2018). Additionally, synthetic nanoparticles known as nanozymes are now being explored to develop enzyme-based diagnostics. Nguyen et al. developed a rapid test for colorimetric detection of cholesterol with nanoceria (cerium oxide nanoparticles) and cholesterol oxidase (ChOx) from *Streptomyces*. In this setup, cholesterol oxidase (ChOx) enzymes were immobilized on nanoceria and chitosan-soaked paper strips. When cholesterol is present, it undergoes oxidation by ChOx, generating H_2O_2, which then reacts with nanoceria to produce an intense color, signaling the presence of the analyte (Nguyen et al. 2020).

4.5 Limitations of Enzyme-Based Diagnostic Technologies

Enzyme-based assays offer precision but have challenges. Enzymes are sensitive to temperature and pH, requiring careful control. Selectivity can limit compatibility with specific targets, necessitating validation. Stability issues and susceptibility to interference are concerns. Achieving precision at low analyte levels often needs standard curves and controls. Costs can limit scalability, and multi-target assays are technically demanding. Detection of ultra-low analyte levels can be challenging for early disease diagnosis. Managing variable enzyme activity, potential sample interference, and complex kinetics are crucial. Environmental factors and population variations add complexity. Cross-reactivity and enzyme engineering require attention. Awareness and rigorous management of these issues are crucial for dependable enzyme-based assays.

References

Afshari E, Amini-Bayat Z, Hosseinkhani S, Bakhtiari N (2017) Cloning, expression and purification of pseudomonas putida ATCC12633 Creatinase. Avicenna J Med Biotechnol 9(4):169–175
Akimoto K, Shinmen Y, Sumida M, Asami S, Amachi T, Yoshizumi H, Saeki Y, Shimizu S, Yamada H (1990) Luminol chemiluminescence reaction catalyzed by a microbial peroxidase. Anal Biochem 189(2):182–185
Andreu Y, Galbán J, de Marcos S, Castillo JR (2000) Determination of direct-bilirubin by a fluorimetric-enzymatic method based on bilirubin oxidase. Fresenius J Anal Chem 368(5):516–521
Aoki K, Suzuki H, Ishimaru Y, Toyama S, Ikariyama Y, Iida T (2005) Thermophilic glucokinase-based sensors for the detection of various saccharides and glycosides. Sensors Actuators B Chem 108(1–2):727–732
Aydin S (2015) A short history, principles, and types of ELISA, and our laboratory experience with peptide/protein analyses using ELISA. Peptides 72:4–15
Aymerich J, Márquez A, Terés L, Muñoz-Berbel X, Jiménez C, Domínguez C, Serra-Graells F, Dei M (2018) Cost-effective smartphone-based reconfigurable electrochemical instrument for alcohol determination in whole blood samples. Biosens Bioelectron 117:736–742

Bozkurt AG, Buyukgoz GG, Soforoglu M, Tamer U, Suludere Z, Boyaci IH (2018) Alkaline phosphatase labeled SERS active sandwich immunoassay for detection of *Escherichia coli*. Spectrochim Acta A Mol Biomol Spectrosc 194:8–13

Cardador CM, Muehlmann LA, Coelho CM, Silva LP, Garay AV, Carvalho AMDS, Bastos IMD, Longo JPF (2023) Nucleotides entrapped in liposome nanovesicles as tools for therapeutic and diagnostic use in biomedical applications. Pharmaceutics 15(3):873

Chandra P, Enespa, Singh R, Arora PK (2020) Microbial lipases and their industrial applications: a comprehensive review. Microb Cell Factories 19:1–42

Chattopadhyay DK (2021) Ratio of serum superoxide dismutase and whole blood glutathione peroxidase: a noteworthy parameter for tuberculosis diagnosis. Indian J Med Biochem 25(3):100–104

Chauhan N, Pundir CS (2011) Co-immobilization of cholesterol esterase, cholesterol oxidase and peroxidase on PVC strip for serum cholesterol determination. Anal Methods 3(6):1360–1365

Chen A, Gui GF, Zhuo Y, Chai YQ, Xiang Y, Yuan R (2015) Signal-off electrochemiluminescence biosensor based on Phi29 DNA polymerase mediated Strand displacement amplification for MicroRNA detection. Anal Chem 87(12):6328–6334

Chen J, Xu Y, Yan H, Zhu Y, Wang L, Zhang Y, Lu Y, Xing W (2018) Sensitive and rapid detection of pathogenic bacteria from urine samples using multiplex recombinase polymerase amplification. Lab Chip 18(16):2441–2452

Chong R, Rho JER, Yoon HJ, Rho THD, Park PS, Kim YH, Lee JH (2012) 1,1′-oxalyldiimidazole chemiluminescent enzyme immunoassay capable of simultaneously sensing multiple markers. Biosens Bioelectron 32(1):19–23

Christensen JE, Reynolds CE, Shukla SK, Reed KD (2004) Rapid molecular diagnosis of lactobacillus bacteremia by terminal restriction fragment length polymorphism analysis of the 16S rRNA gene. Clin Med Res 2(1):37–45

D'Auria S, DiCesare N, Staiano M, Gryczynski Z, Lakowicz JR (2002) A novel fluorescence competitive assay for glucose determinations by using a thermostable glucokinase from the thermophilic microorganism *Bacillus stearothermophilus*. Anal Biochem 303(2):138–144

Dahiya D, Nigam PS (2021) An overview of three biocatalysts of pharmaceutical importance synthesized by microbial cultures. AIMS Microbiol 7(2):124–137

Davis MA, Hancock DD, Besser TE, Call DR (2003) Evaluation of pulsed-field gel electrophoresis as a tool for determining the degree of genetic relatedness between strains of Escherichia coli O157:H7. J Clin Microbiol 41(5):1843–1849

Ding X, Yin K, Li Z, Lalla RV, Ballesteros E, Sfeir MM, Liu C (2020) Ultrasensitive and visual detection of SARS-CoV-2 using all-in-one dual CRISPR-Cas12a assay. Nat Commun 11(1):4711

Doumas BT, Perry B, Jendrzejczak B, Davis L (1987) Measurement of direct bilirubin by use of bilirubin oxidase. Clin Chem 33(8):1349–1353

Duong HD, Appiah-Kwarteng C, Takashima Y, Aye KM, Nagayasu E, Yoshida A (2020) A novel luciferase-linked antibody capture assay (LACA) for the diagnosis of toxoplasma gondii infection in chickens. Parasitol Int 77:102125

Dworzański J, Strycharz-Dudziak M, Kliszczewska E, Kiełczykowska M, Dworzańska A, Drop B, Polz-Dacewicz M (2020) Glutathione peroxidase (GPx) and superoxide dismutase (SOD) activity in patients with diabetes mellitus type 2 infected with Epstein-Barr virus. PLoS One 15(3):e0230374

Eckmann CM, De Laaf RT, Van Keulen JM, Van Mourik JA, De Laat B (2007) Bilirubin oxidase as a solution for the interference of hyperbilirubinemia with ADAMTS-13 activity measurement by FRETS-VWF73 assay. J Thromb Haemost 5(6):1330–1331

Eed HR, Abdel-Kader NS, El Tahan MH, Dai T, Amin R (2016) Bioluminescence-sensing assay for microbial growth recognition. J Sens 2016:1

Fang BY, Wang CY, Li C, Wang HB, Zhao YD (2017) Amplified using DNase I and aptamer/graphene oxide for sensing prostate specific antigen in human serum. Sensors Actuators B Chem 244:928–933

Ferri S, Kojima K, Sode K (2011) Review of glucose oxidases and glucose dehydrogenases: a bird's eye view of glucose sensing enzymes. J Diabetes Sci Technol 5(5):1068–1076

Flenker KS, Burghardt EL, Dutta N, Burns WJ, Grover JM, Kenkel EJ, Weaver TM, Mills J, Kim H, Huang L, Owczarzy R, Musselman CA, Behlke MA, Ford B, McNamara JO 2nd (2017) Rapid detection of urinary tract infections via bacterial nuclease activity. Mol Therapy 25(6):1353–1362

Forootan A, Sjöback R, Björkman J, Sjögreen B, Linz L, Kubista M (2017) Methods to determine limit of detection and limit of quantification in quantitative real-time PCR (qPCR). Biomol Detect Quantif 12:1–6

Freije CA, Myhrvold C, Boehm CK, Lin AE, Welch NL, Carter A et al (2019) Programmable inhibition and detection of RNA viruses using Cas13. Mol Cell 76(5):826–837

Gamella M, Campuzano S, Manso J, González de Rivera G, López-Colino F, Reviejo AJ, Pingarrón JM (2014) A novel non-invasive electrochemical biosensing device for in situ determination of the alcohol content in blood by monitoring ethanol in sweat. Anal Chim Acta 806:1–7

Gulinaizhaer A, Yang C, Zou M, Ma S, Fan X, Wu G (2023) Detection of monkeypox virus using helicase dependent amplification and recombinase polymerase amplification combined with lateral flow test. Virol J 20(1):274

Gurung N, Ray S, Bose S, Rai V (2013) A broader view: microbial enzymes and their relevance in industries, medicine, and beyond. Biomed Res Int 2013:329121

Hasanah U, Sani NDM, Heng LY, Idroes R, Safitri E (2019) Construction of a hydrogel pectin-based triglyceride optical biosensor with immobilized lipase enzymes. Biosensors 9(4):135

Hicks JM (1984) Fluorescence immunoassay. Hum Pathol 15(2):112–116

Hinnebusch CJ, Nikolai DM, Bruckner DA (1991) Comparison of API rapid STREP, Baxter Microscan rapid Pos ID panel, BBL Minitek differential identification system, IDS RapID STR system, and Vitek GPI to conventional biochemical tests for identification of viridans streptococci. Am J Clin Pathol 96(4):459–463

Huang Y, Chen J, Zhao S, Shi M, Chen ZF, Liang H (2013) Label-free colorimetric aptasensor based on nicking enzyme assisted signal amplification and DNAzyme amplification for highly sensitive detection of protein. Anal Chem 85(9):4423–4430

Ito Y, Fujimura M, Ohta M, Hirano T (2011) Development of a homogeneous assay for measurement of small dense LDL cholesterol. Clin Chem 57(1):57–65

Jackson AA, Hinkley TC, Talbert JN, Nugen SR, Sela DA (2016) Genetic optimization of a bacteriophage-delivered alkaline phosphatase reporter to detect Escherichia coli. Analyst 141(19):5543–5548

Jimenez Valencia AM, Kryszak LA, Goheen J, Van Roy W, Woronoff G, Mathieu E, Gudjonsdottir H, Stakenborg T, Clarke WA (2021) Rapid quantification of plasma creatinine using a novel kinetic enzymatic assay. J Appl Lab Med 6(3):688–701

Kang EB, Mazrad ZAI, Robby AI, In I, Park SY (2018) Alkaline phosphatase-responsive fluorescent polymer probe coated surface for colorimetric bacteria detection. Eur Polym J 105:217–225

Kaushal DC, Kaushal NA (2002) Diagnosis of malaria by detection of plasmodial lactate dehydrogenase with an immunodot enzyme assay. Immunol Investig 31(2):93–106

Kim HH, Zhang Y, Heller A (2004) Bilirubin oxidase label for an enzyme-linked affinity assay with O2 as substrate in a neutral pH NACL solution. Anal Chem 76(8):2411–2414

Kim EM, Jeon HS, Kim JJ, Shin YK, Lee YJ, Yeo SG, Park CK (2016) Uracil-DNA glycosylase-treated reverse transcription loop-mediated isothermal amplification for rapid detection of avian influenza virus preventing carry-over contamination. J Vet Sci 17(3):421–425

Kossen K, Vaish NK, Jadhav VR, Pasko C, Wang H, Jenison R, McSwiggen JA, Polisky B, Seiwert SD (2004) High-throughput ribozyme-based assays for detection of viral nucleic acids. Chem Biol 11(6):807–815

Lagacé-Wiens PR, Alfa MJ, Manickam K, Karlowsky JA (2007) Thermostable DNase is superior to tube coagulase for direct detection of *Staphylococcus aureus* in positive blood cultures. J Clin Microbiol 45(10):3478–3479

Lei J, He MY, Li J, Li H, Wang W, Gopinath SCB, Xu LZ (2022) miRNA identification by nuclease digestion in ELISA for diagnosis of osteosarcoma. Biotechnol Appl Biochem 69(4):1365–1372

Li XM, Yang XY, Zhang SS (2008) Electrochemical enzyme immunoassay using model labels. TrAC Trends Anal Chem 27(6):543–553

Li Y, Ma X, Xu Z, Liu M, Lin Z, Qiu B, Guo L, Chen G (2016) Multicolor ELISA based on alkaline phosphatase-triggered growth of au nanorods. Analyst 141(10):2970–2976

Li J, Lei J, He L, Fan X, Yi F, Zhang W (2019) Evaluation and monitoring of superoxide dismutase (SOD) activity and its clinical significance in gastric cancer: a systematic review and meta-analysis. Med Sci Monitor 25:2032–2042

Li Y, He X, Zhu W, Li H, Wang W (2022a) Bacterial bioluminescence assay for bioanalysis and bioimaging. Anal Bioanal Chem 414:1–9

Li P, Wang ZY, Li Y, Liu LZ, Qiu JG, Zhang CY (2022b) Bsu polymerase-mediated fluorescence coding for rapid and sensitive detection of 8-oxo-7,8-dihydroguanine in telomeres of cancer cells. Talanta 243:123340

Li X, Zhong J, Li H, Qiao Y, Mao X, Fan H et al (2023) Advances in the application of CRISPR-Cas technology in rapid detection of pathogen nucleic acid. Front Mol Biosci 10

Liang B, Han L (2020) Displaying of acetylcholinesterase mutants on surface of yeast for ultra-trace fluorescence detection of organophosphate pesticides with gold nanoclusters. Biosens Bioelectron 148:111825

Liang Y, Yan H, Huang L, Zhao J, Wang H, Kang M et al (2021) A luciferase immunosorbent assay for quantitative detection of IgG antibodies against SARS-CoV-2 nucleoprotein. J Virol Methods 292:114141

Liao AM, Pan W, Benson JC, Wong AD, Rose BJ, Caltagirone GT (2018) A simple colorimetric system for detecting target antigens by a three-stage signal transformation-amplification strategy. Biochemistry 57(34):5117–5126

Liu H, Li L, Duan L, Wang X, Xie Y, Tong L et al (2013) High specific and ultrasensitive isothermal detection of microRNA by padlock probe-based exponential rolling circle amplification. Anal Chem 85(16):7941–7947

Liu H, He P, Meng F, Jiang M, Xiong J, Li J et al (2021) A semiautomated luciferase immunoprecipitation assay for rapid and easy detection of African swine fever virus antibody. J Clin Microbiol 59(10):10–1128

Lucarelli F, Marrazza G, Mascini M (2006) Dendritic-like streptavidin/alkaline phosphatase nanoarchitectures for amplified electrochemical sensing of DNA sequences. Langmuir 22(9):4305–4309

Ma L, Zhang X, Peng Y, Chen W, Xiao Y, Fang H, yang, H., & Zhou, Y. (2023) Based on intervening PCR for detection of alkaline phosphatase and zearalenone. Microchem J 186:108314

Maloney S, Engler C, Norton R (2014) Evaluation of the Remel RapID NF plus rapid biochemical method for identification of Burkholderia pseudomallei. J Clin Microbiol 52(6):2175–2176

Mano N (2012) Features and applications of bilirubin oxidases. Appl Microbiol Biotechnol 96(2):301–307

Marchenko SV, Kucherenko IS, Hereshko AN, Panasiuk IV, Soldatkin OO, El'Skaya AV, Soldatkin AP (2015) Application of potentiometric biosensor based on recombinant urease for urea determination in blood serum and hemodialyzate. Sensors Actuators B Chem 207:981–986

Martín-Alonso S, Frutos-Beltrán E, Menéndez-Arias L (2021) Reverse transcriptase: from transcriptomics to genome editing. Trends Biotechnol 39(2):194–210

Melo IMA, Viana MRP, Pupin B, Bhattacharjee TT, de Azevedo Canevari R (2021) PCR-RFLP and FTIR-based detection of high-risk human papilloma virus for cervical cancer screening and prevention. Biochem Biophys Rep 26:100993

Mina S, Marot-Leblond A, Cimon B, Fleury MJ, Larcher G, Bouchara JP, Robert R (2015) Purification and characterization of a mycelial catalase from *Scedosporium boydii,* a useful tool for specific antibody detection in patients with cystic fibrosis. Clin Vaccine Immunol 22(1):37–45

Nalini P, Ellaiah P, Prabhakar T, Girijasankar G (2015) Microbial alkaline phosphatases in bioprocessing. Int J Curr Microbiol Appl Sci 4(3):384–396

Narwal V, Pundir CS (2017) An improved amperometric triglyceride biosensor based on co-immobilization of nanoparticles of lipase, glycerol kinase and glycerol 3-phosphate oxidase onto pencil graphite electrode. Enzym Microb Technol 100:11–16

Nguyen PT, Kim YI, Kim MI (2020) Reagent-free colorimetric cholesterol test strip based on self-color-changing property of nanoceria. Front Chem 8:798

Nigam PS (2013) Microbial enzymes with special characteristics for biotechnological applications. Biomol Ther 3(3):597–611

Nord CE, Lindberg AA, DahlbÄck A (1974) Evaluation of five test-kits—API, AuxoTab, Enterotube, PathoTec and R/B—for identification of Enterobacteriaceae. Med Microbiol Immunol 159(3):211–220

Notomi T, Okayama H, Masubuchi H, Yonekawa T, Watanabe K, Amino N, Hase T (2000) Loop-mediated isothermal amplification of DNA. Nucleic Acids Res 28(12):e63–e63

Ohkusu K (2000) Cost-effective and rapid presumptive identification of gram-negative bacilli in routine urine, pus, and stool cultures: evaluation of the use of CHROMagar orientation medium in conjunction with simple biochemical tests. J Clin Microbiol 38(12):4586–4592

Oishee MJ, Ali T, Jahan N, Khandker SS, Haq MA, Khondoker MU et al (2021) COVID-19 pandemic: review of contemporary and forthcoming detection tools. Infect Drug Resist Volume 14:1049–1082

Oscorbin I, Filipenko M (2023) Bst polymerase—a humble relative of Taq polymerase. Computational and structural. Biotechnol J 21:4519

Perbal B (2008) Avian myeoloblastosis virus (AMV): only one side of the coin. Retrovirology 5(1):1–5

Pollegioni L, Piubelli L, Molla G (2009) Cholesterol oxidase: biotechnological applications. FEBS J 276(23):6857–6870

Porstmann T, Kiessig ST (1992) Enzyme immunoassay techniques an overview. J Immunol Methods 150(1–2):5–21

Rahimi Esboei B, Fallahi S, Zarei M, Kazemi B, Mohebali M, Shojaee S, Mousavi P, Teimouri A, Mahmoudzadeh R, Salabati M, Keshavarz Valian H (2022) Utility of blood as the clinical specimen for the diagnosis of ocular toxoplasmosis using uracil DNA glycosylase-supplemented loop-mediated isothermal amplification and real-time polymerase chain reaction assays based on REP-529 sequence and B1 gene. BMC Infect Dis 22(1):89

Raj SB, Ramaswamy S, Plapp BV (2014) Yeast alcohol dehydrogenase structure and catalysis. Biochemistry 53(36):5791–5803

Rani K, Garg P, Pundir CS (2004) Measurement of bile acid in serum and bile with arylamine-glass-bound 3alpha-hydroxysteroid dehydrogenase and diaphorase. Anal Biochem 332(1):32–37

Ray S, Kumari A, Sonkar M, Kumar P (2023) Polyhydroxyalkanoate-based sensors and their applications. In: Biomaterials-based sensors: recent advances and applications. Singapore, Springer Nature Singapore, pp 223–243

Regalado C, García-Almendárez BE, Duarte-Vázquez MA (2004) Biotechnological applications of peroxidases. Phytochem Rev 3:243–256

Roberts RJ (2005) How restriction enzymes became the workhorses of molecular biology. Proc Natl Acad Sci USA 102(17):5905–5908

Román-Pizarro V, Écija-Arenas Á, Fernández-Romero JM (2023) An integrated microfluidic-based biosensor using a magnetically controlled MNPs-enzyme microreactor to determine cholesterol in serum with fluorometric detection. Mikrochim Acta 190(8):303

Roy S, Vivoli Vega M, Harmer NJ (2019) Carbohydrate kinases: a conserved mechanism across differing folds. Catalysts 9(1):29

Santiago LM, Bejarano-Nosas D, Lozano-Sanchez P, Katakis I (2010) Screen-printed micro-systems for the ultrasensitive electrochemical detection of alkaline phosphatase. Analyst 135(6):1276–1281

Schormann N, Ricciardi R, Chattopadhyay D (2014) Uracil-DNA glycosylases—structural and functional perspectives on an essential family of DNA repair enzymes. Protein Sci 23(12):1667–1685

Scognamiglio V, Staiano M, Rossi M, D'Auria S (2004) Protein-based biosensors for diabetic patients. J Fluoresc 14:491–498

Shanmugakani RK, Wu M (2022) An isothermal amplification-coupled dipstick for the rapid detection of COVID-19. J Med Microbiol 71(4). https://doi.org/10.1099/jmm.0.001519

Shanmugakani RK, Bonam W, Erickson D, Mehta S (2021) An isothermal amplification-based point-of-care diagnostic platform for the detection of mycobacterium tuberculosis: a proof-of-concept study. Curr Res Biotechnol 3:154–159

Sharma P, Garg N, Sharma A, Capalash N, Singh R (2019) Nucleases of bacterial pathogens as virulence factors, therapeutic targets and diagnostic markers. Int J Med Microbiol 309(8):151354

Shen J, Zhou X, Shan Y, Yue H, Huang R, Hu J, Xing D (2020) Sensitive detection of a bacterial pathogen using allosteric probe-initiated catalysis and CRISPR-Cas13a amplification reaction. Nat Commun 11(1):267

Shih A, Bockman JM, George ST (1994) Inventors; Innovir laboratories, Inc, assignee. Ribozyme amplified diagnostics. WO1994013833A1 (Patent)

Silva FSR, Erdogmus E, Shokr A, Kandula H, Thirumalaraju P, Kanakasabapathy MK, Hardie JM, Pacheco LGC, Li JZ, Kuritzkes DR, Shafiee H (2021) SARS-CoV-2 RNA detection by a cellphone-based amplification-free system with CRISPR/CAS-dependent enzymatic (CASCADE) assay. Adv Mater Technol 6(12):2100602

Smith MW, Ghindilis AL, Seoudi IA, Smith K, Billharz R, Simon HM (2014) A new restriction endonuclease-based method for highly-specific detection of DNA targets from methicillin-resistant *Staphylococcus aureus*. PLoS One 9(5):e97826

Song MJ, Yun DH, Hong SI (2009) An electrochemical biosensor array for rapid detection of alanine aminotransferase and aspartate aminotransferase. Biosci Biotechnol Biochem 73(3):474–478

Song Y, Zhang Y, Bernard PE, Reuben JM, Ueno NT, Arlinghaus RB, Zu Y, Qin L (2012) Multiplexed volumetric bar-chart chip for point-of-care diagnostics. Nat Commun 3:1283

Sooch BS, Kauldhar BS, Puri M (2014) Recent insights into microbial catalases: isolation, production and purification. Biotechnol Adv 32(8):1429–1447

Śpibida M, Krawczyk B, Olszewski M, Kur J (2017) Modified DNA polymerases for PCR troubleshooting. J Appl Genet 58:133–142

Staden RISV, Bokretsion RG (2006) Simultaneous determination of creatine and creatinine using monocrystalline diamond paste–based amperometric biosensors. Anal Lett 39(11):2227–2233

Staerck C, Gastebois A, Vandeputte P, Calenda A, Larcher G, Gillmann L, Papon N, Bouchara JP, Fleury MJJ (2017) Microbial antioxidant defense enzymes. Microb Pathog 110:56–65

Stakenborg T, Vicca J, Butaye P, Maes D, De Baere T, Verhelst R, Peeters J, de Kruif A, Haesebrouck F, Vaneechoutte M (2005) Evaluation of amplified rDNA restriction analysis (ARDRA) for the identification of mycoplasma species. BMC Infect Dis 5:46

Sun Y, Gao G, Cai T (2021) Enzymatic characterization of D-lactate dehydrogenase and application in alanine aminotransferase activity assay kit. Bioengineered 12(1):6459–6471

Ueda S, Sakasegawa SI (2016) Enzymatic cycling method using creatine kinase to measure creatine by real-time detection. Anal Biochem 506:8–12

Vacca Smith AM, Scott-Anne KM, Whelehan MT, Berkowitz RJ, Feng C, Bowen WH (2007) Salivary glucosyltransferase B as a possible marker for caries activity. Caries Res 41(6):445–450

Vaquero ME, Barriuso J, Martínez MJ, Prieto A (2016) Properties, structure, and applications of microbial sterol esterases. Appl Microbiol Biotechnol 100(5):2047–2061

Wang H, Chen H, Huang Z, Li T, Deng A, Kong J (2018) DNase I enzyme-aided fluorescence signal amplification based on graphene oxide-DNA aptamer interactions for colorectal cancer exosome detection. Talanta 184:219–226

Wang X, Shang X, Huang X (2020) Next-generation pathogen diagnosis with CRISPR/Cas-based detection methods. Emerg Microb Infect 9(1):1682–1691

Wang F, Wang Y, Liu X, Wang L, Wang K, Xu C, Huang G, Gao X (2022) Rapid, simple, and highly specific detection of *Streptococcus pneumoniae* with visualized recombinase polymerase amplification. Front Cell Infect Microbiol 12:878881

Wong SS, Schenkel OJ (1995) Quantification of plasma hemoglobin in the presence of bilirubin with bilirubin oxidase. Ann Clin Lab Sci 25(3):247–251

Wong KY, Chuan YC, Aggarwal A, Tham L, Kong WM, Tan P (2004) Identifying patterns of DNA for tumor diagnosis using capillary electrophoresis-amplified fragment length polymorphism (CE-AFLP) screening. J Bioinforma Comput Biol 2(3):569–587

Wong CM, Wong KH, Chen XD (2008) Glucose oxidase: natural occurrence, function, properties and industrial applications. Appl Microbiol Biotechnol 78:927–938

Yamket W, Sathianpitayakul P, Santanirand P, Ratthawongjirakul P (2023) Implementation of helicase-dependent amplification with SYBR green I for prompt naked-eye detection of bacterial contaminants in platelet products. Sci Rep 13(1):3238

Yang S, Rothman RE (2004) PCR-based diagnostics for infectious diseases: uses, limitations, and future applications in acute-care settings. Lancet Infect Dis 4(6):337–348

Yoshimi K, Takeshita K, Yamayoshi S, Shibumura S, Yamauchi Y, Yamamoto M et al (2022) CRISPR-Cas3-based diagnostics for SARS-CoV-2 and influenza virus. Iscience 25(2):103830

Zeng Y, Liu M, Xia Y, Jiang X (2020) Uracil-DNA-glycosylase-assisted loop-mediated isothermal amplification for detection of bacteria from urine samples with reduced contamination. Analyst 145(21):7048–7055

Zhang GH, Cong AR, Xu GB, Li CB, Yang RF, Xia TA (2005) An enzymatic cycling method for the determination of serum total bile acids with recombinant 3alpha-hydroxysteroid dehydrogenase. Biochem Biophys Res Commun 326(1):87–92

Zhu B (2014) Bacteriophage T7 DNA polymerase–sequenase. Front Microbiol 5:181

Chapter 5
Organic Acids of Microbial Origin as Nutraceuticals

Hina Maniya, Mehul Chauhan, and Vijay Kumar

Abstract Organic acids are substances with acidic characteristics that are present naturally in a variety of foods. Due to hydrolysis, biochemical metabolism, and microbial activity, they are primarily found in fermented foods. Despite the fact that they are not regarded as nutrients, these are what gives food its distinctive flavor. To naturally extend shelf life and prevent spoilage, many foods are preserved with organic acids. These acids work by creating an acidic environment that inhibits the growth of harmful bacteria and fungi. Determining the amount of organic acids in food products is crucial as they are responsible for the unique flavor and signature aroma of the food. Currently, both artificial and natural preservatives are extensively used by food and beverage industries to mitigate product deterioration. Organic acids are not only effective in the food industries but also pragmatic in the health-care sector, pharmaceutical industries, and cosmetic industries. In this chapter, an overview of organic acids of microbial origin suitable for utilization in food and their implementation as nutraceuticals for human health has been elaborated.

Keywords Acidulant · Anti-aging · Fermentation · Short-chain fatty acid · Traditional nutraceuticals

5.1 Introduction

5.1.1 Overview of Organic Acids and Their Importance

Organic acids are mostly weak acids. They contain carbon, hydrogen, and oxygen. They are widely spread in nature and can be found in both plants and animals. In addition, microorganisms are also capable of producing organic acids through

H. Maniya · M. Chauhan · V. Kumar (✉)
Postbiotics and Foodomics Lab, Department of Microbiology, School of Science, RK University, Rajkot, Gujarat, India
e-mail: vijay.kumar@rku.ac.in

V. Kothari et al. (eds.), *Microbial Products for Health and Nutrition*,
https://doi.org/10.1007/978-981-97-4235-6_5

fermentation (Kumar et al. 2016) that too in a more effective manner with particular characteristics, such as purity and acidity of organic acids. The method of producing organic acid by fermentation is efficient and sustainable as compared to conventional methods, which produce a high number of byproducts that can be harmful to the environment (Kumar et al. 2014a). Another disadvantage of the conventional method is the racemization of the end product, whereas, in microbial fermentation, only one type of isomer is produced, which enhances the purity of the organic acid.

As organic acids are mostly weak in nature, they help to maintain pH in the body by acting as buffering agents. For some of the organisms, it works as an energy source as in the case of short-chain fatty acids (SCFAs) that act as a source of energy for gut microbiota. For the preservation of food, it works as an antimicrobial agent, hence its use in many food and beverage industries (Chahardoli et al. 2020).

5.1.2 Classification of Organic Acids Based on Carboxyl Group

Organic acids are capable of releasing a proton (H^+) to the base. Carboxylic acids, which have the carboxyl group (-COOH), are the most prevalent kind of organic acids. Sulfonic acids (methanesulfonic acid and benzenesulfonic acid), phenols (salicylic acid), and enols (citric acid and succinic acid) are other categories of organic acids. The broad classification of organic acid based on the carboxyl group (Chahardoli et al. 2020) is presented in Fig. 5.1 and further description is given below:

A. Monocarboxylic acids: Each molecule of these acids only has one carboxyl group (-COOH). Acetic acid, formic acid, and lactic acid are a few examples.
B. Dicarboxylic acid: Two carboxyl groups are present in each molecule of a dicarboxylic acid. Succinic acid, malonic acid, and oxalic acid are among the examples.

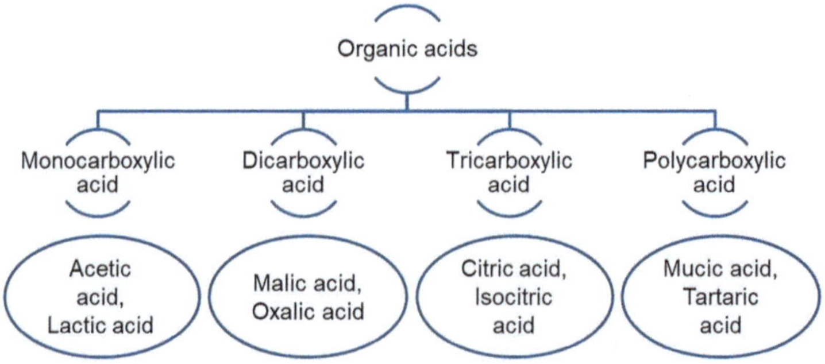

Fig. 5.1 Classification of organic acid based on the carboxylic acid group

C. Tricarboxylic acids: Such molecules contain three carboxyl groups. Citric acid, aconitic acid, and isocitric acid are a few examples.
D. Polycarboxylic acid: Four or more carboxyl groups are present in each polycarboxylic acid molecule. Examples include phytic acid (C-6), mucic acid (C-6), and tartaric acid (C-4).

5.1.3 Occurrence and Distribution of Organic Acids in Nature

Organic acids are abundant in nature and come from a variety of sources, such as plants, animals, and microorganisms. Organic acids make up a significant portion of plant tissues and are crucial to their metabolism. The plant-derived organic acids act as flavoring agents, preservatives, and provide unique aromas and fragrances. In animals, they are present in tissues and are crucial for the elimination of waste and creation of energy. The microorganisms are among the richest sources of naturally occurring organic acids, many of which have crucial roles to play in our health and curing illnesses. Ascorbic acid, butyric acid, lactic acid, and propionic acid are used in the treatment of various illnessess. As a result of their metabolic processes, microorganisms produce a wide range of organic acids (Yadav et al. 2022; Kumar et al. 2014b).

The type of organism, the environment, and the stage of development are a couple of variables that have an impact on the distribution of organic acids in nature. For instance, fruits often contain more organic acids as they ripen. In nature, organic acids have a variety of significant functions. Certain organic acids have the potential to significantly improve soil fertility, as observed in the case of humic and fulvic acids, which can aid in increasing soil fertility by altering the pH of soil as well as enriching it with minerals and elemental contents (Anyasi et al. 2017; Costa and Conte-Junior 2016).

5.1.4 Microorganisms Involved in Organic Acid Production

The microorganism's metabolic pathways determine the type of organic acid synthesized by that organism. For instance, lactic acid-producing bacteria are used in the fermentation process to turn glucose into lactic acid in order to obtain energy. Another example is citric acid which is a primary metabolite of the Krebs cycle, used by the fungi that make citric acid to obtain energy. Various organic acids are produced industrially by utilizing microorganisms. For instance, members of the genus *Acetobacter* are used to ferment carbohydrates and generate acetic acid. Sugar fermentation is carried out using the fungus *Aspergillus niger* that produces citric acid. A few organic acids and their respective microbial producers are depicted in Fig. 5.2.

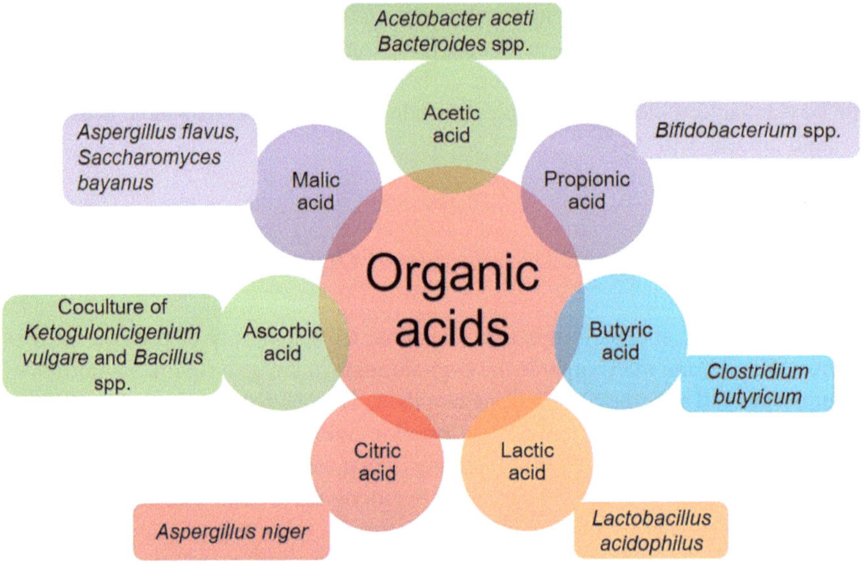

Fig. 5.2 Organic acids production by various microorganisms

5.1.5 Chemical Structures and Properties of Carboxylic Acids

Carboxylic acids, which have the general formula R-COOH, are the most prevalent kind of organic acids. The carboxyl group (COOH) contains an acidic hydrogen. This functional group consists of a carbonyl group (C=O) linked to a hydroxyl group (-OH). The electron-withdrawing nature of the carbonyl group weakens the O-H bond in the hydroxyl group, making the hydrogen more acidic. As per the functional group and the number of replaceable hydrogen they contain, there are certain unique characteristics that attribute to the variety of organic acids (Chahardoli et al. 2020) that are discussed below:

A. Acidity and solubility: Because organic acids may give a proton (H^+), they are acidic. The stability of an organic acid's conjugate base determines how potent the acid is. The stronger the acid, the more stable the conjugate base. The presence of different H^+ ions in the same molecule determines the acidity of molecules. The strongest organic acid is acetic acid. The majority of organic acids dissolve in water. While lower molecular mass organic acids, like lactic acid and formic acid, are miscible in water, higher molecular mass organic acids, like benzoic acid, are insoluble in water. An organic acid becomes more soluble as the number of hydroxyl groups (-OH) in the molecule increases.
B. Melting and boiling points: Organic acids frequently have greater melting and boiling points when compared to other organic compounds with similar molecular mass. The neutral solids are known as dicarboxylic acids, and their melting points are above 100 °C (ascorbic acid and malic acid). Higher boiling points

are indicated by stronger dipole–dipole interactions and intermolecular hydrogen bonding. Very high boiling points are due to the dimerization of carboxylic acids. The strong hydrogen bonds between the carboxylic acid groups are responsible for this.

C. Chemical reactivity: Organic acids are chemically active substances that can react in a number of ways such as acid–base reaction, nucleophilic substitution reaction, elimination reaction, and redox reaction.

5.2 Applications of Organic Acids

Organic acids are used in various fields like healthcare, pharmaceuticals, cosmetics, food, and the bioconversion industry. Figure 5.3 reflects various organic acids and their applications in the respective sectors.

It has a tangy flavor and is used in the food and beverage sector as an emulsifier (stabilizes an emulsion), an acidulant (maintains acidity), a stabilizer (maintains or enhances test and texture), and a food preservative (antimicrobial agent). Organic acids like citric acid, lactic acid, and tartaric acid are the most commonly used food preservatives and acidulants in the food industry. Lactic acid is used for the enhancement of taste in candies and other foods; it also helps clarify the candy base. For making fruit jam, acidulant is used to maintain the pH. In the production of effervescent tablets, jams, and gelatins, citric acid is used to stabilize the pH value. Acids

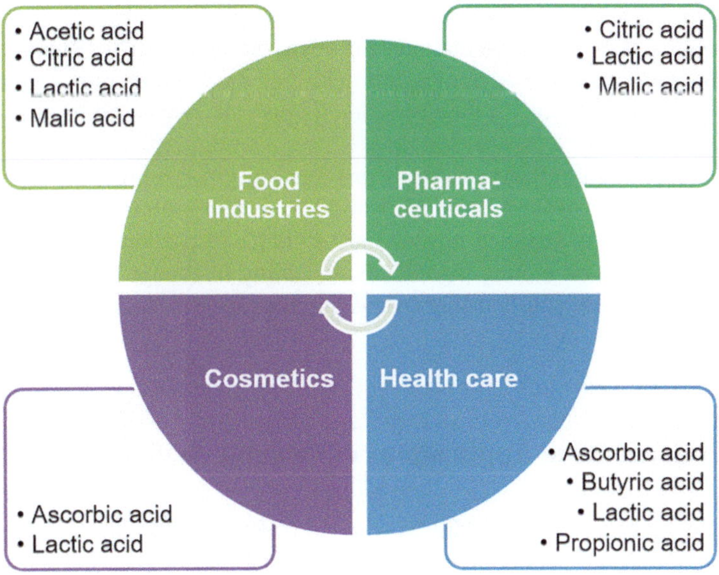

Fig. 5.3 Application of organic acids in various sectors

such as citric acid, lactic acid, and acetic acid are used as emulsifiers in ice cream making, as well as in the cheese-making process (Soccol et al. 2006).

In the pharmaceutical and healthcare sectors, organic acids are used in the formulation of drugs or supplements using various salts of acids. Organic acids are also useful for enhancing the taste of bitter drugs. Some acids such as lactic acid and its derivatives are used as electrolytes while others like citric acid and polymalic acid are used in coating materials (Chi et al. 2016; Abd Alsaheb et al. 2015).

The largest organ in the body, the skin requires attention as it acts as the body's first line of defense against foreign particles. Acids like ascorbic acid (vitamin C), lactic acid, and itaconic acid are mostly used to prevent skin damage caused by sunburn or UV rays. These acids are also useful to maintain skin health (Al Niaimi and Chiang 2017).

5.3 What Are Nutraceuticals?

"Nutraceutical" was first defined and coined by (DeFelice 1995). He used this word by combining the terms "nutrition and pharmaceutics." He defined nutraceuticals as "any substance that is a food or part of a food and provides medical or health benefits, including the prevention and treatment of disease" (Dima et al. 2020). It is classified based on the food items used for the nutraceuticals (nutrients, dietary supplements, herbals, etc.) and whether the item is traditional or nontraditional. Further, dietary supplements are categorized as nutrient and non-nutrients wherein proteins, carbohydrates, lipids, and vitamins are considered as nutrient components of nutraceuticals, while probiotics, prebiotics, synbiotics, postbiotics, and enzymes as non-nutrient components (Pant and Joshi 2023). Nutraceuticals may have two proposed ways of action: (1) they have the ability to reduce the symptoms of diseases or (2) they can directly provide benefits to health (Dudeja and Gupta 2017).

The global nutraceuticals market is projected to grow at a CAGR (compound annual growth rate) of 9.3% from 2021 to 2028, reaching a total value of $658.11 billion. Its market is expected to grow post COVID-19 pandemic, with increased demand for health-promoting ingredients, flexibility in storage options, increased consumer education, and boosting overall immune system health (Nutraceuticals Market Size 2021).

5.4 Nutraceutical Applications of Organic Acids

As a major category of food additives with a broad range of applications including preservative agents, acidity regulators, antioxidant substances, and even more, organic acids play an essential role in protecting the nutritional value and flavor profile of food. Because taste and scent have a cross-modal link, the release of organic acids during oral food processing also affects taste perception. When it comes to nutritional roles, organic acids are a class of nutrients that have been shown to have several positive health effects, including the ability to reduce inflammation, prevent osteoporosis, regulate host immune function, produce intestinal hormones, prevent obesity, promote calcium absorption, and inhibit platelet aggregation (Shi et al. 2022). Organic acids are essential for human health because they are crucial for maintaining human physiology. The organic acids that are part of food or produced after having the food by the gut microflora are discussed here. Information on molecular weight, molecular formula, and molecular structure of organic acids are presented in Table 5.1.

Table 5.1 Molecular formula, weight, and structure of organic acids

Acids	Molecular formula	Molecular weight	Molecular structure
Acetic acid	$C_2H_4O_2$	60.05 g/mol	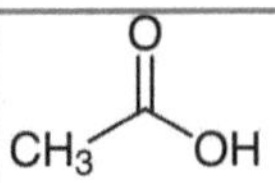
Propionic acid	$C_3H_6O_2$	74.08 g/mol	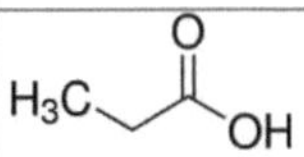
Butyric acid	$C_4H_8O_2$	88.11 g/mol	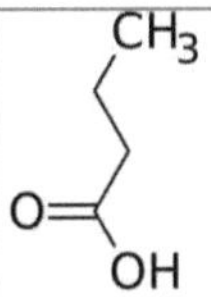
Lactic acid	$C_3H_6O_3$	90.08 g/mol	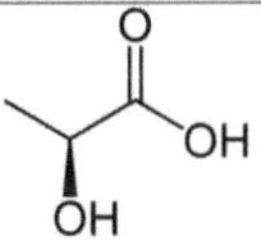
Citric acid	$C_6H_8O_7$	192.12 g/mol	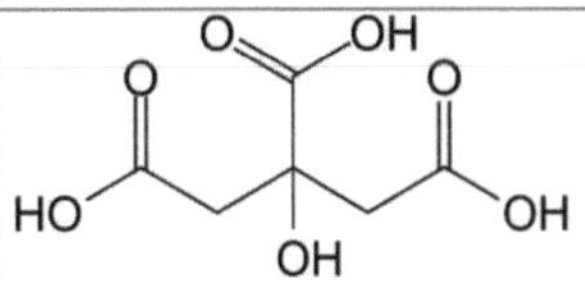
Ascorbic acid	$C_6H_8O_6$	176.12 g/mol	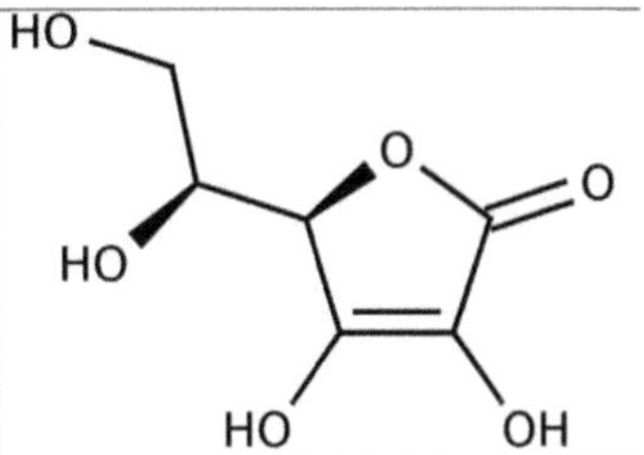
Malic acid	$C_4H_6O_5$	134.09 g/mol	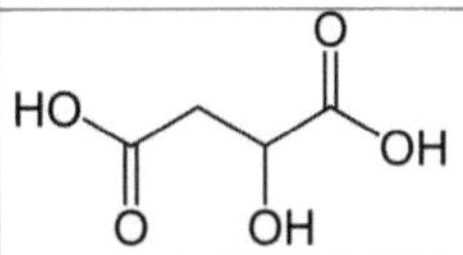

5.4.1 Short-Chain Fatty Acids (SCFAs)

Short-chain fatty acids (SCFAs) are the final byproduct of fermentation by the gut microflora. It is made up of up to six carbon chains or branched aliphatic chains. SCFAs that are more abundant in the gut environment and more studied are propionate, acetate, and butyrate. Apart from these, formate, valerate, and capronate are also present, but in very small amounts. More than 90–95% of the SCFAs are composed of the first three SCFAs mentioned above. They are present in 60:20:20 of acetate, butyrate, and propionate in the gut. From these three, acetate is more concentrated as compared to the rest of the two. The human gut produces approximately 500–600 mM SCFAs per day, depending on the type of fermentative organisms, the composition of the diet, and the residence time of food in the intestinal tract (Markowiak-Kopeć and Śliżewska 2020). Due to their abundance, available literature, and studies, we will be discussing these three SCFAs here in detail.

The organisms that can produce these types of acids are present in the gut as a common microflora. Gut microflora has polymicrobial ecology with active ecosystems that ferment the dietary fibers and convert them into different metabolites. The microbial community present in the intestine digests the polysaccharides such as nonstarch polysaccharides, arabinoxylan, β-glucan, inulin, resistant oligosaccharides, pectin, and gum, which are partially digested or not digested by the human gastrointestinal enzymes. These polysaccharides are known as nondigestible polysaccharides and are found in plant-based fibers. Among these plant fibers, some are soluble fibers that are exceptionally consumed by the microbial community (also known as microbiota-accessible carbohydrates) in the gut and produce SCFAs (Portincasa et al. 2022). Table 5.2 shows the organisms producing different types of SCFAs in the gut lumen. The majority of the SCFAs play a role in the intestinal cells (as energy source for other microflora), but a few SCFAs (mainly butyrate and

Table 5.2 Microbial origin short-chain fatty acids

Short-chain fatty acids	Carbon chain	Producer microorganisms	References
Acetic acid	Two carbon chain	*Akkermansia muciniphila*, *Bacteroides* spp., *Bifidobacterium* spp., *Blautia hydrogenotrophica*, *Clostridium* spp., *Coprococcus* spp. *Prevotella* spp., *Ruminococcus* spp., *Streptococcus* spp.	Mirzaei et al. (2021); Markowiak-Kopeć and Śliżewska (2020)
Butyric acid	Four carbon chain	*Anaerostipes* spp., *Coprococcus catus*, *C. comes*, *C. eutactus*, *Clostridium butyricum*, *C. symbiosum*, *Eubacterium hallii*, *Eu. rectal*, *Faecalibacterium prausnitzii*, *Lactobacillus* spp., *Roseburia* spp., *Ruminococcus* spp.	
Propionic acid	Three carbon chain	*Bacteroides* spp., *Bifidobacterium* spp., *Clostridium* spp., *Coprococcus catus*, *Dialister* spp., *Eubacterium* spp., *Lactobacillus* spp., *Megasphaera elsdenii*, *Phascolarctobacterium succinatutens*, *Roseburia inulinivorans*, *Ruminococcus obeum*, *Salmonella* spp., *Veillonella*	

acetate) at higher concentrations can cross the intestinal cell lining and reach the liver via the hepatic portal vein. The transportation of the SCFAs from the gut lumen to intestinal cells occurs through the monocarboxylate transporter and the sodium-coupled monocarboxylate transporter (Yao et al. 2020). It also plays a crucial role in the host immune system, glucose homeostasis, and cell apoptosis, and it can cross the blood–brain barrier to enhance memory and satiety, with some evidence to restore the autism spectrum diseases, Parkinson's disease, and Alzheimer's disease (Silva et al. 2020).

5.4.1.1 Application of SCFAs in Healthcare

Various studies have identified two major sites for the action of SCFAs. These include G-protein-coupled receptors (GPCR) and histone deacetylases (HDAC). These two sites, by knock-on effect, can affect other pathways to prevent the disease. GPCRs are involved in regulating many physiological roles in the host and are mostly present in all cells. They are able to bind with many extracellular moieties like sugar, lipids, proteins, hormones, and other signaling molecules. Two GPCRs, GPR41 and GPR43, have more tendencies toward the SCFAs, serving as receptors. These receptors are also known as free fatty acid activated receptors 3 (FFAR3) and FFAR2, respectively. If we are talking about the affinity of different SCFAs for FFAR3, propionate, butyrate, and valeroate have the same affinity, followed by acetate, and lastly, formate. Acetate and propionate have a higher affinity than butyrate, and valeroate and formate have the least affinity for FFAR2 (He et al. 2020). The above-mentioned two receptors are highly engaged with metabolic diseases and metabolic syndrome.

SCFAs enhance glucose uptake in the skeletal muscle and adipose tissues through increased expression of glucose transporter type 4 (GLUT4). A reduction in glycolysis and an increased rate of glycogen synthesis are also associated with SCFAs. The two hormones that cause satiety, peptide YY (PYY) and glucagon-like peptide-1 (GLP-1), are secreted by SCFAs in the intestine. Indirectly, SCFAs reduce appetite and food intake, thereby preventing body weight and associated metabolic disorders like type 2 diabetes, hypertension, hypercholesterolemia, and many more (Portincasa et al. 2022).

A diet rich in fiber, especially soluble fibers, increases anti-inflammatory effects. The production of pro-inflammatory mediators interleukin 6 (IL-6), interleukin 8 (IL-8), and tumor necrosis factor (TNF) was observed to be decreased while on the soluble fiber-containing diet for more than 2 weeks. Regulation of cytokine production in various immune cells is regulated by SCFAs, which are responsible for the management of inflammation. The polymicrobial communities present in the gut produce lipopolysaccharides (LPS), which induce inflammation by activating TNF secretion. This reaction is inhibited by activating the FFAR pathway due to the SCFAs.

Another site of target for SCFAs is histone deacetylases (HDACs), which are responsible for the epigenetic effect on the cells of the host. Histone deacetylases

(HDACs) and histone acetyltransferases (HAT) work in coordination with each other. Both enzymes work for gene expression and gene transcription. Particularly, HDACs work to unwind the DNA from the histone octamer and allow various transcription factors and DNA binding factors to attach to the DNA strand, allowing for gene transcription. SCFAs can interfere in this process and cease some of the transcription, which is responsible for the diseases they cause. At higher concentrations ($\geq$10 mM), SCFAs are used as natural inhibitors for HDACs. Among all SCFAs, butyrate was reported to have more inhibitory activity (2 mM) for HDACs (van der Hee and Wells 2021).

In brief, SCFAs are produced by the indigenous healthy gut microbiota and play an important role in maintaining various metabolic pathways like glucose uptake, glycogen synthesis, mucin synthesis, hormone production, and many more. It also involves strong anti-inflammatory activity by enhancing the production of anti-inflammatory cytokines and diminishing the formulation of pro-inflammatory mediators. It has the ability to pass through the blood–brain barrier and improves nervous system-related abnormalities. SCFAs can also arrest the progression of bladder cancer, breast cancer, colorectal cancer, gastric cancer, and lung cancer (Mirzaei et al. 2021).

5.4.2 Lactic Acid

Lactic acid is also known as 2-hydroxypropanoic acid. Due to the presence of a hydroxyl group adjacent to the carboxyl group, it has two isomers (L-lactic acid and D-lactic acid). Lactic acid, produced by bacteria in yogurt, buttermilk, and the gut, is a crucial organic acid resulting from anaerobic respiration. It is classified as homofermentative or heterofermentative and is a waste product of anaerobic respiration in many animals. Organisms used for the production of lactic acid are given in Table 5.3. The human body can easily accept L-lactic acid, and it can easily be metabolized by the body. As bacterial production of lactic acid is very specific in chirality, the production of L-lactic acid by bacterial fermentation is suitable in contrast to the chemically synthesized lactic acid. Lactic acid is mostly produced by lactic acid bacteria (LAB), which include organisms from the genus *Lactobacillus*, *Lactococcus*, *Bifidobacterium*, *Bacillus*, etc. These microbes frequently inhabit the colon of rodents and also humans (Idler et al. 2015).

The production of L-lactic acid by microbes can be achieved by various approaches of fermentation such as batch, fed-batch, repeated batch, and continuous fermentation. All types of fermentation have their advantages and disadvantages. As per the type of substrates used in the fermentation, investment in the instruments used for the production decides the mode and mortgages of fermentation for good yield (Rawoof et al. 2020).

Table 5.3 Lactic acid producing microorganisms

Type of fermentation	Producer microorganisms	References
Obligate Homo fermentative	*Lactobacillus acidophilus*, *Lacticaseibacillus casei*, *L. delbrueckii*, *L. amylophilus*, *L. plantarum*, *L. manihotivorans*, *L. curvatus*, *Ligilactobacillus salivarius* ssp. *salivariusd*, *L. salivarius* ssp. *Saliciniu*, *L. delbrueckii* ssp. *Delbrueckii*, *L. delbrueckii* ssp. *Bulgaricus*, *L. delbrueckii* ssp. *Lactis*, *L. amylovorus*, *L. helveticus*, *Sporolactobacillus inulinus*, *S. laevolacticus*, *Streptococcus faecium*, *Lactococcus lactis* ssp. *lactis*, *L. lactis* ssp. *cremoris*, *L. lactis* ssp. *Hordniae*, *Pediococcus lindneri*, *P. damnosus*, *P. cerevisiae*, *Streptococcus faecalis*, *S. cremoris*, *S. lactis*	Idler et al. (2015); Vijayakumar et al. (2008)
Obligate Hetero fermentative	*Lactobacillus brevis*, *L. fermentum*, *L. reuteri*, *Bifidobacterium adolescentis*, *B. animalis*, *B. bifidum*, *B. breve*, *B. longum*, *B. thermophilum*, *Bacillus. calidolactic*, *B. thermoacidurans*, *B. dextrolacticus*, *B. thermo acidificans*, *Lb. cereal*, *Rhizopus oryzae*, *Leuconostoc mesenteroides*, *L. dextranicum*	
Facultative Hetero fermentative	*Lacticaseibacillus casei*, *L. rhamnosus*, *L. paracasei* ssp. *Paracasei*, *L. coryniformis* ssp. *Coryniformis*, *L. coryniformis* spp. *Torquens*, *Lactiplantibacillus plantarum* ssp. *Plantarum*	

5.4.2.1 Application of Lactic Acid in the Food Industry

The food industry widely uses lactic acid as a preservative for improving food flavor and maintaining the pH of the food. It also has antimicrobial properties, which help in preserving the food for a longer time from pathogenic food-borne microbes (Krishna et al. 2018). As an antimicrobial agent, it also extends the shelf life of food like pickles, yogurt, and sauerkraut. It is also used to make jellies, jams, and puddings; as lactic acid functions as a gelling agent. As a curative agent, lactic acid helps to enhance the flavors and texture of foods like sausages and salamis (Rawoof et al. 2020).

5.4.2.2 Application of Lactic Acid in Pharmaceuticals

Salts of lactic acid are very useful in curing several diseases, as lactate and its salts are easily accepted by the body. In addition, they also have a pleasant to neutral taste and good tolerance. It is widely used in formulations that can be applied or ingested topically. For treating anemia, ferrous lactate is widely used. Calcium lactate is used for osteoporosis, hypertension, and tooth decay treatment. To prevent skin disorders, zinc lactate is used as an ointment. Manganese lactate and magnesium lactate help with glucose intolerance and muscle weakness, respectively. Ammonium lactate is extensively used in severe dry skin or irritating skin disorders. For preventive effects on xerosis and callus formation, ammonium lactate is frequently used. Patients suffering from kidney failure get treatment with CAPD (continuous ambulatory peritoneal dialysis). Research suggests the use of lactate salts rather than acetate salts in this treatment to metabolize it easily by the human body (Cubas Cano et al. 2018; Krishna et al. 2018; Abd Alsaheb et al. 2015).

5.4.2.3 Application of Lactic Acid in Cosmetics

In the cosmetic industry, lactic acid is used as antioxidant and anti-aging agents. It is also used to reduce the pigmentation on the skin. Lactic acid is also reported to retain moisture and hence used as a moisturizer. Also used as a humectant in many cosmetic products and as in food products. It also works against acne by reducing acne-producing bacteria on the skin. For removing tartar formation on teeth, lactic acid is frequently used as an anti-tartar agent (Abd Alsaheb et al. 2015). When consumed in certain amounts lactic acid has many beneficial aspects and provides good health. If the dose of lactic acid is more than 3730 mg/kg (in rats) then it could be lethal and may cause health hazards (Chahal and Starr 2012).

5.4.3 Citric Acid

Citric acid is also known as tricarboxylic acid or 2-hydroxy-1,2,3-propanetricarboxylic acid. It is a typical, naturally occurring metabolite that both plants and animals produce. Citric acid is produced and metabolized approximately 1.5 g/day by a healthy adult human body (Vandenberghe et al. 2017). In 1784, Carl Scheele isolated citric acid from lemon juice. It was industrially possible to isolate citric acid from lemon juice in that era. Because of its greater applicability in the food and beverage industries, an alternative source was needed. Wehmer has shown the accumulation of citric acid by the fungus *Penicillum glaucum* in sugar and inorganic salt-containing media. After these, Currie in 1917 discovered citric acid production from the fungus *A. niger* in high-sugar and mineral salt-containing media. After 1919, the industrial production of citric acid by the fungus *A. niger* commenced, and further research in the production and yield aspects was carried out. The research on the various kinds of fermenters (solid-state fermentation and submerged fermentation) was also carried out. There are various other genera also known to accumulate citric acid, but *Aspergillus* remains a good option for the same (Nangare et al. 2021). Most commonly used moulds for citric acid production are *A. aculeatus*, *A. awamori*, *A. carbonarius*, *A. niger*, *A. wentii*, *A. foetidus*, and *Penicillium janthinelum*, while *Candida citroformans*, *C. guilliermondii*, *C. oleophila*, *C. parapsilosis*, *C. tropicalis*, *Hansenula anamola*, and *Saccahromicopsis lipolytica* (also known as *Yarrowia lipolytica*) are few examples of yeast citric acid producers. The bacteria which are well known to produce citric acid are *Arthrobacter paraffinens*, *Bacillus licheniformis*, and *Corynebacterium* spp. Furthermore, substrates used in the fermentation and their yield as per various strains are given in Table 5.4. The demand for citric acid is high as it is easily metabolized by the

Table 5.4 Citric acid producing microorganisms

Substrate	Producer microorganisms	Yield (g/L)	References
Apple pomace	*A. niger* BC1	124	Soccol et al. (2006); Książek (2023)
Apple waste	*A. niger* NRRL 567	8.3	
Beet molasses	*A. niger* A20	29.7	
	A. niger W78B	110	
	A. niger ATCC 9142	109	
	Y. lipolytica A101	54	
Black strap molasses	*A. niger* GCM 7	86	
Brewery wastes	*A. niger* ATCC 9142	19	
Carob pod extract	*A. niger*	86	
	A. niger ATCC 9142	264	
Cassava	*A. niger* FUO-2	88.73	
	A. niger NRRL 2001	88	
Cooking oil	*Y. lipolytica SWJ-1b*	31	
	A. niger IBO-103MNB	337.94	
Corncob	*A. niger NRRL 2001*	250	
	A. niger NRRL 2270	603.5	
Date syrup	*A. niger J4*	56.7	
Glucose	*Y. lipolytica*	121–129	
	Y. lipolytica VKM Y-2373	80–85	
	Y. lipolytica NRRL Y-1094	30.31	
	A. niger PM-1	121	
Glycerol	*Y. lipolytica NG40/UV7*	115	
	Y. lipolytica SKY7	18.70	
	Y. lipolytica A101	75.9	
	A. niger W78B	150	
	A. niger PD66	100	
Inulin	*Y. lipolytica* SWJ-1b	85	
	Y. lipolytica AWG7 INU 8	200	
Peat	*A. niger* NRRL 567	82	
Pineapple waste	*A. niger* ATCC 1015	132	
	A. niger ACM 4942	194	
Rice bran	*A. niger* CFTRI 30	127	
Starch	*A. niger* GCB-47	45.1	
	A. niger GCMC	69.5	
	A. niger ATCC 9142	2.7	
Starch hydrolysates	*A. niger* UE-1	74	
Sucrose	*A. niger NCIM705*	30.7	
	A. niger CFTRI 30	174	
Sugarcane bagasse	*A. niger ATCC 9142*	97.81	
	A. niger DS1	198	
Sugarcane molasses	*A. niger ATCC 9142*	106.65	
	A. niger GCMC-7	96.1	

(continued)

Table 5.4 (continued)

Substrate	Producer microorganisms	Yield (g/L)	References
Wheat bran	*A. niger CFTRI 30*	85	
Whey	*A. niger ATCC 9642*	2.43	
Wood hemicellulose	*A. niger IMI-41874*	27	
	Y. lipolytica IFO 1658	9	
Xylan hydrolysate	*A. niger YANG No. 2*	72	

human body and does not have harmful effect on the human body. The majority of the citric acid is used in the food industry, followed by pharma companies (Behera et al. 2021).

5.4.3.1 Application of Citric Acid in Food Industries

Citric acid, a weak organic acid found in citrus fruits like lime, lemon, and orange, is used as a flavoring, preservative, stabilizer, and disinfectant. Because it functions as a buffer, it is utilized as an acidulant and pH regulator in the food industry. This characteristic of citric acid is also used to preserve food for a long time. It also exhibits the characteristics of antimicrobial activities against food-borne pathogens and enhances the shelflife of food. To store fruit at a low temperature, citric acid is used to protect the ascorbic acid and nonfunctionalize the enzymes responsible for the oxidation (work as antioxidant agents) (Soccol et al. 2006).

5.4.3.2 Application of Citric Acid in Pharmaceuticals and Healthcare

Citric acid is also applied for efficient drug delivery; this ability is due to the presence of three carboxylic groups and one hydroxyl group. These functional groups allow citric acid to crosslink with other biopolymers via various reactions, like condensation. Biopolymers like glycerol react with citric acid in certain conditions (90–150 °C) to form biodegradable copolymers, which are used for the delivery of gentamicin for efficient antibacterial activity. Citric acid is also cross-linked with polyethylene glycol 200 and maleic acid to entrap smaller molecular drugs and control the release of the drug molecules for up to 7 h.

The metabolite of the Krebs cycle is transported to various organs via receptors like the solute carrier family 13 (SLC13) transporter. In the kidney, SLC13A2 expresses and transports the citrate (II) ions in the urine. Calcium present in the kidney and urine tends to form stones, which are inhibited by the presence of citrate ions. Citrate ions in the urine inhibit crystal formation by various stone-forming ions. Another important SLC13A5 (also known as sodium-coupled citrate transporter, NaCT) is expressed in the liver and brain. This receptor is responsible for various metabolic activities like energy production, glycolysis, and the synthesis of lipids. The gene *SLC13A5* is also linked with some diseased conditions, like type 2 diabetes and nonalcoholic fatty liver disease. Transcription of these genes is high in the liver, and they are homologs of the *Indy* (I am not dead yet) gene in mammals. Less expression of these kinds of genes increases the chances of anti-aging, which is also associated with less deposits in the liver and vice versa (pH 7.4).

As discussed above, citrate ions chelate calcium ions. This property is used in drug delivery to disrupt tight junction integrity and eventually increase paracellular absorption. The drugs, which are not absorbed by the body easily, are mostly mixed with the citrate to enhance their absorption. Citric acid also serves the purpose of being a proteolytic inhibitor. For protein drugs, the main hindrance is proteolytic enzymes (trypsin and chymotrypsin). To enhance the absorption of protein drugs, the inhibition of these enzymes is a must, which is achieved by reducing the local pH near the drug and the concentration of calcium ions. Citrate is used for these purposes as an oral formulation to give protein and peptide drugs. Citrate in concentrations of 5–50% is used to remove inorganic matter from the root canal for ease of filling material in the root canal (Lambros et al. 2022).

Citric acid is also a part of the bone and takes part in bone formation. An estimation shows that around 80% of the citrate in the human body is present in the bones. During bone formation, osteoblasts secrete citrate ions and incorporate them into the bone. The strength, stability, and fracture resistance of bone are due to the citrate ions. To enhance the taste of bitter drugs, citric acid is used to enhance the flavor, regulate acidity, and provide a citrus aroma, which are inversely proportional to the bitter taste. It is also used to make effervescent formulations. In addition to citric acid, carbonates and bicarbonates are added, and gas formation occurs. This property increases the bioavailability of the drug in the gastrointestinal (GI) tract. It also provides rapid dissolution and dispersal in the water for oral dosage. This helps people who have difficulties swallowing capsules and tablets (Nangare et al. 2021; Lambros et al. 2022).

5.4.4 Ascorbic Acid

Ascorbic acid is also known as vitamin C. Vitamins are organic chemicals that are usually obtained from diet and are required by the body in lesser amounts to maintain normal functioning. They can help with a wide range of skin conditions because they are necessary nutrients and some of them are involved in several different biochemical processes. They are more likely to have a major local impact when applied topically for direct delivery to the skin than when taken orally, when the vitamin's dispersion is constrained by circulation to the intended spot (Martínez-Valverde et al. 2022).

Ascorbic acid is produced by the two-step fermentation process. These two steps require different microbes. The first step requires *Gluconobacter oxydans*, which convert D-sorbitol to L-sorbitol. Another step requires the coculture of *Ketogulonicigenium vulgare* and *Bacillus* spp. to convert L-sorbose to L-sorbosone and further to 2-keto-L-gulonic acid (2-KLG). Finally, L-ascorbic acid is produced after the esterification of 2-KLG. Here, *Bacillus* spp. provides various growth factors to *K. vulgare* (Zhang and Lyu 2022).

5.4.4.1 Application of Ascorbic Acid in Healthcare and Pharmaceuticals

Vitamin C has a very well-known antioxidant moiety. This property of ascorbic acid is due to its electron-donating tendency; it easily gets reduced while reacting with reactive oxygen species (ROS). In the cosmetic industry, it is used to rejuvenate the skin and for anti-aging purposes because of its excellent antioxidant capacity. Ascorbic acid also helps in collagen formation by acting as a cofactor for many enzymes that take part in the process. It helps enzymes like prolyl hydroxylase and lysyl hydroxylase, which are involved in the post-translational stage of collagen formation (Al Niaimi and Chiang 2017). Deficiency of the same can lead to defects in collagen maturation. Ascorbic acid is not synthesized by humans; hence, supplements from food are mandatory. It is easily produced by plants and very few animals (Boo 2022).

Vitamin C helps not only in the synthesis but also in the stabilization of collagen fibrins. It also improves wound healing by enhancing fibroblast proliferation and migration toward the wound site. Very essential micronutrients and vitamin C are required after surgery to improve the wound-healing process wherein the neutrophils and monocyte-derived macrophages play a crucial role. They migrate neutrophils to the wounded site and sterilize the site via the release of various antimicrobial proteins and ROS (Carr and Maggini 2017). The apoptotic neutrophils are removed by the macrophages and create an inflammatory response in the body. Vitamin C plays an important role in all the steps of wound healing (Boo 2022; Ravetti et al. 2019).

5.4.4.2 Application of Ascorbic Acid in Cosmetics

For the skin, it works as an antioxidant to rejuvenate it. It helps in the neutralization of oxidative stress that occurs during electrotransfer and donation. Due to the high exposure to sunlight (solar radiation, especially UV and infrared), the formation of ROS occurs, which is responsible for damaging protein, nucleic acid, and other crucial cell organelles. These generated ROSs also provoke a signal transduction cascade, which results in the upregulation of matrix metalloproteinases (MMPs). The high amount of MMPs is responsible for degrading the collagen, or it also reduces the production of collagen and is ultimately responsible for the high accumulation of elastin. This cascade of reactions directs various clinical phenomena like deep wrinkles, solar elastosis, coarse texture of skin, photoaging, and pigmentation. In these reactions, vitamin C works in such a way that the production of MMPs is reduced, and the following reaction gets stopped, reducing collagen damage (Al Niaimi and Chiang 2017).

Table 5.5 Malic acid producing microorganisms

Producer microorganisms		Substrate	Yield (g/L)	References
Bacteria	*Bacillus subtilis*	Glucose	2.1	Chi et al. (2016), Liu et al. (2018), Dai et al. (2018), Wei et al. (2021)
	Escherichia coli	Glucose	36.05	
	Revundimonas diminuta	Pyruvate	1.59	
	Streptomyces lividans	–	–	
Fungi	*A. flavus*	Glucose	113	
	A. niger	Thin stillage	19	
	A. oryzae	Glucose	165	
	Aureobasidium pullulans	Glucose	142.2	
	Penicillium rubens	Glucose	92	
	Plasmopara viticola	–	–	
	Rhizopus delemar	Corn straw hydrolyte	120	
	Schizophyllum commune	Glucose	18	
Yeast	*Saccharomyces cerevisiae*	Glucose	59	
	Zygosaccharomyces rouxii	Glucose	74.9	

5.4.5 Malic Acid

Malic acid is a dicarboxylic acid with four carbon atoms and is found as an intermediate in the TCA (Tricarboxylic Acid) cycle. Commercially, malic acid is obtained by treating fumaric acid under high-temperature and high-pressure conditions. By this method, the racemic mixture of the malic acid (D- and L-malic acid) is obtained. To overcome this, microbial cells containing fumarase enzymes or enzymes themselves can be used. However, nowadays, polymers made up of L-malyl units, that is, polymalic acid, are getting more attention because of their wide range of applicability, and one can produce L-malic acid by acid hydrolysis of the polymalic acid (Chi et al. 2014). Therefore, there are three ways in which the microbial production of L-malic acid is possible, and other producer organisms are presented in Table 5.5.

1. Direct microbial production of L-malic acid (Ex. production by *Aspergillus flavus*)
2. Conversion of fumaric acid to L-malic acid by fumarase enzymes containing microbes (Ex. production by *Saccharomyces bayanus*)
3. Acid hydrolysis by microorganisms that produce polymalic acid (Ex. production of PMLA by *Aureobasidium* sp. P6)

5.4.6 Application of Malic Acid

In the food industry, it is used as a preservative because it has proven antimicrobial activity. Derivatives of malic acid are used in various drugs, like citrulline malate, which serves as a nonessential amino acid. The Food and Drug Administration

(FDA) approved drug sunitinib malate for the treatment of renal cell carcinoma and GI stromal tumors as it has antiangiogenic and antitumor activities. To cure the hyperammonemia, calcium citrate malate is used. It also helps with liver dysfunction, reduces the risk of kidney stone formation, and boosts the strength of bones (Liu et al. 2021). Recent studies are also showing that 1% malic acid can reduce the problem of xerostomia (dry mouth). One such study done by Liu and coworkers in 2021 showed an improvement in xerostomia. With xylitol and fluorides, malic acid has the capability to reduce the demineralization of the enamel; in addition, it also helps in the enhancement of saliva secretion. Another review done by Manfredini and his team has shown that malic acid is also used for the treatment of renal stones. The high amount of malic acid present in pears has been proven to have this efficacy. The presence of other organic acids (citric acid, oxalic acid) in pears also has this kind of activity, but a higher amount of malic acid enhances this kind of activity (Manfredini et al. 2016).

5.5 Future Perspectives

The field of nutraceutical applications of microbiological production of organic acids is rapidly growing. As researchers learn more about the health benefits of organic acids, new and innovative applications are emerging. One challenge in the field is developing efficient and sustainable methods for the production of organic acids. Another challenge is ensuring the quality and safety of organic acid-based nutraceuticals. Despite these challenges, the future of nutraceutical applications of microbiological production of organic acids is bright. Organic acids have the potential to revolutionize the way of preventing and treating diseases.

Considering the increased demand for dietary supplements from the higher and middle classes, the Indian nutraceuticals industry is predicted to rise from $4 billion in 2017 to $18 billion in 2025. Due to the tremendous changes in lifestyle, incidences of non-communicable diseases like diabetes, obesity, hypertension, cancer, cardiovascular diseases, and many more are rising. These can be prevented, treated, or cured by diet intervention; hence, nutraceuticals play an important role in this regard. Indian companies like Dabur, Amway, gsk (GlaxoSmithKline), Hamdard, ITC limited (Imperial Tobacco Company India Limited), Danone, and many others in the list are providing nutraceutical products that contain a good amount of organic acids that are used as preservatives or supplements. These companies are providing nutrients or dietary supplements in the form of tablets, capsules, soft gels, powders, and soft cups. The demand for the same is increasing as lifestyle-oriented diseases increase (Bhatia et al. 2020).

The global nutraceutical market is highly competitive, with leading players like PepsiCo, Nestle, The Kellogg Company, Abbott, Herbalife, and General Mills. Leading nutraceutical manufacturers strategically position their products by addressing consumer needs for specific ingredient content and targeting defined consumer segments within the vast nutraceutical market, considering the market size and growth potential of those segments. New launches in line with consumer trends can help nutraceutical companies to better position themselves in the market (Nutraceuticals Market Size 2021).

According to The National Institute for Occupational Safety and Health (NIOSH) and the reports of Centers for Disease Control and Prevention (CDC), the dose of 1470 μg/kg of consumption of acetic acid through the oral route can cause adverse effects on humans (CDC/NIOSH 2018). The effect includes differences in the structure and function of esophagus as well as ulceration and bleeding in the intestine. Hence, excessive consumption of organic acids beyond their respective recommended dose or lethal dose must be avoided in order to avert any harmful effect on human health while providing nutraceuticals.

5.6 Challenges and Opportunities

The nutraceutical bioactive compounds should be assessed based on their bioavailability, bioaccessibility, bioactivity, bioconversion, and bioequivalence. Bioavailability could be defined as the absorption of nutraceutical molecules in the GI tract, which are then metabolized and distributed to various tissues and organs. These phenomena cover the main three steps: bioaccessibility, absorption, and transformation. The first step focuses on the availability of a nutraceutical molecule for absorption in the GI tract, particularly within the intestinal epithelium. The second step is the absorption of nutraceutical molecules in the epithelial cells of the intestine. The main obstacle in this step is the efflux transporter. Bioactive components are reversed in the lumen of the intestine by the efflux transporter. The last step is transformation, which is a limiting step as it transforms the bioactive molecules. Biochemical and chemical transformation of molecules in the GI tract may degrade the compound or inactivate the molecules. The pH variation in the stomach and the presence of many proteolytic, lipolytic, and various kinds of enzymes and salts make transformations of the bioactive components of functional food.

As discussed in the previous section, the bioavailability of different salts and derivatives of acids should be analyzed in vivo for their efficacy. To enhance its bioavailability, one should focus on it. To improve the bioavailability of the bioactive compound in nutraceutical products, encapsulation is a new emerging technique. Many studies have shown the increased bioavailability of compounds due to encapsulation. Encapsulation prevents the last step, which is transformation, due to the various pH in the GI tract, the presence of enzymes, and the release of molecules in a controlled manner. It also enhances the chances of the molecules being available for absorption in the intestine (Dima et al. 2020).

5.7 Conclusion

Microbial production of organic acids is a promising technology for the development of new and innovative nutraceuticals. Organic acids have a wide range of health benefits, and they can be produced in a sustainable and efficient manner. Microbial fermentation offers a sustainable and versatile approach to producing organic acids with nutraceutical applications. These acids, produced by various microorganisms, possess a range of functionalities with promising health benefits. This chapter has explored the diverse landscape of microbial organic acids with nutraceutical potential. Looking ahead, research on microbial organic acids as nutraceuticals holds immense promise. Future efforts should focus on optimizing fermentation processes for cost-effective production, elucidating the mechanisms behind their health effects, and conducting robust clinical trials to substantiate their efficacy. Furthermore, exploring synergistic effects between different microbial organic acids and other nutraceuticals could lead to the development of even more potent health-promoting products. By unraveling the full potential of these microbial metabolites, we can pave the way for a new generation of natural, safe, and efficacious nutraceuticals to promote health and well-being.

References

Abd Alsaheb RA, Aladdin A, Othman NZ, Abd Malek R, Leng OM, Aziz R, El Enshasy HA (2015) Lactic acid applications in pharmaceutical and cosmeceutical industries. J Chem Pharm Res 7(10):729–735

Al Niaimi F, Chiang NYZ (2017) Topical vitamin C and the skin: mechanisms of action and clinical applications. J Clin Aesthet Dermatol 10(7):14. https://pubmed.ncbi.nlm.nih.gov/29104718

Anyasi TA, Jideani AIO, Edokpayi JN, Anokwuru CP (2017) Application of organic acids in food preservation. In: Vargas C (ed) Organic acids, characteristics, properties and synthesis, pp 1-47. https://www.researchgate.net/publication/312087014

Behera BC, Mishra R, Mohapatra S (2021) Microbial citric acid: production, properties, application, and future perspectives. Food Front 2(1):62–76. https://doi.org/10.1002/fft2.66

Bhatia A, Mehta R, Verma K, Saigal K (2020) Indian food processing sector: the untapped growth opportunity. Invest India.gov.in. Indian food processing: an untapped growth opportunity (investindia.gov.in)

Boo YC (2022) Ascorbic acid (vitamin C) as a cosmeceutical to increase dermal collagen for skin antiaging purposes: emerging combination therapies. Antioxidants 11(9):1663. https://doi.org/10.3390/antiox11091663

Carr AC, Maggini S (2017) Vitamin C and immune function. Nutrients 9(11):1211. https://doi.org/10.3390/nu9111211

CDC/NIOSH, Acetic acid-The Registry of Toxic Effects of Chemical Substances, Centers for disease control and prevention, The National Institute for Occupational Safety and Health (2018) Registry of toxic effects of chemical substances-acute toxicity data and references. RTECS:AF1225000

Chahal SP, Starr JN (2012) Lactic acid, Ullmann's encyclopedia of industrial chemistry. https://doi.org/10.1002/14356007.a15_097.pub2

Chahardoli A, Jalilian F, Memariani Z, Farzaei MH, Shokoohinia Y (2020) Analysis of organic acids. In: Recent advances in natural products analysis. Elsevier, pp 767–823. https://doi.org/10.1016/B978-0-12-816455-6.00026-3

Chi Z, Liu GL, Liu CG, Chi ZM (2016) Poly (β-L-malic acid) (PMLA) from *Aureobasidium* spp. and its current proceedings. Appl Microbiol Biotechnol 100:3841–3851. https://doi.org/10.1007/s00253-016-7404-0

Chi Z, Wang ZP, Wang GY, Khan I, Chi ZM (2014) Microbial biosynthesis and secretion of L-malic acid and its applications. Crit Rev Biotechnol 36(1):99–107. https://doi.org/10.3109/07388551.2014.924474

Costa MP, Conte-Junior CA (2016) Chromatographic analysis of organic acids in food from animal origin. Organic Acids:47–72. https://www.researchgate.net/publication/312087014

Cubas Cano E, González Fernández C, Ballesteros M, Tomás-Pejó E (2018) Biotechnological advances in lactic acid production by lactic acid bacteria: lignocellulose as novel substrate. Biofuels Bioprod Biorefin 12(2):290–303. https://doi.org/10.1002/bbb.1852

Dai Z, Zhou H, Zhang S, Gu H, Yang Q, Zhang W, Dong W, Ma J, Fang Y, Jiang M, Xin F (2018) Current advance in biological production of malic acid using wild type and metabolic engineered strains. Bioresour Technol 258:345–353. https://doi.org/10.1016/j.biortech.2018.03.001

DeFelice SL (1995) The nutraceutical revolution: its impact on food industry R&D. Trends Food Sci Technol 6(2):59–61. https://doi.org/10.1016/S0924-2244(00)88944-X

Dima C, Assadpour E, Dima S, Jafari SM (2020) Bioavailability of nutraceuticals: role of the food matrix, processing conditions, the gastrointestinal tract, and nanodelivery systems. Compr Rev Food Sci Food Saf 19(3):954–994. https://doi.org/10.1111/1541-4337.12547

Dudeja P, Gupta RK (2017) Nutraceuticals. In: Food safety in the 21st century. Academic Press, pp 491–496. https://doi.org/10.1016/B978-0-12-801773-9.00040-6

He J, Zhang P, Shen L, Niu L, Tan Y, Chen L, Zhao Y, Bai L, Hao X, Li X, Zhang S (2020) Short-chain fatty acids and their association with signalling pathways in inflammation, glucose and lipid metabolism. Int J Mol Sci 21(17):6356. https://doi.org/10.3390/ijms21176356

Idler C, Venus J, Kamm B (2015) Microorganisms for the production of lactic acid and organic lactates. Microorganisms in biorefineries, pp 225–273. https://doi.org/10.1007/978-3-662-45209-7_9

Krishna BS, Nikhilesh GSS, Tarun B, Saibaba N, Gopinadh R (2018) Industrial production of lactic acid and its applications. Int J Biotechnol Res 1(1):42–54. https://www.researchgate.net/publication/330292057

Książek E (2023) Citric acid: properties, microbial production, and applications in industries. Molecules 29(1):22. https://doi.org/10.3390/molecules29010022

Kumar P, Singh M, Mehariya S, Patel SK, Lee JK, Kalia VC (2014a) Ecobiotechnological approach for exploiting the abilities of *bacillus* to produce co-polymer of polyhydroxyalkanoate. Indian J Microbiol 54:151–157

Kumar P, Pant DC, Mehariya S, Sharma R, Kansal A, Kalia VC (2014b) Ecobiotechnological strategy to enhance efficiency of bioconversion of wastes into hydrogen and methane. Indian J Microbiol 54:262–267

Kumar P, Ray S, Kalia VC (2016) Production of co-polymers of polyhydroxyalkanoates by regulating the hydrolysis of biowastes. Bioresour Technol 200:413–419

Lambros M, Tran T, Fei Q, Nicolaou M (2022) Citric acid: a multifunctional pharmaceutical excipient. Pharmaceutics 14(5):972. https://doi.org/10.3390/pharmaceutics14050972

Liu J, Li J, Shin HD, Du G, Chen J, Liu L (2018) Biological production of L-malate: recent advances and future prospects. World J Microbiol Biotechnol 34:1–8. https://doi.org/10.1007/s11274-017-2349-8

Liu G, Qiu X, Tan X, Miao R, Tian W, Jing W (2021) Efficacy of a 1% malic acid spray for xerostomia treatment: a systematic review and meta-analysis. Oral Dis 29(3):862–872. https://doi.org/10.1111/odi.14116

Manfredini R, De Giorgi A, Storari A, Fabbian F (2016) Pears and renal stones: possible weapon for prevention? A comprehensive narrative review. Eur Rev Med Pharm Sci 20(3):414–425. https://pubmed.ncbi.nlm.nih.gov/26914114

Markowiak-Kopeć P, Śliżewska K (2020) The effect of probiotics on the production of short-chain fatty acids by human intestinal microbiome. Nutrients 12(4):1107. https://doi.org/10.3390/nu12041107

Martínez-Valverde T, Crespo N, Suñer E (2022) Open-label study to evaluate the efficacy of a topical anhydrous formulation with 15% pure ascorbic acid and ginger as a potent antioxidant. Cosmetics 9(4):74. https://doi.org/10.3390/cosmetics9040074

Mirzaei R, Afaghi A, Babakhani S, Sohrabi MR, Hosseini-Fard SR, Babolhavaeji K, Akbari SKA, Yousefimashouf R, Karampoor S (2021) Role of microbiota-derived short-chain fatty acids in cancer development and prevention. Biomed Pharmacother 139:111619. https://doi.org/10.1016/j.biopha.2021.111619

Nangare S, Vispute Y, Tade R, Dugam S, Patil P (2021) Pharmaceutical applications of citric acid. Future J Pharm Sci 7(1):1–23. https://doi.org/10.1186/s43094-021-00203-9

Nutraceuticals Market Size, Share & COVID-19 Impact Analysis, By Product Type (Dietary Supplements, Functional Foods, and Functional Beverages), Distribution Channel (Hypermarkets/Supermarkets, Convenience Stores, Online Retail, and Others) and Regional Forecast, 2021–2028 (2021) Infographics-Nutraceuticals market

Pant SP, Joshi S (2023) Emerging insights of nutraceuticals as anticancer. Int J Plant Based Pharm 3(2):176–182. https://doi.org/10.29228/ijpbp.30

Portincasa P, Bonfrate L, Vacca M, De Angelis M, Farella I, Lanza E, Khalil M, Wang DQH, Sperandio M, Di Ciaula A (2022) Gut microbiota and short chain fatty acids: implications in glucose homeostasis. Int J Mol Sci 23(3):1105. https://doi.org/10.3390/ijms23031105

Ravetti S, Clemente C, Brignone S, Hergert L, Allemandi D, Palma S (2019) Ascorbic acid in skin health. Cosmetics 6(4):58. https://doi.org/10.3390/cosmetics6040058

Rawoof SAA, Kumar PS, Vo DVN, Devaraj K, Mani Y, Devaraj T, Subramanian S (2020) Production of optically pure lactic acid by microbial fermentation: a review. Environ Chem Lett 19:539–556. https://doi.org/10.1007/s10311-020-01083-w

Shi Y, Pu D, Zhou X, Zhang Y (2022) Recent progress in the study of taste characteristics and the nutrition and health properties of organic acids in foods. Food Secur 11(21):3408. https://doi.org/10.3390/foods11213408

Silva YP, Bernardi A, Frozza RL (2020) The role of short-chain fatty acids from gut microbiota in gut-brain communication. Front Endocrinol 11:25. https://doi.org/10.3389/fendo.2020.00025

Soccol CR, Vandenberghe LP, Rodrigues C, Pandey A (2006) New perspectives for citric acid production and application. Food Technol Biotechnol 44(2):141–149. https://hrcak.srce.hr/109832

van der Hee B, Wells JM (2021) Microbial regulation of host physiology by short-chain fatty acids. Trends Microbiol 29(8):700–712. https://doi.org/10.1016/j.tim.2021.02.001

Vandenberghe LPS, Rodrigues C, De Carvalho JC, Medeiros ABP, Soccol CR (2017) Production and application of citric acid. In: Current developments in biotechnology and bioengineering. Elsevier, pp 557–575. https://doi.org/10.1016/B978-0-444-63662-1.00025-7

Vijayakumar J, Aravindan R, Viruthagiri T (2008) Recent trends in the production, purification and application of lactic acid. Chem Biochem Eng Q 22(2):245–264. https://hrcak.srce.hr/24811

Wei Z, Xu Y, Xu Q, Cao W, Huang H, Liu H (2021) Microbial biosynthesis of L-malic acid and related metabolic engineering strategies: advances and prospects. Front Bioeng Biotechnol 9:765685. https://doi.org/10.3389/fbioe.2021.765685

Yadav P, Chauhan AK, Singh RB, Khan S, Halabi G (2022) Organic acids: microbial sources, production, and applications. In: Functional foods and nutraceuticals in metabolic and non-communicable diseases. Academic Press, pp 325–337. https://doi.org/10.1016/B978-0-12-819815-5.00053-7

Yao Y, Cai X, Fei W, Ye Y, Zhao M, Zheng C (2020) The role of short-chain fatty acids in immunity, inflammation and metabolism. Crit Rev Food Sci Nutr 62(1):1–12. https://doi.org/10.1080/10408398.2020.1854675

Zhang Q, Lyu S (2022) Microbial interactions in a vitamin C industrial fermentation system: novel insights and perspectives. Appl Environ Microbiol 88(18):e01212–e01222. https://doi.org/10.1128/aem.01212-22

Chapter 6
Microbial Production of r-DNA Products

Ishita Modasiya, Priya Mori, and Swati Goswami

Abstract Recombinant DNA (rDNA) technology has transformed human healthcare by enabling mass manufacturing of safe, clean, and effective rDNA expression products. Currently, various kinds of rDNA products are being created for human application, including hormones of therapeutic interest, therapeutic enzymes, thrombolytic agents, blood coagulation products, interleukins, anticoagulants, interferons, and hemopoietic growth factors. The current chapter discusses several aspects of rDNA products, such as expression systems, production, purification, analysis, and medicinal applications, as well as regulatory aspects and the market potential of various products.

Keywords Recombinant DNA (rDNA) technology · Hormones · Therapeutic enzymes · Thrombolytic agents · Blood coagulation products · Interleukins · Anticoagulants

6.1 Introduction

A new technology was implemented between 1971 and 1973, which later led to a significant scientific turning point. It was genetic engineering, also referred to as recombinant DNA (rDNA) technology, which is still the foundation for many biotechnological procedures today (Stryjewska et al. 2013). Paul Berg of Stanford University, who won the 1980 Nobel Prize in chemistry, synthesized rDNA for the first time in 1972. A group supervised by Herbert Boyer (University of California, San Francisco), who in 1973 introduced recombinant plasmids to transform *Escherichia coli* cells and later established Genetic Engineering Technology, followed his lead (Pavlou and Reichert 2004).

I. Modasiya · P. Mori · S. Goswami (✉)
Postbiotics and Foodomics Lab, Department of Microbiology, School of Science, RK University, Rajkot, Gujarat, India
e-mail: swati.goswami@rku.ac.in

V. Kothari et al. (eds.), *Microbial Products for Health and Nutrition*,
https://doi.org/10.1007/978-981-97-4235-6_6

Three things have a big impact on human life: lack of food, health challenges, and environmental problems. In addition to a clean and safe environment, basic human needs include food and health. Improper management of one or many of these factors leads to numerous deaths worldwide every year. According to the website http://GlobalIssues.org/, noncommunicable and communicable diseases like cancer, diabetes, AIDS/HIV, tuberculosis, malaria, and several others cause the deaths of almost 36 million people worldwide every year (Khan et al. 2016). Strategies for gene therapy have emerged as essential tools in the medication and control of human diseases throughout the preceding decades. According to current estimates, a single gene mutation can cause more than 10,000 human illnesses (Leonova and Gainetdinov 2020). A more recent advancement in rDNA technology known as clustered regularly interspaced short palindromic repeats (CRISPR) has played a pivotal role in providing remedies to numerous health conditions (Khan et al. 2016). Although CRISPR technology is anticipated to be a secure and efficacious treatment strategy, its implementation as a therapeutic is still in its early stages, and there are still many obstacles and unsolved issues. This technology ought to be more thoroughly researched and tested before being used to cure genetic illnesses; otherwise, unrestricted usage of this method could have detrimental effects on future generations (Leonova and Gainetdinov 2020). These health issues are overcome through the development of novel medications, vaccines, and treatment strategies (diagnostic kits, monitoring devices, and new therapeutic approaches) are also improved by using modern genetic engineering methods and rDNA technology. rDNA technology is one type of artificial DNA development method that involves combining DNA molecules from two different sources and introducing them into a host organism to generate products for human use, as per Fig. 6.1. A variety of microorganisms, including *Escherichia coli*, *Bacillus subtilis*, *Saccharomyces cerevisiae*, *Pichia pastoris*, *Hansenula polymorpha*, and *Aspergillus niger*, have been employed as hosts and in the yielding of therapeutic rDNA proteins (Vieira Gomes et al. 2018).

Due to the sudden emergence of coronavirus diseases (COVID 19), for which the world was not prepared, biotechnology has had a significant opportunity for the development of drugs and vaccines to save human lives (Prajapat and Jain 2022). For example, the emergence of vaccines (Pfizer–BioNTech, Oxford–AstraZeneca, Sinopharm BIBP, Moderna, Janssen, CoronaVac, Covaxin, Novavax, Convidecia, and Sanofi–GSK) that are based on protein, DNA, and mRNA. It is rDNA technology that helped in the development of vaccines for COVID, and the same method was given the 2023 Nobel Prize to Katalin Karikó and Drew Weissman for their discoveries concerning nucleoside base modifications that enabled the development of effective mRNA vaccines against COVID-19. Many kinds of therapeutic rDNA products are now accessible for human use as a result of the development of rDNA technology and its implementation in the healthcare industry (Bhopale and Nanda 2005). This chapter contains general information on healthcare uses of specific rDNA products as a result of the expanding importance of rDNA products such as hormones of therapeutic interest, therapeutic enzymes, thrombolytic agents, blood coagulation products, interleukins, anticoagulants, interferons, and hemopoietic

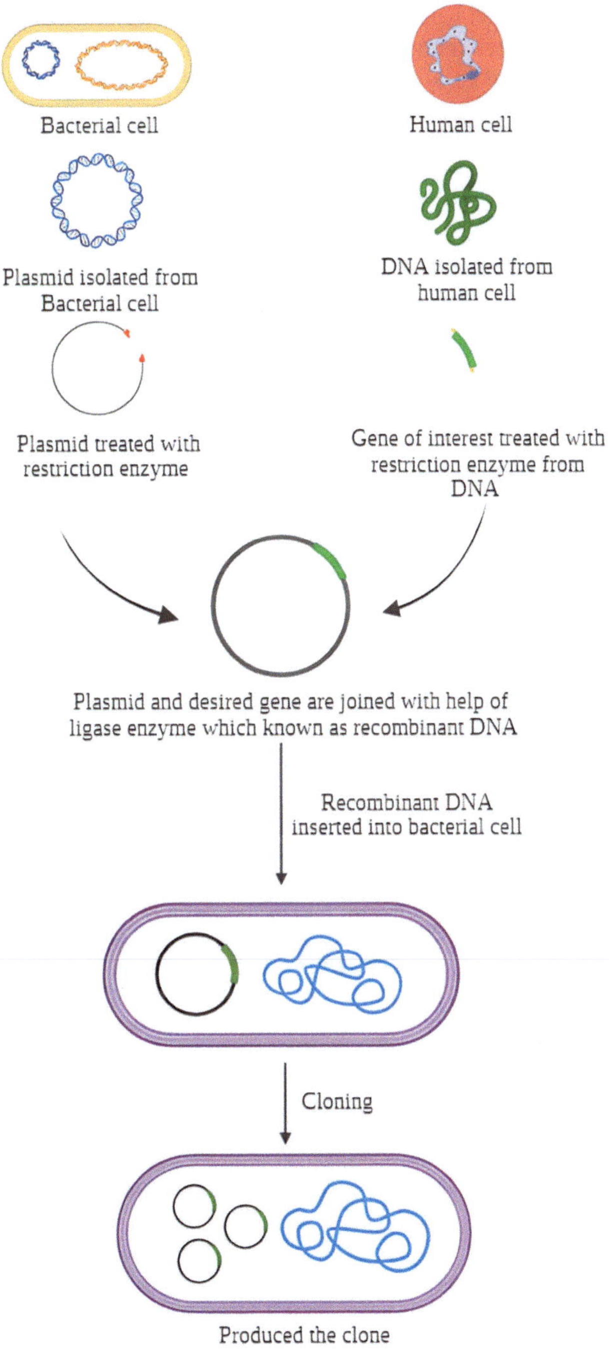

Fig. 6.1 Steps of recombinant DNA technology

growth factors for human healthcare. Recombinant therapeutic products of several different kinds have been authorized for commercialization in India, as mentioned in Table 6.1.

6.2 Therapeutic r-DNA Products

6.2.1 Hormones of Therapeutic Interest

Growth hormone or human growth hormone (hGh) is a nonglycosylated 22 kDa protein with 191 amino acid residues and 2 disulfide bridges (Paramasivam et al. 2022). The master gland, also known as the pituitary gland, produces growth

Table 6.1 Recombinant therapeutics products and its application in healthcare

Therapeutic category	r-DNA products	Expression host	Healthcare applications
Hormones of therapeutic interest	Growth hormone Somatotropin Somatostatin Ghrelin	*E. coli*	Treatment of growth hormone deficiency (GHD), Turner syndrome (TS), Prader–Willi syndrome (PWS), and chronic renal insufficiency (CRI)
	Insulin	*E. coli*, *S. cerevisiae*	Treatment of diabetics
Therapeutic enzymes	L-asparaginase	*E. coli*, *Erwinia carotovora*, *P. vulgari*, *S. cerevisiae*, *Aspergillus*, and *Penicillium*	Treatment of cancer
Thrombolytic agents	Streptokinase	*P. pastoris* *E. coli* *B. subtilis*	Treatment of the blood clot in the lungs
	Urokinase	CHO cells	Treatment for acute arterial occlusion of the legs, anti-thromboembolic drug
Blood and coagulation products	Factor VIII Factor IX	pGEX-4T1(*E. coli*), pMrnF9	Treatment of hemophilia A Treatment of hemophilia B
Interleukins (IL)	IL-2	*E. coli*	Therapeutic drug for the treatment of anemia, cancer, and multiple sclerosis
Anticoagulants	Hirudin	*E. coli*, *S. cerevisiae*	Chronic thromboembolic diseases
Interferons (IFN)	IFN-α2b	*E. coli*	Treatment of over 14 diseases globally, such as hairy-cell leukemia, condyloma acuminatum, Kaposi's sarcoma, and hepatitis
Hemopoietic growth factors	Erythropoietin	CHO cells	Treat anemia linked to conditions such as renal failure, HIV infection, cancer, and surgery

hormones, also known as somatotropin. This growth hormone serves a variety of functions, including maintaining normal body composition and transformation. Formerly, pituitary glands of cadavers derived hGh were used for therapeutic purposes, which were discontinued due to safety concerns like Creutzfeldt–Jakob disease. This issue has been addressed by recombinant hGh discovered by the Genentech team in 1981, which was approved in 1985. This strategy is now employed for various medical applications (Hardin et al. 2007). Mammalian, yeast, and bacterial cells have all been utilized as expression hosts to produce recombinant hGh. Because the nonglycosylated hGh proved a bio-effector, *Escherichia coli* was utilized mostly to produce hGh due to its number of benefits, including the potential for simple bacterial modification, low production costs, quick cell growth, and the ability to cultivate high-density cells (Ghavim et al. 2017). These factors, all together, make it a worthwhile system, especially from an economic standpoint when compared to other expression systems (Choi et al. 2006). The hGh gene has been cloned and expressed as an inclusion body in the cytoplasm of *E. coli*, predominantly (Song et al. 2017). In general, complementary DNA (cDNA) produced from pituitary mRNA is cloned with chemically synthesized DNA during cloning to produce the bacterial form of hGh. A chemically synthesized DNA fragment encoding the first 24 amino acids of hGh is being designed that retains the HaeIII restriction site at amino acid residues 23 and 24, mirroring the natural sequence found in hGh cDNA. This design facilitates future assembly with the complete hGh cDNA. A 551-base pair (bp) long hGh-specific cDNA produced by HaeIII contains the coding sequences for amino acids 24–191 of hGh. The ATG start codon and the coding sequences for residues 1–23 of hGh were included in the chemically synthesized DNA "adaptor" fragment that clones independently from this cDNA fragment. These combinations of DNA fragments would create a synthetic, natural hybrid gene. This gene can regulate the synthesis of fMet-hGh when introduced into the plasmid downstream from the appropriate bacterial promoter and ribosome binding site. Since most bacterial proteins lack N-terminal methionine residues, the fMet should be effectively eliminated, leading to the direct production of hGh (Rigi et al. 2021). This process established the framework for the industrial manufacturing of rhGh. Essentially, original rhGh acts the same way as the major form (22 kDa) of growth hormone made by the pituitary gland.

It is crucial to remember that large-scale rhGh production necessitates not only the cultivation (fermentation) of the synthesizing cells but also a subsequent downstream processing step focused on eliminating all nonhormonal constituents. There are two types of purification methods for rhGh products, which include structural analyses by SDS-PAGE, chromatography (e.g., HI-HPLC), laser densitometry, and protein sequencing. Additionally, immunological assays, radioreceptor assays, and bioassays function as instruments for verifying the chemical and biological equivalence of the product to the pituitary-derived 22-kD hGh. There are strict regulations, like proof of efficacy and safety for the production and approval of rhGh-based drugs, in the pharmaceutical industry. Drug approval follows a two-step process: preclinical testing (in labs and animals) and clinical trials (in humans) (Ranke 2007). The use of rhGh was given authorization by the Food and Drug Administration

(FDA) of the United States in 1985 for the application of rhGh in therapeutics such as growth hormone deficiency (GHD) (e.g., reduced muscle strength, altered body composition, low energy expenditure, and reduced growth, even in the presence of obesity), idiopathic short stature, Turner syndrome, Prader–Willi syndrome, chronic renal insufficiency, and human immunodeficiency virus (HIV), dwarfism, bone fractures, skin burns, bleeding ulcers, and AIDS (Hardin et al. 2007).

Diabetes has become one of the most common metabolic disorders in present-day society and is continuing to rise consistently and substantially. The World Health Organization (WHO) estimates that 171 million people worldwide have diabetes mellitus, a number that is expected to double by 2030 (Walsh 2005). The necessity to address the unsolved problems of diabetes treatments is essential for maintaining human health. Since insulin's discovery in the1920s first by Banting and Best, one of the greatest achievements of pharmaceutical science is its application in the therapy of diabetes mellitus (Gowda et al. 2023).

Insulin is one type of pancreatic hormone consisting of 51 amino acids and it can control the glucose level in the body because diabetes may be triggered by low insulin levels (Gowda et al. 2023). The production of recombinant insulin was first achieved successfully in 1978, when *E. coli* was employed as a host for the production of recombinant human insulin. In the bacterial system, insulin is produced as a precursor and stored in inclusion bodies, from where it is solubilized and refolded into active insulin (Alyas et al. 2021). The precursor insulin, also called proinsulin, consists of three polypeptide chains: A, B, and C. Chains A and B are linked together by the C chain. Proinsulin is broken down into its active form, insulin. This happens by separating two chains (A with 21 amino acids and B with 30 amino acids) and linking them with a disulfide bond. An extra chain (C-chain) is removed during this process before mature insulin is released (Gupta et al. 2017). *E. coli* or *S. cerevisiae* usually play a crucial role in the commercial manufacturing of recombinant human insulin (Alyas et al. 2021). The DNA sequence for the A and B chains is chemically generated and individually placed into two pBR322 plasmids alongside the β-galactosidase gene. These plasmids are individually inserted into two host systems, such as *E. coli*, where they are expressed and secreted as fusion chains of β-galactosidase-A and β-galactosidase-B. A and B insulin polypeptide chains are then separated chemically from the β-galactosidase polypeptide and combined with pure polypeptide chains linked by disulfide bonds by sodium dithionite and sodium sulfite to produce functional recombinant insulin as per Fig. 6.2 (Gupta et al. 2017). The remaining impurities or reagents are eliminated using RP-HPLC. The development of 20 unique *E. coli* strains and the creation of the innovative pIBAINS expression vector, which increases the production of recombinant insulin per liter of media, were both reported by Zielinski et al. (2019). The pharmaceutical industry can then pack and commercialize the purified, active human recombinant insulin. Recombinant insulin has various medicinal applications, which are approved by the FDA. Recombinant products like insulin glulisine have been found effective for treating diabetes, while insulin aspart provides more rapid absorption than regular human insulin, insulin glargine, long-acting recombinant insulin, and insulin lispro, whose main practical benefit is the instant onset of action (Bhopale and Nanda 2005).

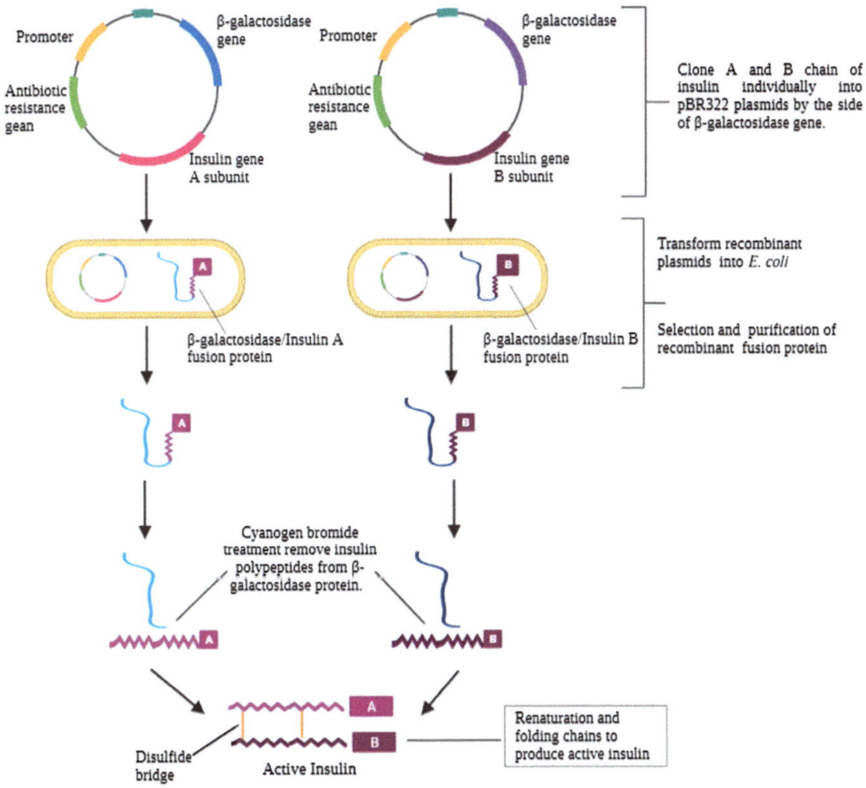

Fig. 6.2 Microbial expression of recombinant human insulin by the chain method

6.2.2 *Therapeutic Enzymes*

Enzymes are important for many kinds of cellular respiration, which are carried out in our body and many diseases are caused by an enzyme deficiency or insufficiency of enzyme activity. These kinds of illnesses and disorders are treated with therapeutic enzymes, which are biocatalysts (Tandon et al. 2021). Various applications of therapeutic enzymes in the treatment of diseases are presented in Fig. 6.3. Therapeutic enzymes are derived by two methods: first, from natural sources and second, by using rDNA technology (Vimal and Kumar 2019). Over the past century, several therapeutic enzymes have found significant uses in the pharmaceutical industry, including L-asparaginase, streptokinase, collagenase, ribonuclease, uricase, glucosidase, etc. (Tandon et al. 2021). Here, we will be discussing L-asparaginase among these four major enzymes.

The enzyme L-asparaginase, also known as L-asparaginase amidohydrolase, which is used in chemotherapy (treatment of cancer), has been long recognized. This enzyme functions as an asparaginase and glutaminase, catalyzing the hydrolysis of both L-asparagine and L-glutamine. It specifically removes the amino group

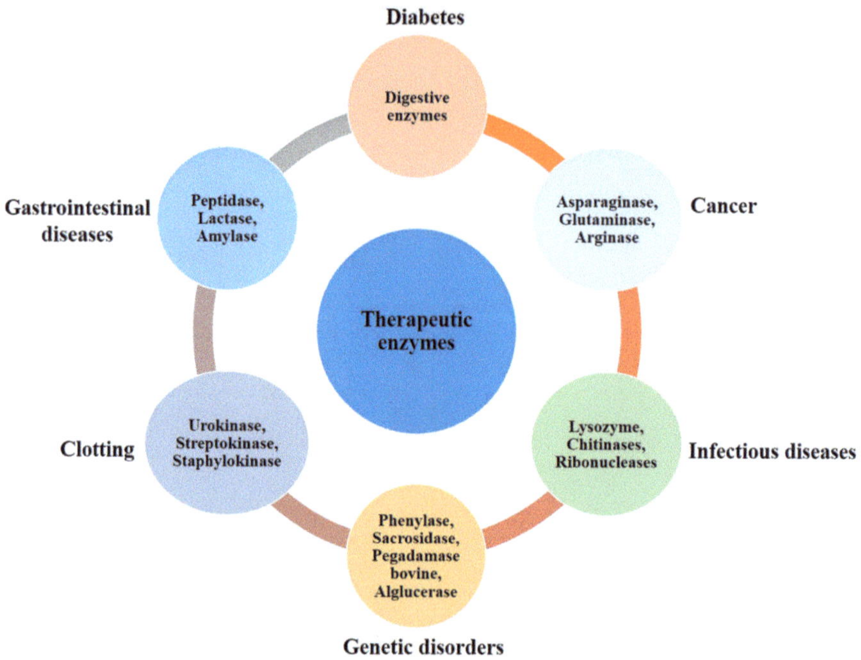

Fig. 6.3 Therapeutic enzymes and their health applications in the treatment of various diseases

from the L-asparagine side chain, converting it to L-aspartate and releasing ammonia. The loss of amino acids and asparagine leads to low nutritional value in tumor cells and ultimately creates death. Most L-asparaginases derive from naturally occurring microorganisms, but some are also designed and produced intracellularly, except for a small number that are secreted extracellularly. This enzyme is produced by a variety of bacteria, including *E. coli*, *Erwinia carotovora* (Kozak and Jurga 2002), and *P. vulgaris*. Building upon earlier findings by Tandon et al. (2021), research has identified several bacterial L-asparaginases that were produced on a large scale and received FDA approval for treating tumor-related disorders. Furthermore, L-asparaginase derived from *S. cerevisiae* has also been shown to be effective in treating acute lymphoblastic leukemia (Kozak and Jurga 2002). Scientists employed *Saccharomyces cerevisiae* (baker's yeast) to isolate the ASP3 gene, which encodes asparaginase II. This gene was then incorporated into the expression vector pET28a, fused with a 6x histidine tag for easier purification. Finally, the construct was introduced into *E. coli* BL21 (DE3) cells to express the recombinant asparaginase II protein. The purification procedure for the enzymes generated from microbial strains involves chromatography techniques like ion exchange and size exclusion with precipitation methods using ammonium sulfate (Mahajan et al. 2014). The enzyme L-asparaginase is isolated from the supernatant using nickel affinity chromatography after being expressed at a high level (225.6 IU/g) as an intracellular and soluble molecule. Recombinant asparaginase is used to treat a variety of illnesses, including chronic lymphocytic leukemia, Hodgkin's disease, acute myelogenous leukemia, critical lymphoblastic leukemia, and myelogenous leukemia. In addition, it has been shown to treat infectious disorders caused by harmful microorganisms such as *Clostridium botulinum*, *Staphylococcus aureus*, *Listeria monocytogenes*, and *Streptococcus pyogenes*. For treating conditions like autoimmune diseases and inflammation, L-asparaginase is a common choice due to its immunosuppressive and anti-inflammatory effects (Tandon et al. 2021).

6.2.3 Thrombolytic Agents

Thrombolytic therapy is critical in the treatment of diverse circulatory disorders such as cardiovascular disease, pulmonary embolism, deep vein disease, thrombosis, and myocardial infarction. Globally, circulatory problems are progressively emerging as a primary reason for death in modern-day life (Kunamneni et al. 2007). Blood clots, also known as thrombi, form inside blood vessels when blood components clump together. These obstructions impede oxygen delivery to the surrounding tissue, causing it to degrade and die. To minimize tissue damage and improve survival rates, successful blood clot treatments prioritize the rapid breakdown of clots (Banerjee et al. 2004). According to Adivitiya and Khasa (2017), thrombolytic therapy offers four main treatment approaches: first-generation plasminogen activators, second-generation thrombolytics, third-generation plasminogen activators, and other plasminogen activators. The first generation of plasminogen activators includes streptokinase and urokinase-type plasminogen activators (urokinase, for short). While both activate plasminogen in the bloodstream, they can also break down fibrinogen and other clotting factors besides fibrin itself. Thrombolytic treatment is chosen based on a number of parameters, including medication cost, adverse effects, stability, half-life, fibrin specificity, and immunogenicity (Table 6.2).

Streptokinase (SK) is a protein produced by certain strains of bacteria. This protein can activate plasminogen, a substance in the blood that can dissolve blood clots. SK (47 kDa, single-chain, 414 amino acids) acts indirectly as a fibrinolytic agent by activating plasminogen in circulation. The first medication for treating acute myocardial infarction was SK, which was established almost 40 years ago. Moreover, it can dissolve blood clot fibrin that forms in shunts during dialysis of the kidneys. This medication, which dissolves clots most effectively when they are newly formed, is released at the site of the clot through a catheter that is placed into an artery (El-Dabaa et al. 2022). Some of the drawbacks of inherent SK generated in *Streptococcus*, especially limited yield, cost-effectiveness, and the existence of additional immunogenic molecules, can be avoided by producing recombinant streptokinase (rSK) in the *E. coli* expression system. Researchers have been focusing on optimizing rSK by making it last longer in the body, reducing side effects, and making it more effective (Assiri et al. 2014). The production of rSK protein utilized *E. coli* BL21 (DE3) as the host organism. To achieve efficient expression, a

Table 6.2 Characteristics of FDA-approved plasminogen activators (Adivitiya and Khasa 2017)

Plasminogen activator	Molecular weight (kDa)	Mode of plasminogen activation	Immunogenicity	Fibrin specificity	Half-life (mins)	Production host	Clinical applications
First generation							
Streptokinase	47.0	Indirect	Yes	No	18	*Streptococci*	AMI, pulmonary embolism, DVT, arterial thrombosis
Urokinase	33.0	Direct	No	No	6–18	Human neonatal kidney cells	Pulmonary embolism
Second generation							
Alteplase	70.0	Direct	No	Yes	6	CHO cell line	AMI, acute ischemic stroke, pulmonary embolism
Anistrep lase	131.0	Direct	Yes	No	90–105	Anisoylated complex	AMI
Third generation							
Reteplase	39.0	Direct	No	Yes	15	*E. coli*	AMI
Tenecteplase	65.0	Direct	No	Yes	10–24	CHO cells	AMI

Note: *AMI* acute myocardial infarction; *DVT* deep vein thrombosis; *LMW* low molecular weight modified gene that codes for the mature SK protein, optimized for uncommon codons and mRNA secondary structure, was cloned into the pET-3a expression vector and then inserted into *E. coli* BL21 (DE3). The quick dilution approach was used to purify, denature, and refold the insoluble rSK inclusion antibodies in 6 M guanidinium chloride. By using anion exchange chromatography, the refolded rSK protein was isolated and assessed by ELISA (El-Dabaa et al. 2022). To increase the half-life of thrombolytic treatment, Wu et al. (1998) produced plasmin-resistant muteins that achieved a 21-fold boost in half-life in the protease-deficient *B. subtilis* WB600 by mutating the Lys^{59} and Lys^{386} residues that are vulnerable to proteolytic activity by plasmin. Along with that, they were using *P. pastoris* to produce rSK by glycosylation, which emerged to form molecules tolerant to proteolysis (Pratap et al. 2000). The drug rSK, which is derived from β-hemolytic streptococci, is now marketed under the trade names Kabikinase® and Streptase®. The same drug, Shankinase (Shantha Biotech), has been approved in India, using *E. coli* as an expression host for production (FDA 2017).

One potent plasminogen activator (PA) that is involved in numerous physiological processes is urokinase (UK). The Arg-Val bond was cleaved by the UK to activate the plasminogen in plasmin. Now, the fibrin polymers in blood clots are broken down by the active plasmin that formed earlier. This active plasmin has a role in cell migration, tissue remodeling, complement activation, prohormone conversion, and wound healing. It also plays an important role in carcinogenesis and metastasis and acts as a very popular cardiovascular agent. The UK was initially found and separated from human urine. The process of separating the UK from urine is time-consuming, expensive, and laborious. Hosts including bacteria *(E. coli, Pseudomonas* sp., *B. subtilis, Halobacillus* sp., and *Enterococcus* sp.), fungi (*Aspergillus* sp., *Penicillium* sp., *Trichothecium* sp., and *S. cerevisiae*), yeast, mammalian cells, insect cells, and plants have been explored for the production of rSK since the development of rDNA technology (Nawaz et al. 2022). Although it is extremely simple to cultivate bacteria or fungi on a large scale, the biggest drawback of employing a prokaryotic system for UK manufacture is that these organisms lack the mechanisms required for post-translational modification. As a result, the nonglycosylated UK that is thus produced lacks the pharmacodynamic qualities and effectiveness of natural UK. Thus, mammalian cell lines are recommended for UK manufacture (Kunamneni et al. 2008).

For almost 20 years, the two-chain version of uPA (tc-uPA), which is derived from human urine or kidney cell culture, has been utilized as a thrombolytic drug. Because it is fibrin-specific and inert in plasma, the single-chain version of uPA has recently been created for thrombolysis. Sc-uPA has been generated by recombinant technology from a few sources, including *E. coli*, mammalian cells, and yeast (Tang et al. 1997). Tang et al. (1997) developed a method to produce single-chain urokinase plasminogen activator (sc-uPA) from *E. coli*. Their approach utilized a pET expression system, which incorporated a histidine tag attached to the C-terminus of the sc-uPA protein. This strategy resulted in high expression levels of the histidine-tagged sc-uPA (ch-sc-uPA) and facilitated its purification and refolding. Presently, human neonatal kidney cells are used in the commercial production of urokinase

(Abbokinase®, Kinlytic™), which was first approved by the FDA in 1978. As the active ingredient, these drugs have a low molecular weight form of u-PA. However, after an investigation revealed issues with virus contamination, the FDA stopped selling the medication in December 1998 because it was not adhering to current good manufacturing practices. As of 2002, urokinase once again became a viable option for treating patients with pulmonary embolism (Adivitiya and Khasa 2017; FDA 2017).

6.2.4 Blood and Coagulation Products

Blood coagulation occurs in mammals when fibrin and activated platelets polymerize together. In this process, many enzyme activation events take place in which proenzymes and pro-cofactors are activated by a serine protease (Smith et al. 2015). This process helps to prevent excessive blood loss during injury. If this blood clot forms inside the blood vessel, it is called pathologic thrombosis, which leads to various disabilities in humans and can even lead to death. To trigger blood clotting, the tissue factor pathway (extrinsic) and contact pathway (intrinsic) play a major role (Norris 2003). These pathways are associated with hemophilia types A and B because if blood clotting factors or proteins are unable to clot the hemorrhage in blood vessels, then hemophilia occurs. Hemophilia A is associated with a deficiency in factor VIII (F8). Bashar and Jeong (2023) investigated the production of recombinant A2 and A3 domains of coagulation factor VIII (F8). The deficiency of F8 causes a congenital bleeding disorder called Hemophilia A. The synthesis of A2 and A3-encoding genes was carried out based on human F8 gene sequences (Takeyama et al. 2018). The expression of rF8-A2 and rF8-A3 was archived using pGEX-4T1, an *E. coli*-expressing vector. The protein was purified from the supernatant using Ni-NTA beads and the concentration was measured using SDS-PAGE. Indirect ELISA was performed to check the antigen–antibody binding capacity of A2 and A3 antigens. The GST-tagged rF8-A2 and rF8-A3 proteins of size 50.5 kDa and 51.6 kDa, respectively, were purified in soluble form. The yield of recombinant protein was increased by introducing GST to the 5′ of F8, which in turn switches the pET28a (+) to the pGETX-4T1 vector. High binding efficiency was observed while using anti-F8 polyclonal antibodies, indicating the presence of epitopes on rF8-A3, which binds with anti-F8 antibodies. Hence, rF8-A2 and rF8-A3 can be used to generate recombinant antibodies against A2 and A3, which helps in recognizing A2/A3 domains in the patients with hemophilia A.

6.2.5 Interleukins

Interleukins (IL) are a major group of naturally produced immunomodulatory proteins that regulate cell-to-cell communication. Although it was previously believed that only leukocytes produced these proteins, now it has been discovered that many other kinds of body cells also produce them. They help to improve the activation and differentiation of immune cells, along with proliferation, maturation, migration, and adhesion. Additionally, they also have both pro- and anti-inflammatory attributes (Commins et al. 2010). By adhering to high-affinity receptors on cell surfaces by using both functions autocrine and paracrine, ILs may induce a variety of responses in cells and tissues. There are now 55 identified IL and IL-related genes in the human IL gene family. Using distinctive structural characteristics, the majority of ILs may be divided into seven different categories and subgroups (Brocker et al. 2010). Recombinant cytokines are one of the techniques that have revolutionized the treatment of many significant diseases. There are currently a variety of cytokine immunotherapies available on the market (Sengupta et al. 2008). There are many human recombinant interleukins like rIL-2, rIL-3, rIL-4, rIL-5, rIL-7, and rIL-11 that have hemopoietic activities. Among all, due to its central role in the immune system, rIL-2 deserved thorough exploration.

Devos et al. in 1983 provided the first description of the expression of human IL-2 in *E. coli* (Halfmann et al. 1993). There is a minor difference between natural human IL-2 (nIL-2) and recombinant human IL-2 (rIL-2). The nIL-2, which is glycosylated, is made up of a polypeptide chain of 133 amino acids, which are arranged into four antiparallel, amphipathic α helices. In contrast, rIL-2 generated from *E. coli* is nonglycosylated, the first amino acid alanine is absent, and the cysteine residue at position 125 has been replaced by a serine residue (Knuver-Hopf and Mohr 1995). Halfmann and colleagues expressed a gene that codes for nIL-2 in *E. coli* to produce a significant amount of rIL-2 (Met-IL-2). When they expressed the mature IL-2 gene in pGF25 at a high level, the produced rIL-2 was accumulated in inclusion bodies. This problem was also faced by Devos et al. (1983), who discovered the first rIL-2, which accumulated in the cytoplasm of *E. coli.* To address this challenge, Halfman and colleagues developed a strategy to fuse IL-2 with either bacterial signal peptides or the periplasmically located maltose-binding protein by targeting IL-2 to the periplasm of *E. coli.* For that, they fused IL-2 with precursor maltose-binding protein (preMBP), which caused a significant amount of the preMBP-IL-2 protein to be appropriately digested and delivered to the periplasmic region. Now we can easily purify the synthesized rIL-2. Aldesleukin (Proleukin, Cetus Oncology Corporation, Emeryville, CA) and teceleukin (Tecin, Hoffmann-La Roche, Nutley, NJ) are two types of recombinant IL-2 available in the market that has undergone significant research. A specific variant of rIL-2 serves as a medicinal agent for managing illnesses such as anemia, cancer, and multiple sclerosis. Metastatic renal cell carcinoma and melanoma have both been successfully treated using rIL-2 as a therapeutic drug to promote T-cell proliferation (Sengupta et al. 2008).

6.2.6 *Anticoagulants*

Anticoagulant therapy has been widely applied for the last 70 years to treat acute thromboembolism, and it has more recently been accepted to cure chronic thromboembolic diseases (Zhang and Lan 2018). Thromboembolism leads to many types of cardiovascular disease (CVDs) because heart and blood vessel problems are the root causes of CVDs. In 2030, CVDs are expected to continue to be the leading cause of death accounting for 23.3 million deaths according to the World Health Organization (Mathers and Loncar 2006). The best treatment for this problem can be thrombolytic therapy (Krishnan et al. 2015), which we have discussed in thrombolytic agents previously, and anticoagulant therapy (Alquwaizani et al. 2013), which helps to reopen blocked blood vessels. These anticoagulant drugs have different aspects of the clotting process. Vitamin K antagonists, direct oral anticoagulants (DOACs), and low molecular weight heparins (LMWH) are the three main types of anticoagulant medications. Examples of anticoagulant drugs are apixaban, dabigatran, edoxaban, rivaroxaban, warfarin, and hirudin (Schulman and Bijsterveld 2007). Among them, hirudin is the original direct thrombin inhibitor and a protease inhibitor, which is a 65-amino-acid protein naturally excreted from the salivary gland of the bloodsucking leech, *Hirudo medicinalis* (Rai et al. 2020).

Bacterial strains like *S. cerevisiae* and *E. coli* are used to produce recombinant anticoagulants. Compared to *E. coli*, *S. cerevisiae* offers several advantages, including a superior extracellular expression efficiency and glycosylation system more closely resembling that of humans and other eukaryotes. *S. cerevisiae* is a popular choice for therapeutic protein production due to its Generally Recognized as Safe (GRAS) status (Zhang and Lan 2018). The US Food and Drug Administration (FDA) approved *S. cerevisiae* for the production of recombinant proteins (Zhang and Lan 2018). Several studies have focused on the expression and secretion of the hirudin gene in the yeast, *S. cerevisiae*. The hirudin gene has been synthesized using the amino acid sequence. This synthesized gene is now cloned into the YEGα-1 expression vector of yeast. This vector also contained the galactose-inducible promoter, GAL10, and the α-mating factor pre-pro leader sequence. It has been found that recombinant *S. cerevisiae* secretes biologically active hirudin into the extracellular medium. After being extracted from the culture medium, the secreted recombinant hirudin (rH) is refined using reverse-phase high-performance liquid chromatography and ultrafiltration. Under less than ideal culture circumstances, 1 mg of hirudin per liter can be formed, and after two stages of purification, the purity increases to almost 90% (Choi et al. 1994). Lipurin, a recombinant derivative of hirudin, which is identical to natural hirudin, is derived from yeast. This molecule closely resembles hirudin, with two key differences: a leucine replaces isoleucine at the beginning (N-terminus) of the chain and the tyrosine at position 69 lacks a sulfate group. This lepirudin, rH, is used for the treatment of heparin-induced thrombocytopenia (Greinacher 2004). Desirudin is also a type of rH, and acts as a thrombin inhibitor, and prevents deep vein thrombosis and thromboembolic diseases in patients (Matheson and Goa 2000).

6.2.7 Interferons

Interferons (IFNs) are the natural defenses of humans, which were discovered by Isaacs et al. (1957). Initially, interferons were believed to have potential applications in anticancer therapy and as a broad antiviral drug. It is a group of proteins involved in many processes, including regulating the immune system and cell growth (Taylor and Taylor 2014). IFNs are presently classified into three major groups: type I, type II, and type III. Type I interferons encompass IFN-α, IFN-β, IFN-ω, IFN-τ, and IFN-κ. These interferons play critical roles in halting viral replication and cell proliferation. Type II IFNs consist solely of IFN-γ, which is essential for stimulating macrophages and natural killer (NK) cells. Type III is comprised of IFN-λ and interleukin (IL)-28/29, which also exhibit antiviral effects in a similar way to type I IFNs. The most potent antiviral action among the three IFNs is exhibited by type I interferons, even though multiple IFNs share some properties (Yu et al. 2021). The successful cloning of the IFN-α gene in *E. coli* in 1980 established it as a model system for mammalian gene cloning. The laboratory of Charles Weissmann at the University of Zurich used an *E. coli* plasmid, which is pBR 322, for expression systems, and biotech company Biogen supported the university for this research. The expression system was established by first isolating mRNA from interferon-producing leukocytes stimulated by the Sendai virus. This mRNA was then reverse-transcribed into double-stranded complementary DNA (cDNA). Finally, the cDNA was cloned into the pBR322 plasmid vector. As a result, *E. coli*, yeast, and mammalian cells were able to produce interferon (Taylor and Taylor 2014).

Rajabi Memari et al. (2010) expressed IFN-α2b in the *E. coli* system using an apET 26b (+) vector with a T7 lac promoter and peIB signal sequence. For analysis of expression, western immunoblotting was used to establish that protein was expressed in the *E. coli* system, and immunoassay was used to measure the concentration of protein. IFN-α2b has immunomodulatory and antiproliferative properties across several cell types. The approval of recombinant IFN-α2b for the treatment of over 14 diseases globally, such as hairy-cell leukemia, condyloma acuminatum, Kaposi's sarcoma, and hepatitis, indicates the practical significance of these multiple benefits (Rajabi Memari et al. 2010). Recombinant human interferon-γ (rhIFN-γ) was established by using the *E. coli* BL21 (DE3) plasmid. rhIFN-γ was purified from inclusion bodies using a novel chromatographic purification technique that combines anion and cation exchange chromatography. HPLC was used to assess the purity of the rhIFN-γ. The rhIFN-γ has the capability to reduce the frequency and severity of serious infections associated with chronic granulomatous diseases (Khalilzadeh et al. 2004).

6.2.8 Hemopoietin Growth Factors

Currently, there are three synthetic hematopoietic growth factors—G-CSF (filgrastim), GM-CSF (sargramostim), and erythropoietin (epoetin; EPO)—that are readily accessible for medical use in the market. Erythropoietin (EPO) is indeed unique among hematopoietic growth factors as it functions as a hormone (Vose and Armitage 1995). Recombinant human erythropoietin (rhuEPO) is extensively utilized to treat anemia linked to conditions such as renal failure, HIV infection, cancer, and surgery. There are three main varieties of rhuEPO: epoetin alfa, epoetin beta, and epoetin omega. EPO, a hematopoietic growth factor, plays a critical role in regulating the final stages of erythroid cell maturation into functional red blood cells (Gabrilove 2000).

Epoetin alfa is a synthetic version of erythropoietin, the primary hormone that regulates the production of red blood cells. Initially used as a treatment for anemia associated with renal failure, epoetin alfa has become an established therapy. It is also employed to address anemia in various other patient groups. In the late 1970s, the isolation and purification of endogenous erythropoietin, followed by the isolation and cloning of the gene responsible for its production, paved the way for the development of rhuEPO. This synthetic version of the hormone is typically produced by expressing the erythropoietin gene in Chinese hamster ovary cells. Epoetin alfa, a form of recombinant human erythropoietin, has been employed as a treatment for anemia in conditions beyond renal failure. It is also used to facilitate autologous blood donation before surgical procedures (Markham and Bryson 1995).

A breakthrough in genetic engineering led to the creation of epoetin beta, a synthetic version of the human protein erythropoietin (r-HuEPO). Scientists first isolated the gene that codes for erythropoietin, then introduced this gene into Chinese hamster ovary cells, enabling them to produce epoetin beta. This glycoprotein hormone plays a crucial role in the growth and maturation of red blood cell precursors in the bone marrow. Treating anemia with epoetin beta has been associated with improvements in various cardiovascular health markers. This includes reductions in cardiac output and resting heart rate. Additionally, patients undergoing hemodialysis due to chronic renal failure (CRF) experienced decreases in left ventricular hypertrophy and exercise-induced cardiac ischemia, along with increased exercise capacity, after 12 months of epoetin beta therapy. However, it is important to note that increased blood pressure might occur due to elevated hematocrit and peripheral vascular resistance as a result of this treatment (Dunn 1996).

Following the creation of epoetins alpha and beta, scientists developed a third version of recombinant human erythropoietin (rHuEPO) and labeled it epoetin-omega. It is produced in BHK cells. It is also used as a treatment for anemia associated with renal failure and cardiac-related diseases (Sikole et al. 2002).

6.2.9 Monoclonal Antibodies

Antibodies are parts of our immune system that eradicate antigens (foreign particles). Monoclonal antibodies, more commonly referred to as mAbs or moAbs, are synthetic proteins that function similarly to the antibodies in our body. These recombinant antibodies (mAbs) have become the gold standard for diagnostic assays due to their high specificity, sensitivity, and reproducibility (Bayer 2019). Early in the 1980s, therapeutic monoclonal antibodies were first developed commercially, and in 1986, Orthoclone OKT3, the first therapeutic monoclonal antibody, was approved to prevent kidney transplant rejection. As the pioneers in producing copious amounts of mAbs in vivo, Milstein and Kohler were awarded the 1984 Nobel Prize in Physiology or Medicine. This was possible by the "hybridoma" technique (fusions of a myeloma cell line with plasma B cells), which permanently altered the way mAbs are produced. Early mAbs, generated using mouse protein (murine), proved nonideal for long-term human use. This incompatibility triggered a re-evaluation of production methods, eventually leading to the development of fully humanized antibodies through technological advancements. Muromonab-CD3, an anti-rejection drug, was the first humanized mAbs to receive FDA approval in the USA (Singh et al. 2018).

The ability to produce antibodies in bacteria has made it more straightforward for researchers to select the desired antibodies depending on how successfully they function, which has led to fascinating novel opportunities for antibody development. The microbial cells, namely *S. cerevisiae*, *P. pastoris*, and *E. coli*, are used as an expression system for producing the recombinant antibodies (Spadiut et al. 2014). There are still numerous benefits to be discovered about recombinant antibodies, which have far more potential than traditional monoclonal antibodies. Advantages stem from two key features: firstly, the immunoglobulin gene can be tested within an *E. coli* system, enabling the use of well-developed methods in genetic manipulation. Secondly, shorter fragments of the antibody, instead of entire IgG molecules, are employed due to their reduced size and absence of the Fc region. As disulfide bridges are critical for the complete conformation and binding capacity of antibodies, the molecule needs to fold under oxidizing conditions (Knappik and Brundiers 2009). Even with competition from alternative expression systems, the Gram-negative bacterium *E. coli*, which was the first to produce recombinant biopharmaceuticals, continues to be the leader in the field, accounting for around 40% of all biopharmaceuticals that are currently in the market (Walsh 2010).

The expression of recombinant proteins in bacterial hosts is typically triggered by self-replicating multi-copy plasmids that carry potent promoters, such as the promoter of bacteriophage T7, the lactose operon (lac) in *E. coli*, or the synthetic tryptophan operon (trp), along with a ribosome binding site that permits high gene dosages. While multi-copy plasmids enhance volumetric productivity, their metabolic burden on *E. coli*, causing cell growth inhibition and death, is a major concern. This has led to the development of plasmid-free expression systems, which integrate the desired DNA directly into the *E. coli* chromosome using site-directed techniques. These systems offer a more favorable alternative with reduced metabolic burden (Spadiut et al. 2014). Once triggered to produce antibodies, *E. coli* cells are broken open (achieved by either adding lysozyme or using a mild osmotic shock technique to extract the periplasmic content). After selection, the antibodies can be individually manufactured and refined in larger quantities. Bacterial colonies that secrete soluble antibody fragments are typically purified using a single-step affinity chromatography method that employs peptide tags fused to the C-terminus of the antibody fragment. Similar to traditional monoclonal antibodies, the purified material can be mostly divided into portions and stored frozen (Knappik and Brundiers 2009). The development and application of monoclonal antibodies (MAbs) have altered the treatment of infections, cancers, organ transplants, autoimmune diseases, and cardiovascular, respiratory, and other neurological disorders.

6.3 Regulatory Aspects for r-DNA Products

Comprehensive risk assessment frameworks are implemented to safeguard the integrity of fermented feed and consumables through established regulatory aspects. In this regard, under Regulation (EC) No. 1831/2003, the regulated product(s) and/or producing organism(s) are only permitted if they are considered safe for use by humans, animals, and the environment. Ensuring that the chosen organisms are both nonpathogenic and nontoxigenic is a top priority for maintaining food and feed safety. The majority of well-known fermentation species comply with the above-mentioned criteria. On the other hand, genome sequencing provides crucial insights for newcomers in the fermentation industry. It allows for the accurate taxonomic identification of the strain and the characterization of functional features that may be of concern, such as the production of recognized toxins, virulence factors, and resistance to antibiotics (Lensch et al. 2022).

In 1989, the Indian government established regulations for handling hazardous microorganisms, genetically engineered organisms (GMOs), and cells. These rules, published under Notification No. G.S.R. 1037 (E) by the Ministry of Environment and Forests, fall under the Environment Protection Act of 1986. This formed the basis of the clinical trials and the Drugs and Cosmetics Rules 1988 (eighth Amendment). The Ministry of Environment and Forests and the DBT of the Ministry of Science and Technology, Government of India, are responsible for implementing these regulations, also known as "Rules 1989." In 1999, India's Department of Biotechnology established a clearer path for approving biopharmaceuticals made with recombinant DNA (rDNA) technology. Their guideline, "Generating pre-clinical and clinical data for r-DNA based vaccines, diagnostics, and other biologicals," requires companies to first obtain approval from the Review Committee on Genetic Manipulations (RCGM) for drugs similar to existing ones (similar biological products or SBPs). To make sure that appropriate safety precautions and containment conditions are followed by the Guidelines and Standard Operating Procedures issued by DBT, RCGM keeps an eye on the safety-related aspects of ongoing recombinant DNA projects and activities involving genetically engineered organisms/hazardous microorganisms. Afterward, the data are sent to the Drugs Controller General (India) [DCG (I)] to obtain final commercial licensure and clearance for clinical trials. The Central and State Drug Control Administrations jointly inspect the facility following the DCG's (I) approval of the product, and the Central Licensing Approval Authority issues the production license (Ghagne et al. 2016).

6.4 Market Size of rDNA Products

Recombinant DNA (rDNA) protein manufacturing is the backbone of biotechnology, which is and most likely will continue to be an R&D-intensive sector. Since the early 1970s, tremendous progress has been made in knowing about the biomolecular roots of human disease due to developments in molecular biology and genetic engineering. The rDNA therapeutics sector is a thriving field, with over 110 companies like Johnson & Johnson, Amgen, Novo Nordisk, and Biogen IDEC actively engaged in the commercialization of innovative rDNA products. These organizations possess a robust development pathway exceeding 80 potential therapeutic agents presently in clinical trials. Their goal is to augment their established collection of 73 currently marketed products (Pavlou and Reichert 2004).

The Department of Pharmaceuticals of the Government of India set an ambitious goal for the biopharmaceutical sector in 2010: by 2020, the sector should evolve into a major worldwide manufacturer of cost-effective biotherapeutic drugs. India's pharmaceutical industry shines in 2023, ranking third globally in production volume. They are also a leader in providing affordable generic medicines and vaccines. The bio-therapeutics are now being manufactured by over 20 Indian companies; the leading of these companies are Dr. Reddy's Laboratories, Ranbaxy, Biocon, Shanta Biotech, Reliance Life Sciences, Panacea Biotech, Bharat Biotech, and Intas Biopharmaceuticals. Approximately 50 bio-therapeutics have already been introduced to the Indian market at prices that are 85% less than those of their competitors, making them accessible to the majority of Indians. In the 2009–2010 financial year, India had increases in sales of $22 million for erythropoietin, $11 million for granulocyte-colony stimulating factor (G-CSF or GCSF), $22 million for interferon, and $15 million for streptokinase. Despite the worldwide downturn in 2009 and the gradual recovery in 2010, the Indian biotechnology industry sustained revenues above $150 billion. Throughout the last 10 years, the sector has seen sustained expansion, exceeding 10% growth annually, with average revenue growth rates above 20%. By 2020, its turnover was estimated to reach $450 billion, indicating it has to expand at a rate of almost 30% annually (Ghagne et al. 2016).

6.5 Downstream Processing Challenges for r-DNA Products

Fermentation occurs under controlled conditions in confined systems where microorganisms designed for use as such or as producing organisms are cultured. "Downstream" procedures (DSP) are used to recover the product after this "upstream" procedure. The needed product purity, which varies based on the application and authorization, which determines the complexity of DSP. The following processes are typically included in DSP: biomass separation, product recovery/purification, and formulation (Lensch et al. 2022). However, there are a number of issues with DSP that can seriously affect the cost and productivity of rDNA production.

Fueled by surging consumer demand, companies are ramping up production of rDNA products. However, achieving large-scale rDNA production efficiently is no easy feat. Here are some key challenges they face: (Palomares et al. 2004)

1. Loss of expression: The target gene must produce enough proteins to have a high output. Several factors can silence the recombinant gene, including mutations within the gene itself or its disappearance from the target cell. Maintaining the ideal temperature, increasing the number of insert copies, and the toxicity of the expressed protein to the host are possible additional factors influencing the yield of cloning. The issue of expression stability may be resolved by chromosomal integration of the foreign gene; however, in extrachromosomal element systems, excessive recombinant genes result in a higher yield.
2. Post-translation processing: Glycosylation—Complex post-translation modification (PTM), a multi-step enzymatic process; how well it holds its shape, interacts with water, interacts with the immune system, folds correctly, gets to the right place in the cell, works properly, and stays active. To produce recombinant proteins, scientists often engineer various organisms like *E. coli*, yeast, or insect (SF9), mice (SP2/0), and hamster cells; nonetheless, it is difficult to produce entirely human PTMs. Commonly utilized rodent-derived mammalian cell lines may carry out glycosylation similar to that of humans. Although recombinant proteins closely resemble their natural counterparts, they may lack certain sugar molecules present in humans (like 2,6-linked sialic acid). Conversely, they might contain additional sugars not found in humans (like terminal galactose-galactose or terminal sialic acids). These structural differences can trigger the immune system, potentially leading to unwanted immune reactions. Since therapeutic proteins often lack the correct sugar structures found in humans, researchers are increasingly using human cell lines and even engineering new ones. This allows us to produce fully glycosylated proteins, potentially leading to more effective treatments.
3. In the prokaryotic system, overexpression of a therapeutic protein may lead to inclusion body formation. Aggregation is more likely when large proteins are expressed and when intracellular proteins aggregate rapidly. Proteins that have accumulated are protected from proteolysis and may be easier to recover. The

presence of clumping proteins together in cells tends to protect the host when the produced protein is toxic to it.

6.6 Patents and Clinical Trials Relevant to the Microbiological Production of rDNA Products

Bringing new therapies, like recombinant proteins (rDNAs), to market involves a rigorous three-phase clinical development process. Each step gets tougher, takes longer, and costs more than the last one. Phase I involves the initial testing of new therapies on humans. Phase I trials prioritize safety, assessing a product's initial tolerability, and basic pharmacokinetics/pharmacodynamics at various doses. These data guide the progression to Phase II, which focuses on efficacy and further safety evaluation in patients. The crucial phase III, which offers exact assessments of the safety of the product and its efficacy in a specified patient population, is designed using the data from these pilot studies. The phase III trial data indicate the precise indication that the FDA or another regulatory body will take into consideration for approval (Spilker 1991).

The clinical development and regulatory review data for 271 rDNA medicines that entered clinical investigation between 1980 and 2002 were gathered by the Tufts Centre for the Investigation of Drug Development from company surveys and publicly available records. rDNA products created as "therapeutic vaccines," therapeutic monoclonal antibodies, and modifications to current products (such as new dosages or reformulations) were not included. Fifty-nine American corporations financed 271 of the products, 222 of which were created there or entered clinical trials. Ten multinational companies obtained FDA marketing approval for one or more therapeutic rDNA products; these companies were the source of the remaining 49 items (Reichert and Paquette 2003a, b).

The U.S. Patent Office published the first major patent in the emerging field of biotechnology on December 2, 1980. This innovation was one of three patents that came to be known as the Cohen-Boyer recombinant DNA cloning patents. It stemmed from a key molecular biology procedure pioneered in 1973–1974 by Stanley N of Stanford and Herbert W. Boyer of Cohen and University of California, San Francisco (UCSF). This procedure, later termed recombinant DNA technology, revolutionized the field (Hughes 2001).

6.7 Future Aspects of rDNA Technology

In the manipulation of target gene expression, rDNA technology currently makes it possible to introduce desired traits that can be applied to improve present technologies. Therapeutic proteins are manufactured in enormous quantities using rDNA technology and have a high degree of stability and consistency. Multidisciplinary uses of this technology could address significant aspects of life, such as boosting health and food resources (Khan et al. 2016). The global rDNA technology market is poised for explosive growth, with analysts estimating it will hit US$145.5 billion by 2022. This momentum is expected to continue, with the market projected to climb at a robust 7.9% annually between 2022 and 2023. Major successes, along with the wide variety of uses of rDNA technology to cure serious human illnesses like cancer, diabetes, and infectious diseases, are anticipated to be the main drivers of market development (Rigi et al. 2021). The development of long-acting therapeutic proteins for drug development and gene therapy, which is known as novel molecular medicine, has been made feasible by advancements in recombinant technologies. This has positively impacted the treatment of genetic diseases and supported the growth of the rDNA technology market. Additionally, scientists now have a powerful tool to precisely alter the DNA of microbes, plants, mice, dogs, and human cells, which is possible through a novel gene editing technique known as Clustered Regularly Interspaced Short Palindromic Repeats or CRISPR. This tool can be used to target the eradication of abnormal genes in human cells (Rigi et al. 2021). The World Health Organization (WHO) issued a public health emergency on January 30, 2020, in response to the COVID-19 outbreak that began in December 2019 and has since spread to over 100 nations worldwide. On the other hand because of the innovative recombinant technologies employed in the development of the COVID-19 vaccine, the COVID-19 pandemic ended up having a favorable effect on the international sector for rDNA technology (Yadav et al. 2022).

Microbial biotechnology has advanced tremendously, and within 4 years after the discovery of rDNA technology, human insulin and human growth hormone were produced and are available on the market through the genetic engineering of bacteria. The process of producing vaccines is another fascinating one, like the hepatitis B vaccine, which was produced in yeast by using virus surface antigen (Maurya et al. 2021).

6.8 Conclusion

Recombinant DNA technology is a historic innovation in science that has greatly improved human welfare. Nowadays, microbial drugs that are produced by rDNA technology are an internationally recognized area of medicine that allows for the therapy of numerous chronic disorders. In this chapter, we discussed the expression systems, manufacture, purification, analysis, and medicinal applications of hormones of therapeutic interest, therapeutic enzymes, thrombolytic agents, blood coagulation products, interleukins, anticoagulants, interferons, and hemopoietic growth factors that have been transformed by rDNA technology. There are several benefits of microbial rDNA products over conventional methods, namely adaptability, reproducibility, purity, safety, cost-effectiveness, and high yields of products. A variety of these microbial products focus on treatments that have positive effects on many different diseases, and they will probably continue to achieve this in the future.

References

Adivitiya, Khasa YP (2017) The evolution of recombinant thrombolytics: current status and future directions. Bioengineered 8:331–358. https://doi.org/10.1080/21655979.2016.1229718

Alquwaizani M, Buckley L, Adams C, Fanikos J (2013) Anticoagulants: a review of the pharmacology, dosing, and complications. Curr Emerg Hosp Med Rep 1:83–97. https://doi.org/10.1007/s40138-013-0014-6

Alyas J, Rafiq A, Amir H, Khan SU, Sultana T, Ali A, Hameed A, Ahmad I, Kazmi A, Sajid T, Ahmad A (2021) Human insulin: history, recent advances, and expression systems for mass production. Biomed Res Ther 8:4540–4561. https://doi.org/10.15419/bmrat.v8i9.692

Assiri AS, El-Gamal BA, Hafez EE, Haidara MA (2014) Production of recombinant streptokinase from *Streptococcus pyogenes* isolate and its potential for thrombolytic therapy. Saudi Med J 35:1482–1488

Banerjee A, Chisti Y, Banerjee UC (2004) Streptokinase—a clinically useful thrombolytic agent. Biotechnol Adv 22:287–307. https://doi.org/10.1016/j.biotechadv.2003.09.004

Bashar S, Jeong HJ (2023) Bacterial production of recombinant coagulation factor VIII domains. Medicina (Kaunas) 59:694. https://doi.org/10.3390/medicina59040694

Bayer V (2019) An overview of monoclonal antibodies. Semin Oncol Nurs 35:150927. https://doi.org/10.1016/j.soncn.2019.08.006

Bhopale GM, Nanda RK (2005) Recombinant DNA expression products for human therapeutic use. Curr Sci 89:614–622

Brocker C, Thompson D, Matsumoto A, Nebert DW, Vasiliou V (2010) Evolutionary divergence and functions of the human interleukin (IL) gene family. Hum Genomics 5:1–26. https://doi.org/10.1186/1479-7364-5-1-30

Choi ES, Sohn JH, Rhee SK (1994) Optimization of the expression system using galactose-inducible promoter for the production of anticoagulant hirudin in Saccharomyces cerevisiae. Appl Microbiol Biotechnol 42:587–594. https://doi.org/10.1007/bf00173925

Choi JH, Keum KC, Lee SY (2006) Production of recombinant proteins by high cell density culture of *Escherichia coli*. Chem Eng Sci 61:876–885. https://doi.org/10.1016/j.ces.2005.03.031

Commins SP, Borish L, Steinke JW (2010) Immunologic messenger molecules: cytokines, interferons, and chemokines. J Allergy Clin Immunol 125:53–72. https://doi.org/10.1016/j.jaci.2009.07.008

Devos R, Plaetinck G, Cheroutre H, Simons G, Degrave W, Tavernier J, Remaut E, Fiers W (1983) Molecular cloning of human interleukin 2 cDNA and its expression in *E. coli*. Nucleic Acids Res 11:4307–4323. https://doi.org/10.1093/nar/11.13.4307

Dunn CJ (1996) Epoetin beta: a review of its pharmacological properties and clinical use in the management of anaemia associated with chronic renal failure. Drugs 51:299–318. https://doi.org/10.2165/00003495-199651020-00008

El-Dabaa E, Okasha H, Samir S, Nasr SM, El-Kalamawy HA, Saber MA (2022) Optimization of high expression and purification of recombinant streptokinase and in vitro evaluation of its thrombolytic activity. Arab J Chem 15:103799. https://doi.org/10.1016/j.arabjc.2022.103799

Gabrilove J (2000) Overview: erythropoiesis, anemia, and the impact of erythropoietin. Semin Hematol 37:1–3. https://doi.org/10.1016/s0037-1963(00)90060-x

Ghagne S, Nerli R, Baligar P, Hiremath M (2016) Status and application of Recombinant DNA therapeutic products used in clinical practices: an Indian scenario. Available at SSRN 2783940. https://doi.org/10.2139/ssrn.2783940

Ghavim M, Abnous K, Arasteh F, Taghavi S, Nabavinia MS, Alibolandi M, Ramezani M (2017) High level expression of recombinant human growth hormone in *Escherichia coli*: crucial role of translation initiation region. Res Pharm Sci 12:168–175. https://doi.org/10.4103/1735-5362.202462

Gowda AP, Schaefer AD, Schuck TK (2023) The impact of EDTA and selenite on the stability of insulin in cell culture media. Int J Biochem Adv 1:31–51

Greinacher A (2004) Lepirudin: a bivalent direct thrombin inhibitor for anticoagulation therapy. Expert Rev Cardiovasc Ther 2:339–357. https://doi.org/10.1586/14779072.2.3.339

Gupta V, Sengupta M, Prakash J, Tripathy BC, Gupta V, Sengupta M, Prakash J, Tripathy BC (2017) Production of recombinant pharmaceutical proteins. In: Basic applied aspects of biotechnology, pp 77–101. https://doi.org/10.1007/978-981-10-0875-7_4

Halfmann Y, Brailly H, Bernadac A, Montero-Julian FA, Lazdunski C, Baty D (1993) Targeting of interleukin-2 to the periplasm of *Escherichia coli*. J Gen Microbiol 139:2465–2473. https://doi.org/10.1099/00221287-139-10-2465

Hardin DS, Kemp SF, Allen DB (2007) Twenty years of recombinant human growth hormone in children: relevance to pediatric care providers. Clin Pediatr (Phila) 46:279–286. https://doi.org/10.1177/0009922806293924

Hughes SS (2001) Making dollars out of DNA: the first major patent in biotechnology and the commercialization of molecular biology 1974-1980. Isis 92:541–575. https://doi.org/10.1086/385281

Isaacs A, Lindenmann J, Valentine RC (1957) Virus interference. II. Some properties of interferon. Proc R Soc Lond B Biol Sci 147:268–273. https://doi.org/10.1098/rspb.1957.0049

Khalilzadeh R, Shojaosadati SA, Maghsoudi N, Mohammadian-Mosaabadi J, Mohammadi MR, Bahrami A, Maleksabet N, Nassiri-Khalilli MA, Ebrahimi M, Naderimanesh H (2004) Process development for production of recombinant human interferon-γ expressed in *Escherichia coli*. J Ind Microbiol Biotechnol 31:63–69. https://doi.org/10.1007/s10295-004-0117-x

Khan S, Ullah MW, Siddique R, Nabi G, Manan S, Yousaf M, Hou H (2016) Role of recombinant DNA technology to improve life. Int J Genomics 2016:2405954. https://doi.org/10.1155/2016/2405954

Knappik A, Brundiers R, (2009) Recombinant antibody expression and purification. In: The protein protocols handbook. pp 1929–1943. ISBN 978-1-60327-474-6. https://doi.org/10.1007/978-1-59745-198-7_203

Knuver-Hopf J, Mohr H (1995) Differences between natural and recombinant interleukin-2 revealed by gel electrophoresis and capillary electrophoresis. J Chromatogr A 717:71–74. https://doi.org/10.1016/0021-9673(95)00557-7

Kozak M, Jurga S (2002) A comparison between the crystal and solution structures of *Escherichia coli* asparaginase II. Acta Biochim Pol 49:509–513

Krishnan R, Chong CS, Goh KM, Wahab FA, Jamaluddin H (2015) Expression and enzymatic properties of a unique recombinant anticoagulant and fibrinolytic enzyme from *Acinetobacter baumannii* TU04. Int J 7:88–94

Kunamneni A, Abdelghani TTA, Ellaiah P (2007) Streptokinase-the drug of choice for thrombolytic therapy. J Thromb Thrombolysis 23:9–23. https://doi.org/10.1007/s11239-006-9011-x

Kunamneni A, Ravuri BD, Saisha V, Ellaiah P, Prabhakhar T (2008) Urokinase-a very popular cardiovascular agent. Recent Pat Cardiovasc Drug Discov 3:45–58. https://doi.org/10.2174/157489008783331670

Lensch A, Duwenig E, Dederer HG, Karenlampi SO, Custers R, Borg A, Wyss M (2022) Recombinant DNA in fermentation products is of no regulatory relevance. Food Control 141:109170. https://doi.org/10.1016/j.foodcont.2022.109170

Leonova EI, Gainetdinov RR (2020) CRISPR/Cas9 technology in translational biomedicine. Cell Physiol Biochem 54:354–370. https://doi.org/10.33594/000000224

Mahajan RV, Kumar V, Rajendran V, Saran S, Ghosh PC, Saxena RK (2014) Purification and characterization of a novel and robust L-asparaginase having low-glutaminase activity from *Bacillus licheniformis*: *in vitro* evaluation of anti-cancerous properties. PLoS One 9:e99037. https://doi.org/10.1371/journal.pone.0099037

Markham A, Bryson HM (1995) Epoetin alfa: a review of its pharmacodynamic and pharmacokinetic properties and therapeutic use in nonrenal applications. Drugs 49:232–254. https://doi.org/10.2165/00003495-199549020-00008

Mathers CD, Loncar D (2006) Projections of global mortality and burden of disease from 2002 to 2030. PLoS Med 3:442–450. https://doi.org/10.1371/journal.pmed.0030442

Matheson AJ, Goa KL (2000) Desirudin: a review of its use in the management of thrombotic disorders. Drugs 60:679–700. https://doi.org/10.2165/00003495-200060030-00012

Maurya DK, Kumar A, Chaurasiya U, Hussain T, Singh SK (2021) Modern era of microbial biotechnology: opportunities and future prospects. In: Microbiomes and plant health. pp 317–343

Nawaz A, ul Haq I, Amin H, Shah Z, Javed A, Mukhtar H, Akram F (2022) Production and market comparison of urokinase and streptokinase as effective and cheap fibrinolytic agents for treatment of cardiovascular diseases. Acta Scientiarum Technol 44:e56088–e56088

Norris LA (2003) Blood coagulation. Best Pract Res Clin Obstet Gynaecol 17:369–383. https://doi.org/10.1016/s1521-6934(03)00014-2

Palomares LA, Mondaca SE, Ramirez OT (2004) Production of recombinant proteins. In: Balbas P, Lorence A (eds) From methods in molecular biology, vol 267. Humana Press, Totowa, pp 15–51

Paramasivam R, Murugananthan Y, Mohammed E, Mohd Fauzi NA (2022) Development of human growth hormones. pp 75–88. http://eprints.uthm.edu.my/id/eprint/7526

Pavlou AK, Reichert JM (2004) Recombinant protein therapeutics-success rates, market trends and values to 2010. Nat Biotechnol 22:1513–1519. https://doi.org/10.1038/nbt1204-1513

Prajapat R, Jain S (2022) Advancement in medical biotechnology: a review. Int J Med Rev 9:217–226. https://doi.org/10.30491/IJMR.2021.279672.1197

Pratap J, Rajamohan G, Dikshit KL (2000) Characteristics of glycosylated streptokinase secreted from Pichia pastoris: enhanced resistance of SK to proteolysis by glycosylation. Appl Microbiol Biotechnol 53:469–475. https://doi.org/10.1007/s002530051643

Rai K, Chu X, Bao Z, Liang Y, Wang X, Yang J, Xian M, Sun Y, Nian R (2020) Enhanced anticoagulant activity of hirudin-i analogue co-expressed with arylsulfotransferase in periplasm of *E. coli* BL21 (DE3). J Biotechnol 323:107–112. https://doi.org/10.1016/j.jbiotec.2020.08.003

Rajabi Memari H, Ramanan RN, Ariff AB (2010) Comparison of expression systems for the production of human interferon-α2b. Cent Eur J Biol 5:446–455. https://doi.org/10.2478/s11535-010-0036-y

Ranke MB (2007) New preparations comprising recombinant human growth hormone: deliberations on the issue of biosimilars. Horm Res 69:22–28. https://doi.org/10.1159/000111791

Reichert JM, Paquette C (2003a) Clinical development of therapeutic recombinant proteins. BioTechniques 35:176–185

Reichert JM, Paquette C (2003b) Therapeutic recombinant proteins: trends in U.S. approvals 1982 to 2002. Curr Opin Mol Ther 5:139–147

Rigi G, Rostami A, Ghomi H, Ahmadian G, Mirbagheri VS, Jeiranikhameneh M, Vahed M, Rahimi S (2021) Optimization of expression, purification and secretion of functional recombinant human growth hormone in *Escherichia coli* using modified staphylococcal protein a signal peptide. BMC Biotechnol 21:1–18. https://doi.org/10.1186/s12896-021-00701-x

Schulman S, Bijsterveld NR (2007) Anticoagulants and their reversal. Transfus Med Rev 21:37–48. https://doi.org/10.1016/j.tmrv.2006.08.002

Sengupta P, Meena K, Mukherjee R, Jain SK, Maithal K (2008) Optimized conditions for high-level expression and purification of recombinant human interleukin-2 in *E. coli*. Indian J Biochem Biophys 45:91–97

Sikole A, Spasovski G, Zafirov D, Polenakovic M (2002) Epoetin omega for treatment of anemia in maintenance hemodialysis patients. Clin Nephrol 57:237–245

Singh S, Tank NK, Dwiwedi P, Charan J, Kaur R, Sidhu P, Chugh VK (2018) Monoclonal antibodies: a review. Curr Clin Pharmacol 13:85–99. https://doi.org/10.2174/15748847126661708091247 28

Smith SA, Travers RJ, Morrissey JH (2015) How it all starts: initiation of the clotting cascade. Crit Rev Biochem Mol Biol 50:326–336. https://doi.org/10.3109/10409238.2015.1050550

Song H, Jiang J, Wang X, Zhang J (2017) High purity recombinant human growth hormone (rhGh) expression in *Escherichia coli* under phoA promoter. Bioengineered 8:147–153. https://doi.org/10.1080/21655979.2016.1212137

Spadiut O, Capone S, Krainer F, Glieder A, Herwig C (2014) Microbials for the production of monoclonal antibodies and antibody fragments. Trends Biotechnol 32:54–60. https://doi.org/10.1016/j.tibtech.2013.10.002

Spilker B (1991) Establishing clinical trial objectives. In: Guide to clinical trials. Raven Press, New York, pp 10–14

Stryjewska A, Kiepura K, Librowski T, Lochyński S (2013) Biotechnology and genetic engineering in the new drug development. Part I. DNA technology and recombinant proteins. Pharmacol Rep 65:1075–1085. https://doi.org/10.1016/s1734-1140(13)71466-x

Takeyama M, Nogami K, Sasai K, Shima M (2018) Contribution of factor VIII A3 domain residues 1793-1795 to a factor IXa-interactive site. Blood 132:1173. https://doi.org/10.1182/blood-2018-99-113260

Tandon S, Sharma A, Singh S, Sharma S, Sarma SJ (2021) Therapeutic enzymes: discoveries, production and applications. J Drug Deliv Sci Technol 63:102455. https://doi.org/10.1016/j.jddst.2021.102455

Tang W, Sun ZY, Pannell R, Gurewich V, Liu JN (1997) An efficient system for production of recombinant urokinase-type plasminogen activator. Protein Expr Purif 11:279–283. https://doi.org/10.1006/prep.1997.0800

Taylor MW, Taylor MW (2014) Interferons. In: Viruses and man: a history of interactions, pp 101–119. https://doi.org/10.1007/978-3-319-07758-1

US Food and Drug Administration (2017) US food and drug administration home page. http://www.fda.Gov/

Vieira Gomes AM, Souza Carmo T, Silva Carvalho L, Mendonça Bahia F, Parachin NS (2018) Comparison of yeasts as hosts for recombinant protein production. Microorganisms 6:38. https://doi.org/10.3390/microorganisms6020038

Vimal A, Kumar A (2019) Transforming the healthcare system through therapeutic enzymes. In: Enzymes in food biotechnology, pp 603–625. https://doi.org/10.1016/B978-0-12-813280-7.00035-9

Vose JM, Armitage JO (1995) Clinical applications of hematopoietic growth factors. J Clin Oncol 13:1023–1035. https://doi.org/10.1200/jco.1995.13.4.1023

Walsh G (2005) Therapeutic insulins and their large-scale manufacture. Appl Microbiol Biotechnol 67:151–159. https://doi.org/10.1007/s00253-004-1809-x

Walsh G (2010) Biopharmaceutical benchmarks 2010. Nat Biotechnol 28:917–924. https://doi.org/10.1038/nbt0910-917

Wu XC, Ye R, Duan Y, Wong SL (1998) Engineering of plasmin-resistant forms of streptokinase and their production in *Bacillus subtilis*: streptokinase with longer functional half-life. Appl Environ Microbiol 64:824–829. https://doi.org/10.1128/aem.64.3.824-829.1998

Yadav R, Tiwari SP, Galgano J (2022) Recent challenges in COVID 19 vaccines and its future advancements. IP Int J Comprehensive Adv Pharmacol 7:1–7

Yu HY, Gao DM, Zhou W, Xia BB, He ZY, Wu B, Jiang MZ, Wang ML, Zhao J (2021) Expression, purification, and bioactivity of a soluble recombinant ovine interferon-tau in. J Vet Res 65:101–108. https://doi.org/10.2478/jvetres-2021-0011

Zhang J, Lan N (2018) Hirudin variants production by genetic engineered microbial factory. Biotechnol Genet Eng Rev 34:261–280. https://doi.org/10.1080/02648725.2018.1506898
Zielinski M, Romanik-Chruścielewska A, Mikiewicz D, Lukasiewicz N, Sokolowska I, Antosik J, Sobolewska-Ruta A, Bierczynska-Krzysik A, Zaleski P, Plucienniczak A (2019) Expression and purification of recombinant human insulin from *E. coli* 20 strain. Protein Expr Purif 157:63–69. https://doi.org/10.1016/j.pep.2019.02.002

Chapter 7
Microbes in Production of Non-alcoholic Beverages

Deevena Jemima and Sheila John

Abstract Fermented foods are garnering much attention in today's world driven by numerous health claims. While many alcohol containing beverages such as different types of beers and wines are studied extensively, less research is available on non-alcoholic fermented beverages prepared from cereals, fruits, vegetables, and dairy. The microorganisms that take part in the production of these beverages are LAB (lactic acid bacteria), moulds, yeasts, and AAB (acetic acid bacteria). Cereal fermentation is a long-recognized technique used for creating refreshing beverages, weaning gruels, and porridge in numerous areas of Africa, Europe, and Asia. Milk which undergoes fermentation, occupies a vital function in the diets of numerous inhabitants around the world, especially in Eastern Mediterranean countries. Lactobacilli have been observed to be the dominant bacteria in many of the products. Fermented non-dairy beverages made of vegetables and fruits are preferred for many reasons these days such as lactose intolerance, food trends such as vegan diets, and awareness about the health benefits of fermented foods. Non-alcoholic fermented beverages also contribute to health as they contain probiotics, antioxidant properties, and a high content of vitamins and minerals. Many of these fermented beverages have a good shelf life, thus benefiting consumers who do not have access to many storage facilities. This chapter focuses on the manufacture of non-alcoholic drinks and the organisms responsible for their fermentation.

Keywords Non-alcoholic · Fermented beverages · LAB · Fermentation · Yeast · Probiotics

D. Jemima (✉) · S. John
Department of Home Science, Women's Christian College, Chennai, Tamil Nadu, India

V. Kothari et al. (eds.), *Microbial Products for Health and Nutrition*,
https://doi.org/10.1007/978-981-97-4235-6_7

7.1 Introduction

Throughout history, mankind has used fermentation to prolong storage life, improve the taste, and nutritional value of food. Many countries have an array of fermented food and liquid refreshments as a part of their culture and culinary tradition (Pérez-Armendáriz and Cardoso-Ugarte 2020). The phrase 'fermented foods' is usually used to depict various types of food whose individual components such as carbohydrates, proteins, fats, and other compounds are broken down causing desirable biochemical changes and undergoing a series of complex modifications in the presence of a variety of microorganisms and enzymes. Certain end products formed as a result of food fermentation such as acids and alcohols inhibit the action of pathogens that may contaminate food. Fermentation of food has many known advantages such as increased shelf life, improved nutritional properties, enhanced organoleptic assets, the removal of anti-nutrients, and reduction of cooking time. The organisms used predominantly in fermenting cereal, dairy, vegetable, fruit, and meat products are coryneforms, LAB, acetic acid bacteria, yeasts, and moulds. Presently, the phrase 'non-alcoholic' is subject to regulations and the rules concerning it differ across countries. For instance, no 1169/2011 of the EU Regulations states that, if a beverage contains an alcoholic potency of above 1.2%, it must be categorized as an alcoholic beverage (Patra et al. 2023).

Fermented beverages are prepared using many ingredients such as dairy, fruits and vegetables, herbs, and leaves. Fermentation of such liquids is affected by many factors such as time taken for fermenting, temperature, pH, moisture content, along with others. Most of these fermented beverages are affordable and have health benefits due to the presence of beneficial microorganisms.

7.2 Cereal Based Non-alcoholic Beverages

Cereals constitute a large section of human food and animal feed with many traditional cereal-based foods being passed down through generations. A wide range of cereals are consumed either singly or processed in combinations to produce germinated and fermented doughs and beverages (Adinsi et al. 2014). Fermentation of cereal leads to an overall enhancement in taste, texture, aroma, digestibility, and storage quality of the food. It is a deep-rooted style utilized for producing edibles such as beverages, gruels, and porridge in many countries in Asia, Africa, and Europe. Cereal fermentation is mainly caused due to the action of microbial species such as *Weissella*, *Leuconostoc*, *Enterococcus*, *Lactococcus*, *Pediococcus*, *Lactobacillus*, *and Streptococcus*. The species of yeast in cereal fermentation include native strains of *Saccharomyces cerevisiae* and others such as *Saccharomycopsis*, *Candida*, *Yarrowia*, *Pichia*, *Hansenula*, *Debaryomyces*, *Kazachstania*, *Trichosporon*, *and Torulopsis*. Other moulds used for fermenting cereals are *Aspergillus* spp. and *Monascus* spp. (Tamang et al. 2016). A summary of microorganisms present in cereal-based non-alcoholic beverages is shown in Table 7.1.

Table 7.1 Microorganisms present in cereal-based non-alcoholic beverages

Name of the beverage	Microbes responsible for fermentation	References
Borş	*Lactobacillus amylolyticus*, *Lactobacillus plantarum*, *Lactobacillus brevis*, *Lactobacillus buchneri*, *Lactobacillus fermentum*, *Lactobacillus casei*, *Lactobacillus oris*, *Enterococcus durans*	Constantin et al. (2023), Zamfir et al. (2022)
Boza	*Lactobacillus acidophilus*, *Leuconostoc paramesenteroides*, *L. mesenteroides*, *L. sanfrancisco*, *Lactobacillus plantarum*, *Lactobacillus pentosus*, *Lactobacillus fermentum*, *Lactobacillus rhamnosus*, *Lactococcus lactis* subsp. *lactis*, *Leuconostoc lactis* and *Pediococcus pentosaceus*, *L. confusus*, *Saccharomyces uvarum*, *Candida* spp., *Saccharomyces*, *Trichosporan*, *Torulospora*, *Rhodotorula*, *Pichia fermentans*	Altay et al. (2013), Hancioğlu and Karapinar (1997), Zorba et al. (2003)
Bushera	*Lactobacillus brevis*, *Lactobacillus fermentum*, *Lactobacillus delbrueckii*, *Lactobacillus plantarum*, *Lactobacillus paracasei*, *Weissella confusa*, *Lactococcus lactis*, *Leuconostoc mesenteroides* subsp. *dextranicun*, *Leuconostoc mesenteroides* subsp. *Mesenteroides*	Muyanja et al. (2003, 2012)
Gowe	*Lactobacillus mucosae*, *Lactobacillus fermentum*, *Kluyveromyces marxianus*, *Pediococcus acetolactic*, *Pediococcus pentosaceus*, *Weissella kimchii*, *Pichia anomala*, *Weissella confusa*, *Clavispora lusitaniae*, *Candida krusei*, *Candida tropicalis*	Greppi et al. (2013), Vieira-Dalodé et al. (2007)
Koji amazake	*Aspergillus oryzae*, *Aspergillus kawachii*, *L. sakei* UONUMA, *Aspergillus luchuensis* MEM-C	Kurahashi (2021), Saigusa and Ohba (2007), Penna et al. (2023), Yamamoto et al. (2011)
Koozh	*Lactobacillus fermentum*, *Weissella paramesenteroides*, *Lactobacillus plantarum*	Ilango and Antony (2014)
Kunun-zaki	*Saccharomyces cerevisiae*, *Lactobacillus lactis*, *Lactobacillus acidophilus*, *Lactobacillus casei*, *Lactobacillus fermentum*, *Lactobacillus delbruecki*, *Lactobacillus plantarum*, *Lactobacillus brevis*	Efiuvwevwere and Akona (1995), Victor-Aduloju et al. (2018)
Kvass	*Saccharomyces cerevisiae*, *Pediococcus pentosaceus*, *Lactobacillus sakei*, *Lactobacillus casei*, *Leuconostoc mesenteroides*	Dlusskaya et al. (2008), Basinskiene et al. (2016)
Mahewu	*Candida glabrata*, *Lactococcus lactis* subsp. *lactis*, *Klebsiella pneumonia*, *Cronobacter sakazakii*, *Saccharomyces cerevisiae*	Bvochora et al. (1999), Pswarayi and Gänzle (2019)

(continued)

Table 7.1 (continued)

Name of the beverage	Microbes responsible for fermentation	References
Munkoyo	*Lactobacillus delbrueckii*, *Saccharomyces cerevisiae*, Streptococcaceae, Leuconostocaceae, Enterobacteriaceae, Lactabacillales, Bacillaceae, Aeromonadaceae	Foma et al. (2012)
Obiolor	*Bacillus* spp., *Lactobacillus* spp., *Bacillus stearothermophilus*, *B. subtilis*, *L. plantarum*, *Lactococcus lactis*	Achi (1990)
Olewonyo	*Lactobacillus fermentum*, *Lactobacillus brevis*, *Lactobacillus plantarum*, *Lactobacillus acidophilus*	Madilo et al. (2022)
Pozol	*S. bovis*, *S. macedonicus*, *L. lactis*, *E. sulfureus*, *Bacillus* sp. strain CS93	Phister et al. (2004), Díaz-Ruiz et al. (2003)
Togwa	*L. plantarum*, *L. brevis*, *Weissella confusa*, *Candida tropicalis*, *L. fermentum*, *L. cellobiosus*, *Kluyveromyces marxianus*, *Pediococcus pentosaceus*, *Issatchenkia orientalis*, *Pichia anomala*, *Pichia norvegensis*, *Pichia burtonii*, *Pichia guilliermondii*, *Hanseniaspora guilliermondii*, *Candida glabrata*, *Saccharomyces cerevisiae*, *Candida pelliculosa*	Hellström et al. (2010), Mugula et al. (2003)

7.2.1 *Borş*

Borş is an acidic liquid used in Romania for preparing a customary soup called *borş* or ciorbă. It is also drunk as an energizing drink. It is also described as 'white *borş*' to differentiate it from other beverages that are red coloured and are termed as borsh or borsch, produced using beetroot extract in Eastern and Central European countries. This beverage is obtained through the spontaneous fermentation of bran that is acquired from wheat processing. It has been used since olden days as a cure for gastric conditions. *Borş* starter is obtained through spontaneous fermentation. The preparation of *borş* differs from area to area in Romania. The bran, corn flour, and water are mixed thoroughly and placed in a loosely secured vessel. This is kept for 2 or 3 days in a cool spot. The mixture is stored at 4 °C after filtration and bottling. The huşte, which is part of the solid sediment, is kept aside to be used as a natural starter for the next batches of *borş*. Other ingredients such as dill, lovage, and leaves of the sour cherry tree are also added for flavouring (Pasqualone et al. 2018). Several beneficial microorganisms have been extracted from *borş* such as *L. amylolyticus, L. buchneri, L. brevis, L. casei, L. fermentum, L. oris, L. plantarum, and Enterococcus durans* (Constantin et al. 2023; Zamfir et al. 2022). During laboratory fermentation, the pH was observed to have dropped from 5.0 to about 4.0 in the first 3 h. LAB were observed in elevated amounts (about 8.0 log cfu/ml) from the beginning of fermentation. Lactic acid bacteria count rose in the course of the initial 6 h of fermentation. The protein, amino acid, and phenolic contents increased during fermentation. *L. brevis* shows great potential as a probiotic additive for dairy products. *Enterococcus durans* was efficient in bioaccumulating selenium and has probiotic aptitude (Pieniz et al. 2014).

7.2.2 *Boza*

Boza is a local Turkish drink prepared by fermenting millets, wheat, rice, or corn through the activity of yeast and LAB. The term '*boza*' is obtained from the term '*buze*' meaning millet in Persian. *Boza* is a viscous drink that has a sour and sweet taste, coloured light yellow, and has a tart odour. Many variations of this drink are prepared around the world, such as, '*braga* or *brascha*', '*busa*', and '*bouza*' in countries of East Europe, Egypt, and the Balkans. Some of these varieties contain a high concentration of alcohol, up to 7%. Two kinds of fermentation were noted, i.e. alcohol fermentation and lactic acid fermentation. The microorganisms isolated from the beverage were LAB and yeast species. After 24 hours of fermentation, the LAB and yeast count showed an increase from 7.6×10^6 to 4.6×10^8 cfu/ml and 2.25×10^5 to 8.10×10^6 cfu/ml, respectively. Among LAB, *Lactobacillus acidophilus, Leuconostoc paramesenteroides, L. mesenteroides, L. sanfrancisco, L. fermentum, Lactobacillus pentosus, Lactobacillus rhamnosus, Lactococcus lactis* subsp. *lactis, L. plantarum, Leuconostoc lactis*, and *P. pentosaceus* were separated from various specimens of the beverage. Yeast isolates that were found were *S. uvarum*, *Candida* spp., *Saccharomyces*, *Trichosporan*, *Torulospora*, and *Rhodotorula*. *Pichia fermentans* was observed to be the most common in Turkish *Boza* (Altay et al. 2013; Hancioğlu and Karapinar 1997). The best culture combination to produce *Boza* was found to be a mixture of *Lactobacillus confusus*, *S. cerevisiae*, and *L. mesenteroides* subsp. *mesenteroides* (Zorba et al. 2003). It was observed that the predominant biogenic amine was tyramine. Boza samples had tyramine levels ranging from 13 to 65 mg/kg. The complete amount of biogenic amines in the beverage samples stretched from 25 to 69 mg/kg. Therefore, the intake of this beverage may pose a health risk to individuals receiving treatment with medications that contain monoamine oxidase inhibitors (MAOI) (Yeğin and Üren 2008). The metallic nanoparticles produced with *Pichia fermentans* JA2 have a good suppressive effect and may be of much use to the drug industry (Chauhan et al. 2015).

7.2.3 Bushera

Bushera is a common beverage consumed in Uganda, where women prepare it for home consumption and sale. It is produced from sorghum or millets that are germinated and pounded into flour. This is blended with boiling water and fermented at room temperature. Sweet *Bushera* is produced by fermenting the mixture for 12–24 h and is usually preferred by women and is used as a weaning food. Sour *Bushera* is fermented for 2–4 days and is usually consumed by men. Samples collected from households showed LAB quantities between 8.4 and 9.0 log cfu/ml. The biggest rise in lactic acid bacteria was observed in the initial 24 h of fermenting the product. Five genera of LAB were identified in *bushera*, namely *Lactococcus*, *Enterococcus*, *Lactobacillus*, *Leuconostoc*, *and Streptococcus*. Strains such as *Lactobacillus paracasei*, *Lactobacillus brevis*, *L. delbrueckii*, *L. fermentum*, and *L. plantarum* were observed to live predominantly in household fermented *bushera*. Eight isolates such as *Lactococcus lactis*, *Weissella confusa*, *Leuconostoc mesenteroides* subsp. *mesenteroides*, *Lactobacillus plantarum*, and *Leuconostoc mesenteroides* subsp. *dextranicun* could produce acid from starch. The major acid varying between 0.34% and 0.66% was lactate. Organic acids including pyruvic, citric, succinic, acetic, lactic, and pyroglutamic acids were identified in household *bushera*. The chief volatile organic compound in *bushera* was ethanol (Muyanja et al. 2003, 2012).

7.2.4 Gowe

Gowe is traditionally produced by malting cereal flour, which is then fermented and cooked, producing a pleasant, sweet, and sour paste. This is drunk as a refreshing brew after diluting with water, and then adding sugar, milk, or ice as desired. *Gowe* varieties are different based on the processing techniques and the grains used for fermenting it. The cereals used are sorghum (*Sorghum bicolor*), maize (*Zea mays*), or a combination of both sorghum and maize, the preparation steps of which are shown in Fig. 7.1. It is produced and drunk commonly in the cities of Benin (Adinsi et al. 2014). The beverage production is indicated by a mixed fermentation of dominant LAB and yeast species such as *Lactobacillus mucosae*, *Lactobacillus fermentum*, *Pediococcus acidilactici*, *Weissella confusa*, *Pediococcus pentosaceus*, *Candida krusei*, *Candida tropicalis*, *Weissella kimchii*, *Kluyveromyces marxianus*,

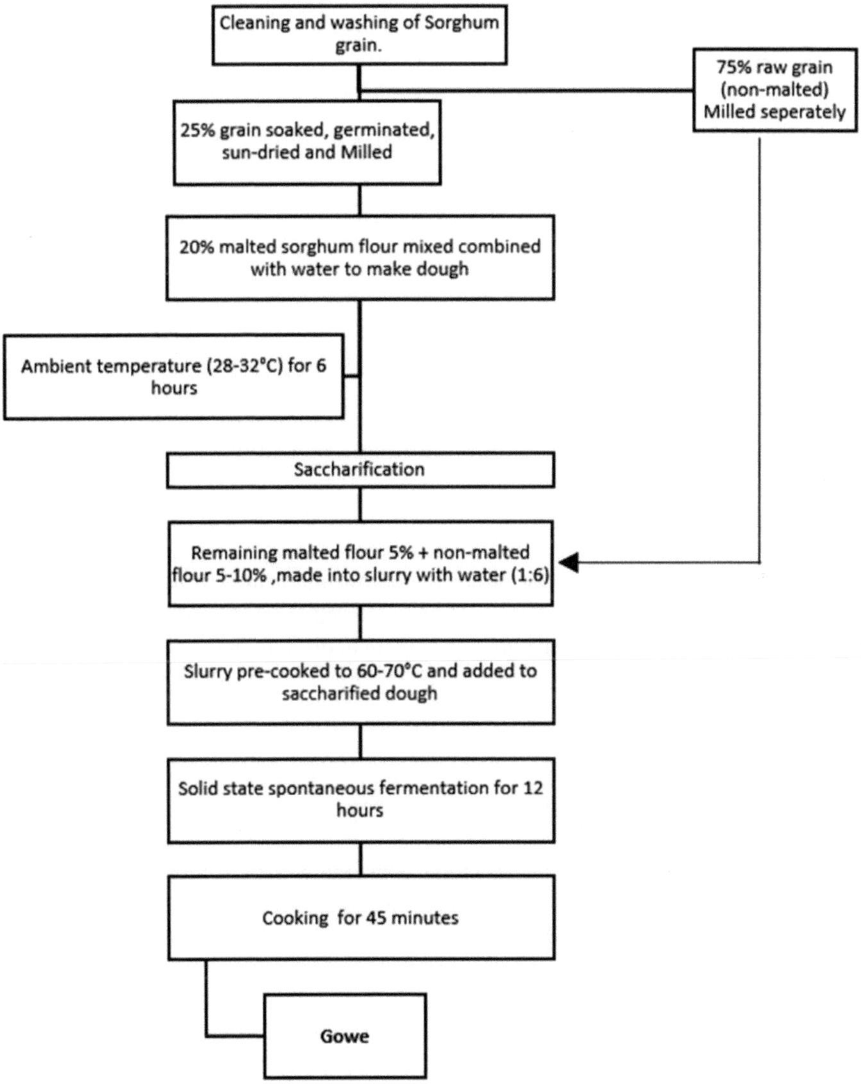

Fig. 7.1 Preparation of gowe

Clavispora lusitaniae, and *Pichia anomala*. The chief hazards of gowe prepared at homes are potential mycotoxin contamination and the occurrence of cyanogenic components that may occur due to the contamination of raw materials and the existence of *E. coli* and Enterobacteriaceae. When stored at room temperature after processing, the beverage often goes bad due to the growth of moulds and a rise in acidity within 2–3 days (Vieira-Dalodé et al. 2007; Adinsi et al. 2017). *Pediococcus acidilactici* has been observed to have antioxidant and anti-cancer properties (Kumar et al. 2020).

7.2.5 Koji Amazake

Koji Amazake is a Japanese conventional non-alcoholic drink which has a sweet taste and appears white in colour. It is made with rice and *Aspergillus oryzae* products. Yellow koji and white koji are *A. oryzae* and *A. kawachii*. Water is mixed with rice koji and kept in a container at 50–60 °C. This causes *A. oryzae* to secrete amylase to break down starch into molecules of glucose. It is a distinctive formulation because its chief sugar is glucose as contrasted to fruit juice where the primary components are fructose and sucrose (Kurahashi 2021; Saigusa and Ohba 2007). Oguro et al. (2017) detected organic acids, amino acids, sugars, and Vitamin B complex in lactic acid fermented amazake and koji amazake. In lactic acid fermented amazake, lactic acid, Vitamin B2, B3, and B6 were considerably improved compared to *Koji amazake*, however, the glutamine and malate content decreased. *L. sakei* UONUMA species made these vitamins. The antioxidant activity of brown rice amazake was higher compared to white rice and sake cake amazake (Yamamoto et al. 2011). Black *Koji amazake*, obtained from *Aspergillus luchuensis* MEM-C strain, showed a greater quantity of citric acid, ferulic acid, and isomaltooligosaccharides compared to yellow *koji amazake*, derived from *Aspergillus oryzae* and showed a protective effect on the gut in reducing endotoxin intrusion and inflammation (Penna et al. 2023). The preparation of *koji amazake* is shown in Fig. 7.2.

7.2.6 Koozh

Ethnic populations in Tamil Nadu, India, consume *Koozh*, a fermented beverage prepared with rice and millet flour. It is either made from finger millet (*Eleusine coracana*) or pearl millet (*Pennisetum glaucum*) and flour made from broken rice. Preparation of *Koozh* is unique in the way that it involves two fermentation stages, before and after cooking. The primary stage is done for 12–15 h and forms the basis for the *koozh's* quality and sensory properties. On day one, flour obtained from millets and water is made into a loose paste by hand mixing. This mixture is fermented overnight, around 12–15 h. The next day, a large quantity of water is used to cook broken rice, added to the previously fermented mixture and heated till it forms a

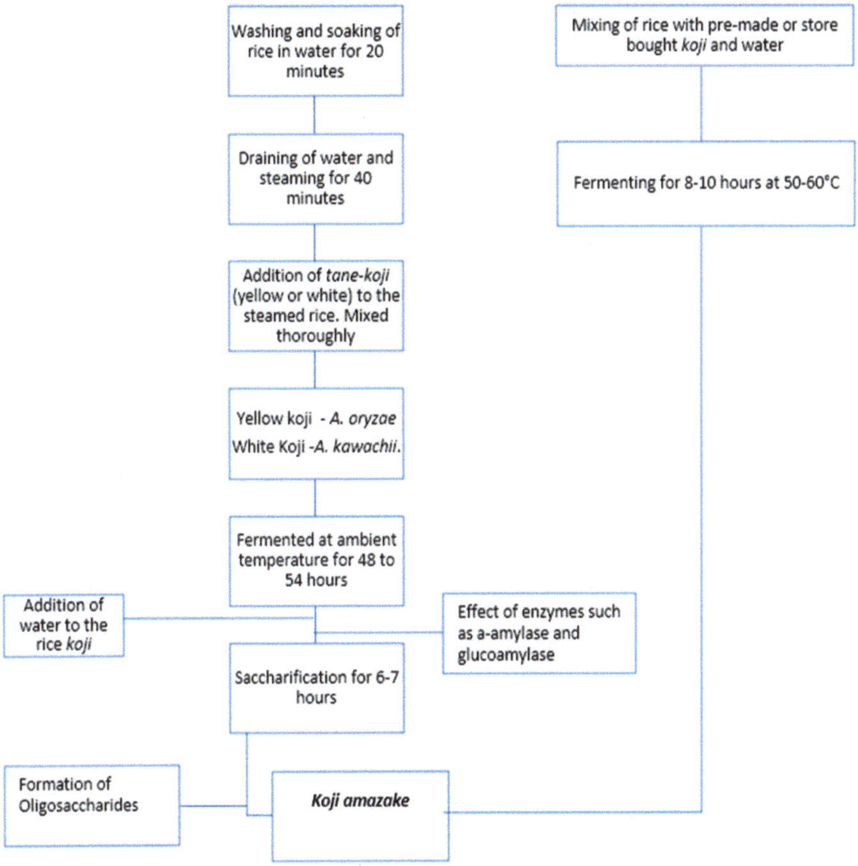

Fig. 7.2 Preparation of Koji amazake

thick porridge (*noyee*). *Noyee* fermentation forms a semi-solid gruel called *kali*. Water and salt are then added to form *koozh*. The traditional method of *koozh* preparation was applied to other common millets and it was observed that *koozh* prepared from pearl millet and sorghum had a sharp flavour and was acceptable to consumers, whereas *koozh* prepared from maize was not well accepted. Bioactive compounds such as *p*-coumaric and ascorbic acid were observed in *koozh* (Thirumangaimannan and Gurumurthy 2013; Ilango and Antony 2014; Manovina et al. 2022). Ilango et al. (2016) isolated 69 strains of presumptive LAB, which were categorized based on their structural features and biochemical investigations as *Streptococcus* sp., *Lactobacillus* sp., *Leuconostoc* sp., and *Enterococcus* sp. Eight strains showed probiotic features. In another study by Kumar et al. (2010), 25 lactic acid bacteria strains were isolated from traditional dishes of South India, namely *Koozh*, *Kallappam* batter, and *Mor Kuzhambu*. They disclosed the occurrence of *Lactobacillus fermentum*, *Weisella paramesenteroides*, and *Lactobacillus plantarum*. Pathogens such as *Bacillus cereus*, *Enterobacteria* sp., *Clostridium* sp.,

Salmonella sp., and *Shigella* species coliforms were observed in some tested samples. This could have been due to the use of contaminated water and poor hygiene practices during hand mixing (Ilango and Antony 2014). From *kambu koozh* (pearl millet porridge) feruloyl esterase-producing LAB were isolated and their probiotic qualities, including their ability to overcome biological barriers, have been observed (Palaniswamy and Govindaswamy 2016).

7.2.7 Kunun-Zaki

Kunun-zaki is a North Nigerian local fermented drink. As shown in Fig. 7.3, the drink is made by steeping millet, sorghum, guinea-corn, or maize and wet milling them. Honey or sugar is used as sweetener along with other ingredients like sweet potatoes, fresh ginger, pepper, and cloves. The technique of preparation is rudimentary without any specific measurements as it is largely a family art. The cereals are steeped in everyday utensils such as earthenware vessels and plastic buckets. The soaked grains are ground into a paste and combined with spices. This mixture is divided into two portions and mixed with hot water and liquefying agents separately. They are then combined and fermented at ambient temperatures for 18–24 h. After fermentation, the mixture is strained using a muslin cloth and sieve. The addition of sweeteners concludes the process and the liquid is ready to be consumed (Elmahmood and Doughari 2007). Organic spices, especially garlic, were found to increase the nutrient value of different varieties of *Kunun-zaki* and may possibly be used as stabilizers for *Kunun-zaki*. There is a wide range of microflora initially, but after the process of steeping, the moulds are destroyed, with Gram (−) and Gram (+) bacteria in a ratio of 1:1. However, by the end of fermentation, all bacteria that are Gram negative are eliminated, leaving the *Lactobacillus* spp. to dominate. LAB such as *L. lactis*, *L. casei*, *L. brevis*, *L. acidophilus*, *L. fermentum*, *L. delbruecki*, and *L. plantarum* have been identified from *kunun-zaki* made with different thickening agents such as paddy rice and sweet potatoes. Growth of *Saccharomyces cerevisiae* was also promoted because of the increase in acidity (Efiuvwevwere and Akona 1995; Victor-Aduloju et al. 2018).

7.2.8 Kvass

Kvass is a fermented drink made with sucrose, rye flour, malt, and stale rye bread among additional grains. Eastern Europe is where people mostly consume this drink. After fermentation, *kvass* is not heated, which results in a high concentration of live yeast and LAB. The organisms usually consist of *Saccharomyces cerevisiae* and LAB, but their composition varies due to the different methods of fermentation and fermenting media. *Kvass* made from fermented squeezed-out rye using

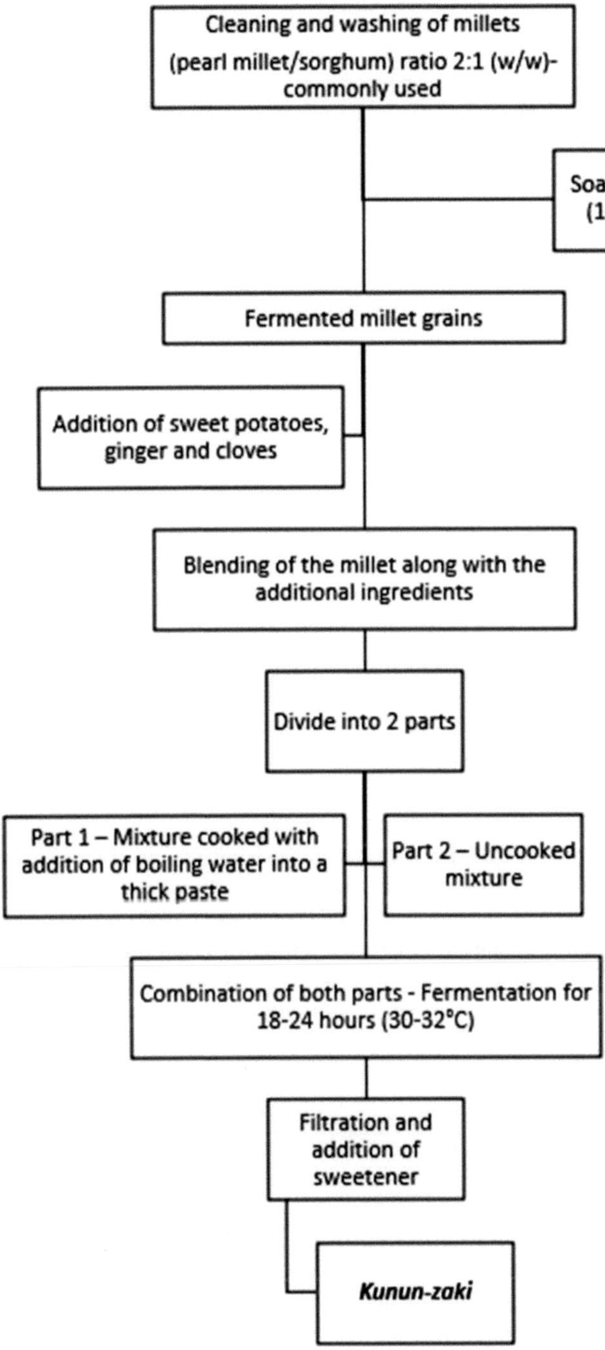

Fig. 7.3 Preparation of Kunun-zaki

xylanolytic enzymes and lactic acid bacteria, demonstrated that the LAB fermented had lesser pH values than the one fermented by yeast. When compared to *Pediococcus pentosaceus*, *Lactobacillus sakei* fermented *kvass* was more palatable, however, *Pediococcus pentosaceus* produced twice as many organic acids. Further observation indicated that after 30 days, the live cells of *L. sakei* lessened in the drink, but the viability of LAB remained high (Basinskiene et al. 2016). In a study conducted by Dlusskaya et al. (2008), a commercial *Kvass* sample was tested for the different microflora that promoted fermentation and the types of microbial metabolites. *Lactobacillus casei* and *Leuconostoc mesenteroides* were observed having counts of 7.3×10^7 and 6.0×10^7 cfu/ml, respectively, whereas *Saccharomyces cerevisiae* was found in cell counts of 3.0×10^7 cfu/ml.

7.2.9 Mahewu

Mahewu is a fermented drink prepared with sorghum, maize, and millet, among others. It is a thin, gritty gruel common among the South African Bantu community. It is also eaten as a thick gruel by adults, whereas for infants between 4 and 18 months, it is diluted and used as weaning foods. *Mahewu* is prepared by cooking cornmeal flour and water together to gelatinize the starch. After which, flour malted from sorghum or millet is combined with the concoction and fermented for 18 h at ambient temperatures. It is consumed cold within 1 or 2 days (Simango 1997). Microbial counts during fermentation showed a rise in numbers of mesophilic and aerobic bacteria, yeasts, and lactic acid bacteria. The primary organisms in the natural fermentation of *Mahewu* are *Lactococcus (lact.) lactis* subsp. *lactis*. Pswarayi and Gänzle (2019) showed that the microorganisms present in *Mahewu* samples usually had 3–7 major varieties of lactobacilli and 2 strains of yeasts, namely *Candida glabrata* and *Saccharomyces cerevisiae*. *Candida glabrata* was found to have increased cell counts in samples that were compiled in summer, however, they did not have similar cell counts in the samples collected in winter. It was observed that model *mahewu* fermentation performed with an inoculum comprising of *Lactobacilli*, *Klebsiella pneumonia*, and *Cronobacter sakazakii*, outcompeted Enterobacteriaceae which showed a sharp decrease in the first 24 h.

7.2.10 Munkoyo

Munkoyo is a traditional fermented cereal drink popular in Zambia and the Katanga province of Congo. It has an acidic taste and contains very little alcohol. It is usually consumed locally in households as a nourishing beverage. *Munkoyo* is different from other cereal-based beverages as it uses the roots of *Rhychosia*, *Eminia*, and *Vigna* species, also known as *munkoyo*. These roots of these plants contain large amounts of amylolytic enzymes which help in the hydrolysis of starch. They also contribute to the colour, flavour, and texture of the beverage. *Munkoyo* is prepared in three steps, i.e. cooking the cereal in water, combining it with *munkoyo* roots to break down the complex carbohydrates, and fermenting the mixture. The rise of acidity in the beverage is because of the formation of lactic acid. Thermophilic LAB promote lactic fermentation. *Lactobacillus delbrueckii* is the representation LAB. Production of alcohol begins only after 15 h of fermenting, due to the occurrence of *Saccharomyces cerevisiae* (Foma et al. 2012). Foma et al. (2012) characterized the bacterial populations of more than 90 samples of *Munkoyo* and observed the presence of 6 dominant families, i.e. Lactabacillales, Streptococcaceae, Bacillaceae, Leuconostocaceae, Enterobacteriaceae, and Aeromonadaceae. The yeast ratio (cfu/ml) to *Lactobacillus* ratio in the starter culture should be 1:1000 or less, and the fermentation process should only last for 24 h, to achieve the intended low concentration of alcohol in *munkoyo* (<100 mmol/l).

7.2.11 *Obiolor*

Obiolor is an indigenous Nigerian drink which is prepared by fermentation. It is slightly brown coloured, has a low viscosity, and tastes sweet. Depending on the locality, it is concocted with ingredients such as maize, millet, and malted sorghum. Similar beverages such as *Pito* and *Burukutu* are also prepared from these cereals. However, these are alcoholic in nature unlike *obiolor*. It is consumed commonly by Igala people who credit this beverage for their good health. In the *obiolor* prepared using the household method, the pH of the mash was found to be 6.81. The chief organisms were *Lactobacillus* and *Bacillus* species. *Streptococcus faecalis* and *Enterobacter* spp. were found in the samples and may have been present as a result of contamination. Isolates from the boiled mash were recognized as *Bacillus stearothermophilus* and *B. subtilis*. During fermentation, *L. plantarum* (TC of 4.8×10^8 microorganisms/g after 22 h) dominated followed by *Bacillus* spp. and *Lactococcus lactis* was observed (Achi 1990). Ajiboye et al. (2016) observed that a diet high in fat, mediated increases in the intensities of conjugated dienes, DNA fragmentation, and lipid hydroperoxides were lowered by the beverage in the livers of rats.

7.2.12 *Olewonyo*

Olewonyo or corn wine, also known as *Aliha* or *Asaana*, is an indigenous non-alcoholic beverage formulated from maize (*Zea mays*) and drunk in Ghana's Volta region and other parts. The manufacturing of this beverage is typically done in batches, weekly once or twice. The cereals are placed on sacks, sprayed with water, and allowed to germinate. After 2 days of drying, grains that are germinated are coarsely crushed, cooked by boiling in water, and allowed to settle overnight. After decanting, brown sugar is added to the creamy liquid and consumed. The highest counts were recorded for *Lactobacillus* species and lactic acid cocci. Lactic acid bacteria such as *L. fermentum*, *L. plantarum*, *L. acidophilus*, and *L. brevis* were extracted from *olewonyo* samples. Fungi and Enterobacteriaceae were found to have significantly increased in the first 24 h of fermentation but drastically declined soon afterwards, between 24 and 72 h. The utensils and vessels used for filtering, mixing, and fermenting as well as compromised sanitary conditions could have influenced the buildup of Enterobacteriaceae in the beverage (Madilo et al. 2022).

7.2.13 *Pozol*

Pozol is a popular mayan beverage made by fermenting a non-sticky paste formed by alkaline cooked maize (nixtamal). It has been eaten as a food or refreshment, especially by low-income groups since the pre-Columbian times (Rizo et al. 2020). Soaking the grains treated with alkali all night, allowed lactic acid bacteria, aerobic mesophiles, and Enterobacteriaceae to flourish, constituting the chief microbiota of *pozol*. LAB are the major cluster in *pozol* fermentation in all its phases. Forty percent of LAB found in the preliminary dough were amylolytic, decreasing to 3% after 72 h of fermentation. However, a large quantity of non-amylolytic LAB were found after the completion of fermentation, explaining the use of mono and disaccharides which were initially present and the other sugars that were formed during starch hydrolysis by the amylolytic LAB. The amylolytic isolates were homofermentative LAB. The isolates belong to the closely linked species such as *Streptococcus*, *Enterococcus*, and *Lactococcus* genera, with the dominant strain being *S. bovis* during fermentation. Another three species of amylolytic LAB that were observed were *E. sulfureus*, *S. macedonicus*, *and L. lactis* and (Díaz-Ruiz et al. 2003). Rizo et al. (2020) showed that the isolated organism from *pozol* shared similarities with three Enterobacteriaceae family members, namely *Klebsiella*, *Kosakonia*, and *Enterobacter*. A strain isolated from *pozol*, *Bacillus* sp. strain CS93 was observed to produce bioactive lipoproteins and antimicrobial compounds such as iturin, bacilysin, and chlorotetaine, thus accounting for its medicinal properties (Phister et al. 2004).

7.2.14 *Togwa*

Togwa is a traditional Tanzanian sweet gruel made by malting and fermenting cereal grains. The four major grains used for this process are sorghum (*Sorghum bicolor*), bulrush millet (*Pennisetum typhoideum*), maize (*Zea mays*), and finger millet (*Eleusine coracona*). The fermentation of *togwa* has been found to be predominated by LAB and yeasts. It was established that within 24 h, the yeasts, the mesophiles, and LAB increased and there was a reduction of Enterobacteriaceae. *L. plantarum*, *L. brevis*, *L. fermentum*, *L. cellobiosus*, *P. pentosaceus*, *W. confusa*, *Issatchenkia orientalis*, *P. anomala*, *P. norvegensis*, *Pichia burtonii*, *Pichia guilliermondii*, *Kluyveromyces marxianus*, *Hanseniaspora guilliermondii*, *C. glabrata*, *Saccharomyces cerevisiae*, *C. pelliculosa*, and *C. tropicalis* were identified in naturally fermented *togwa*. Proteinase and aminopeptidase activities have been found to be considerably more in naturally fermented *togwa* compared to the beverages prepared with starters. Glutamic acid and ornithine have also been found to increase after natural fermentation. Analysis of trace elements and protein content shows that good amounts of nutrients are present in *togwa* (Mugula et al. 2003; Hellström et al. 2010). Although *togwa* is high in minerals, natural inhibitors like phytate frequently prevent these minerals from being absorbed in the human gut (IP (6)), however, the concentration of folate was found to be high when fermented with *C. glabrata* after fermenting for 46 h (Hellström et al. 2012).

7.3 Fruits, Vegetables, and Aromatic Plant Based Non-alcoholic Beverages

Fruits are ideal substrates for producing fermented beverages through microbial transformations due to the presence of sugars and fermentative possibility. New, unconventional, and exotic fruits are being used these days to open prospects for reducing food wastage, improving sustainability, and increasing nutritional value. Fruit extracts, tea leaves, and brine-soaked vegetables that are fermented are extensively utilized across Asia and many parts of Europe. Natural antimicrobial components are found in microbial fermented tea extracts and they show an inhibitory effect against pathogenic bacteria. As fermentation does not always result in beverages with high alcohol, Beverages with low alcohol or no alcohol content are valued because of their quality, flavour, and shelf life in ambient temperatures (Baschali et al. 2017). A summary of the organisms present in non-alcoholic beverages made from vegetables and fruits is shown in Table 7.2.

7.3.1 Cauim

Cauim is a fermented drink consumed by the Tapirapé Amerindians of the Brazilian Tapi'itãwa tribe as a weaning food for babies until 2 years old. Several ingredients such as cassava, corn, groundnuts, and rice are used. However, cassava is used frequently due to its availability in the region. The roots of the cassava plant are submerged in running water for 3–5 days to enable fermentation and softening of the outer layer. The roots are then pared and cut into little pieces, after which they are sun-dried. These are crushed into flour and boiled for 2 h. After the temperature decreases, an inoculum acquired by chewing sweet potatoes is added to aid fermentation which takes around 24–48 h. During fermentation, progressive acidification was observed as the pH diminished from 5.5 to 3.4. Ethanol, lactic acid, and acetic

Table 7.2 Microorganisms present in fruits, vegetables, and aromatic plant-based non-alcoholic beverages

Name of the beverage	Microorganisms involved	References
Cauim	*L. pentosus*, *C. amylocolatum*, *L. plantarum*, *B. pumilus*, *Cornybacterium xerosis*, *Paenibacillus macerans*, *C. vitarumen*, *Bacillus cereus*, *B. licheniformis*, *B. circulans*	Almeida et al. (2007)
Gilaburu juice	*Lactobacillus plantarum*, *Leuconostoc pseudomesenteroides*, *Lactobacillus casei*, *Lactobacillus brevis*, *L. buchneri*, *L. parabuchneri*, *L. pantheris*, *L. delbrueckii*, *Lactobacillus harbinensis*	Sagdic et al. (2014), Sevindik et al. (2022)
Ginger beer	*Saccharomyces cerevisiae*, *Micrococcus* sp., *Streptococcus* sp., *Lactobacillus* sp.	Madden (2008), Pujiati et al. (2022)
Hardaliye	*L. plantarum*, *L. paracasei* subsp. *paracasei*, *L. casei* subsp. *pseudoplantarum*	Bil Der et al. (2017)
Kombucha	*K. rhaeticus*, *K. intermedius*, *Acetobacter*, *K. xylinus*, *Tanticharoenia*, *Gluconobacter*, *K. kombuchae*, *Lactobacillus*, *Lactococcus*, *Oenococcus*, *Enterobacter*, *K. saccharivorans*, *Bifidiobacterium*, *Kluyvera*, *Cellulosimic*, *Pediococcus pentosaceus*, *Schizosaccharomyces pombe*, *Brettanomyces bruxellensis*, *Candida stellata*, *Dekkera bruxellensis*, *Hanseniaspora valbyensis*, *Torulaspora delbrueckii*, *Gluconacetobacter europaeus*, *Zygosaccharomyces bailii*, *Gluconobacter oxydans*, *Oenococcus oeni*, *G. saccharivorans*, *Acetobacter peroxydans*	Teoh et al. (2004), Coton et al. (2017), Reva et al. (2015), Gaggìa et al. (2018), Arıkan et al. (2020)
Shalgam	*Pediococcus pentosaceaceus*, *L. delbrueckii*, *L. plantarum*, *L. paracasei*, *L. brevis*, *L. fermentum*, *L. buhneri*, *L. paralimentarius*, *L. acidophilus*, *L. caseii*, *L. helveticus*, *Leuconostoc mesenteroides*, *P. fermentans*, *Saccharomyces cerevisiae*, *Candida oleophila*, *Kazachstania bulderi*, *Pichia kudriavzevii*, and *Geotrichum candidum*, *Liquorilactobacillus nagelii* AGA58, *Limosilactobacillus fermentum* AGA52	Tanguler and Erten (2012), Kahve et al. (2022)
Sima	*Saccharomyces cerevisiae*	Tozetto et al. (2019), Hotti (2021)
Tepache	*Lactobacillus pentosus*, *Leuconostoc*, *Zygosaccharomyces*, *Acetobacter*, *Epicoccum*, *Lactococcus*, *Gibberella*, *Saccharomyces*, *Candida*, *Meyerozyma*, *Talaromyces*	Gutiérrez-Sarmiento et al. (2022), Escobar-Ramírez et al. (2020)
Water kefir	*L. harbinensis*, *Lactobacillus casei/paracasei*, *Bifidobacterium psychraerophilum/crudilactis*, *L. hilgardii*, *Saccharomyces cerevisiae*, *Dekkerra bruxellensis*, *L. nagelii*, *L. hordei*, *Leuconostoc mesenteroides*, *Acetobacter orientalis*, *Acetobacter fabarum*, *Leuconostoc citreum*, *Lachancea fermentati*, *Zygotorulaspora florentina*, *Hanseniaospora valbyensis*	Laureys and De Vuyst (2014), Gulitz et al. (2011)

acid were found. About 6.8 log cfu/ml of bacteria and 3.7 log cfu/ml of yeast population were observed at the beginning of fermentation. The strains identified were *Bacillus cereus*, *Paenibacillus macerans*, *L. pentosus*, *L. plantarum*, *Corynebacterium xerosis*, *C. vitarumen*, *C. amylocolatum*, *B. licheniformis*, *B. pumilus*, and *B. circulans* (Almeida et al. 2007).

7.3.2 Gilaburu Juice

Gilaburu fruit (*Viburnum opulus*) is a red fruit with a sharp taste, excessive acidity, and pectin content. Other names of this plant include water elder, guelder rose, cherry wood, European guelder, snowball, and crampbark. Viburmine, a compound found in fresh fruit, causes a bitter taste and health risks, so it can only be consumed in processed form. Fermentation also removes the bitterness. *Gilaburu* juice which undergoes fermentation is a non-alcoholic local Turkish beverage produced by the natural fermentation of LAB (Sevindik et al. 2022). The LAB counts of this beverage have been found to be in the range of 3.90–8.30 log cfu/g. A sum of 332 organisms from *Leuconostoc* and *Lactobacillus* species have been identified in traditional fermented *Gilaburu* juice. The LAB strains responsible are *Leuconostoc pseudomesenteroides*, *Lactobacillus plantarum*, *Lactobacillus casei*, *Lactobacillus brevis*, *Lactobacillus buchneri*, *L. pantheris*, *L. parabuchneri, and L. harbinensis* (Sagdic et al. 2014). Juices fermented with *Lactobacillus plantarum* and *L. delbrueckii* displayed better organoleptic acceptability (Sevindik et al. 2022). Sönmezdağ et al. (2017) found that isovaleric acid was a largely plentiful volatile aroma component in fermented GR juice. It was observed that it is safe and effective to treat Medical Expulsive Therapy in distal ureteral calculi with *V. opulus* (Gok et al. 2021). Research also showed that long-term Gilaburu ingestion protects male rats' reproductive systems from injury produced by taxanes (Sarıözkan et al. 2017).

7.3.3 Ginger Beer

Ginger beer, sometimes called *Ginger Ale* is manufactured in two ways, the conventional method of fermentation using the native microflora present on the rhizome or using *Saccharomyces cerevisiae* in a controlled fermentation atmosphere. In the traditional method, the most used ingredients are fresh root ginger, lemon, sugar, and yeast (Madden 2008). Constituents such as chilli (*Capsicum anuum*), liquorice (*Glycyrrhiza glabra*), juniper (*Juniperus communis*), and yarrow (*Achilleas millefolium*) are added for flavour and for the inhibition of bacterial growth. Ground malt is combined with heated water, around 67 °C for 60 min. This is stirred to convert the complex sugars into simple fermentable sugars after which the heat is raised to inactivate the malt enzymes. This mash is cooled and filtered to obtain wort. Wort is boiled along with bitter hops for 45 min. For non-alcoholic beverages, the wort is

Fig. 7.4 Non-alcoholic beverages from fruits and vegetables

watered down to reduce the gravity. Ginger extract is obtained by boiling or pressing sorted and cleaned ginger. The juice is added to aerated wort with lower density. The needed concentration of yeast is combined with the wort and fermented for a specific time at 20 °C. The fermentation sediment or trub is rejected and the green ginger beer is shifted to another fermenter along with the needed sugar for secondary fermentation (Tozetto et al. 2019; Nutakor et al. 2020). A study of various kinds of Ginger beer produced with varieties of ginger showed the presence of *Micrococcus*, *Streptococcus*, and *Lactobacillus* bacteria (Pujiati et al. 2022). A sample representation of the beverage is shown in Fig. 7.4.

7.3.4 Hardaliye

Hardaliye is a non-alcoholic, indigenous Turkish drink prepared with juice extracted from grapes, squashed grapes, and various densities of mustard and leaves from the sour cherry fruit tree. The colour of the beverage indicates the primary colour of the fruits and has a unique scent. Fermentation happens at ambient room temperatures for 7–10 days and can extend up to 20 days if the temperature is cold. After fermentation, the beverage is stored at 4 °C upto 4 months (Coskun and Liu 2017). During the course of fermentation, the complete bacterial count changed from 2.1×10^5 to 1.3×10^2, the lactic acid bacterial count shifted from 6.0×10^4 to 1.2×10^3, and the yeast count from 1.2×10^5 cfu/ml to 1.1×10^3 cfu/ml. *Lactobacillus* species, such as *L. casei* subsp. *pseudoplantarum* and *L. paracasei* subsp. *paracasei*, were observed in the course of fermentation. Bil Der et al. (2017) isolated LAB from 28 samples of *Hardaliye* and observed that *L. plantarum* was the prevailing organism. A high amount phenolic and antioxidant activity values are present in the fresh beverage along with good levels of resveratrol. The serum concentrations of homocysteine, dien conjugate, and malondialdehyde are impacted by *hardaliye* in the diet, potentially because of the antioxidant compounds present (Amoutzopoulos et al. 2013).

7.3.5 Kombucha

Sweetened black or green tea is fermented to create Kombucha, a mildly fizzy beverage. The drink is created by the actions of a 'symbiotic colony of yeast and bacteria', or SCOBY. SCOBY typically manifest as a solid-phase, cellulosic pellicle that is continuously relocated from a finished set of tea to a new set. The presence of at least one *Komagataeibacter* genus acetic acid-producing bacterium (AAB) is what causes this solid phase to form. *K. xylinus*, *K. rhaeticus*, *K. intermedius*, *K. saccharivorans*, and *K. kombuchae* are among the species that have been observed. *Gluconobacter*, *Tanticharoenia*, and *Acetobacter* were the other AAB found. Through studies based on sequencing and culture, LAB such as *Lactobacillus*, *Lactococcus*, and *Oenococcus* have been identified among the SCOBY microbiota. Additional minor organisms include *Enterobacter*, *Bifidobacterium*, *Kluyvera*, and *Cellulosimic*. Strains of *Pediococcus pentosaceus* have also been observed (Reva et al. 2015; Gaggìa et al. 2018; Arıkan et al. 2020). The yeasts that were identified from the beverage included *Schizosaccharomyces pombe*, *Brettanomyces bruxellensis*, *Torulaspora delbrueckii*, *Candida stellata*, *Dekkera bruxellensis*, *Hanseniaspora valbyensis*, and *Zygosaccharomyces bailii*. Species richness declined over an 8-day fermentation. *Gluconacetobacter europaeus*, *Gluconobacter oxydans*, *G. saccharivorans*, and *Acetobacter peroxydans* were found to be the dominant organisms among AAB. *Oenococcus oeni*, the main LAB, was clearly linked with the fermentation of green tea. Even though it is marketed as a beverage with no alcohol, due to the preparation procedure used, it is likely that the beverage contains small amounts of ethanol (Teoh et al. 2004; Coton et al. 2017). It has been shown that kombucha has an immunomodulatory effect and successfully prevents the disorders of cellular immune functions in mice during the early stages of infection. Supplementing with kombucha changed the composition of microflora in the gut and encouraged the increase of bacteria that produce butyrate, which has been shown to have anti-inflammatory properties (Wang et al. 2021). It was observed that kombucha fermented with bamboo leaf by-products had a notable improvement in its sensory acceptability, concentrations of specific compounds, overall phenolic content, and antioxidant capacity (Xiong et al. 2023). A sample representation of the beverage is shown in Fig. 7.4.

7.3.6 Shalgam

Shalgam is an indigenous drink made by fermenting black or purple carrots, salt, turnips, yeast, bulgur, water and chickpea, or bulgur sourdough (Kahve et al. 2022). It is a cherry-coloured beverage that has a cloudy appearance and tastes sour. It is popularly consumed as a soft drink along with meals and by itself. It is made on a small scale in households, villages, and some industries in regions of South Turkey such as Adana, Mersin, Hatay, and Kahramanmaras. Organisms such as LAB, yeasts, mesophilic aerobic bacteria, and coliform bacteria were observed in the counts of 6.97, 6.72, 5.64, 4.47, and 1.49 log cfu/ml, respectively. Total acidity (TA) as lactic acid was attained in the span of 6.54–7.25 g/l and pH values in the span of 3.28–3.48. The LAB found in *shalgam* are *L. delbrueckii*, *L. caseii*, *L. fermentum*, *L. paracasei*, *L. buhneri*, *L. paralimentarius*, *L. acidophilus*, *Pediococcus pentosaceaceus*, *L. helveticus*, *L. brevis*, and *Leuconostoc mesenteroides* (Tanguler and Erten 2012). Some species of endogenous yeasts found in *shalgam* are *Geotrichum candidum*, *Saccharomyces cerevisiae*, *P. kudriavzevii*, *P. fermentans*, *Candida oleophila*, and *Kazachstania bulderi*. It was also noted that the occurrence of *P. kudriavzevii* increased while fermenting, while the other species declined (Kahve et al. 2022). *Liquorilactobacillus nagelii* AGA58 isolated from *shalgam* demonstrates its probiotic characteristics by its tolerance to bile and acid concentrations, which mimic human gut environments (Yetiman and Ortakci 2023). The analysis of the genomic and metabolic characteristics of *Limosilactobacillus fermentum* AGA52, which was extracted from shalgam, showed that it is a viable candidate for dietary supplements that contain probiotics and psychobiotics and that can benefit the host physiologically (Yetiman et al. 2023). A sample representation of the beverage is shown in Fig. 7.4.

7.3.7 Sima

Sima is a traditional fermented low-alcoholic drink, prepared and drunk in Finland. It is sweet to taste and has a murky appearance. It is prepared with water, sugar, or honey, lemon, raisins, and yeast. Due to its very low alcohol content, *sima* is enjoyed by both adults and children as a special beverage to celebrate May Day, also called *Vappu* (Tozetto et al. 2019; Hotti 2021).

7.3.8 Tepache

Tepache is an indigenous Mexican beverage made by the fermentation of pineapple peels or a combination of pulp and peels along with brown cane sugar and spices. The beverage is fermented in barrels made of wood at ambient temperatures for 1–4 days. Since *tepache* is not a distilled beverage like tequila and wine, ethanol is not required to be the major product. Therefore, a low concentration of ethanol is desired along with a partial utilization of sugars, giving a sweet taste to the resulting beverage. *Tepache* production is impacted by many elements including the weather, time, temperature, pH, and the concentration of sugars (Moreno-Terrazas et al. 2016; Romero-Luna et al. 2022). At 72 h of fermentation, *Lactobacillus*, *Leuconostoc*, *Acetobacter*, and *Lactococcus* dominated the microbial community. *Zygosaccharomyces*, *Meyerozyma*, *Saccharomyces*, *Gibberella*,

Candida, *Talaromyces*, and *Epicoccum* were found to be in abundance among the fungal population. *Tepache* fermentation occurred in two successive steps. First, ethanol and lactic acid fermentation and the second an increase in acetogenic bacteria (Gutiérrez-Sarmiento et al. 2022). *Lactobacillus pentosus* isolated from *tepache* was recognized as being unaffected in some in vitro digestive actions and has the potential to be used as a starter in fermenting milk for producing a symbiotic beverage (Escobar-Ramírez et al. 2020).

7.3.9 Water Kefir

Water kefir grains are a distinctive starter culture utilized for producing homemade carbonated, acidic drinks with a low alcohol content. The grains are translucent, grey to white coloured, and have an irregular shape. They are inoculated in water containing sucrose and dried figs. The origin of these grains does not pinpoint to a particular beginning, rather showing more than one origin with names such as *Japanese beer seeds*, *tibi*, *Balm of Gilead*, *California bees*, *African bees*, and *Gingerbeer plants* (Neve and Heller 2002). *Lactobacillus hilgardii* played an important role in *water kefir* grain formation, which consisted of dextran, an α 1–6 linked glucose polymer (Waldherr et al. 2010). The microbiota present in *water kefir* grains were *Lactobacillus casei/paracasei*, *Saccharomyces cerevisiae*, *L. harbinensis*, *L. hilgardii*, *Bifidobacterium psychraerophilum/crudilactis*, and *Dekkerra bruxellensis* (Laureys and De Vuyst 2014). The most abundant strains in *water kefir* were *L. hordei* and *L. nagelii*. Other lactic acid bacteria found were *Leuconostoc mesenteroides* and *L. citreum*. *Acetobacter orientalis* and *A. fabarum* were non-LAB strains that were identified along with *Zygotorulaspora florentina*, *Lachancea fermentati*, and *Hanseniaospora valbyensis* (Gulitz et al. 2011). Diabetic mice were treated with the separated *L. paracasei* from Malaysian Water Kefir Grains, where it was observed that the treatment controlled the expression of multiple genes involved in lipid metabolism and glucose homeostasis. These findings implied that *L. paracasei* might be a viable dietary addition for people with type 2 diabetes (Talib et al. 2023).

7.4 Dairy-Based Non-alcoholic Beverages

Milk is believed to be a complete food for humans and has been used in their diet since the early times. Milk fermentation is carried out as a method to preserve milk along with imparting additional benefits to the product. While fermentation is mostly carried out by the microorganisms naturally present, some kinds of fermentations are done by introducing specific microbes to obtain a desired product. A major beneficial role is played by Lactic Acid Bacteria (LAB) in fermenting milk and associated products (Kumar et al. 2015). A summary of microorganisms present in dairy-based non-alcoholic beverages is shown in Table 7.3.

Table 7.3 Microorganisms present in dairy-based non-alcoholic beverages

Name of the beverage	Microorganisms involved	References
Kefir	*Enterococcus durans*, *Lactobacillus brevis*, *Lactobacillus fermentum*, *L. kefiranofaciens*, *L. kefiranofaciens* subsp. *kefiranofaciens*, *L. acidophilus*, *L. kefiri*, *Leuconostoc mesenteroides* subsp. *cremoris*, *L. casei* subsp. *casei*, *L. paracasei* subsp. *paracasei*, *L. delbrueckii* subsp. *bulgaricus*, *L. helveticus*, *L. kefiranofaciens* subsp. *kefirgranum*, *L. delbrueckii* subsp. *lactis*, *L. parakefiri*, *L. plantarum*, *Lactococcus lactis* subsp. *cremoris*, *Lactococcus lactis* subsp. *lactic*, *Streptococcus thermophilus*, *Leuconostoc mesenteroides* subsp. *mesenteroides* and *Acetobacter aceti*, *K. lodderae*, *K. marxianus*, *Saccharomyces cerevisiae*, *S. pastorianus*, *S. unisporus*, *S. exiguus*, *S. turicensis*, *Dekkera anomala*, *Brettanomyces anomalus*, *Torulaspora delbruecki*, *Candida famata*, *C. friedrichii*, *C. humilis*, *Candida inconspicua*, *Candida kefyr*, *C. lipolytica*, *C. krusei*, *C. maris*, *Candida firmetaria*, *Galactomyces geotrichum*, *Issatchenkia orientalis*, *Cryptococcus humicolus*, *Debaryomyces hansenii*, *Debaryomyces (Schwanniomyces) occidentalis*, *Geotrichum candidum*, *Pichia fermentans*, *Kluyveromyces lactis* var. *lactis*, *Yarrowia lipolytica*, *Zygosaccharomyces rouxii*	Garrote et al. (2001), Simova et al. (2002), Beshkova et al. (2002), Kesenkaş et al. (2017)
Ayran	*L. delbrueckii* subsp. *bulgaricus*	Baschali et al. (2017)
Buttermilk	*Leuconostoc mesenteroides*, *Lactobacillus brevis*, *Lactobacillus kefir*, *Lactobacillus lactis* subsp. *lactis* biovar. *diacetylactis*, *Lactobacillus helveticus*, *Lactobacillus casei*, *Lactobacillus lactis* subsp. *cremoris*, *Lactobacillus lactis* subsp. *lactis*, *Lactobacillus delbruecki* subsp. *lactis*, *Streptococcus thermophilus*, *Lactobacillus delbruecki* subsp. *bulgaricus*, *Lactobacillus acidophilus*	Kumar et al. (2015), Mascarello et al. (2019), Banerjee and Qamar (2022)
Mabisi	*Lactobacillus brevis*, *Lactobacillus plantarum*, *Leuconostoc garlicum*, *Enterococcus durans*, *Lysinibacillus sphaericus*, *Lactococcus lactis*	Schoustra et al. (2013)
Rob	*Saccharomyces exigus*, *Lactococcus lactis*, *Lactobacillus bulgaricus*, *Lactobacillus helveticus*, *Streptococcus thermophilus*, *Saccharomyces globosus*, *Kluyveromyces bulgaricus*, *P. aeruginosa*, *K. lactis*, *L. plantarum*, *A. Baumannii*	Abdelgadir et al. (2001), Soltan Dallal et al. (2017)

7.4.1 *Kefir*

Kefir is a creamy-textured, mildly sour milk that has undergone acidic fermentation. Kefir grains, a tiny mass of microbes bound together with a polysaccharide medium called kefiran, are used to produce it. Kefiran is a glucogalactan that dissolves in water and is produced by *Lactobacillus kefiranofaciens* (Lopitz-otsoa et al. 2006). These grains contain a number of microorganisms of different species that have been identified and isolated. To make kefir, raw, unpasteurized, pasteurized, or UHT-treated milk is combined with kefir grains, which are then allowed to ferment for 24 h at ambient temperatures. The *kefir* granules are retrieved by filtering the cultured milk and mixed with new milk to continue fermenting. *Kefir* is quoted as an example of symbiosis as the existence and development of microbial strains are reliant on one another. Bacterial microbiota such as *Lactobacillus acidophilus*, *Enterococcus durans*, *Leuconostoc mesenteroides* subsp. *cremoris*, *Lactobacillus casei* subsp. *casei*, *Lactobacillus brevis*, *L. paracasei* subsp. *paracasei*, *L. delbrueckii* subsp. *bulgaricus*, *Leuconostoc mesenteroides* subsp. *mesenteroides*, *L. kefiri*, *L. kefiranofaciens* subsp. *kefiranofaciens*, *L. kefiranofaciens* subsp. *kefirgranum*, *L. fermentum*, *L. delbrueckii* subsp. *lactis*, *L. parakefiri*, *L. plantarum*, *Lactococcus lactis* subsp. *lactic*, *Lactococcus lactis* subsp. *cremoris*, *Streptococcus thermophilus*, *L. helveticus*, and *Acetobacter aceti* are found in *kefir* grains. The fungal microbiota included *K. lodderae*, *K. marxianus*, *S. cerevisiae*, *S. pastorianus*, *Saccharomyces unisporus*, *Yarrowia lipolytica*, *D. anomala*, *Brettanomyces anomalus*, *Torulaspora delbruecki*, *C. friedrichii*, *C. firmetaria*, *C. humilis*, *Saccharomyces exiguus*, *C. inconspicua*, *Candida kefyr*, *Zygosaccharomyces rouxii*, *Pichia fermentans*, *Issatchenkia orientalis*, *C. krusei*, *Cryptococcus humicolus*, *Debaryomyces hansenii*, *Candida famata*, *Debaryomyces (Schwanniomyces) occidentalis*, *Galactomyces geotrichum*, *Geotrichum candidum*, *Kluyveromyces lactis* var. *lactis*, and *Saccharomyces turicensis* sp. *nov* (Garrote et al. 2001; Simova et al. 2002; Beshkova et al. 2002; Kesenkaş et al. 2017). Mixed results have been found in human trials investigating kefir's capacity to replicate metabolic health benefits previously noted in rat models. In preventative animal trials, kefir has long been linked to improved health outcomes, including obesity, diabetes, NAFLD, and dyslipidaemia (Bourrie et al. 2020).

7.4.2 *Ayran*

Ayran is a Turkish beverage, fermented by including salt and water to homemade yoghurt or by incorporating culture, *S. thermophilus* and *L. delbrueckii* subsp. *bulgaricus* to homogenized milk. Beverages related to *Ayran* include *ayrani* in Cyprus, *ayryan* in Bulgaria, *ariani* in Greece, and *dhalle* in Albania (Baschali et al. 2017). The microflora that is used to produce ayran made at home is akin to yoghurt's microflora. *Ayran* produced in large-scale industries contains only the needed organisms, whereas, in homemade ones, the risk of contamination is increased resulting in the presence of yeasts such as *Kluyveromyces* and *Saccharomyces* in the final product. Exopolysaccharides (EPS) producing lactic ferments are also employed in the manufacture of *Ayran* (Kocak and Avsar 2010). It is also acquired as a derivative of butter making termed as '*Turkish buttermilk*'. Traditionally, yoghurt is diluted with water before churning. After churning and the transfer of butter, the resulting fluid is consumed as *Ayran* (Altay et al. 2013). It was observed that 1% salt-containing *Ayran* was better liked by consumers compared to 0.5% salt-containing ones (Kocabaş et al. 2022). The addition of concentrated whey and whey protein powder produced *Ayran* with probiotic and functional properties (Ayar and Burucu 2013). The *ayran* filtrates exhibited a significantly higher antioxidant capacity than milk, with high concentrations of lactic and acetic acid and minute amounts of butyric and caproic acid. Therefore *ayran*, possibly because of its antioxidant property, can lessen DNA damage (Grishina et al. 2011).

7.4.3 Buttermilk

Natural *buttermilk* is obtained after churning butter. The leftover liquid with a low-fat content is a famous fermented beverage used widely in India and Middle Eastern countries. Sour *buttermilk* known as *lassi* is a common drink in India. Four types of *buttermilk* have been found, namely cultured *buttermilk*, sweet cream *buttermilk*, sour milk *buttermilk*, and commercial *buttermilk*. Natural *buttermilk* is low in fat and can be used in recipes to replace water. To produce cultured *buttermilk*, the milk is sterilized at 82–85 °C for 2–5 min. After cooling to 22 °C, the starter cultures are added. Sweet cream *buttermilk* is obtained after stirring cream containing a high amount of fat and protein. Sour cream *buttermilk* is the result of the agitating procedure of matured cream. Commercial *buttermilk* is produced by adding lactic ferments to skimmed or low-fat milk and letting it mature. *Buttermilk* holds high quantities of protein, calcium, and riboflavin. *Buttermilk* cultures contain live active organisms such as *L. brevis*, *L. kefir*, *Lactobacillus lactis* subsp. *lactis* biovar. *diacetylactis*, *L. casei*, *L. helveticus*, *Lactobacillus lactis* subsp. *cremoris*, *Lactobacillus lactis* subsp. *lactis*, *L. acidophilus*, *Lactobacillus delbruecki* subsp. *lactis*, *Leuconostoc mesenteroides*, *Lactobacillus delbruecki* subsp. *bulgaricus* and *Streptococcus thermophilus*. The chief role of these bacterial cultures is the creation of organic acids such as lactic acid. When homofermentative bacteria consume or ferment glucose, the final metabolite is lactic acid (Kumar et al. 2015; Mascarello et al. 2019; Banerjee and Qamar 2022). Buttermilk ingestion may be linked to lower cholesterol levels in both men and women, mainly due to its ability to inhibit intestinal cholesterol absorption (Conway et al. 2013). When compared to a placebo, the intake of buttermilk drastically decreased plasma levels of the angiotensin I-converting enzyme, mean arterial blood pressure, and systolic blood pressure (Conway et al. 2014).

7.4.4 *Mabisi*

Mabisi is a brewed dairy beverage made in Zambia's South and Central regions. Raw milk undergoes fermentation for 2 days at ambient temperatures in a calabash. The subsequent brew is a little dense in consistency and has a mild tart taste. After fermentation, the *Mabisi* is mixed to create a uniform consistency. This is the typical 'Tonga type' *Mabisi*. The creamy thick type is prepared in a similar way along with the removal of the top layer of cream and whey. The resulting solution undergoes fermentation on another day. Cooked-type *Mabisi* is made by boiling the milk and then fermenting it. Backslopping is done by leaving a small amount of the *Mabisi* in the brewing vessel for the following batch. The gourd vessel surface offers a chance for the organisms to create biofilms and remain in the crevices so that they are not eliminated while rinsing and drying (Moonga et al. 2019). The microbes responsible for fermenting *Mabisi* were found to be *Lactobacillus brevis*, *Lactobacillus plantarum*, *Leuconostoc garlicum*, *Enterococcus durans*, and *Lysinibacillus sphaericus* (Schoustra et al. 2013). The genera commonly observed were *Streptococcus*, *Buttiauxella*, *Lactococcus*, *Klebsiella*, *Enterobacter*, *Citrobacter*, *Acinetobacter*, *Kluyvera*, *Aeromonas*, and *Lactobacillus* with *Lactococcus* present in *Mabisi* produced in all areas (Moonga et al. 2020). Groenenboom et al. (2020) also observed that *Lactococcus lactis* dominated the microbial community in a sequence of 15 experiments using 2 kinds of brewing containers in combination with different methods of backslopping. It is hypothesized that a bacterial population comprising metabolically active cells from a wide variety of species will produce a product with a greater variety of organoleptic elements and possibly a finer taste than when only some organisms are involved. Additionally, more diverse communities have a higher likelihood of having health-promoting bacteria than a product with a small number of microbial species (Smid and Kleerebezem 2014).

7.4.5 *Rob*

Rob is a well-liked Sudanese fermented milk beverage that is prepared to use up surplus milk. Milk is taken in a vessel and inoculated with a lactic ferment present in the former container used for making *rob*. It is fermented overnight and is churned the next morning. Butter is also obtained from this processing. In some areas, the churning is done in leather containers made of goatskin or dried gourds. Strong gas is emitted in the preliminary phases of fermenting the product, indicating the action of heterofermentative microbes. However, after churning, the gas production subsides. Freshly prepared *rob* has an agreeable taste with a pH of 4.5. Within 1–2 h, it sours considerably. However, this is also consumed with the addition of water and sugar. The microflora in the brewing of *rob* were detected as, *L. fermentum*, *L. bulgaricus*, *L. helveticus*, *Lactococcus lactis*, and *Streptococcus thermophilus*. The yeast observed were *S. globosus*, *K. bulgaricus*, *S. exigus*, and *Kluyveromyces lactis*

(Abdelgadir et al. 2001). *Lb. plantarum* has been found to exhibit adversarial characteristics towards strains of *P. aeruginosa* and *A. baumannii* that cause healthcare-associated infections and significantly inhibit the growth of these strains (Soltan Dallal et al. 2017).

7.5 Conclusion

Non-alcoholic fermented beverages have been used widely around the world for all their different benefits such as ease of preparation, nutritional quality, and shelf life in ambient temperatures. Globally, many non-alcoholic beverages with similar names and ingredients are found, targeting nutritional value, cultural significance, and providing alternates for alcoholic drinks. Although these beverages have different chemical and flavour profiles, all of them contain beneficial components such as probiotics and other therapeutic agents, making them valuable for human health. Studies are also being conducted on improving the nutritional value of these beverages by fortifying them with other nutrient-dense foods. In today's world, the attention has shifted towards foods that are sustainable, locally produced and nutritional. These non-alcoholic beverages cater to these requirements in many ways. As more research is being conducted on these beverages, the scientific knowledge is increasing raising awareness about these beverages. In this review, the formulation of non-alcoholic beverages and the microflora responsible for their fermentation have been summarized. Though some common fermented drinks such as kombucha, kefir, and buttermilk have been studied extensively, many lesser-known and locally available non-alcoholic beverages have much to be studied, especially the activity of the microbes in the fermenting process resulting in the formation of these desired beverages.

References

Abdelgadir WS, Hamad SH, Moller PL, Jakobsen M (2001) Characterisation of the dominant microbiota of Sudanese fermented milk Rob. Int Dairy J 11:63–70. https://doi.org/10.1016/S0958-6946(01)00042-5

Achi OK (1990) Microbiology of 'obiolor': a Nigerian fermented non-alcoholic beverage. J Appl Bacteriol 69:321–325. https://doi.org/10.1111/j.1365-2672.1990.tb01522.x

Adinsi L, Vieira-Dalode G, Akissoe N, Anihouvi V (2014) Processing and quality attributes of gowe: a malted and fermented cereal-based beverage from Benin. Food Chain 4:171–183. https://doi.org/10.3362/2046-1887.2014.016

Adinsi L, Mestres C, Akissoé N, Vieira-Dalode G, Anihouvi V, Durand N, Hounhouigan DJ (2017) Comprehensive quality and potential hazards of gowe, a malted and fermented cereal beverage from West Africa. A diagnostic for a future re-engineering. Food Control 82:18–25. https://doi.org/10.1016/j.foodcont.2017.06.019

Ajiboye TO, Iliasu GA, Adeleye AO, Ojewuyi OB, Kolawole FL, Bello SA, Mohammed AO (2016) A fermented sorghum/millet-based beverage, Obiolor, extenuates high-fat diet-induced dyslipidaemia and redox imbalance in the livers of rats. J Sci Food Agric 96:791–797. https://doi.org/10.1002/JSFA.7150

Almeida EG, Rachid CCTC, Schwan RF (2007) Microbial population present in fermented beverage 'cauim' produced by Brazilian Amerindians. Int J Food Microbiol 120:146–151. https://doi.org/10.1016/J.IJFOODMICRO.2007.06.020

Altay F, Karbancioglu-Güler F, Daskaya-Dikmen C, Heperkan D (2013) A review on traditional Turkish fermented non-alcoholic beverages: microbiota, fermentation process and quality characteristics. Int J Food Microbiol 167:44–56. https://doi.org/10.1016/j.ijfoodmicro.2013.06.016

Amoutzopoulos B, Löker GB, Samur G, Çevikkalp SA, Yaman M, Kose T, Pelvan E (2013) Effects of a traditional fermented grape-based drink "hardaliye" on antioxidant status of healthy adults: a randomized controlled clinical trial. J Sci Food Agric 93:3604–3610. https://doi.org/10.1002/JSFA.6158

Arıkan M, Mitchell AL, Finn RD, Gürel F (2020) Microbial composition of Kombucha determined using amplicon sequencing and shotgun metagenomics. J Food Sci 85:455–464. https://doi.org/10.1111/1750-3841.14992

Ayar A, Burucu H (2013) Effect of whey fractions on microbial and physicochemical properties of probiotic ayran (drinkable yogurt). Int Food Res J 20:1409–1415

Banerjee P, Qamar I (2022) Insights into the technological and nutritional aspects of lactic milk drinks: buttermilk. Adv Dairy Microb Prod:93–103. https://doi.org/10.1016/B978-0-323-85793-2.00002-3

Baschali A, Tsakalidou E, Kyriacou A, Karavasiloglou N, Matalas AL (2017) Traditional low-alcoholic and non-alcoholic fermented beverages consumed in European countries: a neglected food group. Nutr Res Rev 30:1–24. https://doi.org/10.1017/S0954422416000202

Basinskiene L, Juodeikiene G, Vidmantiene D, Tenkanen M, Makaravicius T, Bartkiene E (2016) Non-alcoholic beverages from fermented cereals with increased oligosaccharide content. Food Technol Biotechnol 54:36–44. https://doi.org/10.17113/ftb.54.01.16.4106

Beshkova DM, Simova ED, Simov ZI, Frengova GI, Spasov ZN (2002) Pure cultures for making kefir. Food Microbiol 19:537–544. https://doi.org/10.1006/FMIC.2002.0499

Bil Der T, Arici M, Coskun F, Celikyurt G, Mirik M, Gulcu M, Tokatli N (2017) Some technological and functional properties of lactic acid bacteria isolated from hardaliye. J Agric Sci 23:428–437. https://doi.org/10.15832/ANKUTBD.385870

Bourrie BCT, Richard C, Willing BP (2020) Kefir in the prevention and treatment of obesity and metabolic disorders. Curr Nutr Rep 9:184–192. https://doi.org/10.1007/S13668-020-00315-3

Bvochora JM, Reed JD, Read JS, Zvauya R (1999) Effect of fermentation processes on proanthocyanidins in sorghum during preparation of Mahewu, a non-alcoholic beverage. Process Biochem 35:21–25. https://doi.org/10.1016/S0032-9592(99)00027-8

Chauhan R, Reddy A, Abraham J (2015) Biosynthesis of silver and zinc oxide nanoparticles using Pichia fermentans JA2 and their antimicrobial property. Appl Nanosci 5:63–71. https://doi.org/10.1007/S13204-014-0292-7

Constantin EA, Constantinescu-Aruxandei D, Matei F, Shaposhnikov S, Oancea F (2023) Biochemical and microbiological characterization of traditional Romanian fermented drinks—Socata and Borş—a review. AgroLife Sci J 12:53–61. https://doi.org/10.17930/agl202317

Conway V, Couture P, Richard C, Gauthier SF, Pouliot Y, Lamarche B (2013) Impact of buttermilk consumption on plasma lipids and surrogate markers of cholesterol homeostasis in men and women. Nutr Metab Cardiovasc Dis 23:1255–1262. https://doi.org/10.1016/J.NUMECD.2013.03.003

Conway V, Couture P, Gauthier S, Pouliot Y, Lamarche B (2014) Effect of buttermilk consumption on blood pressure in moderately hypercholesterolemic men and women. Nutrition 30:116–119. https://doi.org/10.1016/J.NUT.2013.07.021

Coskun F, Liu SQ (2017) A traditional Turkish fermented non-alcoholic grape-based beverage, "hardaliye". Beverages 3(1):2. https://doi.org/10.3390/BEVERAGES3010002

Coton M, Pawtowski A, Taminiau B, Burgaud G, Deniel F, Coulloumme-Labarthe L, Fall A, Daube G, Coton E (2017) Unraveling microbial ecology of industrial-scale Kombucha fermentations by metabarcoding and culture-based methods. FEMS Microbiol Ecol 93. https://doi.org/10.1093/FEMSEC/FIX048

Díaz-Ruiz G, Guyot JP, Ruiz-Teran F, Morlon-Guyot J, Wacher C (2003) Microbial and physiological characterization of weakly amylolytic but fast-growing lactic acid bacteria: a functional role in supporting microbial diversity in pozol, a Mexican fermented maize beverage. Appl Environ Microbiol 69:4367–4374. https://doi.org/10.1128/AEM.69.8.4367-4374.2003

Dlusskaya E, Jänsch A, Schwab C, Gänzle MG (2008) Microbial and chemical analysis of a kvass fermentation. Eur Food Res Technol 227:261–266. https://doi.org/10.1007/s00217-007-0719-4

Efiuvwevwere BJO, Akona O (1995) The microbiology of "kunun-zaki", a cereal beverage from northern Nigeria, during the fermentation (production) process. World J Microbiol Biotechnol 11:491–493. https://doi.org/10.1007/BF00286358/METRICS

Elmahmood AM, Doughari JH (2007) Microbial quality assessment of kunun-zaki beverage sold in Girei town of Adamawa State, Nigeria. Afr J Food Sci 1:11–015

Escobar-Ramírez MC, Jaimez-Ordaz J, Escorza-Iglesias VA et al (2020) Lactobacillus pentosus ABHEAU-05: an in vitro digestion resistant lactic acid bacterium isolated from a traditional fermented Mexican beverage. Rev Argent Microbiol 52:305–314. https://doi.org/10.1016/J.RAM.2019.10.005

Foma RK, Destain J, Mobinzo PK, Kayisu K, Thonart P (2012) Study of physicochemical parameters and spontaneous fermentation during traditional production of munkoyo, an indigenous beverage produced in Democratic Republic of Congo. Food Control 25:334–341. https://doi.org/10.1016/J.FOODCONT.2011.10.049

Gaggìa F, Baffoni L, Galiano M, Nielsen DS, Jakobsen RR, Castro-Mejía JL, Bosi S, Truzzi F, Musumeci F, Dinelli G, Di Gioia D (2018) Kombucha beverage from green, black and rooibos teas: a comparative study looking at microbiology, chemistry and antioxidant activity. Nutrients 11(1):1. https://doi.org/10.3390/NU11010001

Garrote GL, Abraham AG, De Antoni GL (2001) Chemical and microbiological characterisation of kefir grains. J Dairy Res 68:639–652. https://doi.org/10.1017/S0022029901005210

Gok B, Tarik Atik Y, Uysal B, Koc E, Tastemur S, Cimen HI (2021) Gilaburu extract (Viburnum opulus Linnaeus) is as effective as Tamsulosin in medical expulsive therapy of distal ureteral calculi. Int J Clin Pract 75(12):e14950. https://doi.org/10.1111/IJCP.14950

Greppi A, Rantsiou K, Padonou W, Hounhouigan J, Jespersen L, Jakobsen M, Cocolin L (2013) Determination of yeast diversity in ogi, mawè, gowé and tchoukoutou by using culture-dependent and -independent methods. Int J Food Microbiol 165:84–88. https://doi.org/10.1016/J.IJFOODMICRO.2013.05.005

Grishina A, Kulikova I, Alieva L, Dodson A, Rowland I, Jin J (2011) Antigenotoxic effect of kefir and ayran supernatants on fecal water-induced DNA damage in human colon cells. Nutr Cancer 63:73–79. https://doi.org/10.1080/01635581.2010.516873

Groenenboom AE, Shindano J, Cheepa N, Smid EJ, Schoustra SE (2020) Microbial population dynamics during traditional production of Mabisi, a spontaneous fermented milk product from Zambia: a field trial. World J Microbiol Biotechnol 36:1–14. https://doi.org/10.1007/S11274-020-02957-5/FIGURES/9

Gulitz A, Stadie J, Wenning M, Ehrmann MA, Vogel RF (2011) The microbial diversity of water kefir. Int J Food Microbiol 151:284–288. https://doi.org/10.1016/J.IJFOODMICRO.2011.09.016

Gutiérrez-Sarmiento W, Peña-Ocaña BA, Lam-Gutiérrez A, Guzmán-Albores JM, Jasso-Chávez R, Ruíz-Valdiviezo VM (2022) Microbial community structure, physicochemical characteristics and predictive functionalities of the Mexican tepache fermented beverage. Microbiol Res 260:127045. https://doi.org/10.1016/J.MICRES.2022.127045

Hancioğlu Ö, Karapinar M (1997) Microflora of Boza, a traditional fermented Turkish beverage. Int J Food Microbiol 35:271–274. https://doi.org/10.1016/S0168-1605(96)01230-5

Hellström AM, Vázques-Juárez R, Svanberg U, Andlid TA (2010) Biodiversity and phytase capacity of yeasts isolated from Tanzanian togwa. Int J Food Microbiol 136:352–358. https://doi.org/10.1016/J.IJFOODMICRO.2009.10.011

Hellström AM, Almgren A, Carlsson NG, Svanberg U, Andlid TA (2012) Degradation of phytate by Pichia kudriavzevii TY13 and Hanseniaspora guilliermondii TY14 in Tanzanian togwa. Int J Food Microbiol 153:73–77. https://doi.org/10.1016/J.IJFOODMICRO.2011.10.018

Hotti P (2021) A Finnish drink with a heroic past—BBC Travel. https://www.bbc.com/travel/article/20210519-a-finnish-drink-with-a-heroic-past. Accessed 6 Nov 2023

Ilango S, Antony U (2014) Assessment of the microbiological quality of koozh, a fermented millet beverage. Afr J Microbiol Res 8:308–312. https://doi.org/10.5897/AJMR2013.6482

Ilango S, Pandey R, Antony U (2016) Functional characterization and microencapsulation of probiotic bacteria from koozh. J Food Sci Technol 53:977–989. https://doi.org/10.1007/S13197-015-2169-5/FIGURES/2

Kahve HI, Akbulut M, Coklar H (2022) Identification and technological characterization of endogenous yeast isolated from fermented black carrot juice, shalgam. LWT Food Sci Technol 154:112823. https://doi.org/10.1016/j.lwt.2021.112823

Kesenkaş H, Gürsoy O, Özbaş H (2017) Kefir. In: Fermented foods in health and disease prevention. Elsevier, pp 339–361. https://doi.org/10.1016/B978-0-12-802309-9.00014-5

Kocabaş H, Ergin F, Aktar T, Küçükçetin A (2022) Effect of lactose hydrolysis and salt content on the physicochemical, microbiological, and sensory properties of ayran. Int Dairy J 129:105360. https://doi.org/10.1016/J.IDAIRYJ.2022.105360

Kocak C, Avsar Y (2010) Development and manufacture of yogurt and other functional dairy products. In: Yildiz F (ed) Development and manufacture of yogurt and other functional dairy products, pp 123–141. https://doi.org/10.1201/9781420082081

Kumar RS, Varman DR, Kanmani P, Yuvaraj N, Paari KA, Pattukumar V, Arul V (2010) Isolation, characterization and identification of a potential probiont from south Indian fermented foods (kallappam, koozh and mor kuzhambu) and its use as biopreservative. Probiotics Antimicrob Proteins 2:145–151. https://doi.org/10.1007/S12602-010-9052-5/METRICS

Kumar R, Kaur M, Garsa AK, Shrivastava B, Reddy VP, Tyagi A (2015) Natural and cultured buttermilk. In: Fermented milk and dairy products. CRC Press, Taylor and Francis, Boca Raton, FL, pp 203–225

Kumar R, Bansal P, Singh J, Dhanda S (2020) Purification, partial structural characterization and health benefits of exopolysaccharides from potential probiotic Pediococcus acidilactici NCDC 252. Process Biochem 99:79–86. https://doi.org/10.1016/J.PROCBIO.2020.08.028

Kurahashi A (2021) Ingredients, functionality, and safety of the japanese traditional sweet drink amazake. J Fungi 7(6):469. https://doi.org/10.3390/jof7060469

Laureys D, De Vuyst L (2014) Microbial species diversity, community dynamics, and metabolite kinetics of water Kefir fermentation. Appl Environ Microbiol 80:2564–2572. https://doi.org/10.1128/AEM.03978-13/ASSET/7F71A806-CE74-46B4-A597-82529B5EE3D0/ASSETS/GRAPHIC/ZAM9991052760005.JPEG

Lopitz-otsoa F, Rementeria A, Elguezabal N, Garaizar J (2006) Kefir: a symbiotic yeasts-bacteria community with alleged healthy capabilities. Rev Iberoam Micol 23:67–74

Madden D (2008) Ginger beer: a traditional fermented low-alcohol drink. Sci Sch

Madilo FK, Kunadu APH, Tano-Debrah K, Mensah GI, Saalia KF, Kolanisi U (2022) Process and product characterization of aliha, a maize-based Ghanaian indigenous fermented beverage. J Food Qual. https://doi.org/10.1155/2022/5604342

Manovina M, Thamarai Selvi B, Prathiviraj R, Selvin J (2022) Potential probiotic properties and molecular identification of lactic acid bacteria isolated from fermented millet porridge or ragi koozh and jalebi batter. Anim Gene 26:200134. https://doi.org/10.1016/J.ANGEN.2022.200134

Mascarello AF, Pinto GI, de Araújo IS, Caragnato LK, da Silva ALL, dos Santos LF (2019) Technological and biological properties of buttermilk: a minireview. Whey Biol Prop Altern Uses. https://doi.org/10.5772/INTECHOPEN.80921

Moonga HB, Schoustra SE, Linnemann AR, Kuntashula E, Shindano J, Smid EJ (2019) The art of mabisi production: a traditional fermented milk. PLoS One 14:e0213541. https://doi.org/10.1371/JOURNAL.PONE.0213541

Moonga HB, Schoustra SE, van den Heuvel J, Linnemann AR, Samad MS, Shindano J, Smid EJ (2020) Composition and diversity of natural bacterial communities in mabisi, a traditionally fermented milk. Front Microbiol 11:1816. https://doi.org/10.3389/FMICB.2020.01816

Moreno-Terrazas R, Reyes-Morales H, Huerta-Ochoa S et al (2016) Note. Consumer awareness of the main sensory attributes of tepache, a traditional fermented fruit beverage. Food Sci Technol Int 7:411–415. https://doi.org/10.1106/BUHF-KMGC-AQ4M-3RXR

Mugula JK, Nnko SAM, Narvhus JA, Sørhaug T (2003) Microbiological and fermentation characteristics of togwa, a Tanzanian fermented food. Int J Food Microbiol 80:187–199. https://doi.org/10.1016/S0168-1605(02)00141-1

Muyanja CMBK, Narvhus JA, Treimo J, Langsrud T (2003) Isolation, characterisation and identification of lactic acid bacteria from bushera: a Ugandan traditional fermented beverage. Int J Food Microbiol 80:201–210. https://doi.org/10.1016/S0168-1605(02)00148-4

Muyanja CMBK, Narvhus JA, Langsrud T (2012) Organic acids and volatile organic compounds produced during traditional and starter culture fermentation of bushera, a Ugandan fermented cereal beverage. Food Biotechnol 26:1–28. https://doi.org/10.1080/08905436.2011.617252

Neve H, Heller KJ (2002) The microflora of water kefir: a glance by scanning electron microscopy. Kieler Milchwirtsch Forschungsberichte 54:337–349

Nutakor C, Essiedu JA, Adadi P, Kanwugu ON (2020) Ginger beer: an overview of health benefits and recent developments. Fermentation 6:102. https://doi.org/10.3390/fermentation6040102

Oguro Y, Nishiwaki T, Shinada R, Kobayashi K, Kurahashi A (2017) Metabolite profile of koji amazake and its lactic acid fermentation product by Lactobacillus sakei UONUMA. J Biosci Bioeng 124:178–183. https://doi.org/10.1016/j.jbiosc.2017.03.011

Palaniswamy SK, Govindaswamy V (2016) In-vitro probiotic characteristics assessment of feruloyl esterase and glutamate decarboxylase producing Lactobacillus spp. isolated from traditional fermented millet porridge (kambu koozh). LWT Food Sci Technol 68:208–216. https://doi.org/10.1016/J.LWT.2015.12.024

Pasqualone A, Summo C, Laddomada B et al (2018) Effect of processing variables on the physico-chemical characteristics and aroma of borş, a traditional beverage derived from wheat bran. Food Chem 265:242–252. https://doi.org/10.1016/J.FOODCHEM.2018.05.095

Patra M, Bashir O, Amin T, Wani AW, Shams R, Chaudhary KS, Mirza AA, Manzoor S (2023) A comprehensive review on functional beverages from cereal grains—characterization of nutraceutical potential, processing technologies and product types. Heliyon 9:e16804. https://doi.org/10.1016/j.heliyon.2023.e16804

Penna B, Casarotti SN, Lelis G et al (2023) Effects of amazake produced with different Aspergillus on gut barrier and microbiota. Foods 12:2568. https://doi.org/10.3390/FOODS12132568

Pérez-Armendáriz B, Cardoso-Ugarte GA (2020) Traditional fermented beverages in Mexico: biotechnological, nutritional, and functional approaches. Food Res Int 136:109307. https://doi.org/10.1016/J.FOODRES.2020.109307

Phister TG, O'Sullivan DJ, McKay LL (2004) Identification of bacilysin, chlorotetaine, and iturin a produced by Bacillus sp. strain CS93 isolated from pozol, a Mexican fermented maize dough. Appl Environ Microbiol 70:631–634. https://doi.org/10.1128/AEM.70.1.631-634.2004

Pieniz S, Andreazza R, Anghinoni T, Camargo F, Brandelli A (2014) Probiotic potential, antimicrobial and antioxidant activities of Enterococcus durans strain LAB18s. Food Control 37:251–256. https://doi.org/10.1016/J.FOODCONT.2013.09.055

Pswarayi F, Gänzle MG (2019) Composition and origin of the fermentation microbiota of mahewu, a Zimbabwean fermented cereal beverage. Appl Environ Microbiol 85:e03130–e03118. https://doi.org/10.1128/AEM.03130-18

Pujiati, Sulistyarsi A, Prafitasari NF (2022) The quality test of fermented ginger drink (ginger Ale) produced from various types of indonesian ginger. Proc 2nd Int Conf Educ Technol (ICETECH 2021) 630:152–159. https://doi.org/10.2991/assehr.k.220103.023

Reva ON, Zaets IE, Ovcharenko LP, Kukharenko OE, Shpylova SP, Podolich OV, de Vera JP, Kozyrovska NO (2015) Metabarcoding of the kombucha microbial community grown in different microenvironments. AMB Express 5:1–8. https://doi.org/10.1186/S13568-015-0124-5/FIGURES/4

Rizo J, Rogel MA, Guillén D, Wacher C, Martinez-Romero E, Encarnación S, Sánchez S, Rodríguez-Sanoja R (2020) Nitrogen fixation in pozol, a traditional fermented beverage. Appl Environ Microbiol 86:1–15. https://doi.org/10.1128/AEM.00588-20

Romero-Luna HE, Peredo-Lovillo A, Dávila-Ortiz G (2022) Tepache: a pre-Hispanic fermented beverage as a potential source of probiotic yeasts. ACS Symp Ser 1406:135–147. https://doi.org/10.1021/BK-2022-1406.CH009

Sagdic O, Ozturk I, Yapar N, Yetim H (2014) Diversity and probiotic potentials of lactic acid bacteria isolated from gilaburu, a traditional Turkish fermented European cranberrybush (Viburnum opulus L.) fruit drink. Food Res Int 64:537–545. https://doi.org/10.1016/J.FOODRES.2014.07.045

Saigusa N, Ohba R (2007) Effects of koji production and saccharification time on the antioxidant activity of amazake. Food Sci Technol Res 13:162–165. https://doi.org/10.3136/fstr.13.162

Sarıözkan S, Türk G, Eken A, Bayram LC, Baldemir A, Dogan G (2017) Gilaburu (Viburnum opulus L.) fruit extract alleviates testis and sperm damages induced by taxane-based chemotherapeutics. Biomed Pharmacother 95:1284–1294. https://doi.org/10.1016/J.BIOPHA.2017.09.057

Schoustra SE, Kasase C, Toarta C, Kassen R, Poulain AJ (2013) Microbial community structure of three traditional Zambian fermented products: mabisi, chibwantu and munkoyo. PLoS One 8:e63948. https://doi.org/10.1371/journal.pone.0063948

Sevindik O, Guclu G, Agirman B, Selli S, Kadiroglu P, Bordiga M, Capanoglu E, Kelebek H (2022) Impacts of selected lactic acid bacteria strains on the aroma and bioactive compositions of fermented gilaburu (Viburnum opulus) juices. Food Chem 378:132079. https://doi.org/10.1016/J.FOODCHEM.2022.132079

Simango C (1997) Potential use of traditional fermented foods for weaning in Zimbabwe. Soc Sci Med 44:1065–1068. https://doi.org/10.1016/S0277-9536(96)00261-4

Simova E, Beshkova D, Angelov A, Hristozova T, Frengova G, Spasov Z (2002) Lactic acid bacteria and yeasts in kefir grains and kefir made from them. J Ind Microbiol Biotechnol 281(28):1–6. https://doi.org/10.1038/SJ/JIM/7000186

Smid EJ, Kleerebezem M (2014) Production of aroma compounds in lactic fermentations. Annu Rev Food Sci Technol 5:313–326. https://doi.org/10.1146/ANNUREV-FOOD-030713-092339

Soltan Dallal MM, Davoodabadi A, Abdi M, Hajiabdolbaghi M, Sharifi Yazdi MK, Douraghi M, Tabatabaei Bafghi SM (2017) Inhibitory effect of Lactobacillus plantarum and Lb. fermentum isolated from the faeces of healthy infants against nonfermentative bacteria causing nosocomial infections. New Microbes New Infect 15:9–13. https://doi.org/10.1016/J.NMNI.2016.09.003

Sönmezdağ AS, Sevindik O, Kelebek H, Selli S (2017) Aroma compounds of non-alcoholic fermented beverage: gilaburu juice. EuroBiotech J 1:226–229. https://doi.org/10.24190/ISSN2564-615X/2017/03.05

Talib N, Mohamad NE, Yeap SK, Ho CL, Masarudin MJ, Abd-Aziz S, Izham MNM, Kumar MR, Hussin Y, Alitheen NB (2023) Anti-diabetic effect of Lactobacillus paracasei isolated from Malaysian water kefir grains. Probiotics Antimicrob Proteins. https://doi.org/10.1007/S12602-023-10159-2/METRICS

Tamang JP, Watanabe K, Holzapfel WH (2016) Review: Diversity of microorganisms in global fermented foods and beverages. Front Microbiol 7:377. https://doi.org/10.3389/fmicb.2016.00377

Tanguler H, Erten H (2012) Chemical and microbiological characteristics of shalgam (şalgam): a traditional Turkish lactic acid fermented beverage. J Food Qual 35:298–306. https://doi.org/10.1111/J.1745-4557.2012.00447.X

Teoh AL, Heard G, Cox J (2004) Yeast ecology of Kombucha fermentation. Int J Food Microbiol 95:119–126. https://doi.org/10.1016/J.IJFOODMICRO.2003.12.020

Thirumangaimannan G, Gurumurthy K (2013) A study on the fermentation pattern of common millets in Koozh preparation—a traditional South Indian food. Indian J Tradit Knowl 12:512–517

Tozetto LT, Nascomento RF, Oloveora M, Beok J, Giovanetti Canteri MH (2019) Production and physicochemical characterization of craft beer with ginger (Zingiber officinale) Luciano. Food Sci Technol (Campinas) 39:962–970

Victor-Aduloju A, Anyamene C, Ogbu K, Ishiwu C (2018) Isolation and identification of probiotic Lactobacillus species from traditional drink kunun-zaki fortified with paddy rice and sweet potatoes. Afr J Food Sci 12:230–237. https://doi.org/10.5897/AJFS2018.1725

Vieira-Dalodé G, Jespersen L, Hounhouigan J, Moller PL, Nago CM, Jakobsen M (2007) Lactic acid bacteria and yeasts associated with gowé production from sorghum in Bénin. J Appl Microbiol 103:342–349. https://doi.org/10.1111/J.1365-2672.2006.03252.X

Waldherr FW, Doll VM, Meißner D, Vogel RF (2010) Identification and characterization of a glucan-producing enzyme from Lactobacillus hilgardii TMW 1.828 involved in granule formation of water kefir. Food Microbiol 27:672–678. https://doi.org/10.1016/J.FM.2010.03.013

Wang P, Feng Z, Sang X et al (2021) Kombucha ameliorates LPS-induced sepsis in a mouse model. Food Funct 12:10263–10280. https://doi.org/10.1039/D1FO01839F

Xiong RG, Wu SX, Cheng J, Saimaiti A, Liu Q, Shang A, Zhou DD, Huang SY, Gan RY, Li HB (2023) Antioxidant activities, phenolic compounds, and sensory acceptability of kombucha-fermented beverages from bamboo leaf and mulberry leaf. Antioxidants (Basel, Switzerland) 12(8):1573. https://doi.org/10.3390/ANTIOX12081573

Yamamoto S, Nakashima Y, Yoshikawa J, Wada N, Matsugo S (2011) Radical scavenging activity of the Japanese traditional food, amazake. Food Sci Technol Res 17:209–218. https://doi.org/10.3136/FSTR.17.209

Yeğin S, Üren A (2008) Biogenic amine content of boza: a traditional cereal-based, fermented Turkish beverage. Food Chem 111:983–987. https://doi.org/10.1016/J.FOODCHEM.2008.05.020

Yetiman AE, Ortakci F (2023) Genomic, probiotic, and metabolic potentials of Liquorilactobacillus nagelii AGA58, a novel bacteriocinogenic motile strain isolated from lactic acid-fermented shalgam. J Biosci Bioeng 135:34–43. https://doi.org/10.1016/J.JBIOSC.2022.10.008

Yetiman A, Horzum M, Bahar D, Akbulut M (2023) Assessment of genomic and metabolic characteristics of cholesterol-reducing and GABA producer Limosilactobacillus fermentum AGA52 isolated from lactic acid fermented shalgam based on "in silico" and "in vitro" approaches. In: Probiotics and antimicrobial proteins. Springer, pp 1–18. https://doi.org/10.1007/S12602-022-10038-2

Zamfir M, Angelescu IR, Voaides C, Cornea CP, Boiu-Sicuia O, Grosu-Tudor SS (2022) Non-dairy fermented beverages produced with functional lactic acid bacteria. Microorganisms 10:2314. https://doi.org/10.3390/microorganisms10122314

Zorba M, Hancioglu O, Genc M, Karapinar M, Ova G (2003) The use of starter cultures in the fermentation of boza, a traditional Turkish beverage. Process Biochem 38:1405–1411. https://doi.org/10.1016/S0032-9592(03)00033-5

Chapter 8
Microbial Pigments: A Potential Substitute of Synthetic Colorants in the Food and Healthcare Sectors

Niralee Patel, Anjisha Maharshi, and Nikitasinh Gohil

Abstract Microbial pigments are gaining attention since they are non-toxic, eco-friendly, and have a wide range of applications, making them a promising alternative to synthetic colorants. Synthetic colorants are still widely used due to high production, intensity, and low cost, but their increased usage has expressed concerns as it directly impacts human health. Algae, fungi, and different bacteria have been utilized recently to produce microbial pigments on a commercial scale. These pigments are then employed as food additives and coloring agents to enhance the biological qualities. Addition of biocolors to food products not only improves the color of the food but also enhances the health benefits by scavenging free radicals from the body. Biological pigments are important sources of natural colors with many therapeutic properties apart from serving as food colorant. They are prominently known for their clinical properties such as antioxidant, anticancer, antimicrobial, anti-allergic, and anti-inflammatory and also used as immunosuppressant. In addition to their industrial and pharmaceutical applicability, they also have higher biodegradability and environmental compatibility. This chapter highlights the significance of bio-pigments, techniques to enhance the stability and emphasize its applications in food and health care sectors.

Keywords Microbial pigments · Therapeutic insights · Food colorants · Pharmacological activities

N. Patel · A. Maharshi · N. Gohil (✉)
Department of Microbiology, Marwadi University, Rajkot, Gujarat, India
e-mail: nikitasinh.gohil@marwadieducation.edu.in

V. Kothari et al. (eds.), *Microbial Products for Health and Nutrition*,
https://doi.org/10.1007/978-981-97-4235-6_8

8.1 Introduction

8.1.1 Microbial Pigments

Pigments, also known as dyes, are a chemical compound capable of inherently light-emitting properties, of selective wavelength, which has many implications. Production of bio-pigments from microbial species is conducted in high demand which is derived from bacteria, fungi, yeasts, and marine bacteria. Organisms producing these pigments through complex mechanism known to help them with their physiological and molecular processes, for the adaptation in various extreme environments, act as a protection against solar radiation, for their survival, sporulation, bioluminescence, biofilm formation, etc. Microorganisms produce varieties of pigments under the influence of various environmental factors having distinct characteristics mainly related with their ecosystem. They are even known for many implications in major industries from agriculture to textiles, cosmetics to pharmaceuticals as they improve the appearance of various products. For example, demand of various carotenoid pigments such as lycopene, zeaxanthin, lutein, beta-carotene, astaxanthin, canthaxanthin is increased by USD 1.53 billion in 2021 and is ever growing (Rana et al. 2021; Kumar et al. 2018). To overcome this demand, microbial production has been increased over plant-derived carotenoid as it is highly expensive (Ram et al. 2020). For this reason, using gene manipulation harvesting of these pigments can be done in large volumes to overcome the demand of consumers for eco-friendly alternatives of pigments (Shindo and Misawa 2014).

8.1.2 Importance of Natural Colorants

The trend to use synthetic colorants or dye were also popular since back from 1850s due to their supercilious coloring abilities, cost effectiveness, easier to enhance large-scale production, consistent in quality and their quick extraction strategies, which remains the equivalently same till this day. Synthetic pigments are commonly used in the various industries such as food, cosmetics, tanning of leather, pharmaceuticals, rubber industries, and textiles. Importance of synthetic dye is still based on the fact that it provides an influencing effect when added to the products, which plays with the preferences of the consumers due to its appealing appearance, and the product apparel decides its fate. The same can be implemented with cosmetic and food industries. Thus, need to use "synthetic pigments" cannot be neglected to fulfill the consumer requirements. The only thing change today is to shift toward naturally derived pigments from synthetic ones for its serious health and environmental issues raised related to its waste generation from the production process (Rodriguez-Amaya 2016). These waste pollutants and untreated dyes are alarming as they are persistent for a longer period of time, cause coloration of water bodies, and cause detrimental effects on the ecosystem which even reduce the light visibility in the surface layer of water. Presence of Azo dye or heavy metal in synthetic dye is of major concern for some health-conscious countries as they are involved in skin allergies, dermatitis, asthma, bladder cancer, and other health-related issues (Numan et al. 2018). So, they made an outcry against the use of synthetic colorants.

Biopigments are eco-friendly and providential as antioxidant, anticancer, and antimicrobial agents. Natural plant-based dyes are available on the market, but they have disadvantages that need to be considered. Growing plants specifically for dyeing purposes can be expensive. In addition, plant-based pigments denature rapidly when subjected to changes in pH and may not guarantee mass reproducibility. Therefore, researchers used another strategy involving microorganisms (Dikshit and Tallapragada 2018). Advantage of using microbial pigments involves easy and efficient extraction techniques, higher yield, and cost-efficient production by optimization for enhancing the growth parameters for efficient extraction of bio-pigments.

8.1.3 Classification

All variety of color hues can be produced by microorganisms such as black, metallic green, yellow, brown, blue, purple, red, gray, green, orange, and indigo based on chemical groups (Velmurugan et al. 2020). Natural pigments are isolated from prokaryotes and eukaryotes which is further categorized into violacein, quinones, carotenoids, indoles, Prodigiosin, rhodopsin, melanin, phycobiliproteins based on their visual, chemical, spectral properties, and source of origin (Fig. 8.1). Monochromatic pigments are easier to isolate compared to that of polychromatic pigments produced by bacteria from different environments such as blue and green pigments by *Pseudomonas aeruginosa* and yellow, orange, and brown pigments by *Streptomyces* sp. etc. In contrast, synthetic pigments are utilized in food, beverages, and even in dyeing industries which is being prepared in laboratories. In general, some microorganisms possess innate abilities and are good pigment producers, but some non-pigmented microbes can acquire the traits from pigmented ones which can classify as acquired pigments. For example, autotrophs bear pigments for photosynthesis; chlorophyll, carotene, and xanthene whereas in heterotrophs they are present as accessory pigments, which help them for their survival such as yellow pigment which is membrane bond produce by *Xanthomonas oryzae* protects them against photodamaging. This accessory pigment biosynthesis is necessary to identify new species and to establish their relatedness (Agarwal et al. 2023).

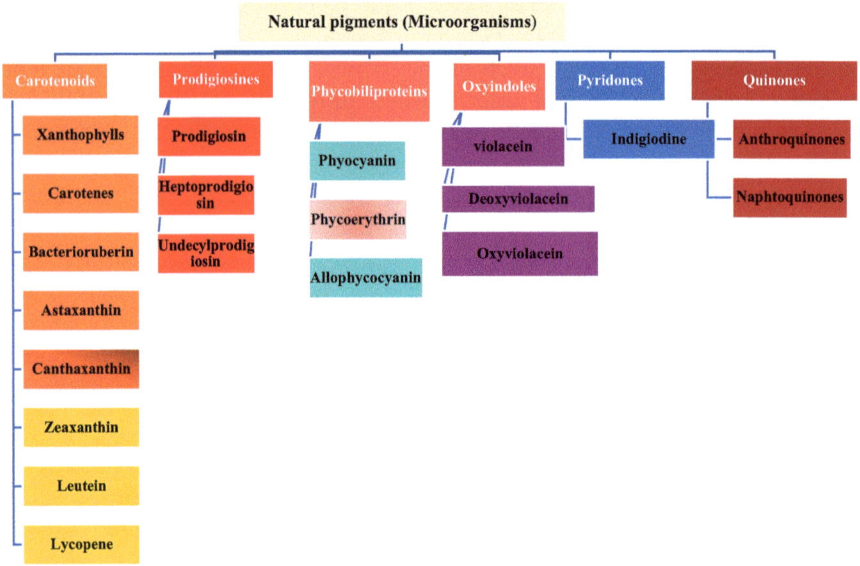

Fig. 8.1 Natural pigments obtained from microbes

8.1.4 Sources

8.1.4.1 Fungi

Pigment producing filamentous fungal growth can be derived from various sources such as mangroves, aquatic source, and soil. Marine fungi are known for its extremes of conditions can produce an axillary metabolite with a remarkable mixture called pigments. Among many microorganisms, fungi show speedy growth and promote higher pigment production, even in a genetically modified manner. Their purification and extraction procedure are easily controlled and handled efficiently during the downstream processing. *Monascus*, *Cordyceps*, *Eurotium*, *Halorosellinia*, *Nigrospora*, *Penicillium*, *Aspergillus*, *Fusarium*, *Talaromyces*, *Phomopsis* are capable of producing carotenoids, melanin, quinones, betalains, anthraquinone, etc. (Lebeau et al. 2017). Some of them also shows antimicrobial, antioxidant, and cytotoxic properties which is more than just a coloring agent. Marine fungi namely *Microsporum* sp. and *Penicillium bilaii* are known to produce yellow colorant fungal polyketide. They are known to promote the fungal growth and provide protection against extreme temperature, desiccation, and photo-oxidation (Meruvu and dos Santos 2021). Isolation of *Talaromyces* sp. and *Trichoderma atroviridae* showed the production of red pigment called azaphilones from maritime sediments. These secondary metabolites are of major importance from biotechnological point of view (Ali et al. 2014). *Trimmatostroma salinum* and *Phaeotheca triangularis* from Adriatic Sea are halophilic strains known to produce melanin pigment.

Mangrove being a specific connectivity between both marine and terrestrial ecosystems have a diversified range of pigment producers (Liu 2012) which can generate the range of pigments. Endophytic fungus, *Alternaria* sp. from mangrove, are known to produce yellow to reddish polyketides namely anthraquinones (Huang et al. 2011).

From terrestrial environments, four bianthraquinone and two monoanthraquinone having orangish-red color compounds were discovered from volcanic ash soil from Japan (Fujitake et al. 1998). Another naphthoquinone pigment was recovered from *Fusarium verticillioides* of Chiang Mai, Thailand (Boonyapranai et al. 2008).

Figure 8.2 depicts chemical structures of secondary metabolites produced by filamentous fungal strains. Ascomycetous fungi are more prominent pigment producers and are mostly of polyketide derivatives.

8.1.4.2 Yeast

Marine yeast is known to produce many bioactive compounds useful in food, pharmaceutical, cosmetic industries (Meruvu and dos Santos 2021). Red yeasts that produce carotenoids, such as Rhodosporidium paludigenum and Sporabolomyces novazealandicus, have been isolated from seawater and marine squid, respectively (Dhaliwal and Chandra 2016). These yeasts have been shown to have antibacterial activity against Shigella sp, Salmonella sp., Escherichia coli, and Vibrio sp. A red yeast named Phaffia rhodozyma is astaxanthin producer showing antioxidant properties (Ye et al. 2019).

8.1.4.3 Bacteria

Among bacteria, several genera are potential pigment producers *Nocardia*, *Kitasatospora*, *Streptomyces*, *Thermomonospora*, *Streptosporangium*, *Rhodococcus* reported for melanin, quinones, indigoidines, and violaceins production which is useful in food, cosmetic, textiles, and even some have good antimicrobial properties with other therapeutic and biomedical importance.

From numerous known species of marine environment only a small fraction has been explored for pigment producers from marine sediments and seawater. Marine bacteria such as *Vibrio* sp., *Bacillus* sp., *Paracoccus* sp., *Pseudomonas* sp., *Pontibacter korlensis*, *Streptomyces* sp. produce compounds such as Prodigiosin, astaxanthin, pyocyanin, melanin, and beta-carotene (Table 8.1).

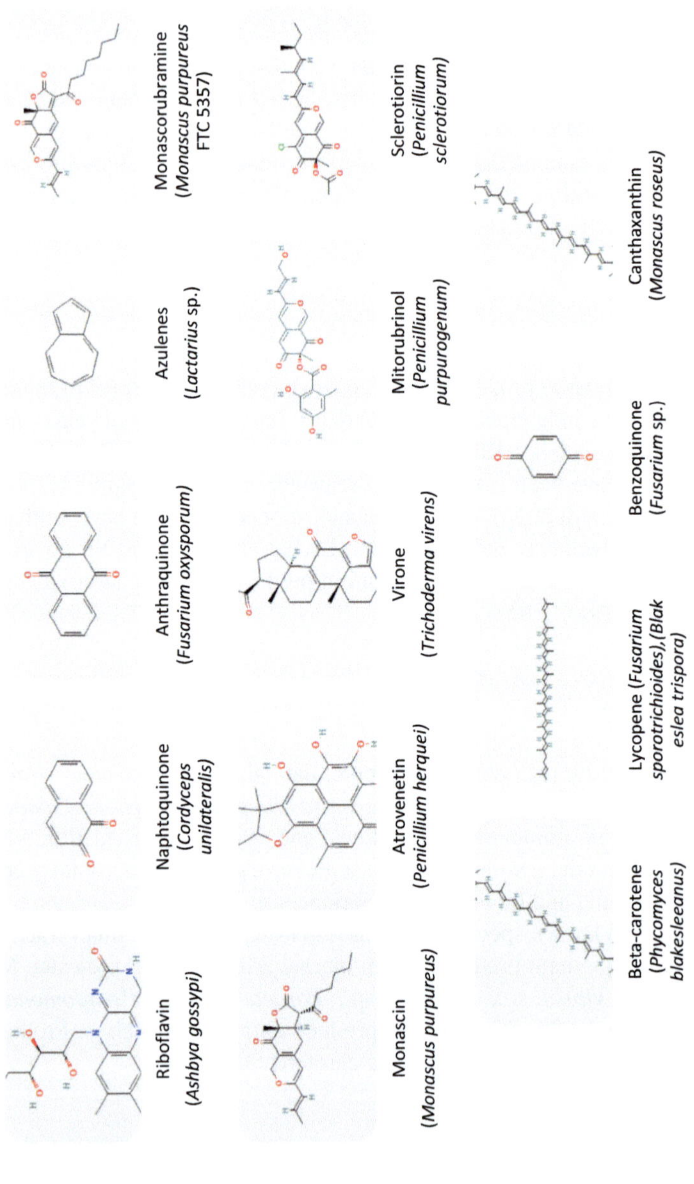

Fig. 8.2 Fungal pigments along with its source

Table 8.1 List of reported genera of bacterial pigment producers

Bacterial producers	Pigment	Structure
Brevundimonas scallop sp. *Corynebacterium glutamicum* sp. *Brevundimonas* sp. strain N-5 *Sphingomicrobium astaxanthinifaciens* sp. nov *Rhodovulum sulfidophilum* sp. *Pontibacter korlensis* sp. AG6 *Exiguobacterium* sp. *Altererythrobacter ishigakiensis* sp. NBRC 107699 *Rhodotorula* sp. *Paracoccus h aeundaensis* sp.	Astaxanthin	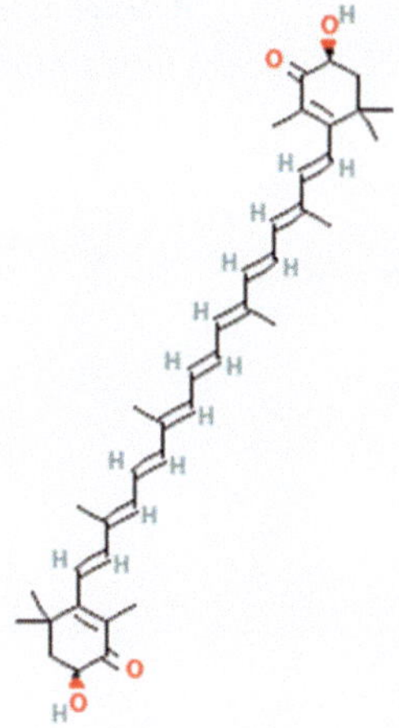
Blastochloris tepida sp. *Salinicoccus roseus* sp.	Lycopene	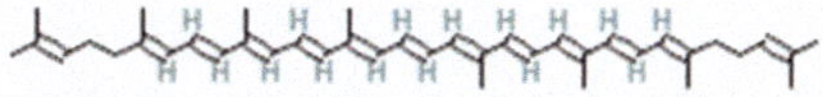

(continued)

Table 8.1 (continued)

Bacterial producers	Pigment	Structure
Cynobacterium *Synechococcus* sp. *Micrococcus* sp. *Vibrio owensii* sp. *Flavicella marina* gen. nov., sp. nov *Gordonia terrae* sp. TWRH01	Beta-carotene	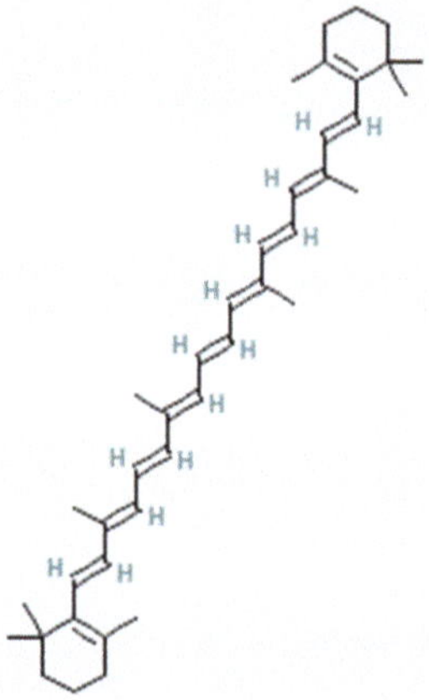
Brevibacterium sp. *Gordonia* sp. *Dietzia* sp.	Canthaxanthin	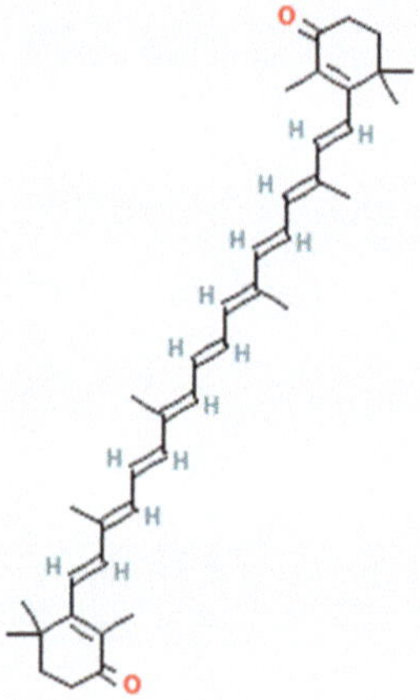

(continued)

Table 8.1 (continued)

Bacterial producers	Pigment	Structure
Pseudomonas aeruginosa sp.	Pyocyanin	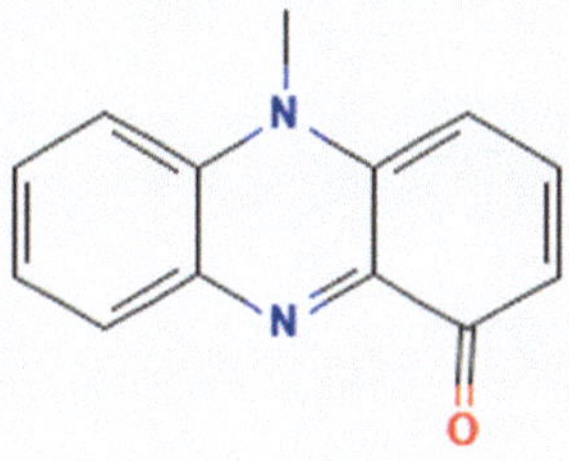
Cyanobacterial sp. *Nostoc punctiforme* sp. ATCC	Scytonemin	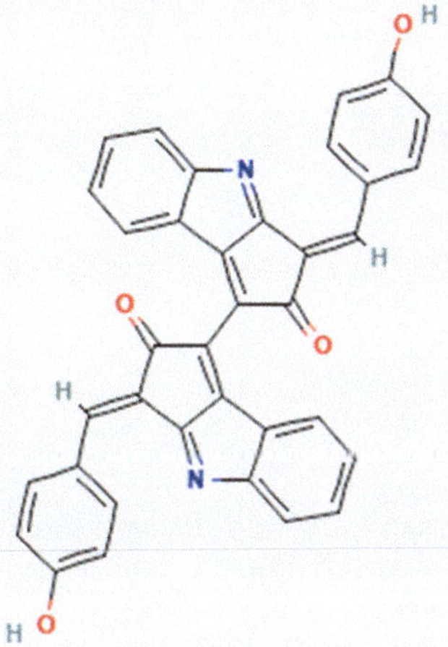

(continued)

Table 8.1 (continued)

Bacterial producers	Pigment	Structure
Pseudoalteromonas luteoviolacea sp. *Pseudoalteromonas ulvae* sp. TC14 *Pseudoalteromonas* sp. 520P1 *Microbulbifer* sp. D250 *Pseudoalteromonas amylolytica* sp. nov *Chromobacterium violaceum* sp. *Collimonas* sp.	Violacein	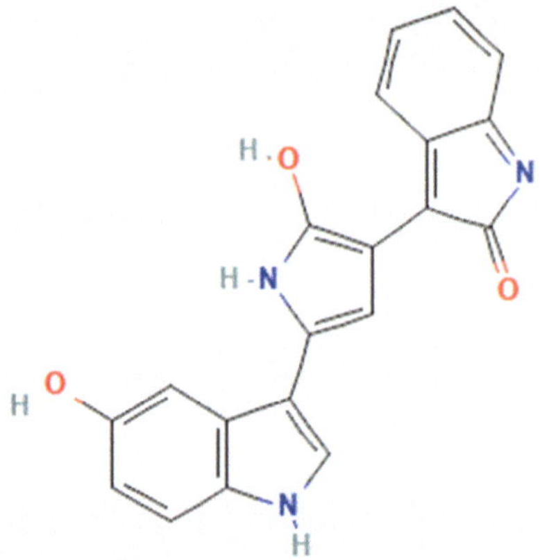
Streptomyces sp. *Pseudomonas* sp. *Marinomonas mediterranea* sp. MMB-1T *Pseudomonas stutzeri* sp. *Bacillus* sp. BTCZ31 *Streptomyces* sp. MVCS13 *Providencia rettgeri* sp. strain BTKKS1 *Marinobacter alkaliphilus* sp. *Leclercia* sp. *Halomonas meridian* sp. *Nocardiopsis dassonvillei* sp. strain JN1 *Vibrio alginolyticus* sp. strain BTKKS3	Melanin	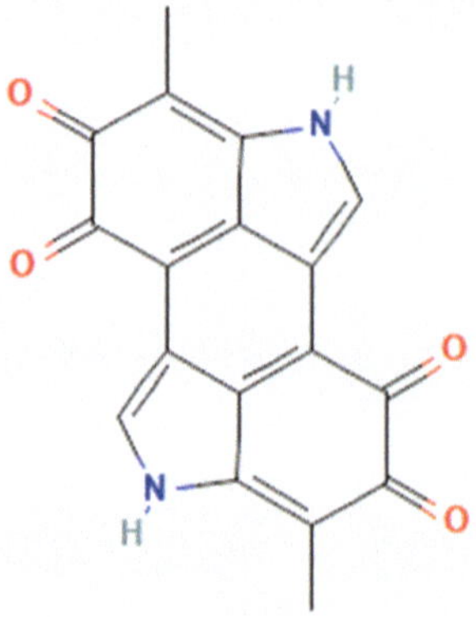

(continued)

Table 8.1 (continued)

Bacterial producers	Pigment	Structure
Sphingomonas phyllosphaerae sp. KODA 19-6 *Mesoflavibacter aestuarii* sp. nov. *Aquibacter zeaxanthinifaciens* gen. nov., sp. nov. *Zeaxanthinibacter enoshimensis* gen. nov., sp. nov *Gramella planctonica* sp. nov *Mesoflavibacter zeaxanthinifaciens* gen. nov., sp. nov *Formosa* sp. KMW *Sphingomonas (Blastomonas) natatoria* sp. DSM 3183^T *Sphingomonas phyllosphaerae* sp. FA2^T *Muricauda lutaonensis* sp. CC-HSB-11T	Zeaxanthin	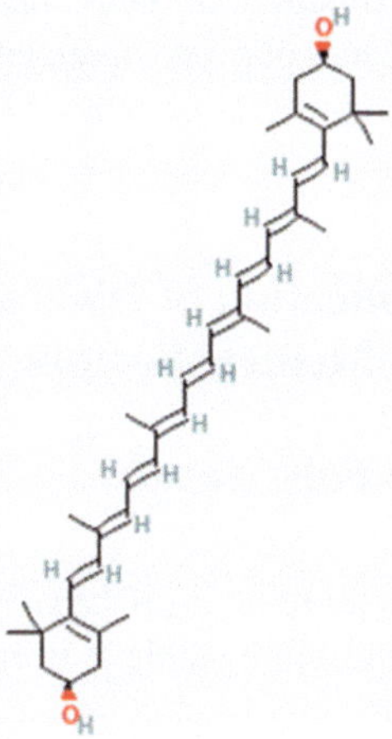

8.1.5 *Pathogenicity*

Despite the numerous applications of pigments produced by different organisms, they are also known for its pathogenicity in humans as well as in aquaculture. An opportunistic pathogens *Chromobacterium violaceum* and a strain of *Serratia marcescens* known for the production of violacein pigment have been reported to cause infection in humans (Sharmin and Kamal 2019). *Janthinobacterium lividum*, another violacein producer's bacterium, are responsible for death of rainbow trout fish *Oncorhynchus mykiss* in Korea. Results showed no involvement of violacein pigments in these virulence activities found in above cases.

Few of the fungi such as *Fusarium* and *Monascus* which are responsible for imparting red pigments, have also shown their involvement in virulence activities and toxins production. Thus, researchers are in search of such microorganisms which show less or no involvement in toxin production or any pathogenic activities to avoid infections in humans and to reduce mortality rate.

8.2 Application of Diverse Bio-Pigments in the Food and Medical Sectors

8.2.1 In Food Sector

The food sector relies more heavily on food coloring. Synthetic food coloring is less expensive and more stable it is encouraged to use it in order to keep costs down. However, artificial food coloring is not good for health because it is created from petroleum waste products. Synthetic pigments have certain drawbacks, for example being less soluble in water, unstable in the presence of heat, light, or acidity, and not always being available. In contrast, microbial pigments are receiving a lot of attention because they are stable, affordable, and resistant to seasonal fluctuations. Microorganisms and plants are the key source of natural pigments; they are easily extracted and can be grown with the help of cultivation technology (Sen et al. 2019).

Mineral pigments like titanium dioxide (E171), calcium carbonate (E170), and iron oxides (E172) have also been permitted by the Food and Drug Administration as food colorants. These pigments are thought to be a good substitute to synthetic food colorants (Rana et al. 2021).

Currently, a range of food color additives are available commercially which are produced by fermentative processes. Pigments that are used as natural colorings are extracted from a variety of microbial sources (Tuli et al. 2015).

8.2.1.1 Red Color Pigments

Carotenoids are lipid-soluble pigments that range greatly in color from red to yellow and are found in many different types of bacteria, algae, fungi, and plants. In nature, carotenoids are widely dispersed and are easily found in vegetables and other plants, which are mainly lipid-soluble. One crucial type of coloring additive utilized in the food sector is carotenoids. It has been documented that a wide range of microorganisms can synthesize carotenoids, including canthaxanthin, zeaxanthin, astaxanthin, and beta-carotene (Elkhateeb and Daba 2023).

Beta-carotene is an orange-red water-insoluble natural pigment and an excellent source of vitamin A. β-Carotene is gaining popularity as a food coloring agent, utilized in cheese, butter, margarine, juices, and beverages at concentrations varying from 2 to 50 parts per million. Since it is the precursor to pro-vitamin A and vitamin

A, it is also served as a dietary supplement. Moreover, one of the main nutritional issues in the world's least developed regions is hypovitaminosis A (Agarwal et al. 2023). Microbial feed stocks are considerably enough to fulfill the need for synthetic derivatives. Beta-carotene is extensively used in ice cream, pastry, crème cheese, etc.

Beta-carotene pigment was primarily derived from an algae *Dunaliella salina* but *Blakeslea trispora* is the most important industrial producer of beta-carotene. It is utilized in many different foods that range in color from red to yellow (Sen et al. 2019).

Prodigiosin is a bacterial secondary metabolite having a tripyrrole group alkaloid structure (Shaikh 2016). It can be used in items like ice cream cones, cereals, canned beverages, and yogurt. It is similar to FD&C Red No. 3 and No. 40. *Hahella chejuensis*, *Vibrio psychroerythrus*, and *Serratia marcescens* are the primary producers of this pigment (Namazkar and Ahmad 2013).

Lycopene, derived from the carotenoid biosynthetic pathway and insoluble in water, is commonly found in red fruits and vegetables. However, the production of lycopene via microbial processes is far more economical and sustainable. This method has been achieved through genetic engineering of host microorganisms such as *Blakeslea trispora*, *E. coli*, and yeasts (Chen et al. 2016). Many nations, including the USA, Australia, and New Zealand, have approved lycopene as a meat coloring agent (Rana et al. 2021).

Arprink red, an extracellular metabolite of the anthraquinone class derived from *Penicillium oxalicum*, can be taken as a dietary supplement and also has an anticancer property. It is used as a food coloring in a range of food products in varying amounts, as per the recommendations made by the Codex Alimentarius Commission.

8.2.1.2 Yellow Color Pigments

Riboflavin, a water-soluble pigment that is also referred to as vitamin B2, is produced by an array of microorganisms. Food additives and supplements containing riboflavin can be found in dairy products, sauces, baby food, fruit, and energy drinks. It helps the body break down dietary polymeric compounds like proteins, fats, and carbohydrates so that the body can use them as fuel and produce energy (Rana et al. 2021).

Zeaxanthin, a kind of xanthophyll produced by *Flavobacterium* sp., has been proposed as an antioxidant, food coloring, and feed additive in poultry. This pigment gives paprika, corn, and wolfberries their characteristic yellow-orange color. After breaking down, zeaxanthin yields safranal and picrocrocin (Yeum and Russell 2002). The taste, aroma, and color of saffron are derived from these two pigments. In general, zeaxanthin is a pigment found in bacteria, including *Staphylococcus aureus*, *Bacillus* sp., and *Corynebacterium michigannise*. Therefore, it is simply used as a biocolorant after extraction. As a food additive, zeaxanthin is marketed under the code E161h (van het Hof et al. 2000).

8.2.1.3 Orange Color Pigments

Canthaxanthin is a powerful antioxidant and lipid-soluble carotenoid with an orange to deep pink color that has shown strong antioxidant qualities (Orlandi et al. 2022). It is approved for use as a food coloring agent in a range of foods, which include poultry feed and fish like salmon, and it is isolated from *Bradyrhizobium* sp. (Surai 2012).

Astaxanthin is a naturally occurring lipid-soluble pigment that is a member of the Xanthophylls class and can be found in yeast, microalgae, marine life, and some feathers of birds (Sen et al. 2019). It is a great antioxidant that, when added to food products as a colorant, not only gives them a pleasing hue but, in addition, lengthens their shelf life, making it a preservative (Surai 2012). *Xanthophyllomyces dendrorhous*, red yeast, contains astaxanthin, a primary carotenoid pigment that is used as a food coloring. Nowadays, astaxanthin is cultivated on a large scale using the red yeast and microalgae. *Haematococcus pluvialis*, *Agrobacterium aurantiacum*, *Paracoccus marcusii*, and *Paracoccus carotinifaciens* are a few examples of bacterial sources (Bandyopadhyay et al. 2019; Kumar and Kim 2019). It is recognized as a colorant for use in fish and animal products (Rana et al. 2021; Sen et al. 2019). Di Salvo et al. (2023) estimate that the market value of astaxanthin is currently US$ 647 million, which by 2026 is expected to have grown at a yearly rate of 8.3–16.8% and will reach US$ 880–968 million (Di Salvo et al. 2023).

8.2.1.4 Blue-Purple-Violet Color Pigments

Phytocyanin is a water-soluble blue-colored accessory pigment for photosynthesis in chlorophyll-A-containing cyanobacteria (Sen et al. 2019). Additionally, produced by *Spirulina* sp. and *Aphanizomenon flos-aquae*, it is also known as "Lina Blue" and is used as an organic coloring ingredient in the food and beverage industry. It is often found in ice cream and other sweet treats (Cuellar-Bermudez et al. 2015). It is also taken as a high-protein dietary supplement.

Indigoidine is one of the organic members of the Azaquinones group, which can replace synthetic food colorings like FD&C Blue NO 1 and 2, which are frequently found in baked goods, cereals, ice cream, yogurt, and snack foods for added color. Indigoidine is extracted from a variety of sources, including bacteria like *Corynebacterium insidiosum*, *Escherichia coli*, and *Streptomyces chromofuscus* (Bandyopadhyay et al. 2019).

Violacein, a secondary metabolite of violet colored vitexin, has a wide range of biological functions. Owing to its color, it can substitute for a number of deadly artificial colors, including FD&C Blue No. 2. It is therefore frequently used in yogurt, ice cream, baked goods, cereals, snack foods, and confections (Bandyopadhyay et al. 2019; Dufossé 2018). Numerous gram-negative bacteria produce violacein, including *Duganella*, *Pseudoalteromonas*, *Collimonas*, *Janthinobacterium lividum*, *Alteromonas luteoviolacea*, and *Chromobacterium violaceum* (Rana et al. 2021; Di Salvo et al. 2023). According to a study, violacein, which is produced by *C. violaceum*, has potential antioxidant properties that also help in strengthening mucosal defense mechanisms and are therefore added to food (Pauer et al. 2018).

Anthocyanins having heterogenous chemical structures mainly extracted from plants, including black rice, black soybeans, blueberries, and raspberries, contain anthocyanins as pigments. However, anthocyanins can also be derived from microbiological sources and offer viable solutions for satisfying industry demand. *E. coli* strain with altered metabolisms is capable of producing anthocyanin. The ability of other modified bacteria to produce this pigment is still being researched (Bandyopadhyay et al. 2019).

8.2.1.5 Monascus Pigments

Vendruscolo et al. (2013) state that these pigments have the following colors: orange (rubropunctatin and monascorubramine), yellow (ankaflavin and monascin), and red (monascorubramine and rubropunctamine). Several fungal species, such as *M. pilosus*, *M. purpureus*, *M. ruberand*, and *M. froridanus*, are the source of these pigments, which have long been used as food colorings in yogurt, red liquor, hams, sausages, and meats (Dufossé et al. 2005). The fungus Monascus produces angkak, which is capable of breaking down starchy substrates into a variety of metabolites, including alcohols, enzymes, fatty acids, flavor substances, organic acids, antihypertensives, pigments, and vitamins. As a result, adding Monascus pigment to food as a coloring agent gave the goods an extra boost of distinct flavor.

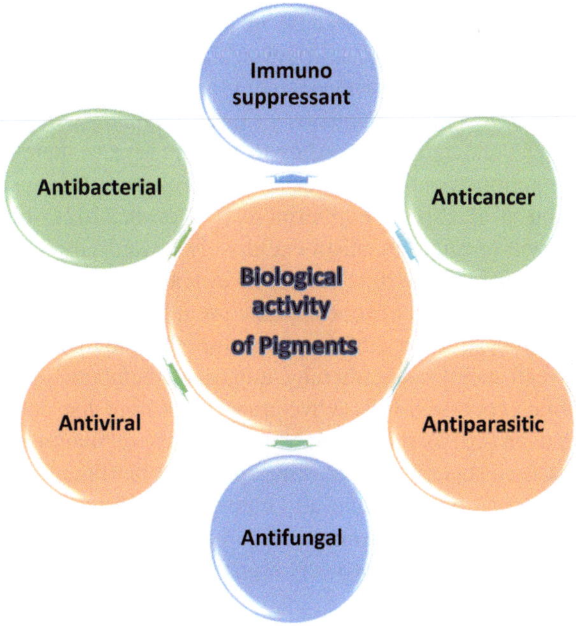

Fig. 8.3 Various biological activity exhibited by the microbial pigments/pharmaceutical importance of microbial pigments

8.2.2 *In Healthcare and Pharmaceutical*

A wide range of synthetic colorants due to their hyper allergenicity, carcinogenicity, and other toxicological problems have been banned or their use is prohibited. So, there is a need to switch to eco-friendly alternatives, moving to the microbial production of pigments will also be helpful in their straightforward downstream processing, easy manipulation, and plentiful availability. Microbial pigments such as prodigiosin, pyocyanin, melanin, carotenoids, zeaxanthin, violacein, astaxanthin, flavins, quinines, monascins, and actinorhodin are subject to have significant properties and potential in modern medicine. There is a need to concentrate on their varied pharmacological features, specifically antimicrobial properties, antioxidant, anticancer, immunoregulator, antiviral, and antiparasitic properties (Fig. 8.3).

8.2.2.1 Antibacterial

Prodigiosin (PG), a red pigment isolated from *Streptomyces* sp. JS520, has been found to give efficient defense against Bacillus and Micrococcus infection (Stankovic et al. 2012). PG acts on cell membrane integrity and diminished *P. aeruginosa* biofilm development (Kimyon et al. 2016). PG, heterocyclic pigments, influenced the permeability and physiology of the cell. PG from *Neisseria* spp. shown potent antifungal action against various fungi (Wagh and Mane 2017). PG from *Chromobacterium prodgiosum* shows antagonism and inhibits the growth of many organisms.

Undecylprodigiosin is efficient toward enteric group of bacteria. On the other hand, Mumtaz et al. (2019) discovered bacterial contagions poses an endless risk to human populations. Because of their rapid spread and prompt mutation, they are often gain resistant to antibiotics. Tagliabue and Rappuoli (2018) estimated that due to antimicrobial resistance each year around ten million mortalities would occur after 2050 and simultaneously cancer frequency will be twice because of population growth and aging. According to a recent review, there is increasing attentiveness in the antimicrobial properties of the red pigment PG against wide range of pathogens (You et al. 2019). Furthermore, Yip et al. (2021) discovered that it suppressed growth of microorganism exclusively toward Gram-positive bacteria. Considering all of this information, further investigation can be conducted to formulate undecylprodigiosin as a novel candidate for antibacterial medications.

Pyocyanin is one of the phenazines thought to interfere with the respiratory chain of cell membranes, thereby impairing membrane-bound, energy-requiring metabolic processes, like active transport within the cell. According to Pierson and Pierson (2010), these pigments control the regulation of cellular genes that promote bacterial survival and the development of biofilms. Pyocyanin showed effectiveness against *C. albicans* and all Gram-positive bacteria, according to El-Shouny et al. (2011). Even in minute quantity, it is able to inhibit Gram-positive bacteria; however, it has no effect at a very high dosage on gram-negative bacteria (Raji El Feghali and Nawas 2018). The blue-green hue of *Pseudomonas* spp. represents its presence. It was previously identified as colicin due to its ability to suppress the growth of *E. coli*, *P. aeruginosa*, *S. aureus*, *S. saprophyticus*, and *E. faecalis* (Venil et al. 2020).

Melanin purified from *P. balearica* has been documented as a phytopathogenic Erwinia biological control agent (Zerrad et al. 2014). Roseoflavin, a potential broad-ranged antibiotic extracted by the bacteria *S. davaonensis* and *S. cinnabarinus*, prevents the development of various pathologically important organisms (Mora-Lugo et al. 2019).

Violacein has antimicrobial effects against Gram-positive bacteria including *S. aureus* and is harmful to bacteriovorus predators like *Caenorhabditis elegans* (Choi et al. 2015). Potential antimicrobial chemicals synthesized by marine pigmented bacteria pertain to marinoazepinone B, marinoqsuinoline I, and marinopyrazinone B (Kalinovskaya et al. 2017). It has extracted from *Chromobacterium violaceum* and demonstrated higher effectiveness as an antibacterial agent against *S. aureus*, perhaps by degradation of membrane integrity. Violacein prevented the biofilm development of *S. epidermidis* and was active against MRSA strains (You et al. 2019).

Other pigments like roseoflavin and toxoflavin purified from *Streptomyces* spp. and *Burkholderia* spp. correspondingly also exhibit inhibitory activity against microorganisms (Mora-Lugo et al. 2019).

The carotenoids obtained from *Halomonas* sp. demonstrated antimicrobial activity against Gram-negative (*Klebsiella* sp., *P. aeruginosa*, *E. coli*) and Gram-positive (*S. aureus*, *S. pyogenes*) organisms. Orange pigment forming *P. flavipulchra* inhibited several multi-drug-resistant bacteria (Ayuningrum et al. 2017). In a research study by Sibero and coworkers, extracted and identified a yellow pigment (carotene) yielding bacteria—*V. owensii* TNKJ.CR.24-7 (MH488980.1) from an unidentified coral, and the extract of pigment suppressed many *MRSA* strains (Sibero et al. 2019).

Cerein, a generated green color pigment by *Bacillus cereus* strains, is bactericidal to their other *Bacillus cereus* bacteria (Naclerio et al. 1993).

Thus, recent research has looked into the antibacterial prospective in reads to use pigments of halophilic bacteria. Although these colorants are natural, they have no adverse impact on the user, showing their candidature more appealing for mass production of antibiotics. Furthermore, some current research has also been signified that these pigments with antimicrobial activity may prevent the development of toxic biofilms (Xie et al. 2023) and have abundant applications as antimicrobial and can be used in the pharmacological and animal fodder industries.

8.2.2.2 Antiviral

Indirubin, pigment which have an indigoid color that can be synthesized by flavin-containing monooxygenase, inhibited the expression and formation of the chemokine (RANTES) in human cells H292 infested with influenza viruses (Mak et al. 2004).

Violacein A heterocyclic compound exhibits inhibition of herpes simplex virus 1, Simian rotavirus SA11, and poliovirus type 2 viral replication. Antiviral action of pigments as violacein generated by *C. violaceum* against the above-mentioned organisms has been demonstrated through minimizing viral replication efficiency (Durán et al. 2010). The antiviral properties of violacein derived from *Janthinobacterium lividum* XT1 have also been demonstrated (Lu et al. 2009).

PG reported in many investigations may represent a novel class of antiviral chemicals. PG, exhibits antiviral properties when examined against cells infested with BmNPV (*Bombyx mori* nucleopolyhedrovirus); furthermore, it hinders viral DNA synthesis and renders anticipation of virus-mediated fusion proteins. A very recent study on *Serratia marcescens* documented that it has produced a separate class of PG, having antiviral efficacy against the herpes simplex virus (HSV). The antiviral properties are because of intensified cell mortality through the viricidal complex destroys the host signaling pathway required for herpes virus existence in the host cell (Suryawanshi et al. 2020). All of the bioactive chemicals discussed above are having great possibilities for additional investigation in order to enter the unexploited market of antiviral compound from microbial pigment origin.

8.2.2.3 Antifungal Activities

Pyocyanin extracted extracellularly from *Pseudomonas aeruginosa* exhibits potential antifungal activity without harmful effect to humans. Additionally, mycotoxin concentrations produced by *F. graminearum* were observed to be considerably reduced by pyocyanin formed by *P. aeruginosa*, hence preventing the development of the allied fungi (Houshaymi et al. 2019). Pyocyanin exhibits antifungal activity against several fungus belonging to Ascomycetes class of via seizure of the electron transport chain (Orlandi et al. 2022).

PG, which binds to chitinase, inhibited the germination of *Mycosphaerella fijiensis* (Gutiérrez-Román et al. 2015). Likewise, *Neisseria* spp. produces a unique red pigment with exceptional antifungal activity against several major fungi which are infectious to humans, i.e., *Candida* spp., *Aspergillus* spp., and *Trichoderma* spp. (Wagh and Mane 2017). *R. psychrophilum* GL8 isolate was isolated from India (Pangong Tso, Leh Ladakh), forming various prodigiosins of having red pigment. These recently exposed red pigments show antifungal properties against yeast and *C. albicans* (MTCC 277, ATCC 90028). Moreover, it has been also validated by Bisht et al. (2020) regarding their synergistic effects as an antifungal compound in combination with certain antifungal medicines, like amphotericin B, vancomycin, luconazole, and erythromycin.

Violacein has been shown to be effective against several species of fungi and yeasts that are pathogenic to plants and humans. Bacterial symbiotic association is known to contribute to the protection of the skin of amphibians. Additionally, analysis of the frog's skin microbiome exposed two inhabitant bacteria producing pigments, first one violacein producing *Janthinobacter lividum* and second one PG producing *S. marcescens* (Woodhams et al. 2018). Separated pigments from these bacteria exhibited efficient antifungal properties against chytrid fungal suspension of *B. salamandrivorans* (Bsal) and *B. dendrobatidis* (Bd), by violacein having major inhibitory ability toward fungal cultures. Recently, yellow pigments, considered to belong to the carotenoid family, have antifungal activity against several economically important fungal pathogens, such as *S. rolfsi* and *R. solani* (Dawoud et al. 2020). In accordance with research study by Bisht et al. (2020), violacein and deoxyviolacein have antimycotic efficacy on pathogenic fungus *Rosellinia necatrix*.

Carotenoid, yellow pigment formed by *B. gibsonii*, purified from the *D. aegialita* (lichen), exhibits peculiar antifungal properties against economically important fungi and, with extensive research and clinical studies, can be in practice as a prescribed antifungal formulation and also formulized as an ointment against fungal infection (Dawoud et al. 2020). As aforementioned, it exemplifies and confirms bioactive pigments have great potential to be used as fungicides if more exploration and testing will lead toward the field of micro-pharmacology (Agarwal et al. 2023).

8.2.2.4 Antiparasitic Activities

Parasitic infections are among the most dangerous and rapidly transmitted vector-borne infections. These include helminthiasis, leishmaniasis, malaria, and encephalitis diseases. These parasites enter into the host and consume essential nutrients, thus harming the host's whole development. The data itself are disturbing, but it doesn't stop there. Approximately one billion populaces has been suffering at least one type of aforesaid parasitic infection. These infections are not just restricted to only human but also extending up to plants and animals, hence triggering an economic agony. Because of the multifaceted pathogenicity and evolutionarily advanced virulence, cultivating formulations that eliminate parasites while being safe for the host is consistently a challenging task. From the continent of Antarctica, an exploration of psychrotolerant bacteria exposed that a species of *Pseudomonas* harvested biologically active pigments revealed its antiparasitic activity (Silva et al. 2018).

PG and violacein exhibiting antimalarial activity and the violacein are active against chloroquine-resistant *Plasmodium falciparum strains* (Lopes et al. 2009). Also noted in a previous study that at concentrations more than pentamidine, violacein showed anti-onchocerciasis activity without prominent side effects (Leon 2001). These kinds of studies reflect the true effectiveness of bacterial pigments and divulge the unexploited potential of psychopaths.

Violacein and deoxyviolacein pigments have large-scale production and usage in the pharmaceutical as antimicrobials, although in current years their antiparasitic activity (especially against *P. falciparum* and *T. cruzi*) has attract the attention. Determinations are being made for production of violacein in bulk scale and its immense usage as drug against plasmodium and trypanosomal cure (Bilsland et al. 2018).

Prodigiosins and their synthetic derivatives are also found to be promising antiprotozoal compounds (Woodhams et al. 2018). In addition, these pigments exhibit uttermost decline in nematode infection in plants. For example, *S. marcescens pigment* inhibits juvenile forms of *R. similis* (LC50 value 83 μl/ml) *and M. javanica* at minute (LC50 value 79 μl/ml) doses (Rahul et al. 2014). Substantiation from these researches ascertains that the novel application of these bacterial pigments assists in battle against the fatal parasites. Furthermore, recently, in bioengineered *Escherichia coli*, Wilkinson et al. (2020) demonstrated antiparasitic activity of violacein extracted from bacteria against the laboratory reference strain *P. falciparum* 3D7 as well as against patient isolates having drug-resistant and drug-susceptibility confirms their potential agent as drug against malarial.

8.2.2.5 Antioxidant

Antioxidants play an important role in protecting mankind against innumerable infestation and disease prognosis by restraining and eliminating free radicals and research workers are pondering on finding substitute antioxidants from natural resources over the synthetic one to remedying the health.

Violacein has antioxidant properties and plays crucial role in membrane fortification against eukaryotic predation, oxidative stress, and fungal diseases. Numerous researches have elaborated defensive properties of biological compounds against damage caused by oxidative damage to cells. For example, violacein extracted by *C. violaceum* (Antonisamy and Ignacimuthu 2010) indicated to have the capability to guard cells via various defense mechanisms. It was witnessed that a consistent consumption of violacein influenced the gut microbiome of rats. A remarkable report validated that violacein synthesis from *C. violaceum* could be boosting mucosal defense mechanism as an antioxidant (Pauer et al. 2018).

Carotenoid extracted from *K. marina* DAGII, *Meiothermus*, and *Thermus* exhibit strong antioxidant activities and inhibit lipid peroxidation (Rezaeeyan et al. 2017). Furthermore, Fiedor and Burda (2014) also demonstrated that it can eradicate the negative effects of ROS in humans. Moreover, the anti-inflammatory and antioxidant properties of bacterial pigments could help in decrease the chance of emerging cancer and can also have straight effects on cancer cells and lead to programmed cell death (Agarwal et al. 2023). Furthermore, an investigation carried out by Majumdar et al. (2019) found *Planococcus* sp. TRCI producing carotenoid has revealed significant antioxidant properties notable to industrial implementation and forthcoming as bioremediates.

These outcomes indicate that the pigments are potential compounds for subsequent application in food and pharmaceutical.

8.2.2.6 Anticancer Activity

Various factors lead to malignant tendencies in somatic cells—cancer such as unhealthy dietary habits, carcinogenic pollutants, cancer-causing viruses and bacteria, and numerous other unknown causes. Chemotherapy is available to treat solid tumors. However, these drugs do not specifically kill cancer cells because they are not morphologically different from healthy cells. So, in addition to killing tumor cells, some nearby healthy cells also get sabotage. Thus, the limitation of current chemotherapy persuades the exploration for a novel anti-tumor drug. Normal cells, during activity, produce free radicals, which, once accumulated in cells, cause cancer. In recent years, the number of cancer patients has drastically increased and become the second chief reason of death worldwide. The main cause is exposure of individuals uncontrolled to carcinogenic compounds (Siegel et al. 2019). Several microbial pigments have shown noteworthy anti-tumor activity.

Current research has validated that deinoxanthin is enhancing apoptosis in various human cancer cell line. The marine microbiome has been the subject of rigorous analyses because several pigments isolated from marine actinomycetes also show cytotoxic activity against numerous cancer cell lines of human. *S. torulosus*, *Actinomycetes* spp., and *Streptomyces* spp. have known to produce blue, yellow, orange, brown, purple, green, and red pigments (Fernandes et al. 2021), among these red is valued as the greatest therapeutic therapy (Chang et al. 2020).

PG has been projected to have therapeutic properties against progressive breast cancers. It stimulates apoptosis and caspase activation, and they block the cell in the late G1 phase of division cycle. From the genus *Streptomyces* voluminous species are known to yield numerous biologically active substances and pigments that exhibited anti-tumor activity. The exploration for an innovative anticancer drug, *S. aburaviensis* and *S. psammoticus* has selected, perhaps as a new novel anticancer drug (Ramirez-Rodriguez et al. 2018).

Violacein prompts apoptosis in hepatocellular carcinoma cell lines and expresses toxicity to cell and apoptosis respective to amount of doses on Huh7 and NNC#2 cells (Tsukimoto et al. 2021). The cytotoxicity synchronized activity of the microbial pigments makes them impeccable model for cancer treatment. In Archaeon, Haloarchaeal carotenoids are formed by the haloalkaliphile *Natrialba* sp. M6 has alternative opulent source of anticancer bioactive substances. An effective anticancer activity was explained by in vitro studies through inhibitory and apoptosis-dependent matrix metalloproteinase (MMP)-9, which was accountable for angiogenesis and cancer metastasis (Hegazy et al. 2020).

Pyocyanin, found in *P. aeruginosa*, was primarily seen to have a cytotoxic effect on human Panc-1 (pancreatic cancer cell lines). Analogous anticancer properties were observed by pyocyanin on 66.34% of glioma cells and human hepatoma cells (HepG2). It is isolated from a similar organism by Koyun et al. (2022) who had found that pyocyanin has an inhibitory effect on AChE (acetylcholinesterase) and its various activities, viz. protective effects against neurodegenerative damage, antioxidant and inhibitory effects, anticoagulant effects. On the basis of this, it may be forecasted as active ingredient of the drug.

8.2.2.7 Immunosuppressant

A pioneer study in this area was implemented by Kawauchi and coworkers (1997), which has reported that cycloprodogiosine hydrochloride selectively inhibits T-cell proliferation. This was one of the first studies performed and opened up the possibility of using biologically active elements as immunosuppressants.

Prodigiosin,2,20-[3-methoxy-10amyl-50-methyl-4-(1″-pyrryl)] dipyrrylmethene (MAMPDM) was recognized to inhibit the proliferation of B and T lymphocytes. MAMPDM induces apoptosis in Con A-stimulated cells, signifying principally inhibition of T cells (Pandey et al. 2003). A strain genetically similar to *Hahella* sp. is MS-02-063 which had been also reported to have immunosuppressive activity. It also inhibits the proliferation of lymphocytes and limits the liberation of superoxide by these cells. Additionally, Anwar et al. (2022) discussed it as an immunomodulator in various cancer cell types such as natural killer cells (NK), T and B lymphocytes, tumor-associated dendritic cells (TADCs), myeloid-derived suppressor cells (MDSC), and tumor-associated macrophages (TAMs).

In one of the case studies of organ transplants, the patient's cells generate immune response habitually considered the organ tissue to be inanimate, leading to rejection of the organ. Immunosuppressive drugs are administered to these patients for reducing their immune reaction, thereby rendering them immunocompromised (Gutierrez-Dalmau and Campistol 2007). However, Kim et al. (2008) stated that the xanthophyll pigment of *H. chejuensis* has immunosuppressive and anticancer abilities. In recent years, plentiful natural pigments have drawn the perceptiveness of researchers for therapeutic approaches because they have established their properties as anti-inflammatory or regulatory agent's immunity (Di Salvo et al. 2023).

8.3 Enhancing Microbial Pigment Stability

Microorganisms can produce vibrant pigments that have a wide range of applications, but under certain conditions (light, pH, oxygen, UV, temperature), their stability decreases, which shortens their shelf life and causes them to become less colorful over time (Agarwal et al. 2023). Utilizing cutting-edge methods such as nano-encapsulation, nano-emulsion preparations, or nanoformulations extends the shelf life of natural microbial pigments. This increases the market potential in an economical way that makes the microbial pigments more durable for industrial applications (Sen et al. 2019).

8.3.1 Encapsulation

When exposed to certain inherent conditions, bacterial pigments struggle to maintain their properties and become highly unstable in the presence of air, daylight, and temperature fluctuations (Narsing Rao et al. 2017). Important considerations when choosing natural colorant for food preparation include the food's chemical and physical properties, its stability and quality throughout the procedure and its storage, and regulatory concerns. Nutraceutical activity-possessing β-carotene is limited in its application as a food colorant by its low bioavailability, low melting point, stability, and solubility (Sen et al. 2019).

Encasing the active ingredients in microparticles through microencapsulation is an alternate method to increase solubility and prolong the self-life of the material. Any solid, liquid, or gas can be condensed into microparticles by microencapsulation, with a diameter of mm to nm. The shelf life of encapsulated colorants is extended due to their enhanced solubility and stability under ambient conditions.

Numerous studies have been published on the encapsulation of bacterial pigments. Two such methods are the spray-drying method (which uses maltodextrin as the wall material) and the freeze-drying method (which uses modified starch as the wall material) for encapsulating anthocyanin and beta-carotene, respectively (Silva et al. 2013). It has been discovered that these encapsulated powders are stable and effective when used in baked products, curd, and carbonated beverages. The non-encapsulated form of *Serratia marcescens* prodigiosin is unstable in the presence of light, pH, and temperature; however, its encapsulated form is more stable and soluble and can serve as a viable alternative to commercially available synthetic colorants (Namazkar and Ahmad 2013).

8.3.2 Nano-emulsion

The three constituents of the nano-encapsulation or nano-emulsion technique are water, soil, and emulsifier in droplet sizes of no more than 100 nm. The most critical step in the creation of a nano-emulsion is the addition of an emulsifier to reduce the surface tension between the oil and water phases. Surfactants are the most commonly used types of emulsifiers, though lipids and proteins can also be used. Nano-emulsion are more useful than micro- or macro-emulsions due to their resistance to chemical or physical change, kinetic stability, and large surface area. These food coloring nano-emulsion are suitable to be used in the food sector due to their non-toxicity and irritability, among their many other advantages (Rana et al. 2021; Sen et al. 2019; Venil et al. 2020).

Bioactive compounds are more soluble when encapsulated in nano-emulsion, which also promote gradual release and absorption in the gut and absorption through cells (Aswathanarayan and Vittal 2019). Because these nano-emulsions are tasteless, non-toxic, and non-irritating, they are suitable for use in the food industry. Food colorant nano-emulsions are a cost-effective solution because they can considerably lower the quantity of colorant needed to produce the preferred food particle color. A nano-emulsion containing β-carotene was reported to form and stabilize with the use of β-lactoglobulin, a biocompatible emulsifier. The reliability and size of β-carotene nano-emulsions have been studied concerning temperature, pH, and surfactant type (Yi et al. 2014).

8.4 Patents and Commercially Available Product of Microbial Pigments

An invention from Japan has established a method for production of algal colorant, phycocyanin, and phycoerythrin. These pigments have broad applicability as a colorant for foods such as ice cream. A patent number US 2011/0250656A1 claimed production of polyketide-based colorant, that is Monascus-like pigment from a non-toxicogenic and non-pathogenic fungal source. It is used as a coloring agent in food items and/or non-food items and in cosmetic compositions (Mapari et al. 2011). A patent number WO2011151426A2, named carotenoid sunscreen, has described photoprotective, coloring, and antioxidant properties of carotenoid. Violacein, isolated from a non-pathogenic, psychrotolerant microorganism is commercially available from Kerafast-an absolute biotech company.

8.5 Global Market and Demand of Microbial Pigments

In recent times, people are aware of the harmful effects of artificial colors; hence, the demand for biopigments is increasing over synthetic colorants. Continuous research and development have played a vital role in the growth of the market share and increasing the revenue for the market players. As listed in Bio-based Pigment and Dyes Market and Organic Pigments Global Market Reports 2023, some of the companies in the global bio-based pigments and dyes market are Sudarshan Chemical Industries Limited, Kiri Industries, Kronos Worldwide Inc., Huntsman Corporation, BASF SE, Cabot Corporation, Heubach GmbH, Ferro Corporation, Trust Chem Co. Ltd., Dominion Colour Corporation, Neochem Industry, Vib Fast Pigment Pvt. Ltd., Toyocolor Co. Ltd.

The global market value of food-grade microbial pigments is not well-documented. Estimating the true global market demand for microbial pigments is becoming more difficult due to a lack of surveys and literature on the cost and demand of microbial pigments (Chatragadda and Dufossé 2021). Carotenoid value as a global commercial product reached \$1.2 billion in 2010, \$1.5 billion in 2014, and \$2.0 billion by 2022 and is projected to grow at a compound annual growth rate (CAGR) of 5.7% from 2022 to 2027 to reach \$2.7 billion as listed by Global markets for carotenoids (Martínez-Cámara et al. 2021). In the market, prodigiosin and violacein fetch approximately \$5000 × 105 per kg. Compared to synthetic carotenoids (76%) with a value of \$250–2000 kg^{-1}, natural carotenoids (24%) are gaining a higher market value of \$350–7500 kg^{-1}. The two most sought-after pigments in the world by far are astaxanthin and β-carotene, with projected market values of \$225 and \$309 million by 2018, respectively (Bhosale and Bernstein 2005; Chatragadda and Dufossé 2021).

8.6 Future Perspectives and Conclusive Remarks

Microbial pigments are multifunctional biomolecules and have very versatile uses. Microorganisms represent a natural source of diverse pigment production as their ability to control growth, which does not get affected by seasonal variations and tolerate environmental factors. Microbial pigments showed a wide range of applications and can replace synthetic pigment which opens up a range of new roles and possibilities in the field of drugs, food, and other industries which fulfill the need of bio-pigments at global level. They also play a crucial role in environmental conservation and securing the health issues of society. Therefore, alternative sources for pigment extraction need to be explored from various environment to produce novel and potential colorants for their multifaceted roles. Therefore, there is a great need to explore the vast microbial diversity for rare, novel, safe pigments using appropriate tools and techniques. Microbial pigments are a growing story, charting a brighter, more colorful, and more sustainable future for industries and ecosystems. In addition to the coloring properties, the biological properties of this indigenous diversity of fungi, bacteria, and marine algae may open new avenues for their use in the production of other valuable resources. Therefore, there is a need to focus more on the chemical structure and structure–function relationship of pigments for biotechnological applications originating from unstudied and understudied microorganisms rescue to develop their diverse applications to satisfy the public interest and market demand.

References

Agarwal H, Bajpai S, Mishra A, Kohli I, Varma A, Fouillaud M, Dufossé L, Joshi NC (2023) Bacterial pigments and their multifaceted roles in contemporary biotechnology and pharmacological applications. Microorganisms 11(3):614. https://doi.org/10.3390/microorganisms11030614

Ali I, Siwarungson N, Punnapayak H, Lotrakul P, Prasongsuk S, Bankeeree W, Rakshit SK (2014) Screening of potential biotechnological applications from obligate halophilic fungi, isolated from a man-made solar saltern located in Phetchaburi Province, Thailand. Pak J Bot 46(3):983–988

Antonisamy P, Ignacimuthu S (2010) Immunomodulatory, analgesic and antipyretic effects of violacein isolated from Chromobacterium violaceum. Phytomedicine 17(3–4):300–304. https://doi.org/10.1016/j.phymed.2009.05.018

Anwar MM, Albanese C, Hamdy NM, Sultan AS (2022) Rise of the natural red pigment 'prodigiosin' as an immunomodulator in cancer. Cancer Cell Int 22(1):1–20. https://doi.org/10.1186/s12935-022-02815-4

Aswathanarayan JB, Vittal RR (2019) Nanoemulsions and their potential applications in food industry. Front Sustain Food Syst 3. https://doi.org/10.3389/fsufs.2019.00095

Ayuningrum D, Kristiana R, Asagabaldan MA, Sabdono A, Radjasa OK, Nuryadi H, Trianto A (2017) Isolation, characterisation and antagonistic activity of bacteria symbionts hardcoral Pavona sp. isolated from Panjang Island, Jepara against infectious multi-drug resistant (MDR) bacteria. IOP Conf Ser Earth Environ Sci 55:012029. https://doi.org/10.1088/1755-1315/55/1/012029

Bandyopadhyay K, Dutta P, Ganguly S, Pramanik A (2019) A review on microbial-pigment: a good source of biocolour. Pharma Innov J 8(12):416–420. http://www.thepharmajournal.com

Bhosale P, Bernstein PS (2005) Microbial xanthophylls. Appl Microbiol Biotechnol 68(4):445–455. https://doi.org/10.1007/s00253-005-0032-8

Bilsland E, Tavella TA, Krogh R, Stokes JE, Roberts A, Ajioka J, Spring DR, Andricopulo AD, Costa FTM, Oliver SG (2018) Antiplasmodial and trypanocidal activity of violacein and deoxyviolacein produced from synthetic operons. BMC Biotechnol 18(1):22. https://doi.org/10.1186/s12896-018-0428-z

Bisht G, Srivastava S, Kulshreshtha R, Sourirajan A, Baumler DJ, Dev K (2020) Applications of red pigments from psychrophilic Rhodonellum psychrophilum GL8 in health, food and antimicrobial finishes on textiles. Process Biochem 94:15–29. https://doi.org/10.1016/j.procbio.2020.03.021

Boonyapranai K, Tungpradit R, Lhieochaiphant S, Phutrakul S (2008) Optimization of submerged culture for the production of naphthoquinones pigment by fusarium verticillioides. Chiang Mai J Sci 35(3):457–466

Chang Y, Xing L, Sun C, Liang S, Liu T, Zhang X, Zhu T, Pfeifer BA, Che Q, Zhang G, Li D (2020) Monacycliones G–K and ent-Gephyromycin A, angucycline derivatives from the marine-derived Streptomyces sp. HDN15129. J Nat Prod 83(9):2749–2755. https://doi.org/10.1021/acs.jnatprod.0c00684

Chatragadda R, Dufossé L (2021) Ecological and biotechnological aspects of pigmented microbes: a way forward in development of food and pharmaceutical grade pigments. Microorganisms 9(3):1–27. https://doi.org/10.3390/microorganisms9030637

Chen Y, Xiao W, Wang Y, Liu H, Li X, Yuan Y (2016) Lycopene overproduction in Saccharomyces cerevisiae through combining pathway engineering with host engineering. Microb Cell Factories 15(1):113. https://doi.org/10.1186/s12934-016-0509-4

Choi SY, Yoon K, Lee J Il, Mitchell RJ (2015) Violacein: Properties and Production of a Versatile Bacterial Pigment. BioMed Research International 1–8. https://doi.org/10.1155/2015/465056

Cuellar-Bermudez SP, Aguilar-Hernandez I, Cardenas-Chavez DL, Ornelas-Soto N, Romero-Ogawa MA, Parra-Saldivar R (2015) Extraction and purification of high-value metabolites from microalgae: essential lipids, astaxanthin and phycobiliproteins. Microb Biotechnol 8(2):190–209. https://doi.org/10.1111/1751-7915.12167

Dawoud TM, Alharbi NS, Theruvinthalakal AM, Thekkangil A, Kadaikunnan S, Khaled JM, Almanaa TN, Sankar K, Innasimuthu GM, Alanzi KF, Rajaram SK (2020) Characterization and antifungal activity of the yellow pigment produced by a Bacillus sp. DBS4 isolated from the lichen Dirinaria agealita. Saudi J Biol Sci 27(5):1403–1411. https://doi.org/10.1016/j.sjbs.2019.11.031

Dhaliwal MK, Chandra N (2016) Isolation of carotenoids producing marine red yeasts. Indian J Geo-Marine Sci 45(8):1029–1034

Di Salvo E, Vecchio GL, De Pasquale R, De Maria L, Tardugno R, Vadal R, Cicero N (2023) Health and other industries

Dikshit R, Tallapragada P (2018) Comparative study of natural and artificial flavoring agents and dyes. In: Natural and artificial flavoring agents and food dyes. Elsevier, pp 83–111. https://doi.org/10.1016/B978-0-12-811518-3.00003-X

Dufossé L (2018) Red colourants from filamentous fungi: are they ready for the food industry? J Food Compos Anal 69:156–161. https://doi.org/10.1016/j.jfca.2017.11.002

Dufossé L, Galaup P, Yaron A, Arad SM, Blanc P, Chidambara Murthy KN, Ravishankar GA (2005) Microorganisms and microalgae as sources of pigments for food use: a scientific oddity or an industrial reality? Trends Food Sci Technol 16(9):389–406. https://doi.org/10.1016/j.tifs.2005.02.006

Durán M, Faljoni-Alario A, Durán N (2010) Chromobacterium violaceum and its important metabolites—review. Folia Microbiol 55(6):535–547. https://doi.org/10.1007/s12223-010-0088-4

Elkhateeb W, Daba G (2023) Fungal pigments: their diversity, chemistry, food and non-food applications. Appl Microbiol 3(3):735–751. https://doi.org/10.3390/applmicrobiol3030051

El-Shouny WA, El-Shouny WA, Al-Baidani ARH, Hamza WT (2011) Antimicrobial activity of pyocyanin produced by Pseudomonas aeruginosa isolated from surgical wound-infections. Int J Pharm Med Sci 1(1):1–07

Fernandes CJ, Doddavarapu B, Harry A, Dilip SPS, Ravi L (2021) Isolation and identification of pigment producing actinomycete Saccharomonospora azurea SJCJABS01. Biomed Pharmacol J 14(4):2261–2269. https://doi.org/10.13005/bpj/2326

Fiedor J, Burda K (2014) Potential role of carotenoids as antioxidants in human health and disease. Nutrients 6(2):466–488. https://doi.org/10.3390/nu6020466

Fujitake N, Suzuki T, Fukumoto M, Oji Y (1998) Predomination of dimers over naturally occurring anthraquinones in soil. J Nat Prod 61(2):189–192. https://doi.org/10.1021/np9703050

Gutierrez-Dalmau A, Campistol JM (2007) Immunosuppressive therapy and malignancy in organ transplant recipients. Drugs 67(8):1167–1198. https://doi.org/10.2165/00003495-200767080-00006

Gutiérrez-Román MI, Holguín-Meléndez F, Dunn MF, Guillén-Navarro K, Huerta-Palacios G (2015) Antifungal activity of Serratia marcescens CFFSUR-B2 purified chitinolytic enzymes and prodigiosin against Mycosphaerella fijiensis, causal agent of black Sigatoka in banana (Musa spp.). BioControl 60(4):565–572. https://doi.org/10.1007/s10526-015-9655-6

Hegazy GE, Abu-Serie MM, Abo-Elela GM, Ghozlan H, Sabry SA, Soliman NA, Abdel-Fattah YR (2020) In vitro dual (anticancer and antiviral) activity of the carotenoids produced by haloalkaliphilic archaeon Natrialba sp. M6. Sci Rep 10(1):5986. https://doi.org/10.1038/s41598-020-62663-y

Houshaymi B, Awada R, Kedees M, Soayfane Z (2019) Pyocyanin, a metabolite of Pseudomonas aeruginosa, exhibits antifungal drug activity through inhibition of a pleiotropic drug resistance subfamily FgABC3. Drug Res 69(12):658–664. https://doi.org/10.1055/a-0929-4380

Huang CH, Pan JH, Chen B, Yu M, Huang HB, Zhu X, Lu YJ, She ZG, Lin YC (2011) Three bianthraquinone derivatives from the mangrove endophytic fungus Alternaria sp. ZJ9-6B from the South China Sea. Mar Drugs 9(5):832–843. https://doi.org/10.3390/md9050832

Kalinovskaya NI, Romanenko LA, Kalinovsky AI (2017) Antibacterial low-molecular-weight compounds produced by the marine bacterium Rheinheimera japonica KMM 9513T. Antonie Van Leeuwenhoek 110(5):719–726. https://doi.org/10.1007/s10482-017-0839-1

Kim D, Jihyun, Kim F, Joung HY, Soon-Kyeong K, Choong HL, Hong KL (2008) J Microbiol Biotechnol 18(10):1621–1629

Kimyon Ö, Das T, Ibugo AI, Kutty SK, Ho KK, Tebben J, Kumar N, Manefield M (2016) Serratia secondary metabolite prodigiosin inhibits Pseudomonas aeruginosa biofilm development by producing reactive oxygen species that damage biological molecules. Front Microbiol 7. https://doi.org/10.3389/fmicb.2016.00972

Koyun MT, Sirin S, Erdem SA, Aslim B (2022) Pyocyanin isolated from Pseudomonas aeruginosa: characterization, biological activity and its role in cancer and neurodegenerative diseases. Braz Arch Biol Technol 65. https://doi.org/10.1590/1678-4324-2022210651

Kumar P, Kim BS (2019) *Paracoccus* sp. strain LL1 as a single cell factory for the conversion of waste cooking oil to polyhydroxyalkanoates and carotenoids. Appl Food Biotechnol 6(1):53–60

Kumar P, Jun HB, Kim BS (2018) Co-production of polyhydroxyalkanoates and carotenoids through bioconversion of glycerol by *Paracoccus* sp. strain LL1. Int J Biol Macromol 107:2552–2558

Lebeau J, Venkatachalam M, Fouillaud M, Petit T, Vinale F, Dufossé L, Caro Y (2017) Production and new extraction method of polyketide red pigments produced by ascomycetous fungi from terrestrial and marine habitats. J Fungi 3(3):34. https://doi.org/10.3390/jof3030034

Leon LL (2001) Antileishmanial activity of the violacein extracted from Chromobacterium violaceum. J Antimicrob Chemother 48(3):449–450. https://doi.org/10.1093/jac/48.3.449
Liu Y (2012) Spectral response to varying levels of leaf pigments collected from a degraded mangrove forest. J Appl Remote Sens 6(1):063501. https://doi.org/10.1117/1.jrs.6.063501
Lopes SCP, Blanco YC, Justo GZ, Nogueira PA, Rodrigues FLS, Goelnitz U, Wunderlich G, Facchini G, Brocchi M, Duran N, Costa FTM (2009) Violacein extracted from Chromobacterium violaceum inhibits plasmodium growth in vitro and in vivo. Antimicrob Agents Chemother 53(5):2149–2152. https://doi.org/10.1128/AAC.00693-08
Lu Y, Wang L, Xue Y, Zhang C, Xing X-H, Lou K, Zhang Z, Li Y, Zhang G, Bi J, Su Z (2009). Production of violet pigment by a newly isolated psychrotrophic bacterium from a glacier in Xinjiang, China. Biochemical Engineering Journal, 43, 135–141. https://doi.org/10.1016/j.bej.2008.09.009
Majumdar S, Priyadarshinee R, Kumar A, Mandal T, Dasgupta Mandal D (2019) Exploring Planococcus sp. TRC1, a bacterial isolate, for carotenoid pigment production and detoxification of paper mill effluent in immobilized fluidized bed reactor. J Clean Prod 211:1389–1402. https://doi.org/10.1016/j.jclepro.2018.11.157
Mak N-K, Leung C-Y, Wei X-Y, Shen X-L, Wong RN-S, Leung K-N, Fung M-C (2004) Inhibition of RANTES expression by indirubin in influenza virus-infected human bronchial epithelial cells. Biochem Pharmacol 67(1):167–174. https://doi.org/10.1016/j.bcp.2003.08.020
Mapari SAS, Meyer AS, Friswad JC (2011) (12) Patent Application Publication (10) Pub. No.: US 2011/0250656A1. 1(19)
Martínez-Cámara S, Ibañez A, Rubio S, Barreiro C, Barredo J-L (2021) Main Carotenoids Produced by Microorganisms Encyclopedia 1(4):1223–1245. https://doi.org/10.3390/encyclopedia1040093
Meruvu H, dos Santos JC (2021) Colors of life: a review on fungal pigments. Crit Rev Biotechnol 41(8):1153–1177. https://doi.org/10.1080/07388551.2021.1901647
Mora-Lugo R, Stegmüller J, Mack M (2019) Metabolic engineering of roseoflavin-overproducing microorganisms. Microb Cell Factories 18(1):146. https://doi.org/10.1186/s12934-019-1181-2
Mumtaz R, Bashir S, Numan M, Shinwari ZK, Ali M (2019) Pigments from soil bacteria and their therapeutic properties: a mini review. Curr Microbiol 76(6):783–790. https://doi.org/10.1007/s00284-018-1557-2
Naclerio G, Ricca E, Sacco M, De Felice M (1993) Antimicrobial activity of a newly identified bacteriocin of Bacillus cereus. Appl Environ Microbiol 59(12):4313–4316. https://doi.org/10.1128/aem.59.12.4313-4316.1993
Namazkar S, Ahmad WA (2013) Spray-dried prodigiosin from Serratia marcescens as a colorant. Biosci Biotechnol Res Asia 10(1):69–76. https://doi.org/10.13005/bbra/1094
Narsing Rao MP, Xiao M, Li W-J (2017) Fungal and bacterial pigments: secondary metabolites with wide applications. Front Microbiol 8. https://doi.org/10.3389/fmicb.2017.01113
Numan M, Bashir S, Mumtaz R, Tayyab S, Rehman NU, Khan AL, Shinwari ZK, Al-Harrasi A (2018) Therapeutic applications of bacterial pigments: a review of current status and future opportunities. 3 Biotech 8(4):207. https://doi.org/10.1007/s13205-018-1227-x
Orlandi VT, Martegani E, Giaroni C, Baj A, Bolognese F (2022) Bacterial pigments: a colorful palette reservoir for biotechnological applications. Biotechnol Appl Biochem 69(3):981–1001. https://doi.org/10.1002/bab.2170
Pandey R, Chander R, Sainis KB (2003) A novel prodigiosin-like immunosuppressant from an alkalophilic Micrococcus sp. Int Immunopharmacol 3(2):159–167. https://doi.org/10.1016/S1567-5769(02)00114-5
Pauer H, Hardoim CCP, Teixeira FL, Miranda KR, Barbirato DS, de Carvalho DP, Antunes LCM, Leitão ÁAC, Lobo LA, Domingues RMCP (2018) Impact of violacein from Chromobacterium violaceum on the mammalian gut microbiome. PLoS One 13(9):e0203748. https://doi.org/10.1371/journal.pone.0203748
Pierson LS, Pierson EA (2010) Metabolism and function of phenazines in bacteria: impacts on the behavior of bacteria in the environment and biotechnological processes. Appl Microbiol Biotechnol 86(6):1659–1670. https://doi.org/10.1007/s00253-010-2509-3

Rahul S, Chandrashekhar P, Hemant B, Chandrakant N, Laxmikant S, Satish P (2014) Nematicidal activity of microbial pigment from Serratia marcescens. Nat Prod Res 28(17):1399–1404. https://doi.org/10.1080/14786419.2014.904310

Raji El Feghali PA, Nawas T (2018) Pyocyanin: a powerful inhibitor of bacterial growth and biofilm formation. Madridge J Case Rep Stud 3(1):101–107. https://doi.org/10.18689/mjcrs-1000125

Ram S, Mitra M, Shah F, Tirkey SR, Mishra S (2020) Bacteria as an alternate biofactory for carotenoid production: a review of its applications, opportunities and challenges. J Funct Foods 67:103867. https://doi.org/10.1016/j.jff.2020.103867

Ramirez-Rodriguez L, Stepanian-Martinez B, Morales-Gonzalez M, Diaz L (2018) Optimization of the cytotoxic activity of three Streptomyces strains isolated from Guaviare river sediments (Colombia, South America). Biomed Res Int 2018:1–14. https://doi.org/10.1155/2018/2839356

Rana B, Bhattacharyya M, Patni B, Arya M, Joshi GK (2021) The realm of microbial pigments in the food color market. In: Frontiers in sustainable food systems, vol 5. Frontiers Media S.A. https://doi.org/10.3389/fsufs.2021.603892

Rezaeeyan Z, Safarpour A, Amoozegar MA, Babavalian H, Tebyanian H (2017) Original Article: High carotenoid production by a Halotolerant bacterium, Kocuria sp. strain qwt-12 and anticancer. EXCLI J 16:840–851

Rodriguez-Amaya DB (2016) Natural food pigments and colorants. Curr Opin Food Sci 7:20–26. https://doi.org/10.1016/j.cofs.2015.08.004

Sen T, Barrow CJ, Deshmukh SK (2019) Microbial pigments in the food industry—challenges and the way forward. In: Frontiers in nutrition, vol 6. Frontiers Media S.A. https://doi.org/10.3389/fnut.2019.00007

Shaikh Z (2016) Biosynthesis of prodigiosin and its applications. IOSR J Pharm Biol Sci 11(6):01–28. https://doi.org/10.9790/3008-1106050128

Sharmin S, Kamal SM (2019) Review on Chromobacterium violaceum, a rare but fatal bacteria needs special clinical attention. Anwer Khan Modern Medical College J 10(2):169–175. https://doi.org/10.3329/akmmcj.v10i2.44131

Shindo K, Misawa N (2014) New and rare carotenoids isolated from marine bacteria and their antioxidant activities. Mar Drugs 12(3):1690–1698. https://doi.org/10.3390/md12031690

Sibero MT, Bachtiarini TU, Trianto A, Lupita AH, Sari DP, Igarashi Y, Harunari E, Sharma AR, Radjasa OK, Sabdono A (2019) Characterization of a yellow pigmented coral-associated bacterium exhibiting anti-bacterial activity against multidrug resistant (MDR) organism. Egypt J Aquatic Res 45(1):81–87. https://doi.org/10.1016/j.ejar.2018.11.007

Siegel RL, Miller KD, Jemal A (2019) Cancer statistics, 2019. CA Cancer J Clin 69(1):7–34. https://doi.org/10.3322/caac.21551

Silva PI, Stringheta PC, Teófilo RF, de Oliveira IRN (2013) Parameter optimization for spray-drying microencapsulation of jaboticaba (Myrciaria jaboticaba) peel extracts using simultaneous analysis of responses. J Food Eng 117(4):538–544. https://doi.org/10.1016/j.jfoodeng.2012.08.039

Silva TR, Duarte AWF, Passarini MRZ, Ruiz ALTG, Franco CH, Moraes CB, de Melo IS, Rodrigues RA, Fantinatti-Garboggini F, Oliveira VM (2018) Bacteria from Antarctic environments: diversity and detection of antimicrobial, antiproliferative, and antiparasitic activities. Polar Biol 41(7):1505–1519. https://doi.org/10.1007/s00300-018-2300-y

Stankovic N, Radulovic V, Petkovic M, Vuckovic I, Jadranin M, Vasiljevic B, Nikodinovic-Runic J (2012) Streptomyces sp. JS520 produces exceptionally high quantities of undecylprodigiosin with antibacterial, antioxidative, and UV-protective properties. Appl Microbiol Biotechnol 96(5):1217–1231. https://doi.org/10.1007/s00253-012-4237-3

Surai PF (2012) The antioxidant properties of canthaxanthin and its potential effects in the poultry eggs and on embryonic development of the chick. Part 2. Worlds Poult Sci J 68(4):717–726. https://doi.org/10.1017/S0043933912000840

Suryawanshi RK, Koujah L, Patil CD, Ames JM, Agelidis A, Yadavalli T, Patil SV, Shukla D (2020) Bacterial Pigment Prodigiosin Demonstrates a Unique Antiherpesvirus Activity That Is Mediated through Inhibition of Prosurvival Signal Transducers. Journal of Virology 94(13). https://doi.org/10.1128/JVI.00251-20

Tagliabue A, Rappuoli R (2018) Changing priorities in vaccinology: antibiotic resistance moving to the top. Front Immunol 9. https://doi.org/10.3389/fimmu.2018.01068

Tsukimoto A, Saito R, Hashimoto K, Takiguchi Y, Miyano-Kurosaki N (2021) Violacein induces various types of cell death in highly differentiated human hepatocellular carcinoma and hepatitis C virus replicon cell lines. Int J Anal Bio-Sci 9(4):46–56

Tuli HS, Chaudhary P, Beniwal V, Sharma AK (2015) Microbial pigments as natural color sources: current trends and future perspectives. J Food Sci Technol 52(8):4669–4678. https://doi.org/10.1007/s13197-014-1601-6

van het Hof KH, Weststrate JA, West CE, Hautvast JGAJ (2000) Dietary factors that affect the bioavailability of carotenoids. J Nutr 130(3):503–506. https://doi.org/10.1093/jn/130.3.503

Velmurugan P, Venil CK, Veera Ravi A, Dufossé L (2020) Marine bacteria is the cell factory to produce bioactive pigments: a prospective pigment source in the ocean. Front Sustain Food Syst 4. https://doi.org/10.3389/fsufs.2020.589655

Vendruscolo F, Luise Müller B, Esteves Moritz D, de Oliveira D, Schmidell W, Luiz Ninow J (2013) Thermal stability of natural pigments produced by Monascus ruber in submerged fermentation. Biocatalysis and Agricultural Biotechnology 2(3):278–284. https://doi.org/10.1016/j.bcab.2013.03.008

Venil CK, Dufossé L, Renuka Devi P (2020) Bacterial pigments: sustainable compounds with market potential for pharma and food industry. Front Sustain Food Syst 4:1–17. https://doi.org/10.3389/fsufs.2020.00100

Wagh P, Mane R (2017) Identification and characterization of extracellular red pigment producing Neisseria spp. isolated from soil sample. Int J Innov Knowl Concept 5:23–25

Wilkinson MD, Lai HE, Freemont PS, Baum J (2020) A biosynthetic platform for antimalarial drug discovery. Antimicrob Agents Chemother 64(5):e02129–e02119. https://doi.org/10.1128/AAC.02129-19

Woodhams DC, LaBumbard BC, Barnhart KL, Becker MH, Bletz MC, Escobar LA, Flechas SV, Forman ME, Iannetta AA, Joyce MD, Rabemananjara F, Gratwicke B, Vences M, Minbiole KPC (2018) Prodigiosin, violacein, and volatile organic compounds produced by widespread cutaneous bacteria of amphibians can inhibit two Batrachochytrium fungal pathogens. Microb Ecol 75(4):1049–1062. https://doi.org/10.1007/s00248-017-1095-7

Xie Q, Liu G, Zhang Y, Yu J, Wang Y, Ma X (2023) Active edible films with plant extracts: a updated review of their types, preparations, reinforcing properties, and applications in muscle foods packaging and preservation. Crit Rev Food Sci Nutr 63(32):11425–11447. https://doi.org/10.1080/10408398.2022.2092058

Ye K, Fan T, Keen LJ, Han B (2019) A review of pigments derived from marine natural products. Israel J Chem 59(5):327–338. https://doi.org/10.1002/ijch.201800154

Yeum K-J, Russell RM (2002) Carotenoid bioavailability and bioconversion. Annu Rev Nutr 22(1):483–504. https://doi.org/10.1146/annurev.nutr.22.010402.102834

Yi J, Lam TI, Yokoyama W, Cheng LW, Zhong F (2014) Cellular uptake of β-carotene from protein stabilized solid lipid nanoparticles prepared by homogenization–evaporation method. J Agric Food Chem 62(5):1096–1104. https://doi.org/10.1021/jf404073c

Yip C-H, Mahalingam S, Wan K-L, Nathan S (2021) Prodigiosin inhibits bacterial growth and virulence factors as a potential physiological response to interspecies competition. PLoS One 16(6):e0253445. https://doi.org/10.1371/journal.pone.0253445

You Z, Zhang S, Liu X, Zhang J, Wang Y, Peng Y, Wu W (2019) Insights into the anti-infective properties of prodiginines. Appl Microbiol Biotechnol 103(7):2873–2887. https://doi.org/10.1007/s00253-019-09641-1

Zerrad A, Anissi J, Ghanam J, Sendide K, El Hassouni M (2014) Antioxidant and antimicrobial activities of melanin produced by a Pseudomonas balearica strain kinetic of bioactive molecules view project ChemTech: a Chemical Library From Natural Sources View project. https://www.researchgate.net/publication/269391661

Chapter 9
Microbial Enzymes: Relevance to Food Industry

Hafsa Sattar, Saira Yahya, and Ayaz Ahmed

Abstract Enzymes are increasingly replacing chemicals in diverse industrial applications due to their environmentally friendly characteristics. Unlike traditional chemicals, enzymes do not release greenhouse gases, contributing to a reduction in energy consumption. Therefore, to accomplish the increasing need for enzymes in industries and overcome production challenges, microbial origins have emerged as the preferred choice. Microbes, known for their abundant availability, rapid growth, and genetic adaptability, can be engineered to function optimally in diverse industrial settings. The long-standing practice of employing enzymes from microorganisms in food preparation has evolved alongside technological progress. Bacteria and fungi, along with their enzymes, considerably improve the taste and texture of various food products, providing substantial economic benefits to industries. Microbial enzymes, distinguished by numerous benefits such as ease of production parameters and cost-effectiveness, outshine their plant- or animal-derived counterparts. Microbial enzymes, known for their robustness and stability, excel in comparison to their plant and animal counterparts, gaining supremacy across various sectors in industrial bioprocesses.

Keywords Microbial enzymes · Food sectors · Bioprocess technology · Food processing enzymes

---

H. Sattar · S. Yahya
Faculty of Life Sciences, Shaheed Zulfikar Ali Bhutto Institute of Science and Technology, Karachi, Pakistan

A. Ahmed (✉)
Dr. Panjwani Centre for Molecular Medicine and Drug Research, International Center for Chemical and Biological Sciences, University of Karachi, Karachi, Pakistan
e-mail: ayaz.ahmed@iccs.edu

V. Kothari et al. (eds.), *Microbial Products for Health and Nutrition*,
https://doi.org/10.1007/978-981-97-4235-6_9

9.1 Introduction

Biocatalysts are steadily replacing toxic chemicals used in the food industry, such as nitriles, phenols, and amine compounds that contribute to several health risks. Unlike chemical treatments, enzymes do not emit any toxic compounds and contribute to lower energy demand (Singh et al. 2016). To address the growing need for enzymes in industries, overcome challenging health concerns, and improve production conditions, microbial sources have become the preferred option. Microbial enzymes possess numerous valuable advantages as they are abundantly available, cost-effective, grow rapidly, and can be genetically engineered to perform optimally under diverse industrial conditions (Sanchez and Demain 2011). Microbial enzymes find extensive application in numerous industries, such as proteases in the textile industry, keratinases in biomedical, pharmaceuticals, and agro-industrial processes (Brandelli et al. 2010), xylanases in the paper manufacturing step (Gupta et al. 2022), pectinases in the agricultural industry, and amylases, β-galactosidases in the food production process (Vasudha et al. 2023). Their versatile nature makes them essential components of industrial processes. Furthermore, recent advancements in enzyme technology for the food industry are explored, accompanied by an extensive inventory of enzymes utilized in food processing. Microorganisms like bacteria and fungi, along with their enzymes, play a vital part in enhancing the taste, texture, aroma, and anti-staling properties of several food products, providing substantial monetary benefits to industries. Microbial enzymes have several advantages over plant- or animal-derived enzymes, including ease of production, cost-effectiveness, and consistency (Sanchez and Demain 2011). Enzymes derived from microbes are particularly significant for advancements in industrial bioprocesses. Nearly 29% of enzymatic protein production is utilized in the food processing industry, with fungi contributing 58% to this total (Singh et al. 2016).

9.2 A Survey of Commercial Products Based on Microbial Enzymes

In response to the escalating global population and the increasing demand for food resources, the utilization of microorganisms has emerged as a promising solution to meet these challenges. Historically, microbes have served as key players in fermented foods, showcasing their longstanding relevance in the culinary world. Beyond their traditional role in fermentation, microorganisms have found extensive applications in the food industry. Enzymes, polysaccharides, organic acids, colorants, flavors, polyhydric alcohols, vitamins, lipids, and glycolipids are among the diverse array of commercial products derived from microbial sources. These microbial-derived compounds not only contribute to the enhancement of food quality and flavor but also serve as essential ingredients in various food products. Harnessing the potential of microorganisms represents a sustainable and innovative approach to addressing the ever-growing demands of the modern food industry (Antony et al. 2024).

Microbial enzymes with unique utilitarian attributes have become a key factor in the food industry (Soufflet Biotechnologies 2017). These enzymes are instrumental in shaping the desired attributes of products, including taste, texture, appearance, and flavor. Moreover, their significance extends to the realm of biofuels, where they contribute to the production processes. For example, there is a high demand for carbohydrates, a subset of industrial enzymes of prime importance to food and biofuel manufacturing, thus suggesting a boom in the market of industrial enzyme. Among the diverse types of industrial enzymes, proteases dominate the market share because of their extensive utilization across various sectors, such as the food and beverage, detergent, and biofuel industries. Globally, the demand for industrial enzymes has increased drastically, making this a multi-millionaire industry. Notably, in 2009, sales in the USA alone reached an impressive $5.1 billion. The 2028 report on the "Industrial Enzymes Market" considering the steep demand for industrial enzymes predicted that demand will reach 10.2 billion by 2028 with an anticipated annual compound growth rate of 6.6%, suggesting a significant increase and widespread adoption of industrial enzymes across diverse industrial sectors.

9.2.1 Overview of Microbial-Based Products

Microbial-based products have become indispensable in the food industry, offering innovative solutions that address key challenges such as food preservation, flavor enhancement, and the promotion of health benefits. A myriad of microbial sources, encompassing a diverse array of taxonomic entities, are evident within the microbial kingdom, comprising bacteria, fungi, and algae. Fermentation is a notable process in the production of microbial-based products. Fermentation holds a key position in the manufacturing of innovative food ingredients and products. As compared to conventional agriculture practices, fermented products hold advantages in terms of reduced reliance on land, emissions of greenhouse gases, and water usage (Teng et al. 2021). Table 9.1 shows the compilation of a few microbial-derived major food product manufacturers (Pivot Food investment 2022).

9.2.2 Nutritional Benefits of Microbial Enzymes

Trouble-free isolation significantly enhanced yield, cheap production costs with a narrow timeline, and durability are the key factors that elevated the rise in popularity of microbial enzymes. Phytic acid is the primary contributor to the poor mineral bioavailability as it chelates with minerals to form an insoluble and non-absorbable mineral-phytate complex in the gut (Greiner and Konietzny 2006). Thus, phytase stands out as a crucial enzyme in the digestive process, playing a pivotal role in

Table 9.1 Microbial-derived food product manufacturer

S. No.	Company	Formulation
1	Better Dairy	Production of dairy proteins using precision fermentation
2	AIIG	
3	Perfect Day	
4	Maya Milk	
5	Change Foods	
6	Legendairy	
7	Novacca	
8	Remilk	
9	Harmony	Human milk proteins using fermentation for infant formula
10	Richcore Lifesciences	Recombinant proteins such as growth factors
11	Peprotech	
12	Arzeda	Customized fermentation strains for ingredient production
13	Provenance Bio	Synbio tools for generating animal protein
14	Air Protein	Alternative fermentative method to transform CO_2
15	Proprotein	Whey and Caseins proteins for cheese production
16	Final Food	
17	New Culture	
18	Biosyntia	Mycelium as a platform for animal muscle proteins in meat alternatives
19	Fumi Ingredients	Egg protein or egg replacement protein via Microbial fermentation
20	Clara Foods	
21	Geltor	Collagen/gelatin using fermentation platform
22	Motif Food Work	Ingredients for meat and dairy alternatives
23	String Bio	Converts methane for alternative protein applications
24	Yali Bio	Fermentation derived fat from animal or animal free products
25	RhYme Biotechnology	
26	Nourish	
27	Melt and Marble	
28	Change Food	
29	C16 Biosciences	Palm oil alternative produced by fermentation
30	Biotork	Microalgae omega-3 fatty acids
31	Cargill-DSM	Steviol glycosides produced by fermentation
32	DuPont	Probiotics for food, beverage, and dietary supplements
33	Fybraworks Foods	Leg hemoglobin for color and flavor in plant-based meats
34	Impossible Foods	Phytolon—Functional ingredients for plant-based meats by fermentation
35	Cargill	Fermentation-derived postbiotics for food applications

human health. The main role of this enzyme is to counteract the nutritional antagonist effects of phytic acid, which chelates minerals and proteins. So, the phytase enzyme improves the availability of vital nutrients, thus improving overall health

(Song et al. 2019). Notably, the enzymatic breakdown of phytic acid eventually leads to the production of non-toxic derivatives such as free inositol, metals, and biomolecules (Rebello et al. 2017). This occurrence could potentially play a major role in exacerbating malnutrition in Asian nations, given that up to 75% of their total calorie intake originates from cereals, which have significantly reduced amounts of protein and minerals with elevated starch and phytate content (Sands et al. 2009; Savita et al. 2017). Thus, microbial phytase can be an efficient means to eliminate the nutritionally adverse content of cereals and legumes.

9.3 Microbial Enzyme in Food Industries

In this context, the following enzymes will be discussed in detail with mechanisms, applications, and advancements in food technological process (Fig. 9.1).

1. α-amylase
2. Glucoamylase
3. Protease
4. Lactase
5. Phytase
6. Pectinase
7. Xylanase

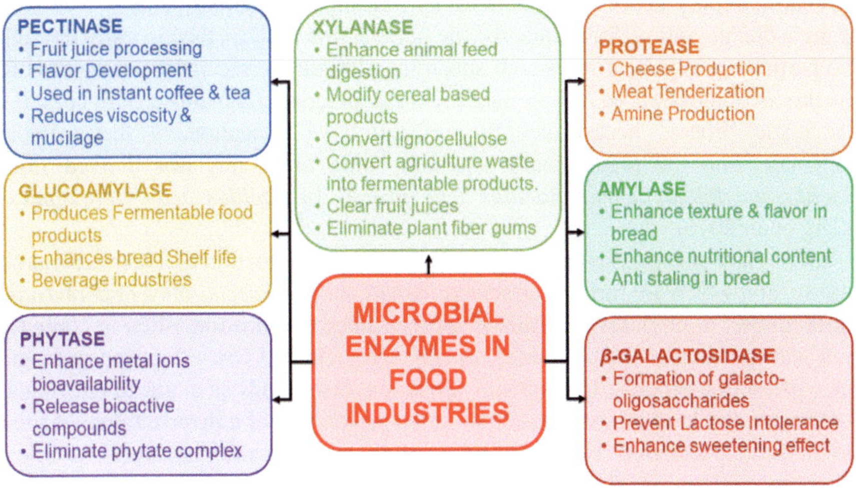

Fig. 9.1 The application of microbial enzymes in food sector

9.3.1 Amylase

The food industry assumes a pivotal role in contemporary society by offering a diverse array of products tailored to a wide spectrum of tastes and dietary preferences. To meet the escalating demand for processed foods that exhibit enhanced texture, flavor, and nutritional characteristics, the food industry extensively relies on a repertoire of enzymes as indispensable instruments for augmenting food manufacturing and quality (Souza 2010).

α-Amylase (EC 3.2.1.1) has the ability to degrade starch by hydrolyzing α-1,4 glycosidic bonds within polysaccharides, ultimately resulting in short-chain dextrins (Raveendran et al. 2018). It is ubiquitously distributed across multifarious living beings. The majority of alpha-amylases fall into the category of metalloenzymes, necessitating the presence of calcium ions for their enzymatic activity, stability, and structural integrity (Couto and Sanromán 2006). α-amylases find extensive applications within the food industry related to brewing, baking, starch liquefaction, and serving as a digestive aid (Van Der Maarel et al. 2002).

Within the bakery sector, α-amylases play a multifaceted role, functioning both as flavor enhancers and anti-staling agents to enhance bread quality. In the course of baking, alpha-amylases are introduced into the dough to get small dextrin from starch to initiate fermentation with the aid of yeast. This enzymatic action not only improves the taste of bread but also enhances its crust color and toasting properties (Aiyer 2005). Additionally, α-amylases serve as glazing agents as they are instrumental in yielding high molecular mass branched dextrins, which aid in the production of powdery foods and rice cakes (Shiv 2015).

In the starch industry, α-amylases are employed in the process of starch liquefaction, a crucial procedure to get glucose and fructose syrups. This conversion is comprised of three sequential steps, including gelatinization, liquefaction, and saccharification. Gelatinization initiates by forming a viscous suspension via starch granule disintegration. Subsequently, the liquefaction reduces the viscosity through the partial hydrolysis of the starch substrate. Ultimately, saccharification leads to the production of desired end products, namely, glucose and maltose. This complex enzymatic process necessitates the utilization of exceptionally thermostable enzymes, with the predominant selection being alpha-amylases derived from *Bacillus amyloliquefaciens, Bacillus licheniformis, or Bacillus stearothermophilus* (Garg et al. 2016).

α-Amylase also plays a significant role in ethanol production as it converts starch into sugars, which are further fermented into alcohol by *Saccharomyces cerevisiae.* Furthermore, α-amylases are utilized for fruit juice clarification, often in conjunction with pectinases and cellulases, to enhance yield and cost-effectiveness in the juice extraction process. In summary, alpha-amylase stands as a crucial enzymatic tool within the food industry, facilitating the production of a diverse range of food products characterized by improved quality, texture, and nutritional profiles. Ongoing exploration of its enzymatic potential, coupled with advancements in enzyme engineering and biotechnology, holds the promise of further expanding its role in shaping the future landscape of food production and innovation (Far et al. 2020).

9.3.2 Glucoamylase

Glucoamylase, scientifically known as α-1,4-glucan glucohydrolase and also referred to as amyloglucosidase, is a multifunctional glycoprotein consisting of 640 amino acids. This glycoprotein comprises a mixture of amylose and amylopectin in a fixed ratio of 1:5. It is an exo-enzyme with a pivotal role in catalyzing starch hydrolysis, specifically targeting the non-oxidizing ends (α-1,4 glycosidic bonds) of starch molecules to form β-glucose. These enzymes are commonly known as saccharifying enzymes and are ubiquitously produced by various living organisms.

However, glucoamylase also possesses the capability to hydrolyze alpha-1,6 linkages within starch, albeit to a lesser extent (Ren et al. 2022). A diverse array of microorganisms have the capacity to produce glucoamylase, including *Paenibacillus amylolyticus*, *Humicola* spp., *Saccharomyces* spp., *Aspergillus niger*, *Alternaria alternate*, and *Bacillus* spp. (Lincoln et al. 2019; Riaz et al. 2007; Prajapati et al. 2014; Kumar et al. 2023; Bretträger et al. 2023; Mendonça et al. 2023). Among these microorganisms, *Aspergillus niger* and *Aspergillus awamori* are the predominant glucoamylase producers, while those derived from *Rhizopus oryzae* find extensive use in industrial applications. It is noteworthy that most glucoamylases exhibit stability at lower temperatures but undergo conformational changes and lose activity at higher temperatures (Almeida et al. 2017).

Glucoamylases are essential in the manufacturing process of fermented goods such as sake, soy sauce, and light beer. They metabolize dextrins, transforming them into fermentable sugars, resulting in beer with lower calorific value and alcohol content. This enzyme holds the distinction of being one of the oldest and most extensively used biocatalysts in the food industry. Its primary function lies in facilitating the saccharification of partially processed starch and dextrin into glucose, a fundamental substrate crucial for several fermentation processes and various sectors within the food and beverage industries (Wayllace et al. 2023). Traditionally, commercial glucoamylase production has relied on filamentous fungi. However, ongoing research endeavors have been dedicated to identifying newer glucoamylases and employing protein engineering techniques to enhance their properties, including thermostability, optimal pH range, and catalytic activity. In beer production, glucoamylase is integral, as it converts brown rice starch into fermentable sugar, a critical step in the brewing process. It also finds applications in soy sauce production and in the hydrolysis of whole grains for alcohol production (Hoang and Toan 2022). Furthermore, glucoamylase plays a major role in the baking industry, contributing to the enhancement of product quality, texture, and nutritional attributes. It plays an important role in improving dough texture, flavor, pasting viscosity, and water content in a wide range of baked products (Struyf et al. 2017). Additionally, glucoamylase reduces the need for added sugar in baked goods while simultaneously preserving the desired color and sweetness. In synergy with other enzymes, it extends the shelf life of bread products by mitigating the effects of staling (Diler et al. 2021). In the realm of beverage industries, glucoamylase assumes a crucial role in the production of fermented beverages like whisky and beer by facilitating the breakdown of starch into fermentable sugars, predominantly glucose. It offers versatility in achieving specific fermentation properties, enabling the production of specialized variants such as light beers or low-calorie beers. However, its stability in beer can pose challenges, potentially leading to increased sweetness over time. Brewers adeptly address this issue through the strategic timing of enzyme addition or the utilization of thermolabile glucoamylases (Blanco et al. 2014). To summarize, glucoamylase stands as a versatile enzyme with expansive applications spanning multiple industries, encompassing food and beverage production, baking, and brewing. Its remarkable ability to efficiently convert starch into glucose and other fermentable sugars renders it indispensable in these sectors, significantly enhancing product quality and process efficiency. Ongoing research endeavors and protein engineering initiatives continue to refine its performance and broaden its spectrum of potential applications.

9.3.3 Protease

Proteases constitute a class of hydrolases that facilitate the hydrolysis of high molecular weight proteins, cleaving peptide bonds to yield smaller peptides. They exhibit broad distribution, encompassing animals and plants, but microorganisms, notably, hold preference for protease production due to their practical and commercial benefits. In the food manufacturing domain, the use of proteases is well acclaimed, either as coagulants, shelf life enhancers, and for the safety of dairy products. Based on their site of action within the polypeptide chain, proteases are categorized as exo- or endo-peptidases. Remarkably, proteases dominate the industrial enzyme market, representing a substantial 60% share. Microbial proteases, mainly those derived from fungal sources, are esteemed for their ease of recovery (Sumantha et al. 2006). Proteases are classified on the basis of their origin (either animal, plant or microorganism), reactive group, and catalytic activity.

Furthermore, proteases are effectively utilized to augment the sensory attributes, nutritional content, solubility, and digestibility of food proteins. Their pivotal role lies in the alteration of the functional characteristics of these proteins, exerting a substantial influence on coagulation and emulsification processes (Miguel et al. 2013). In the realm of the baking industry, proteases are essential for crafting bread, assorted baked goods, waffles, and crackers. Proteases facilitate reduced dough viscosity with reduced mixing time and, moreover, ensure uniformity and strength of gluten, thereby leading to enhancements in both texture and flavor (Deng et al. 2016). For example, the acid protease derived from *Aspergillus usamii* has proven to be highly effective in augmenting the functional attributes of wheat gluten, while *Aspergillus niger* KIBGEIB36 serine metalloprotease has been successfully employed for the degradation of complex casein molecules into smaller peptides (Diler et al. 2021; Sattar et al. 2019).

The dairy industry is yet another domain where proteases hold a prominent position. Naturally occurring proteases substantially influence the flavor of cheese. They accelerate cheese ripening and mitigate the allergenic aspects of milk products (Mamo and Assefa 2018). In cheese production, proteases are instrumental in hydrolyzing specific peptide bonds to generate paracasein and glycomacropeptides (Zaheer and Gupta 2019). Notably, protease-producing microorganisms like Bacillus tequilensis, *Aspergillus niger*, *Mucor miehei*, *A. oryzae*, *B. stearothermophilus*, *B. amyloliquefaciens*, and *M. pusillus* are commonly employed for various applications spanning brewing, baking, meat tenderization, oriental food production (e.g., soy sauce), and cheese manufacturing (Aruna et al. 2014; Aminlari 2022).

Proteases find another significant application in brewing, particularly in achieving proper fermentation. They facilitate the release of amino acids and peptides into the wort, aiding in this crucial process. Acidic proteases from fungal sources are of prime importance as they maintain their working efficiency at an acidic pH, thus enhancing beer fermentation by balancing their amino acid profile.

Within the brewing industry, the catalytic action of proteases assumes a pivotal role in the advancement of beer stability through the reduction of haze formation during cold storage and the augmentation of soluble protein levels. This, consecutively, contributes to the facilitation of yeast fermentation processes. While historical utilization involved papain for such purposes, it has since been replaced by more cost-efficient microbial proteases. Specifically, proteases from *Bacillus subtilis* are employed to heighten amylase activity, elevate the concentration of soluble proteins, and actively participate in the degradation of polypeptide haze precursors within the beer matrix. These enzymatic catalysts exhibit optimal functionality from 45 to 50 °C and maintain a pH environment of approximately 5.6 (Serna-Saldivar and Rubio-Flores 2017).

Microbial proteases find application in enhancing dough extensibility and strength, allowing for precise modulation of dough rheology (Deniz 2019). Wheat flour, a crucial component of baking, contains gluten, composed of gliadin and glutenin proteins, responsible for dough characteristics. These proteins form a viscoelastic network that traps carbon dioxide, vital for dough expansion during baking. Proteases in baking modify gluten's extensional properties, ensuring dough uniformity, reducing consistency, enhancing flavor, and increasing loaf volumes and cookie spread ratios (Dong and Karboune 2021). Notably, acidic protease from *Aspergillus usamii* has been reported to enhance water solubility as well as water and oil retention capacity, thus promoting wheat glutan functional properties (Deng et al. 2016). Similarly, immobilized proteases from *Aspergillus* species KIBGE IB36 have shown improved catalytic behavior at extreme operational conditions, thus further justify the remarkable application of protease enzymes in the food sector (Sattar et al. 2018).

9.3.4 *β-Galactosidase/Lactase*

β-Galactosidase belongs to the hydrolase family and is obtainable from various biological sources, including animals, plants, and microorganisms. Bacteria and fungi are particularly favored for large-scale production due to their better yield and cost-effectiveness. Its function is to cleave the β-1–4 glycosidic bond, which holds glucose and galactose as disaccharide sugar i.e., lactose. The enzyme selection with respect to the source organism is contingent upon their intended application, such as β-galactosidase from yeast, is functional at neutral pH and used to degrade lactose in milk or whey, on the contrary, fungal β-galactosidase with pH optima in the acidic range is used to degrade acidic whey (Panesar et al. 2010). Moreover, the production of this enzyme is dependent on specific ions such as Mn^{2+}, Na^{+}, Mg^{2+}, or K^{+}. These enzymes are used to enhance the solubility and digestibility of milk and its products, thereby facilitating commercial dairy products (Carminati et al. 2010). Additionally, lactase, owing to its transglycosylation activity, also results in galactooligosaccharides (GOS) formation upon lactose hydrolysis, thus adding nutrition value and benefiting prebiotic food (Gibson and Wang 1994). The dual hydrolytic and transgalactosylation properties of β-galactosidases yield several advantages in the realms of the food and dairy industries. These advantages encompass addressing lactose intolerance, the synthesis of prebiotic oligosaccharides, the mitigation of lactose crystallization, and the enhancement of lactose solubility (Fig. 9.2).

Beyond these advantages, lactase's ability to break down lactose into monosaccharides holds the potential to reduce the need for additional sweeteners, ultimately improving the overall sweetness of various products (Zadow 1992; Shah et al. 1993). Apart from this, whey, the primary residual material resulting from cheese production, carries a substantial organic burden of lactose, proteins, and minerals,

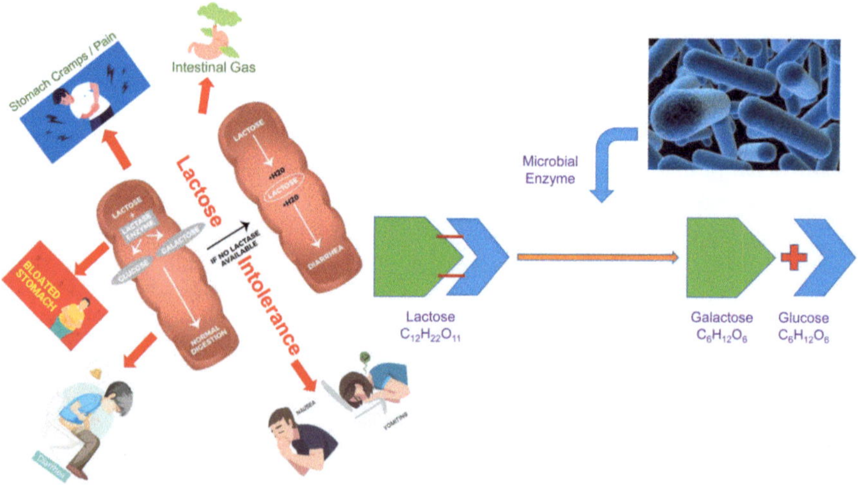

Fig. 9.2 Application of microbial lactase in lactose intolerance patients

contributing to pressing environmental challenges encountered within the dairy industry. These issues revolve around lactose's association with heightened levels of chemical oxygen demand (COD) and biological oxygen demand (BOD) (Khider et al. 2004; Johansen et al. 2002).

An additional significant utilization of lactase lies in the hydrolysis of lactose within whey, a byproduct originating from the cheese manufacturing process. Nevertheless, the superior biodegradation of whey, and the generation of valuable byproducts from whey, such as bioethanol, sweet syrup, and animal feed, is remarkable significance of this enzyme.

9.3.5 *Phytase*

Phytases, a class of hydrolytic enzymes, play a crucial role in degrading phytic acid—an anti-nutritional compound prevalent in plant-based dietary sources like legumes, cereals, grains, and oilseeds (Samtiya et al. 2020). Phytic acid acts as a strong chelator, hindering the absorption of essential metal ions (Fe^{2+}, Cu^{2+}, Fe^{3+}, Ca^{2+}, Zn^{2+}, Mg^{2+}) and leading to micronutrient deficiencies, especially in growth-retarded young children and pregnant women (Raboy 2001; Gibson et al. 2015). Monogastric organisms lack the phytase enzyme needed for phytic acid degradation, resulting in the accumulation of phytate and hindering micronutrient absorption. This unabsorbed mineral content is subsequently excreted in significant amounts, contributing to environmental pollution. To address this global health concern, microbial phytase presents a solution for degrading phytic acid or phytate in plant-based food materials (Joudaki et al. 2023). Phytases are classified based on the sequence through which they liberate phosphate groups from phytic acid, including 3-phosphatases, 4/6-phosphatases, and 5 phosphatases. These enzymes are produced by bacteria and fungi, existing as either cell-bound or intracellular entities (Singh et al. 2020). Fungal phytases, predominantly extracellular, are widely used in commercial applications. Bacteria such as *Bifidobacterium* spp., *Enterobacter* spp., *Lactobacillus casei*, *Serratia*, *Pseudomonas* spp., as well as molds and yeasts like *Aspergillus oryzae*, *Pichia kudriavzevii*, *Sporotrichum thermophile*, *Humicola nigrescens*, *Zygosaccharomyces*, *Aspergillus flavus*, *Penicillium purpurogenum*, *S. cerevisiae*, are known phytase producers (Kłosowski et al. 2018; Kalsi et al. 2016).

Microbial phytases find various applications in the food sector, enhancing phosphorus bioavailability in animal feed formulations and contributing to improved nutrient utilization and growth in monogastric organisms (Fig. 9.3). Additionally, their use can enhance sensory attributes and functional characteristics in food items, releasing bioactive compounds like inositol with potential health benefits (Joudaki et al. 2023).

Phytate negatively impacts mineral absorption in the small intestine. Phytase shows promise in reducing phytin content in bread, improving its quality by elevating nutritional value through phytate reduction (Sandberg et al. 1996). Fungal phytase inclusion in wheat bread rolls resulted in a twofold increase in non-heme iron

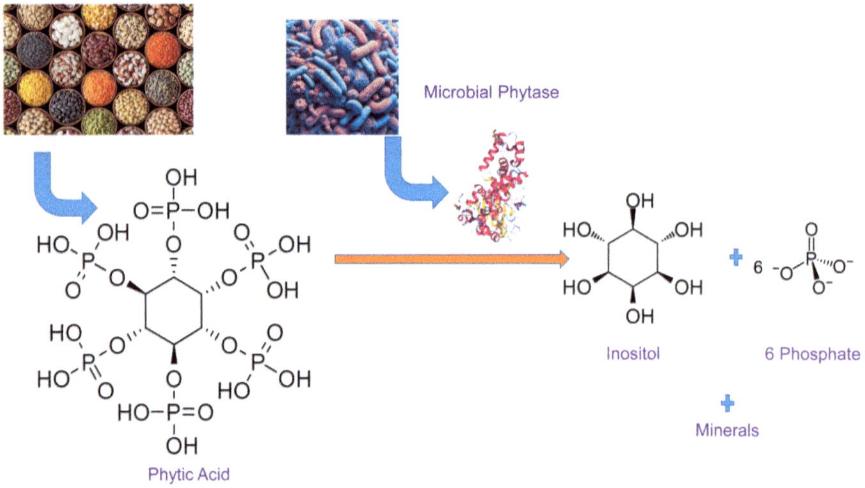

Fig. 9.3 Application of microbial phytase to overcome the bio-availability of Micronutrients

absorption in human subjects compared to non-phytase-treated bread (Simell et al. 1989). Soymilk, with a substantial phytate content (0.56%), can benefit from phytase supplementation to yield phytate-free soybean milk (Khare 1994).

Studies on iron absorption in porridge from various grains revealed improvements, particularly when prepared with water (Caransa et al. 1988). Microbial phytase also has the potential to expedite the steeping process in maize wet milling, increasing corn steep liquor production (Bassi et al. 2021). Traditional Turkish fermented grain meal, Tarhana, exhibited reduced phytic acid content and increased overall mineral and protein levels during fermentation, especially under increasingly acidic conditions. In Indian cuisine, chapathi, a popular whole-wheat flour snack with high phytic acid content, could benefit from using a mutant strain of Candida versatilis as a phytase source, potentially lowering phytate levels by 10–45% (Bassi et al. 2021).

9.3.6 Pectinase

Pectin, an acidic heteropolysaccharide found within the middle lamella of plant cell walls, is subject to enzymatic degradation by pectinases, yielding monomeric sugars, particularly galacturonic acid. Pectinases are extracted from plants, bacteria, and fungi. Numerous microorganisms, such as *Saccharomyces*, *Bacillus*, *Erwinia*, *Aspergillus*, *Penicillium*, *Rhizopus*, and *Fusarium*, are known producers of pectinases (Dwivedi et al. 2023; Yapo et al. 2007). Among these, fungi are often favored for their high production yield and their ability to thrive under varying environmental settings, including pH, temperature, and pressure fluctuations.

The utilization of pectinase dates back to ancient times, with its initial commercial applications observed in 1930, particularly in the fruit and wine beverage industry, to improve the quality of juice (de Souza and Kawaguti 2021; Tapre and Jain 2014). They are applied in the wine industry for macerating fruit tissues, as macerating enzymes in the extraction of oils are used for a range of products, from aromatic goods to cosmetics and the detergent industry. Pectinases are also a fundamental player in various food products, especially in the development of essential flavors for coffee and instant tea preparation to reduce viscosity and mucilage (Kashyap et al. 2001).

Processing fruit to get its juice involves mechanical grinding, resulting in a jelly-like product. Due to the complex pectin structure within fruits, extracting juice from highly viscous, gel-like pulp can be challenging (Shiv 2015). Pectinase enzymes act on pectin to reduce its water-holding capacity and enhance juice yield by cleaving glycosidic bonds between galacturonic acid monomers (Ali et al. 2005). Some juices, like those from apples, berries, and grapes, require intensive pectinase treatment, often in combination with cellulases, to achieve complete clarification, haze reduction, and viscosity control. In contrast, juices like pineapple and orange require milder pectinase treatment to enhance juice extraction and yield without intensive clarification, as these juices are typically consumed with characteristic pulpy clouds. Achieving the desired results involves optimizing process parameters, including enzyme concentration, pH of the solution, incubation temperature, and treatment duration (Ali et al. 2005).

During the fermentation of tea, fungal pectinases target the pectin in tea leaves' cell walls. However, maintaining a specific enzyme concentration is crucial to avoid damaging the tea leaves. Pectinases also serve as antifoaming agents in instant tea powders by breaking down pectin (Schwan and Wheals 2004). In coffee fermentation, pectinases aid in removing the mucilage layer from coffee beans. Enzymatic pectinase preparations offer greater efficiency compared to other methods, as they accelerate the fermentation process (Koshy & De 2019; Collins et al. 2005). Cocoa fermentation plays a vital role in developing chocolate flavor, with pectinolytic microorganisms contributing to the degradation of cocoa pulp. This results in the extraction of premium cocoa beans with superior cocoa flavor (Ouattara et al. 2010).

In the making of jams and jellies, pectin esterase, via the demethylation process, converts high-methoxylated pectins into low-methoxylated pectins. This enzyme demonstrates gelation properties dependent on calcium, which lessens the necessity for additional sugars. As a result, it is utilized in creating compotes, jams, jellies, soups, and sauces (Silva et al. 2000).

9.3.7 *Xylanase*

Xylan, a plant cell wall component, is a complex polymer of xylose subunits connected via 1,4 glycosidic bonds (Collins et al. 2005). Xylanase is an enzyme that catalyzes the xylan complex matrix, generating industrially valued commodities such as xylose, xylobiose, and xylo-oligosaccharides (Chen et al. 2019). Xylanase has impacted various processes in the baking, beverage, paper, textile, and animal feed processing industries (Khanahmadi et al. 2018). Furthermore, it also has a fundamental role in producing bioethanol from various agricultural wastes (Katsimpouras et al. 2019).

The production of xylanase is observed across a diverse range of organisms, including bacteria (Alokika and Singh 2019), molds, actinomycetes, and yeast as well as in protozoans, marine algae, snails, insects, crustaceans, and the seeds of land plants (Nair et al. 2008; Saini et al. 2015; Liu et al. 1998; Harris and Ramalingam 2010). Xylanase holds significant applications in the food industry, contributing to various processes such as refining the digestibility of animal feed, modifying cereal-based products, converting lignocellulosic materials and agricultural waste into fermentable products, clarifying fruit juices, and eliminating plant fiber gums, among other functions (Harris and Ramalingam 2010). In biscuit production, xylanase ensures lightweight cream crackers with better palatability, texture, and uniformity (Christopher and Kumbalwar 2015). When combined with cellulases, amylases, and pectinases, xylanases increase juice yield by liquefying fruits and vegetables, stabilizing fruit pulp, and enhancing the recovery of vitamins, essential oils, aromas, minerals, pigments, edible dyes, etc. They reduce viscosity, hydrolyze substances that impede juice clarification, and prevent cloudiness in juice concentrates (Polizeli et al. 2005). Xylanase, in collaboration with endoglucanase, contributes to the breakdown of starch and arabinoxylan, facilitating the separation of gluten from starch in wheat flour. Additionally, xylanase is utilized in processing coffee bean mucilage (Wong 1993). Characteristics such as its stability and activity at an acidic pH make this enzyme important in the food industry.

Enzymatic processes are widely utilized in fruit juice extraction and clarification. The challenges posed by polysaccharides such as starch, surface-bound lignin, cellulose, hemicellulose, and pectin were easily addressed using enzymes. These components can lead to haziness and increased viscosity, impacting juice quality (Danalache et al. 2018). Enzyme applications, involving centrifugation and filtration, reduction in viscosity and removal of suspended solids, result in enhanced clarity, color, and aroma of the juice (Danalache et al. 2018). Semi-refined xylanase from *Streptomyces* sp. AOA40 has been used in the beverage industry to increase clarity in juices from grapes, oranges, and apples (Adigüzel and Tunçer 2016). Specifically, *Streptomyces* sp. xylanase has demonstrated efficacy in clarifying orange, mousambi, and pineapple juices, achieving clarity percentages of 20.9%, 23.6%, and 27.9%, respectively (Rosmine et al. 2017). Immobilized xylanase from *Bacillus pumilus* VLK-1 has been utilized for enriching orange (29%) and grape juices (26%) (Kumar et al. 2014). Similarly, polyacrylamide *Geobacillus stearothermophilus* xylanase showed remarkable efficiency at 70 °C even after 60 min. This gives a clear example of a bioprocess in a thermal industrial setting (Bibi et al. 2022). In the realm of computational modeling, Hassan et al. (2020) optimized apple juice clarification using immobilized enzymes, achieving approximately 84% clarification through an artificial neural network that maximized the recovery of xylanase and pectinase activity of *Mucor hiemalis* AB1.

9.4 Conclusion

Biocatalysts from microbial sources play a crucial role in diverse industrial sectors, including pharmaceuticals, biomedical, detergent, and paper manufacturing industries. This immense applicability preference over chemical hydrolysis is due to its eco-friendly nature and efficient production at low-cost medium. Unlike enzymes of plant and animal origin, microbial enzymes can be produced with great effectiveness using fermentation techniques. Additionally, large-scale production of microbial enzymes is easily achievable through diverse molecular techniques. An enhanced outcome with specific activity is possible through the overexpression of their genes. Owing to these tremendous properties, microbial enzymes have been used in numerous sectors of food the processing industry.

References

Adigüzel AO, Tunçer M (2016) Production, characterization and application of a xylanase from Streptomyces sp. AOA40 in fruit juice and bakery industries. Food Biotechnol 30(3):189–218

Aiyer PV (2005) Amylases and their applications. Afr J Biotechnol 4(13):1525–1529

Ali A, Williams IH, Martin GB, Sipsas S (2005) Hydrolysis of lupin pectin by pectinases for broilers. In: Proceedings of Australian poultry science symposium, pp 219–222

Almeida PZ, Pereira MG, Carvalho CC, Heinen PR, Ziotti LS, Messias JM, Jorge JA, Polizeli MLTM (2017) Bioprospection and characterization of the amylolytic activity by filamentous fungi from Brazilian Atlantic Forest. Biota Neotropica 17(3):e20170337. https://doi.org/10.1590/1676-0611-BN-2017-0337

Alokika, Singh B (2019) Production, characteristics, and biotechnological applications of microbial xylanases. Appl Microbiol Biotechnol 103:8763–8784

Aminlari M (2022) Plants- and animal-derived enzymes and their potential application in food processing and preservation. In: Novel food grade enzymes: applications in food processing and preservation industries. Springer Nature, Singapore, pp 17–63

Antony C, Selvaraj T, Mohanakrishnan D, Ghodke PK, Sivaprakasam S, Sharma AK (2024) Microbial application in food industry. In: Sustainable production innovations: bioremediation and other biotechnologies. Wiley, pp 273–306

Aruna K, Shah J, Birmole R (2014) Production and partial characterization of alkaline protease from Bacillus tequilensis strains CSGAB 0139 isolated from spoilt cottage cheese. Int J Appl Biol Pharm 5:201–221

Bassi D, Kaur K, Singh TP, Kaur J (2021) Quality attributes and Chapatti making property of biofortified wheat as influenced byparticle size. J Food Sci Technol 58(3):1156–1164

Bibi Z, Sattar H, Nawaz MA, Karim A, Pervez S, Qader SAU, Aman A (2022) Polyacrylamide hydrogel carrier (matrix-type macrogel beads): improvement in the catalytic behavior, stability, and reusability of industrially valuable xylanase from a thermophile Geobacillus stearothermophilus. Curr Res Biotechnol 4:229–237

Blanco CA, Caballero I, Barrios R, Rojas A (2014) Innovations in the brewing industry: light beer. Int J Food Sci Nutr 65(6):655–660

Brandelli A, Daroit DJ, Riffel A (2010) Biochemical features of microbial keratinases and their production and applications. Appl Microbiol Biotechnol 85:1735–1750

Bretträger M, Sacher B, Gastl M, Becker T (2023) The black gap: understanding the potential roles of black fungal-derived enzymes in malting and brewing quality: a review. J Am Soc Brew Chem 82:93–108

Caransa A, Simell M, Lehmussaari A, Vaara M, Vaara T (1988) A novel enzyme application for corn wet milling. Starch-Stärke 40(11):409–411
Carminati D, Giraffa G, Quiberoni A, Binetti A, Suárez V, Reinheimer J (2010) Advances and trends in starter cultures for dairy fermentations. In: Biotechnology of lactic acid bacteria: novel applications. Wiley, pp 177–192
Chen Z, Liu Y, Zaky AA, Liu L, Chen Y, Li S, Jia Y (2019) Characterization of a novel xylanase from Aspergillus flavus with the unique properties in production of xylooligosaccharides. J Basic Microbiol 59(4):351–358
Christopher N, Kumbalwar M (2015) Enzymes used in food industry a systematic review. Int J Innov Res Sci Eng Technol 4(10):9830–9836
Collins T, Gerday C, Feller G (2005) Xylanases, xylanase families and extremophilic xylanases. FEMS Microbiol Rev 29(1):3–23
Couto SR, Sanromán MA (2006) Application of solid-state fermentation to food industry—a review. J Food Eng 76(3):291–302
Danalache F, Mata P, Alves VD, Moldão-Martins M (2018) Enzyme-assisted extraction of fruit juices. In: Fruit juices. Academic, pp 183–200
de Souza TS, Kawaguti HY (2021) Cellulases, hemicellulases, and pectinases: applications in the food and beverage industry. Food Bioprocess Technol 14(8):1446–1477
Deng L, Wang Z, Yang S, Song J, Que F, Zhang H, Feng F (2016) Improvement of functional properties of wheat gluten using acid protease from Aspergillus usamii. PLoS One 11:e0160101
Deniz I (2019) Production of microbial proteases for food industry. In: Green bio-processes: enzymes in industrial food processing. Springer, pp 9–14
Diler G, Rannou C, Guyon C, Prost C, Le-Bail A (2021) Use of amyloglucosidase in a soft wheat dough: impact of process and formulation on glucose production. Appl Food Res 1(2):100007
Dong Y, Karboune S (2021) A review of bread qualities and current strategies for bread bioprotection: flavor, sensory, rheological, and textural attributes. Compr Rev Food Sci Food Saf 20(2):1937–1981
Dwivedi S, Yadav K, Gupta S, Tanveer A, Yadav S, Yadav D (2023) Fungal pectinases: an insight into production, innovations and applications. World J Microbiol Biotechnol 39(11):305
Far BE, Ahmadi Y, Khosroshahi AY, Dilmaghani A (2020) Microbial alpha-amylase production: progress, challenges and perspectives. Adv Pharm Bull 10(3):350
Garg G, Singh A, Kaur A, Singh R, Kaur J, Mahajan R (2016) Microbial pectinases: an ecofriendly tool of nature for industries. 3 Biotech 6:1–13
Gibson GR, Wang X (1994) Regulatory effects of bifidobacteria on the growth of other colonic bacteria. J Appl Microbiol 77(4):412–420
Gibson RS, Wawer AA, Fairweather-Tait SJ, Hurst R, Young SD, Broadley MR et al (2015) Dietary iron intakes based on food composition data may underestimate the contribution of potentially exchangeable contaminant iron from soil. J Food Compos Anal 40:19–23
Greiner R, Konietzny U (2006) Phytase for food application. Food Technol Biotechnol 44(2):13–21
Gupta GK, Dixit M, Kapoor RK, Shukla P (2022) Xylanolytic enzymes in pulp and paper industry: new technologies and perspectives. Mol Biotechnol 64(2):130–143
Harris AD, Ramalingam C (2010) Xylanases and its application in food industry: a review. J Exp Sci 1(7):1–11
Hassan SS, Williams GA, Jaiswal AK (2020) Computational modelling approach for the optimization of apple juice clarification using immobilized pectinase and xylanase enzymes. Curr Res 3:243–255
Hoang VM, Toan ND (2022) Response surface optimization of enzymatic hydrolysis of germinated brown rice for higher reducing sugar production. Vietnam J Food Control 5(4):645
Johansen AG, Vegarud GE, Skeie S (2002) Seasonal and regional variation in the composition of whey from Norwegian Cheddar-type and Dutch-type cheeses. Int Dairy J 12(7):621–629
Joudaki H, Aria N, Moravej R, Rezaei Yazdi M, Emami-Karvani Z, Hamblin MR (2023) Microbial phytases: properties and applications in the food industry. Curr Microbiol 80(12):1–13

Kalsi HK, Singh R, Dhaliwal HS, Kumar V (2016) Phytases from Enterobacter and Serratia species with desirable characteristics for food and feed applications. 3 Biotech 6:1–13

Kashyap DR, Vohra PK, Chopra S, Tewari R (2001) Applications of pectinases in the commercial sector: a review. Bioresour Technol 77(3):215–227

Katsimpouras C, Dedes G, Thomaidis NS, Topakas E (2019) A novel fungal GH30 xylanase with xylobiohydrolase auxiliary activity. Biotechnol Biofuels 12(1):1–14

Khanahmadi M, Arezi I, Amiri MS, Miranzadeh M (2018) Bioprocessing of agro-industrial residues for optimization of xylanase production by solid-state fermentation in flask and tray bioreactor. Biocatal Agric Biotechnol 13:272–282

Khare SK (1994) Entrapment of wheat phytase in polyacrylamide gel and its application in soymilk phytate hydrolysis. Biotechnol Appl Biochem 19:193–198

Khider K, Akretche DE, Larbot A (2004) Purification of water effluent from a milk factory by ultrafiltration using Algerian clay support. Desalination 167:147–151

Kłosowski G, Mikulski D, Jankowiak O (2018) Extracellular phytase production by the wine yeast S. cerevisiae (Finarome Strain) during submerged fermentation. Molecules 23(4):848

Koshy M, De S (2019) Effect of Bacillus tequilensis SALBT crude extract with pectinase activity on demucilation of coffee beans and juice clarification. J Basic Microbiol 59(12):1185–1194

Kumar L, Nagar S, Mittal A, Garg N, Gupta VK (2014) Immobilization of xylanase purified from Bacillus pumilus VLK-1 and its application in enrichment of orange and grape juices. J Food Sci Technol 51:1737–1749

Kumar S, Nath P, Bhattacharyya A, Mazumdar S, Bhattacharjee R, Satyanarayana T (2023) Diversity and biotechnological applications of microbial glucoamylases. In: Glycoside hydrolases. Academic, pp 365–387

Lincoln L, More VS, More SS (2019) Isolation, screening and optimization of extracellular glucoamylase from Paenibacillus amylolyticus strain NEO03. Biocatal Agric Biotechnol 18:101054

Liu W, Zhu W, Lu Y, Kong J, Ma G (1998) Production, partial purification and characterization of xylanase from Trichosporon cutaneum SL409. Process Biochem 33(3):331–336

Mamo J, Assefa F (2018) The role of microbial aspartic protease enzyme in food and beverage industries. J Food Qual. https://doi.org/10.1155/2018/7957269

Mendonça APS, Dos Reis KL, Barbosa-Tessmann IP (2023) Aspergillus clavatus UEM 04: an efficient producer of glucoamylase and α-amylase able to hydrolyze gelatinized and raw starch. Int J Biol Macromol 249:125890

Miguel ÂSM, Martins-Meyer TS, Veríssimo da Costa Figueiredo E, Lobo BWP, Dellamora-Ortiz GM (2013) Enzymes in bakery: current and future trends. In: Muzzalupo I (ed) Food industry. InTech, Rijeka, Croatia

Nair SG, Sindhu R, Shashidhar S (2008) Purification and biochemical characterization of two xylanases from Aspergillus sydowii SBS 45. Appl Biochem Biotechnol 149:229–243

Ouattara HG, Reverchon S, Niamke SL, Nasser W (2010) Biochemical properties of pectate lyases produced by three different Bacillus strains isolated from fermenting cocoa beans and characterization of their cloned genes. Appl Environ Microbiol 76(15):5214–5220

Panesar PS, Kumari S, Panesar R (2010) Potential applications of immobilized β-galactosidase in food processing industries. Enzyme Res 2010:473137

Polizeli MLTDM, Rizzatti ACS, Monti R, Terenzi HF, Jorge JA, Amorim DDS (2005) Xylanases from fungi: properties and industrial applications. Appl Microbiol Biotechnol 67:577–591

Prajapati VS, Trivedi UB, Patel KC (2014) Kinetic and thermodynamic characterization of glucoamylase from Colletotrichum sp. KCP1. Indian J Microbiol 54:87–93

Raboy V (2001) Seeds for a better future:'low phytate'grains help to overcome malnutrition and reduce pollution. Trends Plant Sci 6(10):458–462

Raveendran S, Parameswaran B, Ummalyma SB, Abraham A, Mathew AK, Madhavan A et al (2018) Applications of microbial enzymes in food industry. Food Technol Biotechnol 56(1):16

Rebello S, Jose L, Sindhu R, Aneesh EM (2017) Molecular advancements in the development of thermostable phytases. Appl Microbiol Biotechnol 101:2677–2689

Ren Y, Wang X, Li Y, Li YY, Wang Q (2022) Lactic acid production by fermentation of biomass: recent achievements and perspectives. Sustainability 14(21):14434

Riaz M, Perveen R, Javed MR, Nadeem H, Rashid MH (2007) Kinetic and thermodynamic properties of novel glucoamylase from Humicola sp. Enzym Microb Technol 41(5):558–564

Rosmine E, Sainjan NC, Silvester R, Alikkunju A, Varghese SA (2017) Statistical optimisation of xylanase production by estuarine Streptomyces sp. and its application in clarification of fruit juice. J Genet Eng Biotechnol 15(2):393–401

Saini A, Aggarwal NK, Sharma A, Yadav A (2015) Actinomycetes: a source of lignocellulolytic enzymes. Enzyme Res 2015:279381

Samtiya M, Aluko RE, Dhewa T (2020) Plant food anti-nutritional factors and their reduction strategies: an overview. Food Prod Process Nutr 2:1–14

Sanchez S, Demain AL (2011) Enzymes and bioconversions of industrial, pharmaceutical, and biotechnological significance. Org Process Res Dev 15(1):224–230

Sandberg AS, Hulthén LR, Türk M (1996) Dietary Aspergillus niger phytase increases iron absorption in humans. J Nutr 126(2):476–480

Sands AL, Leidy HJ, Hamaker BR, Maguire P, Campbell WW (2009) Consumption of the slow-digesting waxy maize starch leads to blunted plasma glucose and insulin response but does not influence energy expenditure or appetite in humans. Nutr Res 29(6):383–390

Sattar H, Aman A, Javed U, Qader SAU (2018) Polyacrylamide beads: polymer entrapment increases the catalytic efficiency and thermal stability of protease. Mol Catal 446:81–87

Sattar H, Bibi Z, Kamran A, Aman A, Qader SAU (2019) Degradation of complex casein polymer: production and optimization of a novel serine metalloprotease from Aspergillus niger KIBGE-IB36. Biocatal Agric Biotechnol 21:101256

Savita PD, Yallappa M, Nivetha N, Suvarna VC (2017) Phytate solubilizing microorganisms and enzyme phytase to combat nutritional problems in cereal-based foods. J Bacteriol Mycol Open Access 4(3):86–89

Schwan RF, Wheals AE (2004) The microbiology of cocoa fermentation and its role in chocolate quality. Crit Rev Food Sci Nutr 44(4):205–221

Serna-Saldivar SO, Rubio-Flores M (2017) Role of intrinsic and supplemented enzymes in brewing and beer properties. In: Microbial enzyme technology in food applications. Routledge, p 271

Shah NP, Spurgeon KR, Gilmore TM (1993) Use of dry whey and lactose hydrolysis in yogurt bases. Milchwissenschaft 48(9):494–498

Shiv K (2015) Role of enzymes in fruit juice processing and its quality enhancement. Adv Appl Sci Res 6(6):114–124

Silva CF, Schwan RF, Dias ËS, Wheals AE (2000) Microbial diversity during maturation and natural processing of coffee cherries of Coffea arabica in Brazil. Int J Food Microbiol 60(2–3):251–260

Simell M, Turunen M, Piironen J, Vaara T (1989) Feed and food applications of phytase. Lecture at 3rd meet. In: Industrial applications of enzymes, Barcelona, Spain, pp 33–38

Singh R, Kumar M, Mittal A, Mehta PK (2016) Microbial enzymes: industrial progress in 21st century. 3 Biotech 6:1–15

Singh B, Boukhris I, Kumar V, Yadav AN, Farhat-Khemakhem A, Kumar A, Alghamdi OA (2020) Contribution of microbial phytases to the improvement of plant growth and nutrition: a review. Pedosphere 30(3):295–313

Song HY, El Sheikha AF, Hu DM (2019) The positive impacts of microbial phytase on its nutritional applications. Trends Food Sci Technol 86:553–562

Soufflet Biotechnologies (2017) Enzymes for brewing and distilling. Soufflet Biotechnologies, Nogent-sur-Seine, France. https://www.souffletbiotechnologies.com/wp-content/uploads/2017/11/BrewingDistilling.pdf

Souza PMD (2010) Application of microbial α-amylase in industry—a review. Braz J Microbiol 41:850–861

Struyf N, Van der Maelen E, Hemdane S, Verspreet J, Verstrepen KJ, Courtin CM (2017) Bread dough and baker's yeast: an uplifting synergy. Compr Rev Food Sci Food Saf 16(5):850–867

Sumantha A, Larroche C, Pandey A (2006) Microbiology and industrial biotechnology of foodgrade proteases: a perspective. Food Technol Biotechnol 44(2):211

Tapre AR, Jain RK (2014) Pectinases: enzymes for fruit processing industry. Int Food Res J 21(2):447

Teng TS, Chin YL, Chai KF, Chen WN (2021) Fermentation for future food systems: precision fermentation can complement the scope and applications of traditional fermentation. EMBO Rep 22(5):e52680

Van Der Maarel MJ, Van der Veen B, Uitdehaag JC, Leemhuis H, Dijkhuizen L (2002) Properties and applications of starch-converting enzymes of the α-amylase family. J Biotechnol 94(2):137–155

Vasudha M, Prashantkumar CS, Bellurkar M, Kaveeshwar V, Gayathri D (2023) Probiotic potential of β-galactosidase-producing lactic acid bacteria from fermented milk and their molecular characterization. Biomed Rep 18(3):23

Wayllace NM, Martín M, Busi MV, Gomez-Casati DF (2023) Microbial glucoamylases: structural and functional properties and biotechnological uses. World J Microbiol Biotechnol 39(11):293

Wong KK (1993) Applications of hemicellulases in food, feed, pulp and paper industries. In: Hemicellulose and hemicellulases. Portland Press, pp 127–142

Yapo BM, Lerouge P, Thibault JF, Ralet MC (2007) Pectins from citrus peel cell walls contain homogalacturonans homogenous with respect to molar mass, rhamnogalacturonan I and rhamnogalacturonan II. Carbohydr Polym 69(3):426–435

Zadow JG (ed) (1992) Whey and lactose processing. Springer, Dordrecht, The Netherlands

Zaheer MR, Gupta A (2019) Current development and future perspectives of microbial enzymes in the dairy industry. In: Enzymes in food biotechnology. Academic, pp 287–302

Chapter 10
Microbial Production of Fat-Soluble Vitamins

Amka Nagar, Mohini Singh, Kanu Priya, and Justyna Mozejko-Ciesielska

Abstract Fat-soluble vitamins are micronutrients that solubilize in organic solvent and fat but are insoluble in water. It includes A, D, E, and K which are used in food, pharmaceuticals, and cosmetic industries. With the surge in the demand for vitamins, there is a requirement to invest in unconventional production strategies. Chemical synthesis is costly and uses non-renewable chemicals and toxic compounds which can be hazardous and cause safety concerns. The use of microorganisms for vitamin production is sustainable, economical, and safe for the environment. Biotechnology has a major contribution to the production of vitamins which includes genetic, metabolic, and systematic engineering of microorganisms. In this chapter, the authors attempt to elucidate how and to what extent microbes including bacteria, fungi, and microalgae can be used and modified to produce fat-soluble vitamins.

Keywords Vitamins · Genetic engineering · Metabolic engineering · Fat soluble · β-Carotene · Systems biology

10.1 Introduction

Vitamins are organic substances required in trace amounts (<1 g/day) (Borel and Desmarchelier 2018). They are essential for various biological processes i.e., roles in the development, reproduction, maintenance, and growth of living organisms, and their deficiencies can cause serious health problems (Surana et al. 2019). The

A. Nagar · M. Singh (✉) · K. Priya
Department of Life Sciences, Sharda School of Basic Sciences and Research, Sharda University, Greater Noida, India
e-mail: mohini.singh@sharda.ac.in

J. Mozejko-Ciesielska
Department of Microbiology and Mycology, Faculty of Biology and Biotechnology, University of Warmia and Mazury in Olsztyn, Olsztyn, Poland

V. Kothari et al. (eds.), *Microbial Products for Health and Nutrition*,
https://doi.org/10.1007/978-981-97-4235-6_10

discovery of vitamins started at the beginning of the nineteenth century. Casimir Funk coined the term "vitamin" in the year 1912 (Piro et al. 2010). Funk discovered that the anti-beriberi component of unpolished rice was an amine, a class of chemicals that contains nitrogen (Ostrowski 1986). The term "vitamine" was created by him, combining the words "vita" and "amine," signifying the "amine of life." He believed that amines were essential for life. Nevertheless, it was eventually discovered that all vitamins, including vitamin A, are either "nitrogen" or "amines" (Mozaffarian et al. 2018). Thus, Jack Drummond suggested that the last "e" be eliminated from the word vitamines, leading to the word "vitamin" (Williams 2016).

Subsequent research demonstrated that various vitamins had distinct chemical characteristics and benefits, as long as they are not consumed in excessive amounts beyond what the body requires, as excessive intake can be harmful (Williams 2016; Fletcher and Fairfield 2002).

Vitamins are classified into two classes based on their solubility: water- and fat-soluble vitamins (Fig. 10.1). There are a total of 13 recognized vitamins, vitamin B-complex (B1, B2, B3, B5, B6, B9, B12, biotin) and C has the ability to solubilize in water, whereas A, D, E, and K are fat-soluble vitamins (Acevedo-Rocha et al. 2019). Fat-soluble vitamins are dissolved and stored in the fatty tissues of the body (Stevens 2021). Water-soluble vitamins are not linked to fats and are unaffected by modifications to the absorption of fat. At cooking temperature, the water-soluble vitamins can be degraded whereas fat-soluble vitamins remain stable (Borel and Desmarchelier 2018).

However, many vitamins cannot be synthesized by mammals. So, we get them from outside sources such as feed additives and dietary supplements. As a result, various industries involved in the manufacturing of vitamins have been successfully developed. In the current scenario, vitamins are industrially manufactured and

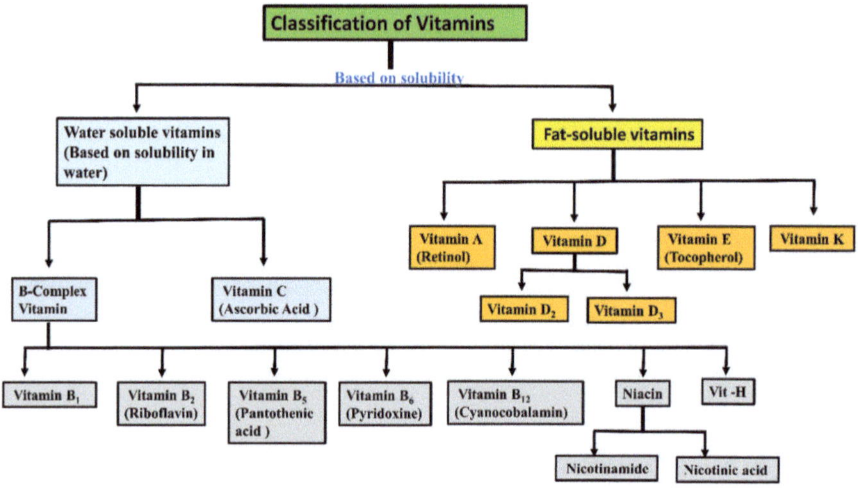

Fig. 10.1 Classification of vitamins based on their solubility

extensively used in food and feed additives, health, pharmaceuticals, technological aids, and also in cosmetics (Tarento et al. 2019).

Fat-soluble vitamins are utilized by the human body for a variety of physiological processes. Vitamin A is present in substances with retinol biological activity. The primary roles of carotenoids are as antioxidants in both animals and plants (Hashemi et al. 2019). Vitamin D2 and D3 are the two forms of vitamin D and are observed in numerous fortified foods and dietary supplements. Vitamin D2 is produced from ergosterol in yeast through UV irradiation, while vitamin D3 is synthesized from cholesterol or lanolin via chemical conversion or through UV irradiation of 7-dehydrocholestrol, respectively (Charoenngam et al. 2019). Vitamin E encompasses a group of compounds, such as four tocopherols and four tocotrienols. Among these, α-tocopherol exhibits the maximum biological activity, while other tocotrienols display 0–50% activity when compared to α-tocopherol (DellaPenna and Last 2006). Vitamin K is naturally present in three forms, vitamin K1, K2, and K3 also referred to as phylloquinone, menaquinone, and menadione, respectively (Shearer and Newman 2008). 2-Methyl-1,4-naphthoquinone core is present in various forms of vitamin K but these forms vary in the length and saturation of their polyisoprenoid side chains at the third position (Schurgers and Vermeer 2000). Microbes have been used for large-scale vitamin production for the past few years, and most of the vitamins that are commercially produced are fed to domestic animals (Ledesma-Amaro et al. 2015).

In the conventional approach, vitamins have been synthesized by organic chemical processes, which may be laborious and expensive due to the need for numerous reactions and the use of solvents that are usually hazardous to the environment. The biological production of vitamins has been explored as a solution to these inadequacies. This method entails to identify natural vitamin-producing organisms, determining the best economic conditions for plate culture, increasing production, and developing downstream methods to extract the pure result. Nevertheless, several vitamin titers cannot compete with chemical synthesis due to the inherent limits of the majority of natural producers (Wang et al. 2021). Thus, biotechnology presents itself as an eco-friendly strategy for enhancing vitamin productivity, either by enhancing existing natural sources of vitamins or developing new sources that are more appropriate for industrial applications. Thus, "white" biotechnology can genetically and metabolically engineer microorganisms, converting them into proficient vitamin producers. Conversely, "green" biotechnology yields crops having high vitamin concentrations, which can be harnessed directly as a source of dietary vitamins (Barcelos et al. 2018). Throughout the chapter, we shall explore each fat-soluble vitamin and their production from the microorganism.

10.2 Vitamin A

Vitamin A is an important fat-soluble vitamin with diverse functions. The molecular structure of vitamin A is comprised of a methylcyclohexane ring, a complex structure with 20 carbons, and a tetraene side chain with an aldehyde group referred to as retinal, with a hydroxyl group referred to as retinol and with carboxylic acid known as retinoic acid (Otten et al. 2006; Ashor et al. 2014) (Fig. 10.2). Vitamin A is important for immune system function, vision, embryonic cell growth and development, and epithelial cell support (Fig. 10.2). Additionally, it is regarded as an essential antioxidant that contributes toward good health (Ashor et al. 2014). It is involved in maintaining the mucosal epithelial cell layer that serves as a barrier against microorganism throughout the body. This lining of the gastrointestinal system can be damaged by some conditions like diarrhea, which may result in poor absorption of vitamin A leading to its deficiency (Maggini et al. 2018).

In our diet, we can obtain both pro-vitamin A and vitamin A from food sources. β-Carotene is the most prominent carotenoid found in pro-vitamin A, followed by α-carotene and β-cryptoxanthin. These pro-vitamins can be converted by our bodies into retinal and retinoic acid, considered as the active forms of vitamin A. Retinol and retinyl ester are the commonly available pre-formed vitamins in daily-consumed food products such as green-leafy vegetables including spinach, carrots, sweet potato, mangoes, red peppers, egg yolk milk, and others. Leafy greens, orange, and yellow vegetables, as well as plant pigments, are often rich sources of pro-vitamin A (Carazo et al. 2021). Pro-vitamin A and vitamin A need to undergo metabolic processes to be converted into retinal and retinoic acids. This conversion is essential for their biological activity (Carazo et al. 2021).

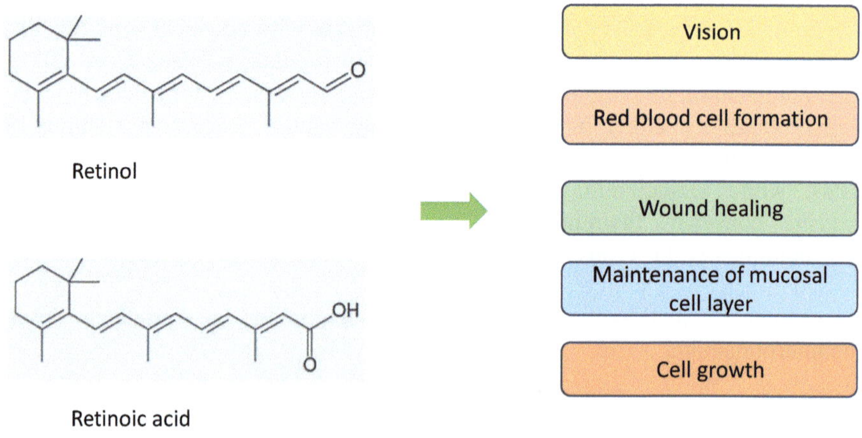

Fig. 10.2 Chemical structures of vitamin A and their biological functions

It is known that the primary components of vitamin A are β-carotene, α-carotene, and β-cryptoxanthin (Xia et al. 2015). Algae, some bacteria, and fungi are the primary producers of β-carotene which acts as a precursor to vitamin A and is separated into all-trans and cis isomers. The biosynthesis of β-carotene occurs by isopentenyl diphosphate (IPP) and dimethylallyl diphosphate (DMAPP) which act as precusor via mevalonate (MVA) pathway (Yoon et al. 2009).

To further enhance vitamin A production, microorganisms such as fungus *Blakeslea trispora* and microalga *Dunaliella* were grown in specific environmental conditions. *Dunaliella* when grown at 20–30% salt, limited nitrogen, and light intensity of 10,000 lux, for 3 months at 25–27 °C resulted in the accumulation of more than 0.1 g/g dry cells (Borowitzka and Borowitzka 1988). Fermentation of *B. trispora* can yield 0.2 g/g dry cells within 7 days and when whole cells are directly used as a feed additive, β-carotene can be directly extracted and purified (Borowitzka and Borowitzka 1989).

Studies indicate that β-carotene production by conventional method is unsatisfactory to meet the population requirement, and thus, there is a need for the use of engineered microorganisms to increase the yield. Thus, engineered strains of *Escherichia coli*, microalgae, and fungi have been used (Ye and Bhatia 2012).

MVA pathway has a crucial role in the synthesis of β-carotene. This pathway involves two segments: the first segment involves the synthesis of mevalonate from acetyl-CoA, and the second segment involves the formation of IPP and DMAPP from mevalonate. The complete MVA pathway was assembled by connecting the first and second segment of the MVA pathway of *Streptococcus pneumoniae* and *Enterococcus faecalis*, respectively, for enhanced β-carotene production (Yoon et al. 2009). In 2009, Yoon et al. introduced a foreign MVA pathway to enhance the availability of IPP and DMAPP for the synthesis of β-carotene. It was observed that approximately 465 mg/L of β-carotene was synthesized by the engineered *E. coli* strain with a complete MVA pathway (Yoon et al. 2009; Ma et al. 2011; Zhao et al. 2013).

There are two crucial cofactors in the β-carotene production pathway: nicotinamide adenine dinucleotide phosphate (NADPH) and adenosine-triphosphate (ATP). Methylerythritol phosphate (MEP) pathway also results in synthesis of IPP and DMAPP in *E. coli* with the help of cofactors including NADPH, ATP, and cytidine triphosphate. Further, β-carotene production was improved by developing and optimizing a central metabolic module in *E. coli* to enhance the amount of ATP and NADPH in the β-carotene production pathway (Zhao et al. 2013). The strain CAR005 (with the ability to produce high levels of ATP and NADPH) produced 2.1 g/L of β-carotene in fed-batch fermentation where the yield was 60 mg/g dry cell weight. Further investigations were done by manipulation of *E. coli* strain ZF43Δ*gdhA*, where the developed strain ECW4/p5C-*nadK* led to the production of beta carotene approximately 266 mg/L and 2579 mg/L in flask culture and 5 L bioreactor, respectively (Wu et al. 2020).

In the year 2018, *Yarrowia lipolytica* was engineered by overexpressing heterologous carotene synthase (*Crt*) gene to increase β-carotene production (Larroude et al. 2018). The engineered strain with the efficient promoter led to the production of 1.5 g/L of β-carotene. Further, using fed-batch fermentation and optimizing the fermentation conditions, the production titer of β-carotene was increased to 90 mg/g DCW and 6.5 g/L (Larroude et al. 2018).

Saccharomyces cerevisiae does not produce any carotenoid but it was genetically engineered to produce β-carotene. Initially, in *S. cerevisiae*, bacterial genes geranyl-geranyl diphosphate synthase (*crtE*), phytoene synthase (*crtB*), phytoene desaturase (*crtI*), and lycopene beta-cyclase (*crtY*) were introduced through episomal expression vector which led to the production of β-carotene but the yield was quite low, approximately 103 μg/g dry wt. (Yamano et al. 1994). Further to maximize the production, in the genome of *S. cerevisiae*, carotenoid genes such as *crtE*, *crtI*, and *crtY* from red yeast *Xanthophyllomyces dendrorhous* were integrated. Also, to enhance the carbon flux, isoprenoid genes were introduced in yeast. Together, this resulted in an enhanced synthesis of β-carotene with 5.9 mg/g dry wt. (Yan et al. 2012; Miura et al. 1998). *Candida utilis* was also genetically engineered to produce carotene including β-carotene. Genes involved in carotenoid biosynthesis from microbes *Erwinia uredovora* and *Agrobacterium aurantiacum* were introduced in *C. utilis*, and their expressions were regulated by constitutive promoter and terminator. This eventually led to the production of 0.4 mg/g dry wt. of β-carotene in *C. utilis* (Miura et al. 1998). *Pichia pastoris* X-33 strain was genetically modified by genomic integration of the plasmid containing genes from *E. uredovora* which encode for proteins phytoene synthase, phytoene desaturase, and geranylgeranyl diphosphate synthase and from *Ficus carica* gene encoding for protein lycopene β-cyclase. This led to the production of 339 μg/g dry wt. of β-carotene (Verwaal et al. 2007; Yan et al. 2012). Further, modulating the conditions of oleaginous red yeast *Sporidiobolus pararoseus* KM281507 by introducing crude glycerol at the concentration of 55 g/L, maintaining the medium pH of 5.53 and growing them at 24 °C (Chaiyaso and Manowattana 2018).

Sun et al. in 2019 constructed a *S. cerevisiae* strain which expresses the heterologous vitamin A biosynthetic pathway and can produce vitamin A in the presence of xylose and/or glucose. Production of vitamin A from xylose as an energy source was four times higher than glucose and upon introducing two-phase in situ extraction using dodecane or olive oil; it was further enhanced twofold. In fed-batch fermentation using xylose and dodecane in situ extraction, a final yield of 3.35 g/L was obtained which included retinal and retinol as 2.1 g/L and 1.256 g/L, respectively (Sun et al. 2019).

Vitamin A is synthesized through chemical processes approximately 2700 tons per annum (t/a) whereas β-carotene is synthesized as 400 t/a at industrial level by Hoffmann-La Roche and BASF (Parker et al. 2016). At present, the major production of vitamin A occurs by chemical synthesis from acetone and acetylene. Production of β-carotene from *E. coli* at industrial scale might not be suitable as endotoxin contamination could be possible when producing at large scale, also there could be bacteriophage contamination. Thus, yeast could be a better option. Use of energy sources other than glucose, such as xylose an abundant nonedible sugar, can be utilized which will eventually result in cost reduction. The work on yeast can result in a sustainable bioprocess for the production of vitamin A industrially.

10.3 Vitamin D

Vitamin D is a fat-soluble vitamin which is crucial for absorbing and retaining phosphorus, magnesium, and calcium and other biological functions, such as neuromuscular function, cell growth modulation, and immune function (Pfotenhauer and Shubrook 2017; Kaiser et al. 1990) (Fig. 10.3). Insufficient concentration of vitamin D results in disorders like rickets and osteoporosis in children and adults, respectively, and also leads to dental complications (Mungai et al. 2021). Two forms of vitamin D: D2 also known as ergocalciferol found mainly in fungi and vitamin D3 referred to as cholecalciferol found mainly in animals (Fig. 10.3). The precursor of vitamin D2 is synthesized from ergosterol upon exposure to the ultraviolet radiation (Papoutsis et al. 2020). In humans and animals, vitamin D3 in active form is present as 25-hydroxyvitamin D3 (25-OH-VD3). Vitamin D2 finds extensive applications in the fields of food, feed industry, and medicine, demonstrating promising market potential (Wei et al. 2023). At industrial level, vitamin D is chemically synthesized from compounds like sterols, ergosterols, and cholesterol using UV radiation (Eggersdorfer et al. 1996).

The ergosterol is primarily found within plant's and certain fungi's cell membrane, with yeast having the highest concentration (often ranging from 0.03% to 4.6% of a DCW). However, in some species like *S. cerevisiae*, the concentration might reach 7–10%. Since fungi are capable of producing ergosterol naturally but exposure to UV radiation is important to synthesize vitamin D2 and thus needs to be optimized. Yeast such as *S. cerevisiae*, *S. uvarum*, and *C. utilis* produces ergosterol. Fungi including *Trichoderma*, *Cephalosporium*, and *Fusarium* also synthesize ergosterol but the concentration is comparatively less than that produced by *S.*

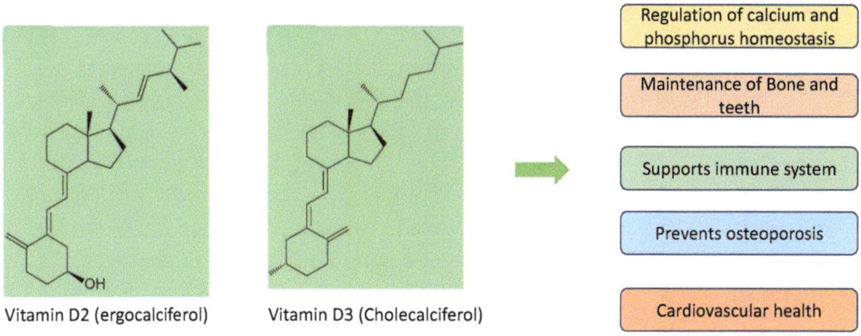

Fig. 10.3 Chemical structures of vitamin D2 and D3 and their biological functions

cerevisiae (Shimizu 2001). In order to enhance the production of ergosterol, *S. cerevisiae* has been genetically modified by overexpressing enzymes involved in the ergosterol synthesis pathway. The enzymes introduced were *Erg4p* and *Are2p* corresponding to sterol C-24(28) reductase and sterol acyltransferase, respectively, and the synthesis was carried out in the medium containing molasses. The yield was 1.71 g/L after 40 h of cultivation in batch fermentation (He et al. 2007).

7-Dehydrocholestrol (7-DHC) is the precursor for vitamin D3. 7-DHC is converted into 25-OH-VD3 followed by 1,25-dihydroxyvitamin D3 [1,25$(OH)_2$D3] by enzymatic activity in the liver and kidney, respectively (Minghetti and Norman 1988). Vitamin D3 hydroxylase (*Vdh*), cytochrome P450 monooxygenase isoforms plays an important role in hydroxylation step to generate 25-OH-VD3 and 1,25$(OH)_2$D3 (Prosser and Jones 2004). In 1991, 300 isolates of streptomyces were screened and two strains were found to synthesize 25-OH-VD3 at the level of 7 mg/L/min (Sasaki et al. 1991). Another isolate from soil, *Actinomyces hyovaginalis* A11-2 has the ability to synthesize 1,25$(OH)_2$D3 (Abbas et al. 2011). Tang et al. in 2020 identified a novel strain *Bacillus cereus* which converts vitamin D3 to 25-OH-VD3. The strain *B. cereus* under optimal conditions results in the production of 796 mg/L and 830 mg/L of 25-OH-VD3 in a shake flask and in a 5 L fermentation tank, respectively (Tang et al. 2020).

During microbial production of vitamin D3, more focus is on the hydroxylation step to enhance the yield. For instance, Yasutake et al. in 2013 used *Rhodococcus erythropolis* with mutant Vdh (T107A) and treated the cells with nisin, an antimicrobial peptide which resulted in the production of approximately 573 μg/mL 25-OH-VD3 by 2-h bioconversion (Yasutake et al. 2013). Additionally, vitamin D3 is hydrophobic in nature, in order to enhance its membrane permeability, Yasutake et al. treated the *R. erythrococcus* membrane with *lactostreptococcus* peptide, which resulted in a significant enhancement in the vitamin D3 (Yasutake et al. 2013). In yeast, genetic and metabolic modification is required in order to produce vitamin D3. It involves introduction of gene encoding for 7-DHC instead of ergosterol. A yeast strain BY4742 was constructed having 7-DHC biosynthetic pathway and by knocking down *ERG5* gene (codes for C-22 desaturase), blocked the endogenous ergosterol pathway. The production of 7-DHC was enhanced by 74% in shake-flask by the strain (Tang et al. 2020). Xiao et al. constructed a strain by overexpressing genes *ERG10, ERG13, tHMG1, ERG12, ERG8, ERG19, IDI1, and ERG20* that are involved in the mevalonate pathway. Further, delta (24)-dehydrocholesterol reductase was introduced from *Gallus gallus* and *ERG6* encoding for delta (24)-sterol C-methyltransferase was blocked to inhibit the ergosterol pathway. This led to the enhancement of 7-DHC to approximately 1.1 g/L in 5 L bioreactor (Guo et al. 2018). Another study uses a system metabolic engineering approach to enhance the production of 7-DHC in yeast. This included genetic modification involving knock out of *ROX1* and *GDH1*, overexpression of essential genes, and increasing the gene copy number of *ERG1* and *DHCR24* with the help of Ty1 transposon and inhibition of ergosterol pathway by using CRISPR interference. This led to the biosynthesis of 1.328 g/L of 7-DHC in a 3 L bioreactor (Qu et al. 2022). Further, study conducted by Amiri et al. showed that besides metabolic engineering culture condition optimization also plays a very important role. Based on their study, the initial concentration of vitamin D3, shaking speed, sucrose, and tryptophan are the key factors (Amiri et al. 2019).

At industrial level, vitamin D is chemically synthesized at a level of 38 t/a (Eggersdorfer et al. 1996). Yeast due to its short generation time, minimal media requirement, common environment conditions, genetic modification, and specific fermentation properties is considered an ideal strain for producing vitamin D. Still, studies are going on to produce vitamin D microbially at an industrial scale. At present, the synthesis of vitamin D is mainly by chemical synthesis but due to the cost effectiveness, safety and sustainability, microbial production of vitamin D by industries will take over in future. A few of the major producers of vitamin D are Solvay-Duphar, Hoffmann-La Roche, BASF, DSM, Fermenta biotech and Zhejiang Medicine Co. Ltd.

10.4 Vitamin E

Vitamin E is a group of fat-soluble molecules with antioxidant properties. It is a tocochromanol known as tocopherols and tocotrienols (Fig. 10.4). There are four forms of tocopherols and tocotrienols and are referred to as α-, β-, δ-, γ-tocopherols and α-, β-, δ-, γ-tocotrienols. These compounds contain a hydrophobic side chain and a chromanol ring. Of all, α-tocopherol is the most biologically active form of vitamin E (Kaiser et al. 1990). α-tocopherol is produced by autotrophic organisms which include green plants, eukaryotic algae, and cyanobacteria such as *Synechocystis* (Taketomi et al. 1983). It is crucial for maintaining the immune system, cell membrane integrity, prevention of platelet aggregation, and formation of red blood cells (Lee and Han 2018) (Fig. 10.4). The source of vitamin E includes

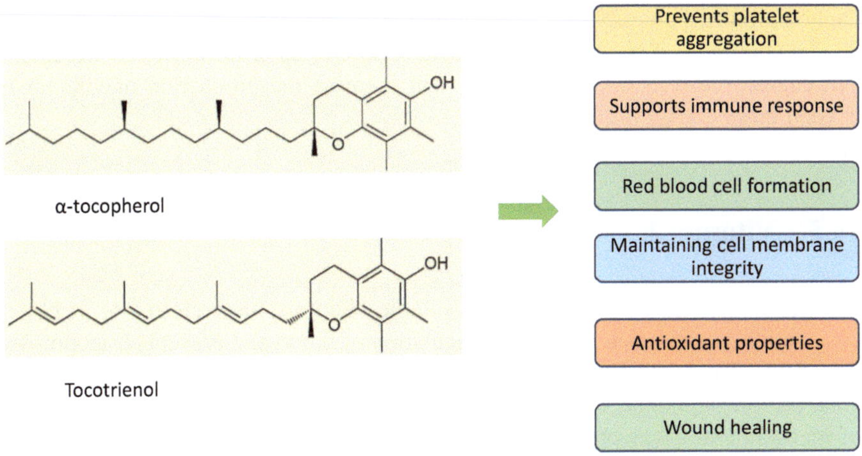

Fig. 10.4 Chemical structures of vitamin E and their biological functions

oils (olive, palm, wheat germ, and other plant oils), almonds, animal fat, asparagus, fish, nuts, peanuts, and its product, pumpkin and sunflower seeds (Lee and Han 2018).

Many studies have been done in plants to improve the synthesis of α-tocopherol but very few studies are available related to genetic and metabolic engineering in context with the microbial production of vitamin E. *Euglena gracilis*, a microalgae, has a high level of α-tocopherol, and thus, studies have been done on their cultivation system to improve the α-tocopherol production (Ogbonna 2009). Researchers have explored the role of nitrogen concentration and its source in *Nannochloropsis oculate* on the production of α-tocopherol. The use of nitrogen source in the form of ammonia and nitrate enhances the production of α-tocopherol (Durmaz 2007). Shen et al. constructed a *S. cerevisiae* strain by introducing heterologous genes from autotrophic organisms along with the endogenous shikimate and MEP pathway to produce tocotrienols. They incorporated a temperature control system that is triggered by cold temperature leading to the production of 0.32 g/L in a 5 L fermentative process (Shen et al. 2020). A study in 2022 by Ye et al. introduced the β-Farnese synthase gene into *S. cerevisiae* via protein and system metabolic engineering to synthesize Farnese. The Farnese is then chemically converted into isophytol to produce vitamin E (Ye et al. 2022). This step has reduced the seven-step reaction into a three-step reaction. Besides introduction of Farnese, researchers have introduced the lycopene gene as well to reduce the cost of production as lycopene is an antioxidant and is used in healthcare products. This vitamin E production is a combination of microbial and chemical synthesis to reduce the cost of production.

The present demand for vitamin E at a global scale is 80,000 tons. BASF, Zhejiang NHU Co. Ltd, DSM, Jilin Beisha Pharmaceutical Co. Ltd are the major producers of vitamin E.

Approximately 20% of the global vitamin E is produced by Nenter and Co. in cooperation with Amyris, a company situated in Brazil. Amyris provides the fermented product Farnese as a raw material for the synthesis of vitamin E (Ye et al. 2022). This method is cost-effective. It is believed that with further advancement and engineering process, this process will increase the production of vitamin E beyond 20%.

10.5 Vitamin K

Vitamin K belongs to the class of fat-soluble vitamins, also referred to as blood coagulation vitamin as it promotes coagulation of blood and prevents osteoporosis (Fig. 10.5). Vitamin K is primarily present as K1 and K2 referred to as phylloquinone and menaquinone, respectively, are present (Fig. 10.5). Based on the number of menaquinone rings present, vitamin K2 is further classified into 14 isoforms.

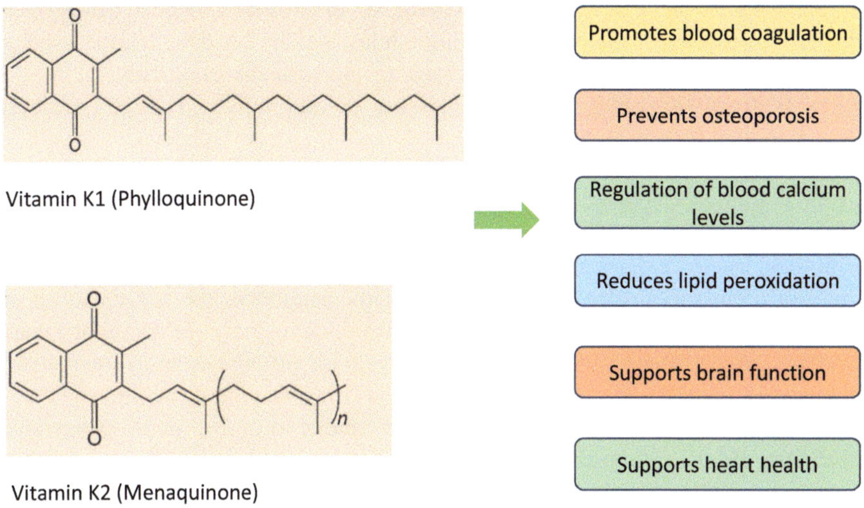

Fig. 10.5 Functions of vitamin K

Vitamin K1 is a naturally occurring plant-based compound that is abundant in green leafy vegetables (Widhalm et al. 2012). Vitamin K2 is produced by animal and bacterial cells (Mahdinia et al. 2019). Menaquinone-7 (MK-7) is generated by a wide variety of bacterial species when compared to other variants of vitamin K2 (Geleijnse et al. 2004). MK-7 in its all-trans forms is biologically active when compared to the cis, trans, and cis/trans forms. The source of vitamin K include dark green vegetables, cruciferous vegetables, asparagus, beef, eggs, fish, liver, peas, and spinach (Simes et al. 2020).

There are many bacteria which have the ability to produce MK-7, for instance, *Bacillus subtilis*, *E. coli*, *Flavobacterium* sp., lactic acid bacteria, and *Bacillus amyloliquefacians* (Sharma et al. 1993; Mahanama et al. 2011; Sato et al. 2001; Xu et al. 2017; Yang et al. 2018; Ma et al. 2019). *B. subtilis* isolated from natto produces a high level of MK-7 and has been considered as food safe by FDA (Mahanama et al. 2011). Further, to enhance the productivity, various *Bacillus* strains have been mutagenized by chemical or physical methods to overproduce vitamin K. *Bacillus* strain D200-41 was generated upon exposure to *N*-methyl-*N*′-nitro-*N*-nitrosoguanidine (NTG) with diphenylamine (DPA) resulting in enhanced production of MK-7 approximately 60 mg/L when cultured with shaking for 1 day at 37 °C followed by static culture for 5 days at 45 °C (Sato et al. 2001). Besides mutagenesis and metabolic engineering has been used to modify the MK-7 metabolic pathway to enhance its production. For instance, in *B. amyloliquefaciens*, the endogenous enzymes involved in MK-7 biosynthetic pathway were overexpressed (Xu et al. 2017). Of all the enzymes that were overexpressed, the expression of heptaprenyl diphosphate synthase (HepS) exhibited maximum production of MK-7, approximately 273 μg/g dry cell weight (Xu et al. 2017). Study conducted in 2018 by Yanget al. categorized the MK-7 biosynthetic pathway into four modules, and the key enzymes involved in the respective module were overexpressed to increase the production of MK-7 (Yang et al. 2018). Later, the overexpression of rate-limiting enzymes such as DHNA polyprenyltransferase (*MenA*), 1-Deoxy-D-xylose-5-phosphate synthase (*Dxs*), Deoxyxylulose Phosphate Reductoisomerase (*Dxr*), and IPP isomerase (*Idi*), resulted in the production of 50 mg/L of MK-7 in *B. subtilis* at lab scale where the cultures were grown at 40 °C, 200 rpm in SYG medium (Ma et al. 2019). In case of lactic acid bacteria, *Lactococcus lactis* produces different isoforms of menaquinone. In order to enhance the production of menaquinone, the key enzymes of menaquinone biosynthetic pathway such as mevalonate kinase (*Mvk*), diphosphate synthase (*preA*), *MenA*, and isochorismate synthase (*MenF*) were overexpressed (Bøe and Holo 2020). *E. coli* was also modified by systematic metabolic engineering to produce MK-7. This was done by overexpression of *Idi*, *MenA*, exogenous *UbiE*, and *HepPPS* from different bacterial species to enhance MK-7 production (Gao et al. 2021). *Flavobacterium meningosepticum* was randomly mutagenized using NTG and HNA to generate strains *F. meningosepticum* HNA 350-22 and IFO 12535 to produce menaquinone (Tani et al. 1986). Further to increase the secretion of vitamin K from the cell, surfactants such as soybean oil, span20 which increases the membrane permeability were used (Ren et al. 2020). For *Bacillus*, the secretion of MK-7 upon addition of soybean oil increased by 61% (Hu et al. 2017). At industrial scale, *B. subtilis* natto is used for vitamin K production as it can secrete vitamin K extracellularly and intracellularly. Liquid fermentation is favored during vitamin K production as static culture result in biofilm and pellicle formation (Mahdinia et al. 2019).

There are mainly six companies which use microbial fermentation techniques to produce vitamin K2. Nattopharma and Gnosis are the major producers of vitamin K with the trademark MenaQ7 and VitaMK7, respectively. There is a requirement to increase the industrialization of the microbial production of Vitamin K which can be achieved by various methods, such as strain improvement by structural analog resistant strategy or genetic modification leading to increased biosynthesis, enhanced production by process optimization including cultivation media, environmental conditions, and use of biofilm bioreactor to enhance the industrialization process (Sun et al. 2019). Microbially produced vitamin K is highly active, environmentally friendly, and safe to use and thus is a requirement to use advanced biotechnology increase the production, to reduce cost of microbial vitamin K production, and speed up the industrialization (Table 10.1).

Table 10.1 Microbial production of fat-soluble vitamins

Vitamin	Microorganism	Strain	Production	Yield	Substrate	Culture conditions	References
Vitamin A	*E. coli*	DH5α BL21(DE3) DSM-17873 VTCC-B-935	β-carotene	65 mg/L	Mevalonate	Ampicillin/kanamycin (50 μg/mL) incubate at 37 °C overnight, mevalonate at 6.6 mM, and β-carotene 65 mg/L	Nam et al. (2017)
	Rhodotorula	NCIM 3168, 3169, 3170, 3353, 3379 NCIM 3171, 3172, 3173, 3174, Mutant-R3, W4, Y and 32	β-carotene	77% and 87% of β-carotene	Sugarcane molasses	Medium with hydrolyzed molasses containing total reducing sugar (TRS) at 20 g/L concentration Fed-batch fermentation with dissolved oxygen. Diluted molasses 40 g/L TRS Feeding rate: 120 and 180 mL/h	Bhosale and Gadre (2001)
	E. coli	DH5α	β-carotene	663 mg/L	Glycerol	Batch culture: 37 °C and 180 rpm 30 g/L glycerol or glucose	Kim et al. (2009)
	Y. lipolytica	ob-CHC^{TEF}C^{TEF}	β-carotene	6.5 g/L	Glucose	Fed-batch fermentation for 122 h. Initially, glucose was added at 5 g/L. Fed Solution glucose at 500 g/L at 6 g/h rate (equivalent to 179.2 g/L)	Yamano et al. (1994)
	Dunaliella salina	DF15, DF17, DF40, CCAP 19/30 and UTEX 2538	β-carotene	–	Johnsons medium	Cultures were maintained at 20 ± 2 °C under a (12/12 LD) by LED lights with ~200 μmol photons/m^2/s of light intensity	Xu et al. (2018)
	Rhodotorula glutinis	KCTC 7989	β-carotene	258.5 μg/g	YM Broth	Inoculating the yeast in 100 mL YM broth, and incubating at 22 °C, 300 rpm for 48 h. Fermentation is carried out for 3 days in a 5 L stirred tank	Park et al. (2007)
Vitamin D	*R. erythropolis*	JCM3201	25-hydroxyvitamin D3	573 μg/mL	Lysogeny broth	Cells were cultured in 100 mL lysogeny broth (LB) medium at 37 °C then inoculated into LB (900 mL) followed by incubation at 60 °C for 15 min	Yasutake et al. (2013)
	Saccharomyces cerevisiae	No. A-3	Ergosterol	1160 mg/L	Glucose	Plate culture: 30 ± 1 °C for 20–24 h Seed culture medium: incubated at 30 ± 1 °C for 20–24 h. *Fed-batch*: 30 °C 1.5 vvm	Tan et al. (2003)

(continued)

Table 10.1 (continued)

Vitamin	Microorganism	Strain	Production	Yield	Substrate	Culture conditions	References
Vitamin E	*Nannochloropsis oculate*	CCAP-8491	α-tocopherol	2325.8 ± 39 μg/g DW	F/2 medium	Kept at 18 ± 1 °C with continuous stirring	Shen et al. (2020)
	Saccharomyces cerevisiae	BY4741	Tocotrienols	7.6 mg/g (DCW)	Tyrosine	Shake-flask culture: 5 mL of culture in appropriate medium, grown at 220 rpm, 30 °C overnight	Ye et al. (2022)
	Euglena gracilis	Z (Klebs SAG 1224-5/25)	α-Tocopherol and paramylon	5.3 ± 0.12 mg/L α-tocopherol- and 9.3 ± 0.1 g/L paramylon	Basal medium	Cultures: grown in 100 mL at 28 °C on a rotary shaker with an amplitude and frequency of 25 mm and 120 per min, respectively	Grimm et al. (2015)
Vitamin K	*Bacillus subtilis natto*	–	Menaquinone-7	226 mg/L	Soybean oil	Cells suspended in a sodium chloride solution (0.9% w/v) followed by heating at 80 °C for 30 min	Berenjian et al. (2014)
	Bacillus subtilis natto, Bacillus licheniformis, and Bacillus amy lolyquifaciens	NRRL NRS-352, NRRL B-14201, DSM 17766, NRRL B-65066, NRRL NRS-745, NRRL B-14394, F1, F2, F3, F4, NF1, NF2, NF3, and NF4	Menaquinone-7	35.5 mg/L	Glycerol		Mahdinia et al. (2017); Kang et al. (2022)
	Bacillus subtilis	MTCC 2756	Menaquinone-7	10.3 μg/mL	Soy powder and glycerol	Culture in flasks: incubated at 37 °C, 16 h and 180 rpm Fermentation: In liquid medium and sterilized medium incubated at 37 °C, 180 rpm for 24 h	Puri et al. (2015)
	E. coli	M01, M22, M23, M25, M28, M29, M41, M44, M45, M49, M50, MK01 –MK23	Menaquinone-7	1.35 g/L	Glucose	Strains grown at 37 °C, 220 rpm in LB (Luria–Bertani) medium For protein, ZYM medium 1% inoculum incubated at 220 rpm at 30 °C	Gao et al. (2021)
	B. amyloliquefaciens	Y-2 and W-21	Menaquinone-7	93.62% and 93.29%	Glucose and glycerol	Seed medium: TYG medium pH approximately 7.2 with 20% (w/v) NaOH	Tani et al. (1986)

10.6 Conclusion

The production of vitamins from microbial sources such as bacteria, microalgae, and fungi has advantages over chemical synthesis including cost, safety, absorption, biological activity, and others. Thus, the production of vitamins at the industrial scale is moving from chemical synthesis to microbial production. This is mainly due to the advancement in biotechnology and the use of metabolic, protein, and genetic engineering. In future, microorganisms such as *E. coli* and *S. cerevisiae* can easily be modified via engineering to produce vitamins by either generating a complete biosynthetic pathway or by regulating the enzyme activity involved in the pathway. Further use of CRISPR technology in microorganisms can enhance the production of vitamins to the next level. Based on our current understanding, vitamin A can be produced both chemically and microbiologically whereas for other fat-soluble vitamins have one or more microbial enzymatic steps. With continuous advancement in metabolic engineering and synthetic biotechnology, more sustainable, economical, and environment-friendly vitamin-producing cell factories in the form of microorganisms can be generated.

References

Abbas AM, Aboshanab KM, Aboulwafa MM, Hassouna NA (2011) Actinomyces hyovaginalis: a novel bacterial isolate with transforming activity of vitamin D3 to 1α, 25-dihydroxyvitamin D3. J Am Sci 7(9):230–237

Acevedo-Rocha CG, Gronenberg LS, Mack M, Commichau FM, Genee HJ (2019) Microbial cell factories for the sustainable manufacturing of B vitamins. Curr Opin Biotechnol 56:18–29. https://doi.org/10.1016/j.copbio.2018.07.006

Amiri MM et al (2019) Production of vitamin D3 enriched biomass of Saccharomyces cerevisiae as a potential food supplement: evaluation and optimization of culture conditions using Plackett–Burman and response surface methodological approaches. Iran J Pharm Res 18(2):974

Ashor AW, Siervo M, Lara J, Oggioni C, Mathers JC (2014) Antioxidant vitamin supplementation reduces arterial stiffness in adults: a systematic review and meta-analysis of randomized controlled trials. J Nutr 144(10):1594–1602

Barcelos MCS, Lupki FB, Campolina GA, Nelson DL, Molina G (2018) The colors of biotechnology: general overview and developments of white, green and blue areas. FEMS Microbiol Lett 365(21). https://doi.org/10.1093/femsle/fny239

Berenjian A, Mahanama R, Talbot A, Regtop H, Kavanagh J, Dehghani F (2014) Designing of an intensification process for biosynthesis and recovery of menaquinone-7. Appl Biochem Biotechnol 172:1347–1357

Bhosale P, Gadre RV (2001) β-Carotene production in sugarcane molasses by a Rhodotorula glutinis mutant. J Ind Microbiol Biotechnol 26:327–332

Bøe CA, Holo H (2020) Engineering Lactococcus lactis for increased vitamin K2 production. Front Bioeng Biotechnol 8:191

Borel P, Desmarchelier C (2018) Bioavailability of fat-soluble vitamins and phytochemicals in humans: effects of genetic variation. Annu Rev Nutr 38(1):69–96. https://doi.org/10.1146/annurev-nutr-082117-051628

Borowitzka MA, Borowitzka LJ (1988) Micro-algal biotechnology. Cambridge University Press, Cambridge

Borowitzka L, Borowitzka M (1989) β-Carotene (provitamin A) production with algae. In: Biotechnology of vitamins, pigments and growth factors. Springer

Carazo A, Macáková K, Matoušová K, Krčmová LK, Protti M, Mladěnka P (2021) Vitamin A update: forms, sources, kinetics, detection, function, deficiency, therapeutic use and toxicity. Nutrients 13(5):1703. https://doi.org/10.3390/nu13051703

Chaiyaso T, Manowattana A (2018) Enhancement of carotenoids and lipids production by oleaginous red yeast *Sporidiobolus pararoseus* KM281507. Prep Biochem Biotechnol 48(1):13–23. https://doi.org/10.1080/10826068.2017.1381620

Charoenngam N, Shirvani A, Holick MF (2019) Vitamin D for skeletal and non-skeletal health: what we should know. J Clin Orthop Trauma 10(6):1082–1093. https://doi.org/10.1016/j.jcot.2019.07.004

DellaPenna D, Last RL (2006) Progress in the dissection and manipulation of plant vitamin E biosynthesis. Physiol Plant 126(3):356–368

Durmaz Y (2007) Vitamin E (α-tocopherol) production by the marine microalgae Nannochloropsis oculata (Eustigmatophyceae) in nitrogen limitation. Aquaculture 272(1–4):717–722. https://doi.org/10.1016/j.aquaculture.2007.07.213

Eggersdorfer M, Adam G, John M, Labler L (1996) Vitamins: biotechnology. Wiley-VCH, Weinheim, pp 114–158

Fletcher RH, Fairfield KM (2002) Vitamins for chronic disease prevention in adults: clinical applications. JAMA 287(23):3127. https://doi.org/10.1001/jama.287.23.3127

Gao Q, Chen H, Wang G, Yang W, Zhong X, Liu J, Huo X, Liu W, Huang J, Tao Y, Lin B (2021) Highly efficient production of menaquinone-7 from glucose by metabolically engineered Escherichia coli. ACS Synth Biol 10(4):756–765

Geleijnse JM et al (2004) Dietary intake of menaquinone is associated with a reduced risk of coronary heart disease: the Rotterdam Study. J Nutr 134(11):3100–3105. https://doi.org/10.1093/jn/134.11.3100

Grimm P, Risse JM, Cholewa D, Müller JM, Beshay U, Friehs K, Flaschel E (2015) Applicability of Euglena gracilis for biorefineries demonstrated by the production of α-tocopherol and paramylon followed by anaerobic digestion. J Biotechnol 215:72–79

Guo X-J et al (2018) Metabolic engineering of Saccharomyces cerevisiae for 7-dehydrocholesterol overproduction. Biotechnol Biofuels 11:1–14

Hashemi SMB, Pourmohammadi K, Gholamhosseinpour A, Es I, Ferreira DS, Mousavi Khaneghah A (2019) Fat-soluble vitamins. In: Innovative thermal and non-thermal processing, bioaccessibility and bioavailability of nutrients and bioactive compounds. Elsevier, pp 267–289. https://doi.org/10.1016/B978-0-12-814174-8.00009-3

He X, Guo X, Liu N, Zhang B (2007) Ergosterol production from molasses by genetically modified Saccharomyces cerevisiae. Appl Microbiol Biotechnol 75(1):55–60. https://doi.org/10.1007/s00253-006-0807-6

Hu X, Liu W, Luo M, Ren L, Ji X, Huang H (2017) Enhancing menaquinone-7 production by Bacillus natto R127 through the nutritional factors and surfactant. Appl Biochem Biotechnol 182(4):1630–1641. https://doi.org/10.1007/s12010-017-2423-6

Kaiser S, Di Mascio P, Murphy ME, Sies H (1990) Physical and chemical scavenging of singlet molecular oxygen by tocopherols. Arch Biochem Biophys 277(1):101–108

Kang MJ, Baek KR, Lee YR, Kim GH, Seo SO (2022) Production of vitamin K by wild-type and engineered microorganisms. Microorganisms 10(3):554

Kim JH, Kim SW, Nguyen DQ, Li H, Kim SB, Seo YG, Yang JK, Chung IY, Kim DH, Kim CJ (2009) Production of β-carotene by recombinant Escherichia coli with engineered whole mevalonate pathway in batch and fed-batch cultures. Biotechnol Bioprocess Eng 14:559–564

Larroude M, Celinska E, Back A, Thomas S, Nicaud J, Ledesma-Amaro R (2018) A synthetic biology approach to transform *Yarrowia lipolytica* into a competitive biotechnological producer of β-carotene. Biotechnol Bioeng 115(2):464–472. https://doi.org/10.1002/bit.26473

Ledesma-Amaro R, Serrano-Amatriain C, Jiménez A, Revuelta JL (2015) Metabolic engineering of riboflavin production in Ashbya gossypii through pathway optimization. Microb Cell Factories 14(1):163. https://doi.org/10.1186/s12934-015-0354-x

Lee G, Han S (2018) The role of vitamin E in immunity. Nutrients 10(11):1614. https://doi.org/10.3390/nu10111614

Ma SM et al (2011) Optimization of a heterologous mevalonate pathway through the use of variant HMG-CoA reductases. Metab Eng 13(5):588–597

Ma Y et al (2019) Metabolic engineering of the MEP pathway in *Bacillus subtilis* for increased biosynthesis of menaquinone-7. ACS Synth Biol 8(7):1620–1630. https://doi.org/10.1021/acssynbio.9b00077

Maggini S, Pierre A, Calder PC (2018) Immune function and micronutrient requirements change over the life course. Nutrients 10(10):1531

Mahanama R et al (2011) Enhanced production of menaquinone 7 via solid substrate fermentation from Bacillus subtilis. Int J Food Eng 7(5). https://doi.org/10.2202/1556-3758.2314

Mahdinia E, Demirci A, Berenjian A (2017) Strain and plastic composite support (PCS) selection for vitamin K (Menaquinone-7) production in biofilm reactors. In: NABEC papers 2004. American Society of Agricultural and Biological Engineers, p 1

Mahdinia E, Demirci A, Berenjian A (2019) Biofilm reactors as a promising method for vitamin K (menaquinone-7) production. Appl Microbiol Biotechnol 103(14):5583–5592. https://doi.org/10.1007/s00253-019-09913-w

Minghetti PP, Norman AW (1988) 1,25 (OH) 2-vitamin D3 receptors: gene regulation and genetic circuitry. FASEB J 2(15):3043–3053

Miura Y, Kondo K, Saito T, Shimada H, Fraser PD, Misawa N (1998) Production of the carotenoids lycopene, β-carotene, and astaxanthin in the food yeast *Candida utilis*. Appl Environ Microbiol 64(4):1226–1229. https://doi.org/10.1128/AEM.64.4.1226-1229.1998

Mozaffarian D, Rosenberg I, Uauy R (2018) History of modern nutrition science—implications for current research, dietary guidelines, and food policy. BMJ 361:k2392. https://doi.org/10.1136/bmj.k2392

Mungai LN, Mohammed Z, Maina M, Anjumanara O (2021) Vitamin D review: the low hanging fruit for human health. J Nutr Metab 2021:6335681

Nam VH, Trang HHT, Huong TLT, Cuong DV (2017) Enhanced β-carotene biosynthesis in recombinant Escherichia coli Harboring the bottom portion of Mevalonate pathway of enterococcus faecium. VTCC-B-935 isolated in Vietnam. J Appl Biotechnol Bioeng 3(4):00073. https://doi.org/10.15406/jabb.2017.03.00073

Ogbonna JC (2009) Microbiological production of tocopherols: current state and prospects. Appl Microbiol Biotechnol 84(2):217–225. https://doi.org/10.1007/s00253-009-2104-7

Ostrowski PF (1986) Who discovered vitamins? Pol Rev 31:171–183

Otten J, Hellwig J, Meyers L (2006) Institute of medicine dietary reference intakes: the essential guide to nutrient requirements. The National Academies Press, Washington, DC

Papoutsis K et al (2020) Recovery of ergosterol and vitamin D2 from mushroom waste—potential valorization by food and pharmaceutical industries. Trends Food Sci Technol 99:351–366

Park PK, Kim EY, Chu KH (2007) Chemical disruption of yeast cells for the isolation of carotenoid pigments. Sep Purif Technol 53(2):148–152

Parker GL, Smith LK, Baxendale IR (2016) Development of the industrial synthesis of vitamin A. Tetrahedron 72(13):1645–1652. https://doi.org/10.1016/j.tet.2016.02.029

Pfotenhauer KM, Shubrook JH (2017) Vitamin D deficiency, its role in health and disease, and current supplementation recommendations. J Osteopathic Med 117(5):301–305. https://doi.org/10.7556/jaoa.2017.055

Piro A, Tagarelli G, Lagonia P, Tagarelli A, Quattrone A (2010) Casimir Funk: his discovery of the vitamins and their deficiency disorders. Ann Nutr Metab 57(2):85–88. https://doi.org/10.1159/000319165

Prosser DE, Jones G (2004) Enzymes involved in the activation and inactivation of vitamin D. Trends Biochem Sci 29(12):664–673

Puri A, Iqubal M, Zafar R, Panda BP (2015) Influence of physical, chemical and inducer treatments on menaquinone-7 biosynthesis by Bacillus subtilis MTCC 2756. Songklanakarin J Sci Technol 37(3)

Qu L et al (2022) Engineered yeast for efficient de novo synthesis of 7-dehydrocholesterol. Biotechnol Bioeng 119(5):1278–1289

Ren L, Peng C, Hu X, Han Y, Huang H (2020) Microbial production of vitamin K2: current status and future prospects. Biotechnol Adv 39:107453. https://doi.org/10.1016/j.biotechadv.2019.107453

Sasaki J, Mikami A, Mizoue K, Omura S (1991) Transformation of 25- and 1 alpha-hydroxyvitamin D3 to 1 alpha, 25-dihydroxyvitamin D3 by using Streptomyces sp. strains. Appl Environ Microbiol 57(10):2841–2846

Sato T, Yamada Y, Ohtani Y, Mitsui N, Murasawa H, Araki S (2001) Production of menaquinone (vitamin K2)-7 by Bacillus subtilis. J Biosci Bioeng 91(1):16–20

Schurgers LJ, Vermeer C (2000) Determination of phylloquinone and menaquinones in food. Pathophysiol Haemost Thromb 30(6):298–307

Sharma V, Meganathan R, Hudspeth M (1993) Menaquinone (vitamin K2) biosynthesis: cloning, nucleotide sequence, and expression of the menC gene from Escherichia coli. J Bacteriol 175(15):4917–4921

Shearer MJ, Newman P (2008) Metabolism and cell biology of vitamin K. Thromb Haemost 100(10):530–547

Shen B, Zhou P, Jiao X, Yao Z, Ye L, Yu H (2020) Fermentative production of Vitamin E tocotrienols in Saccharomyces cerevisiae under cold-shock-triggered temperature control. Nat Commun 11(1):5155. https://doi.org/10.1038/s41467-020-18958-9

Shimizu S (2001) Vitamins and related compounds: microbial production. In: Rehm H-J, Reed G (eds) Biotechnology set, 1st edn. Wiley, pp 318–340. https://doi.org/10.1002/9783527620999.ch11k

Simes DC, Viegas CSB, Araújo N, Marreiros C (2020) Vitamin K as a diet supplement with impact in human health: current evidence in age-related diseases. Nutrients 12(1):138. https://doi.org/10.3390/nu12010138

Stevens SL (2021) Fat-soluble vitamins. Nurs Clin N Am 56(1):33–45. https://doi.org/10.1016/j.cnur.2020.10.003

Sun L, Kwak S, Jin Y-S (2019) Vitamin A production by engineered *Saccharomyces cerevisiae* from xylose *via* two-phase *in situ* extraction. ACS Synth Biol 8(9):2131–2140. https://doi.org/10.1021/acssynbio.9b00217

Surana KR et al (2019) Nutritional advantages and effects of vitamins in human health. Chhattisgarh J Sci Technol 16(3):67–79

Taketomi H, Soda K, Katsui G (1983) Results of screening test in tocopherols in microbial realm. Vitamins (Japan) 57:133–138

Tan T, Zhang M, Gao H (2003) Ergosterol production by fed-batch fermentation of Saccharomyces cerevisiae. Enzym Microb Technol 33(4):366–370

Tang D, Liu W, Huang L, Cheng L, Xu Z (2020) Efficient biotransformation of vitamin D3 to 25-hydroxyvitamin D3 by a newly isolated Bacillus cereus strain. Appl Microbiol Biotechnol 104:765–774

Tani Y, Asahi S, Yamada H (1986) Menaquinone (vitamin K2)-6 production by mutants of Flavobacterium meningosepticum. J Nutr Sci Vitaminol 32(2):137–145

Tarento TD et al (2019) A potential biotechnological process for the sustainable production of vitamin K1. Crit Rev Biotechnol 39(1):1–19

Verwaal R et al (2007) High-level production of beta-carotene in Saccharomyces cerevisiae by successive transformation with carotenogenic genes from Xanthophyllomyces dendrorhous. Appl Environ Microbiol 73(13):4342–4350

Wang Y, Liu L, Jin Z, Zhang D (2021) Microbial cell factories for green production of vitamins. Front Bioeng Biotechnol 9:661562. https://doi.org/10.3389/fbioe.2021.661562

Wei X, Pandohee J, Xu B (2023) Recent developments and emerging trends in dietary vitamin D sources and biological conversion. Crit Rev Food Sci Nutr:1–17. https://doi.org/10.1080/10408398.2023.2220793

Widhalm JR, Ducluzeau A, Buller NE, Elowsky CG, Olsen LJ, Basset GJC (2012) Phylloquinone (vitamin K_1) biosynthesis in plants: two peroxisomal thioesterases of lactobacillales origin hydrolyze 1,4-dihydroxy-2-naphthoyl-coa. Plant J 71(2):205–215. https://doi.org/10.1111/j.1365-313X.2012.04972.x

Williams P (2016) The missing vitamin alphabet. Nutr Dietetics 73(2):205–214. https://doi.org/10.1111/1747-0080.12212

Wu Y et al (2020) Enhancing β-carotene production in Escherichia coli by perturbing central carbon metabolism and improving the NADPH supply. Front Bioeng Biotechnol 8:585. https://doi.org/10.3389/fbioe.2020.00585

Xia W, Chen W, Peng W, Li K (2015) Industrial vitamin B12 production by Pseudomonas denitrificans using maltose syrup and corn steep liquor as the cost-effective fermentation substrates. Bioprocess Biosyst Eng 38(6):1065–1073. https://doi.org/10.1007/s00449-014-1348-5

Xu J-Z, Yan W-L, Zhang W-G (2017) Enhancing menaquinone-7 production in recombinant Bacillus amyloliquefaciens by metabolic pathway engineering. RSC Adv 7(45):28527–28534

Xu Y, Ibrahim IM, Wosu CI, Ben-Amotz A, Harvey PJ (2018) Potential of new isolates of Dunaliella salina for natural β-carotene production. Biology 7(1):14

Yamano S, Ishii T, Nakagawa M, Ikenaga H, Misawa N (1994) Metabolic engineering for production of *β*-carotene and lycopene in *Saccharomyces cerevisiae*. Biosci Biotechnol Biochem 58(6):1112–1114. https://doi.org/10.1271/bbb.58.1112

Yan G, Wen K, Duan C (2012) Enhancement of β-carotene production by over-expression of HMG-CoA reductase coupled with addition of ergosterol biosynthesis inhibitors in recombinant Saccharomyces cerevisiae. Curr Microbiol 64:159–163

Yang S et al (2018) Modular pathway engineering of Bacillus subtilis to promote de novo biosynthesis of menaquinone-7. ACS Synth Biol 8(1):70–81

Yasutake Y, Nishioka T, Imoto N, Tamura T (2013) A single mutation at the ferredoxin binding site of P450 Vdh enables efficient biocatalytic production of 25-hydroxyvitamin D_3. Chembiochem 14(17):2284–2291. https://doi.org/10.1002/cbic.201300386

Ye VM, Bhatia SK (2012) Pathway engineering strategies for production of beneficial carotenoids in microbial hosts. Biotechnol Lett 34:1405–1414

Ye Z et al (2022) Revolution of vitamin E production by starting from microbial fermented farnesene to isophytol. Innovation 3(3).100228

Yoon S-H et al (2009) Combinatorial expression of bacterial whole mevalonate pathway for the production of β-carotene in E. coli. J Biotechnol 140(3–4):218–226

Zhao J et al (2013) Engineering central metabolic modules of Escherichia coli for improving β-carotene production. Metab Eng 17:42–50. https://doi.org/10.1016/j.ymben.2013.02.002

Chapter 11
Microbes in the Baking Industry: Harnessing the Power of Microbes in Baking Products

Disha Dutta, Sombodhi Bhattacharya, and Sisir Nandi

Abstract The baking industry's enduring relationship with microbes and the integration of biotechnological advancements have transformed the production of bakery products. The symbiotic interaction between microbes and baking has been instrumental in creating a vast range of bakery products throughout history. This chapter delves into the historical and contemporary importance of both microbes and biotechnology in various baking processes, demonstrating their role in modern bakery product applications. The discovery of yeast leavening in ancient Egypt revolutionized bread-making and led to the development of various bread types with distinct flavors and textures. Modern times have integrated Biotechnological techniques into the baking industry, leading to innovative techniques and applications. Biotechnological advancements have led to the development of genetically engineered, yeast and bacteria which have enhanced their fermentation capabilities, resulting in improved leavening, texture, and flavor development. Biotechnological techniques have facilitated the use of enzymes in baking, enabling precise control over dough conditioning and product quality. The integration of probiotics and prebiotics through biotechnology has resulted in the production of healthy baked goods. The combined impact of microbes and biotechnology in the baking industry has transformed the production of bakery products, enhancing their quality, diversity, and sustainability.

Keywords Microbes · Baking · Fermentation · Biotechnology · Probiotics · Prebiotics

D. Dutta
Devsthali Vidyapeeth College of Pharmacy, Rudrapur, Uttarakhand, India

S. Bhattacharya · S. Nandi (✉)
Global Institute of Pharmaceutical Education and Research, Kashipur, Uttarakhand, India

V. Kothari et al. (eds.), *Microbial Products for Health and Nutrition*,
https://doi.org/10.1007/978-981-97-4235-6_11

11.1 Introduction

The baking industry has a rich history that spans centuries and is deeply rooted in various cultures worldwide. The origins of baking can be traced back to ancient civilizations, where early forms of bread were made by mixing grains with water and baking them over an open fire. Over time, the process of baking evolved, and the techniques were refined, leading to a diverse range of baked goods we enjoy today. The history of the baking industry is a fascinating journey that spans thousands of years, with roots in ancient civilizations, culminating into the diverse and thriving industry we know today. Throughout history, the baking industry has continuously evolved, adapting to the changes in technology, societal preferences, and cultural influences. Today, baking remains an essential part of daily life and celebrations, with a global market that continues to innovate and delight consumers with a vast array of delicious baked goods.

11.2 Historical Perspective: The Intertwined Journey of Microbes and Baking

11.2.1 Origins of Baking: Early Reliance on Natural Fermentation

The art of baking can be traced back to around 10,000 BCE when humans transitioned from nomadic lifestyles to settled communities. Early civilizations in Egypt, Mesopotamia, and Greece started making simple flatbreads by mixing crushed grains with water and baking them on hot stones or in primitive ovens. These early forms of bread were a staple food and a crucial development in human civilization. Flatbreads have always been the product of sensible and successful solutions to the challenge of converting cereals and non-cereals into digestible and appetizing meals, and they have lasted to the present day due to their flexibility (Pasqualone 2018). Proteomics evidence showed that in China, sourdough bread was made from barley and broomcorn millet by leavening with a renewable starter comprising baker's yeast and lactic acid bacteria, 2500 years ago (Shevchenko et al. 2014).

11.2.2 Egyptian Innovations: The Discovery of Yeast Leavening

Around 3000 BCE, ancient Egyptians discovered the fermentation process, which led to the use of yeast as a leavening agent which allowed bread to rise, producing lighter and more palatable loaves. This marked a significant advancement in baking technology. The ancient Romans further refined baking techniques by building large, communal ovens and developing different types of bread and pastries. Baking became a fundamental aspect of Egyptian society, and they spread their baking knowledge throughout their empire (Gobbetti and Ganzle 2013).

11.2.3 Middle Ages: Artisanal Bakers and the Rise of Sourdough Tradition

During the middle ages, baking became a respected profession, and guilds were established to regulate and protect bakers' interests. These guilds set standards for baking quality, controlled prices, and passed down baking traditions and knowledge from one generation to the next. As trade routes expanded during the Renaissance, baking techniques and recipes spread across continents. European explorers introduced new ingredients like cocoa and sugar, leading to the creation of various pastries and cakes. Baking became more diverse and culturally enriched (Gobbetti and Ganzle 2013).

11.2.4 The Modern Era

With the advent of the Industrial Revolution in the eighteenth and nineteenth centuries, significant changes in the baking industry were made. These involved steam-powered ovens and the mass production of baking powder, which revolutionized baking and enabled large-scale production of bread and pastries. Bakeries expanded their operations and distribution, supplying baked goods to a broader market. The twentieth century saw further innovations in baking technology, such as electric ovens, mechanical mixers, and automated baking processes. Convenience became a priority, leading to the development of ready-made mixes and pre-packaged baked goods. In recent decades, a shift in consumer preferences toward healthier and more authentic foods has led to a resurgence of artisanal baking. Small-scale bakeries and craft bakers focus on traditional methods, natural ingredients, and handmade techniques, offering a more personalized and premium baking experience. The rise of the internet and technology allowed baking enthusiasts and professionals alike to have access to vast resources, online communities, and recipe-sharing platforms. Social media has also played a significant role in popularizing baking trends and showcasing visually appealing creations (Konstantinos et al. 2019).

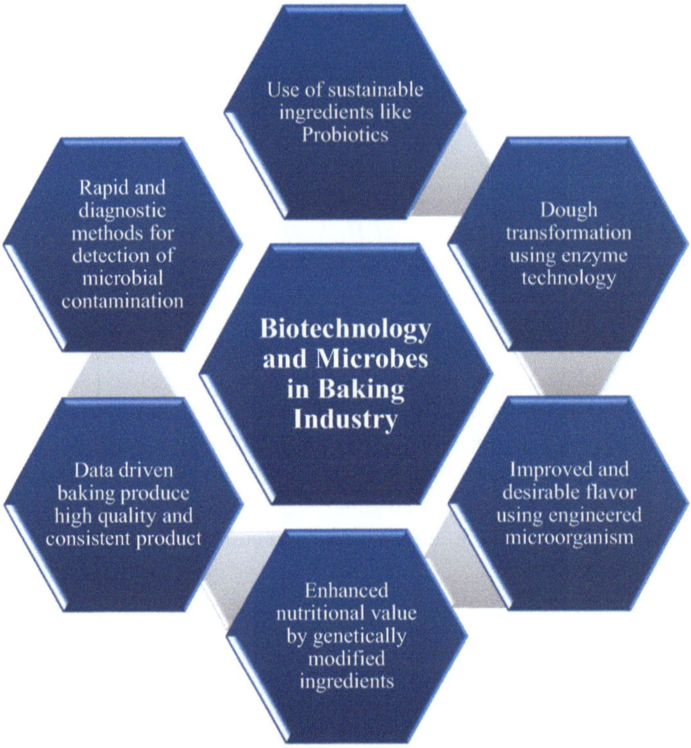

Fig. 11.1 Role of biotechnology and microbes in baking industry

11.3 The Contemporary Impact of Biotechnology in Baking

Biotechnological advancements have sparked a revolution in the baking industry, transforming the way we produce, enhance, and consume baked goods. Biotechnology utilizes biological processes for industrial applications, which has improved the quality, nutritional value, and sustainability of baked products. Some of the key biotechnological advancement applications (Fig. 11.1) that revolutionized baking have been elaborated in Table 11.1.

11.3.1 Improvements in Yeast Strains by Genetic Engineering

The *Saccharomyces* family is the conventional family of yeasts which are used in baking. *Saccharomyces cerevisiae*, also known as the baker's yeast belongs to this family and is the most widely used yeast strain in baking. Genetic engineering techniques have allowed us to engineer and modify yeast strains used in baking. Metabolic pathways have been explored to identify important genes. Numerous

Table 11.1 Biotechnological applications for improvements in baking products

Improvements	Description	Reference
Improvements in yeast strains	Genetic engineering techniques like overexpression and deletion of genes, as well as yeast hybridization methods, have been utilized to offer various benefits, such as faster fermentation, improved flavor profiles, and increased tolerance to environmental stress factors	Struyf et al. (2017)
Enzyme technology	Enzymes such as alpha-amylases, protease, xylanase, lipase, lipoxygenase, glucose oxidase, transglutaminase, cellulase, hemicellulase, pectinase, and phospholipase have been utilized efficiently to target particular aspects of dough transformation, resulting in better volume, crumb structure, and freshness in baked goods	Miguel et al. (2013)
Flavor enhancements	Numerous microorganisms have been identified and engineered to produce natural flavor compounds, enhancing the taste of baked goods without the need for artificial additives. Fermentation techniques involving specific bacteria and fungi can generate unique and desirable flavors in bread and other products	Verma et al. (2022)
Food preservation	New preservation techniques have been explored by biotechnology which extend the shelf life of baked goods without compromising taste and texture. These methods can help reduce food waste and ensure fresher products for consumers	Sridhar et al. (2021)
Sustainable ingredients	Biotechnology has enabled the development of sustainable ingredients for baking. Alternative protein sources like insect proteins have been explored as substitutes for traditional animal-based proteins, reducing the environmental impact of baked goods production	Munialo et al. (2022)
Microbial safety	Biotechnology plays a vital role in ensuring the safety of baked goods by providing tools for pathogen detection and control. Rapid and sensitive diagnostic methods help identify potential microbial contamination, ensuring product safety and consumer protection	Lorenzo et al. (2018)
Data-driven baking	The integration of biotechnology and data analytics has enabled precision baking, optimizing processes, ingredient formulations, and baking conditions to achieve consistent and high-quality results	Sridhar et al. (2023)

attempts have also been made by overexpression and deletion of genes, which contributed to the production of novel yeast strains with improved capabilities. Yeast hybridization is the strategy used to combine two yeast strains with distinctive properties via protoplast fusion. These novel hybrids have characteristics from both of the parents, hence giving an improved strain (Parapouli et al. 2020). The following characteristics have been improved in the genetic engineering of yeast cells:

11.3.1.1 Enhanced Leavening

Heterologous overexpression and deletion of numerous genes can lead to faster and more consistent leavening. This results in improved volume and texture of the baked goods (Zhang et al. 2016). Maltase (alpha-D-glucosidase) (*MAL61*, *MAL62*), AMP-activated S/T protein kinase (AMPK) (*SNF1*), glycogen synthase GSY2, H/ACA box small nucleolar RNA (snoRNA) (*SNR84*) overexpression and *NTH1*, phosphoglucomutase (PGM 2), multicopy Inhibitor of GAL gene expression (*MIG1*) (Lin et al. 2014), and *SUC2* deletion improved dough leavening abilities (Dangi et al. 2017). Mikumo et al. showed that *Saccharomyces cerevisiae* AK 46 which were isolated from cherry fruits were resistant to 2-deoxyglucose (2-DOG). 2-DOG resistance contributed to enhanced dough leavening capabilities (Mikumo et al. 2015).

11.3.1.2 Flavor Enhancement

Genetic engineering enables the introduction of specific genes into microorganisms that produce desirable flavor compounds during fermentation. Numerous yeast strains producing aromatic compounds also have been designed. Gottardi et al. characterized a strain that could synthesize *trans*-cinnamic acid derivatives by inducing heterologous overexpression of phenylalanine ammonia-lyase 2 (*AtPAL2*), aryl carboxylic acid reductase (*acar*), and phosphopantetheinyl transferase (*entD*) (Gottardi et al. 2017). Zha et al. isolated *Saccharomyces cerevisiae* strains from the fermentation of Chinese sesame-flavored Baijiu which generated the 2-furfurylthiol, an aromatic compound. Overexpression of *STR3* and *CYS3* genes leads to a significant increase in the production of the aromatic compound (Zha et al. 2017). An in silico metabolic approach showed that deletion of the *PDC1* gene can increase the production of vanillin β-D-glucoside by two times (Brochado et al. 2010). 3-Dehydroshikimate dehydratase from the dung mold *Podospora pauciseta*, an aromatic carboxylic acid reductase (ACAR) from a bacterium of the *Nocardia* genus, and an *O*-methyltransferase from *Homo sapiens* were introduced into yeast cells by pathway engineering approach which resulted into production of vanillin from *Saccharomyces cerevisiae* (Hansen et al. 2009).

11.3.1.3 Reducing Undesirable Compounds

In microbial fermentation, the selection of appropriate strains is paramount. Diverse microbial taxa generate a range of metabolites, necessitating meticulous selection to minimize undesirable compounds and enhance the production of target flavor profiles. Beyond strain selection, fermentation parameters like temperature, time, and nutrient availability exert a significant influence on the metabolic pathways employed by the chosen microbes. For example, by reducing the production of acetic acid, bakers can achieve a milder flavor in sourdough bread while maintaining its characteristic tang. The deletion of the *PDE2* gene can reduce the anaerobic production of acetic acid (Curiel et al. 2016).

11.3.1.4 Cryostability and Osmostability

Cryopreservation and higher sucrose concentration of dough diminish its quality (Randez-Gil et al. 1999). Phosphodiesterase (*PDE-2*), maltase (alpha-D-glucosidase) (*MAL62*) overexpression, and neutral trehalase (*NTH1*), Proline oxidase (*PUT1*) deletion can give rise to anti-freezing and freeze tolerant effects in yeast (Dangi et al. 2017). Spencer et al. produced a novel yeast strain by crossing *Saccharomyces diastaticus* and *Saccharomyces* (*Zygosaccharomyces*) *rouxii*, by protoplast fusion. This new yeast strain had the sugar fermenting capabilities of *Saccharomyces diastaticus*; however, it was also osmotolerant, which is a characteristic of *Saccharomyces* (*Zygosaccharomyces*) *rouxii*. This new strain had greater leavening abilities and was able to handle dough with a higher sucrose concentration (Spencer et al. 1985). One of the studies revealed that, in *Saccharomyces cerevisiae*, glycerol is the most abundant and suitable solute. When yeast cells are exposed to osmotic stress, such as during semi-solid state bread dough fermentation, they create and store glycerol to prevent dehydration by balancing internal osmolarity against the one found in the surrounding atmosphere. Higher glycerol production, on the other hand, leads to decreased CO_2 generation, which may diminish dough leavening (Aslankoohi et al. 2015).

11.3.1.5 Nutritional Improvements

Genetic engineering can be used to enhance the nutritional content of baked goods. For example, microbes can be engineered to produce vitamins, minerals, or other beneficial compounds, contributing to the overall nutritional value of the final product. Yeast cells were engineered to produce plant terpenoids as precursors for the terpenoid biosynthesis pathway present in yeast sterol, and targeting of sterol biosynthesis pathway genes *HMG2*, *ERG20*, and *IDI1* leads to increased production of plant terpenoids. Plasmid-driven co-expression of *HMG2* and *IDI1* genes leads to strain improvement (Ignea et al. 2011). Processing of HAC1 mRNA, which is an important transcription factor in the Unfolded protein response pathway in yeast cells, was targeted. Targeting HAC1 mRNA splicing, mRNA ligation, mRNA decay, translation, and Hac1p degradation converted the yeast cells into protein factories (Lin et al. 2023b).

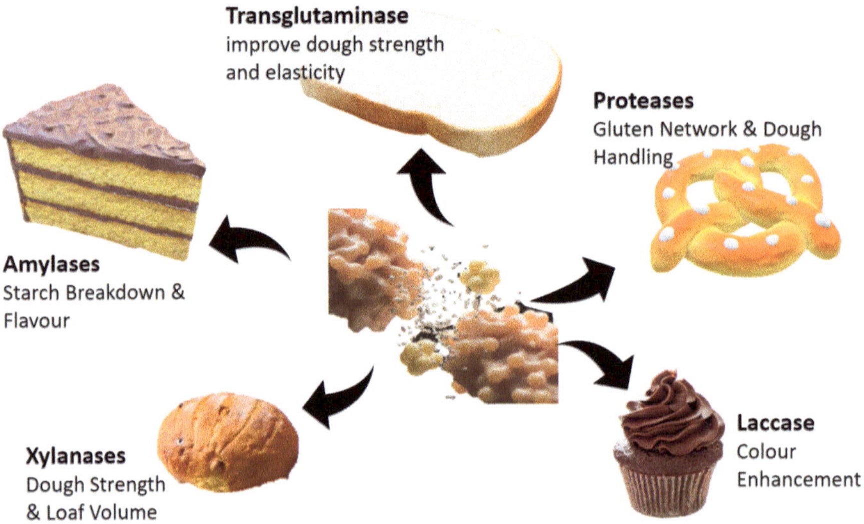

Fig. 11.2 Effect of microbial enzymes in bakery products

11.3.2 Enzyme Applications: Precision Control of Dough Conditioning

Enzymes play a crucial role in the baking industry, contributing to various aspects of dough development, texture improvement, and flavor enhancement (Linko et al. 1997). The role of enzymes has been graphically sketched in Fig. 11.2.

One of the studies tested enzymes (α-amylase, xylanase, lipase, protease) in different ovens (microwave, halogen-microwave combo, conventional) for bread quality improvement. Enzymes reduced initial firmness and increased volume in microwaved and combo-baked bread. In conventional baking, the effects on firmness were seen during storage. Protease notably affected bread color across all ovens (Keskin et al. 2004).

These enzymes are typically derived from microbial sources and are used to modify the properties of dough and the final baked products (Okpara 2022). Common enzymes used in baking and their microbial sources have been tabulated in Table 11.2:

11.3.2.1 Amylases

Amylases are enzymes that catalyze the hydrolysis of starch into sugars. There are several different varieties of amylases, out of them α-amylases used in bakery very often, which are derived from bacterial and fungal sources and also malted barley. It breaks down complex starches into simpler sugars, improving dough fermentation and creating softer bread. The generation of α-amylase via submerged

Table 11.2 Enzymes used in baking industries and their applications

Enzyme	Microbial source	Role in baking industry	Reference
Alpha-amylase	*Bacillusamylo liquefaciens* *Aspergillus oryzae*	Converts starch to sugars, improving fermentation, and crumb texture	Rebholz et al. (2021)
Protease	*Lactobacillus alimentarius* *Aspergillus oryzae*	Weakens gluten, and enhances dough handling	Heredia-Sandoval et al. (2016)
Xylanase	*Bacillus* spp. *Trichoderma* species	Breaks down hemicellulose, and improves dough handling	Babu et al. (2019)
Lipase	*Rhizopus oryzae*	Contributes to flavor development by breaking down lipids	Fereidoon and Hossain (2022)
Lipoxygenase	*Fusarium oxysporym*	Oxidizes lipids, and improves dough strength and volume	Buche et al. (2014)
Glucose oxidase	*Aspergillus niger*	Strengthens dough by improving protein cross-linking	Rasiah et al. (2005)
Transglutaminase	*Streptomyces mobaraensis*	Forms covalent bonds between proteins, improving dough texture	Yang et al. (2022)
Cellulase	*Trichoderma* species	Breaks down cellulose, and improves dough volume and texture	Jayasekara and Ratnayake (2019)
Hemicellulase	*Bacillus* species	Breaks down hemicellulose, and enhances dough extensibility	Tian et al. (2021)
Pectinase	*Aspergillus* species	Improves dough handling and contributes to a softer crumb	Ahmad et al. (2014)
Phospholipase	*Salmonella enterica*	Enhances dough rheology and gas retention	Burak and Unal (2017)

fermentation (SmF) and solid-state fermentation (SSF) has been studied and is affected by several physicochemical variables. Because of the simplicity with which diverse factors like pH, temperature, aeration, oxygen transfer, and moisture can be controlled, SmF has historically been employed for the manufacture of industrially essential enzymes. SSF systems look promising owing to their inherent potential and benefits. Amylases function to enhance the dough's extensibility, resulting in improved crumb structure (De Souza and Magalhães 2010).

The starch industry is the most common application for α-amylases, which are utilized for starch hydrolysis in the starch liquefaction process, which turns starch into fructose and glucose syrups (Nielsen and Borchert 2000). All starch undergoes enzymatic conversion, which includes gelatinization, which involves the dissolution of starch granules, resulting in a viscous suspension; liquefaction, which involves partial hydrolysis and viscosity loss; and saccharification, which involves the production of glucose and maltose through further hydrolysis used in industry (Gupta et al. 2003).

11.3.2.2 Proteases

Bacterial and fungal proteases are routinely used in the baking industry as they reduce the mixing time, decrease dough consistency and uniformity; regulate the gluten strength in bread, and improve the texture and flavor. They are also utilized in gluten-free products. Proteases from *Aspergillus usamii* have been utilized to improve the functional properties of wheat gluten (Raveendran et al. 2018). Bacterial proteases sourced from *Bacillus stearothermophilus* improved bread crumb appearance, increased volume, and softened texture (Kawamura-Konish et al. 2013). Glutenin and gliadin are broken down by proteases, which can be utilized to make gluten-free products (Singh and Bajaj 2017).

11.3.2.3 Xylanases

Xylanases help degrade hemicellulose in flour, improving dough handling and final product quality. Xylanase from *Streptomyces* S27 was cloned and expressed in *P. pastoris*, which resulted in its increase in pH stability and thermostability. This xylanase on treatment with bread flour, an increase in reducing sugars in dough, favoring the fermentation and gas-retention was observed. This also increased the bread volume (Cunha et al. 2018). A recombinant arabinoxylan sourced from *Halolactibacillus miurensis* improved the structure of the dough and increased its air holding capacity. These enhanced leavening abilities also improve the texture of the final product (Zhang et al. 2022). Another recombinant xylanase sourced from *Plectosphaerella cucumerina* also improved bread volume, texture, and taste (Zhan et al. 2014).

11.3.2.4 Laccase

Laccase enzyme can be used to replace chemical additions in bread to increase texture and volume. A laccase-encoding gene from *Phlebia brevispora*, a natural fungus from *Misiones*, *Argentina*, was expressed in the yeast *Kluyveromyces lactis*, which is generally considered harmless. Medium conditions were adjusted to promote laccase activity. The laccase enzyme raises the specific volume and lowers the hardness of bread, making it an alternative to chemical additions in the bakery business (Molina et al. 2023).

11.3.2.5 Lipases

Lipid concentrations are low in flour dough; however, it significantly affects its shelf-life. Lipases can improve bread volume by breaking down galactolipids. It can also help in the formation of amyloid lipid complexes, which can reduce amylopectin degradation and hence increase their shelf life (Reyes et al. 2022).

11.3.2.6 Glucose Oxidase

Glucose oxidase helps to strengthen dough by improving cross-linking of gluten proteins. A recombinant Glucose oxidase produced from *Cladosporium neopsychrotolerans* by cloning into Pichiapastoris showed greater activity than Glucose oxidase sourced from *Aspergillus niger* (Jianzhong et al. 2020).

11.3.2.7 Transglutaminase

Transglutaminase is an enzyme that catalyzes protein crosslinking by forming isopeptide bonds between glutamine and lysine amino acid side chains. In baking, it can be added to doughs and batters to strengthen and improve the texture of products by crosslinking the gluten and other wheat proteins as well as proteins in ingredients like dairy. Specifically, transglutaminase facilitates an acyl transferase reaction that results in the formation of epsilon-(gamma-glutamyl) lysine isopeptide bonds. The amine group of lysine binds to the carboxamide group on the side chain of glutamine residues catalyzed by the enzyme. This crosslinking reinforces the gluten network and overall dough matrix, leading to increased dough elasticity and stability. The tighter protein network is stronger and more resistant to mechanical shearing. This improved handling properties, better gas retention during proofing, and greater volume as well as finer crumb structure in the final baked items (Boukid et al. 2018).

11.3.2.8 Cellulases

Cellulases can modify the dough structure by breaking down cellulose, breaking down complex polysaccharides present in flour, such as hemicellulose and beta-glucans. This breakdown can improve dough handling properties, leading to better dough elasticity and extensibility leading to improved volume and texture. It is also utilized to reduce levels of roughage in dough. By breaking down certain polysaccharides, cellulases can reduce the retrogradation of starch, which helps maintain moisture retention and delays staling, thereby extending the product's freshness (Kumar et al. 2018).

11.3.2.9 Hemicellulases

Hemicellulases break down hemicellulose, improving dough extensibility and gas retention. One of the studies showed an increase in bread volume and firmness of breadcrumb texture after hemicellulase treatment (Zhou et al. 2021). The Gluten network is improved by the addition of disulfide bonds, hydrophobic interactions, and hydrogen bonds (Tian et al. 2021).

11.3.2.10 Pectinases

Pectinases can be generated from both natural and transgenic microorganisms, with attempts to boost their thermostability and production. Sourced from fungi, the pectinase enzyme has the capability of breaking down pectin, a complex polysaccharide found in plant cell walls improving dough handling properties and contributing to a softer crumb. Used for the clarification of fruit juices, improves color and flavor of the product. Using concentrated date syrup treated with pectinase in pan bread compositions resulted in a much greater specific loaf volume (Raveendran et al. 2018).

11.3.2.11 Phospholipases

Phospholipases improve dough rheology and can enhance gas retention. Phytase (EC 3.1.3.8) is a type of phosphatase that catalyzes the hydrolysis of inorganic phosphorus from phytic acid found in grains and oilseeds. Phytase's enzymatic activity promotes the absorption of minerals such as calcium, zinc, and iron, which are sometimes attached to phytic acid in grains and oilseeds. Consequently, phytase's hydrolytic activity on phytic acid boosts calcium, zinc, and iron bioavailability in monogastric animals. Phytase is a naturally occurring enzyme found in fungi, bacteria, plants, and ruminants. *Aspergillus niger*, *Aspergillus fumigatus*, *Aspergillus ficuum*, *Schizosaccharomyces pombe*, *Klebsiella terrigena*, *Klebsiella ooxytoca*, *Escherichia coli*, *Bacillus amyloliquefaciens*, and *Bacillus subtilis* are microbial sources of phospholipases (Okpara 2022).

11.3.3 Microbial Strain Selection: Tailoring Fermentation for Baked Goods

The art and science of baking remain integral to cultures worldwide, providing nourishment and delight to people of all ages. The baking industry produces a diverse range of products, catering to various tastes and preferences.

11.3.3.1 Lactic Acid Bacteria

Some baking processes involve the use of lactic acid bacteria, such as *Lactobacillus* species. These bacteria are responsible for sourdough fermentation, which is a natural leavening method used in traditional bread-making. In a sourdough starter, a mixture of flour and water, lactic acid bacteria, and wild yeast are present. These bacteria produce lactic acid during fermentation, which helps to acidify the dough and contributes to the tangy flavor characteristic of sourdough bread (Perez-Alvarado et al. 2022).

The activity of lactic acid bacteria leads to acidification of the dough, which has several benefits. Acidification improves the quality of bread by inhibiting the growth of spoilage microbes. It also helps break down gluten in the dough, making it easier to handle during shaping and giving the bread a softer crumb. Several factors influence the diversity of LAB in sourdough. Geographical location is one such factor, as different regions may have different naturally occurring LAB strains. The type of flour used in the bread-making process also affects the LAB diversity, as different flours may contain different populations of LAB. Other factors such as temperature, humidity, and the fermentation process itself can influence the growth and development of LAB in sourdough (Arezoo et al. 2024).

The metabolic activities and enzymatic capabilities of different LAB strains further contribute to the wide range of flavor profiles and bread textures observed in sourdough. LAB can produce various flavor compounds, including lactic acid, acetic acid, and other volatile compounds that give sourdough its distinct taste. They also produce enzymes that break down complex carbohydrates, leading to the formation of simpler sugars that contribute to the overall flavor and texture of the bread (Akamine et al. 2023).

LAB (Lactic Acid Bacteria) can be found in various sources within the baking industry. Common sources of LAB found in the industry are given in Table 11.3.

Tailoring the fermentation process for baked products involves manipulating various factors to achieve specific desired characteristics in the final product. Here are some key factors that can be adjusted to tailor the fermentation process:

1. **Type of LAB starter culture**: The choice of LAB starter culture can significantly impact flavor, texture, and fermentation speed. Different strains produce varying levels of acids, gasses, and other byproducts. Selecting the right culture can help achieve the desired taste and texture profile (Ayivi and Ibrahim 2022).
2. **Inoculation size**: The amount of starter culture or sourdough used for inoculation affects fermentation kinetics. Adjusting the inoculation size can control the rate and intensity of fermentation (Celano et al. 2016).

Table 11.3 Common sources of LAB in baking industry

Sources of LAB in the baking industry	Description
Starter cultures	Specific LAB strains are intentionally added to control fermentation, obtained from suppliers as dried or liquid cultures
Spontaneous fermentation	Naturally occurring LAB from flour, grains, and air initiate fermentation, used in traditional baking methods
Backslopping	Using a portion of old dough/starter to jump-start fermentation in the new batch, maintains flavor and consistency
Flour	Some flours (e.g., whole wheat, rye) naturally contain LAB, affecting the dough's LAB content and final bread
Equipment and environment	Baking equipment, surfaces, and air can carry LAB. Proper sanitation prevents unintended contamination

3. **Temperature**: Fermentation temperature influences the activity of LAB. Higher temperatures accelerate fermentation, while lower temperatures slow it down. Choose temperatures that align with the desired fermentation rate and product characteristics (Mazorra-Manzano et al. 2020).
4. **Fermentation duration**: Longer fermentation allows for more complex flavor development, as well as improved dough structure. Shorter fermentation times might be preferred for certain products to achieve specific textures (Perez-Alvarado et al. 2022).
5. **Hydration level**: The hydration level of the dough affects the overall fermentation process. Higher hydration can lead to more extensible dough and greater LAB activity, while lower hydration can yield a tighter crumb structure (Minervini et al. 2019).
6. **Flour types**: Different flours contribute varying levels of nutrients and LAB. Incorporating flours with higher LAB content, like whole wheat or rye, can enhance fermentation and flavor (Perez-Alvarado et al. 2022).
7. **Backslopping ratio**: Adjusting the amount of old dough or starter used in backslopping can influence fermentation speed and intensity. A higher ratio of old dough accelerates fermentation (Ozel et al. 2020).
8. **Acidification rate**: LAB produces lactic acid during fermentation. Controlling the rate of acidification can impact dough extensibility, crumb texture, and flavor (Lonner 1988).
9. **pH control**: Monitoring and adjusting the pH of the dough can influence LAB activity and fermentation characteristics. pH levels can impact enzyme activity, dough structure, and final product quality (Lonner 1988).
10. **Dough handling techniques**: Techniques such as folding and kneading can influence dough development and fermentation. Proper handling ensures even distribution of LAB and gasses.
11. **Use of additives**: Some additives like enzymes or malt can impact LAB activity and fermentation dynamics. Using these additives can help tailor the process to achieve specific results (Da Silva et al. 2015).
12. **Environmental conditions**: Factors such as humidity and air circulation in the baking environment can affect fermentation. Ensure the environment is conducive to desired fermentation outcomes (Mengesha et al. 2022).

Fraberger et al. showed that lactic acid bacteria isolated from sourdough preparations also could inhibit the growth of molds; hence, LABs can show natural preservative properties (Fraberge et al. 2020). Preservative qualities were also shown by the use of lactic acid bacteria *L. casei*, which made a softer final product and reduced staleness (Muhialdin et al. 2018).

11.3.4 Sustainability Considerations: Biotechnology and Eco-Friendly Practices

Certainly, sustainability considerations in the baking industry involve adopting biotechnology and eco-friendly practices to reduce environmental impact and promote more responsible production. Here are some ways to integrate these concepts into baking processes:

(a) **Sourdough and natural fermentation**: Embrace natural fermentation methods like sourdough, which rely on the inherent microflora in flour and the environment. This reduces the need for commercial starter cultures and minimizes resource consumption.
(b) **Local and organic ingredients**: Source local and organic ingredients when possible. Supporting local agriculture reduces transportation emissions, while organic practices promote healthier soils and ecosystems.
(c) **Reducing food waste**: Implement strategies to minimize food waste. Proper inventory management, portion control, and creative use of surplus dough can significantly reduce waste.
(d) **Alternative flours**: Incorporate alternative flours like whole wheat, ancient grains, and legume flours. These can have lower environmental footprints and offer unique flavors and nutritional benefits (Turfani et al. 2017).
(e) **Biotechnology for improved dough**: Explore biotechnological approaches to improve dough quality and reduce resource use. Enzymes and microbial additives can enhance dough characteristics while reducing ingredient quantities.
(f) **Reduced packaging**: Minimize packaging waste by using eco-friendly packaging materials or adopting bulk purchasing for ingredients.
(g) **Waste management and recycling**: Implement comprehensive waste management practices. Recycle, compost, or find innovative ways to repurpose by-products like spent grain.
(h) **Carbon footprint measurement**: Assess the carbon footprint of the baking processes and products. This information can guide efforts to reduce emissions and promote more sustainable practices (Rivera et al. 2014). The carbon footprint (CF) is an indicator that aims to measure the entire quantity of greenhouse gas emissions—expressed in carbon dioxide equivalent—created directly or indirectly by activity or accumulated throughout a product's life or phases. Studies were done to estimate the CF of the baking process, comprising the phases of primary production, milling, baking, consumption, and the associated transit of wheat and flour, as well as to compare the CF of the bread produced.
(i) **Education and outreach**: Educate staff and customers about the environmental benefits of your sustainable practices. This can create awareness and encourage others to support eco-friendly baking.
(j) **Collaboration and partnerships**: Collaborate with local farmers, suppliers, and organizations that share your sustainability goals. This can foster a community of like-minded businesses working toward a common goal.

(k) **Continuous improvement**: Regularly evaluate and refine your sustainable practices. Technology and knowledge are constantly evolving, so staying up-to-date can help you adopt even more efficient and eco-friendly methods.

11.4 Improved Leavening and Texture

11.4.1 Croissants and the Art of Layering

Croissants are renowned for their delicate, flaky layers. Achieving this texture relies on precise control of butter incorporation and dough folding during preparation. Genetic engineering has introduced specialized yeast and enzyme strains designed to enhance the gluten structure in the dough, creating the perfect balance of elasticity and extensibility (Omedi et al. 2019). This results in dough that can be rolled thin without tearing, leading to those signature croissant layers.

Additionally, genetically modified yeasts contribute to improved leavening, creating air pockets that expand during baking, further enhancing the flakiness and volume of croissants.

11.4.2 Novelty Cakes and Confections

Unleashing creative possibilities: genetic engineering plays a remarkable role in the world of novelty cakes and confections. Customized designs and intricate decorations are made possible by genetically engineered microorganisms that produce food-safe pigments and colors. These microorganisms are used to develop edible ink, which can be used in 3D food printers to create detailed designs directly onto cakes, cookies, and other confections (Lin et al. 2023a).

Genetic engineering is also employed to meet consumer demand for unique and personalized confections. By selecting yeast strains and enzymes tailored to produce specific flavor compounds and textures, bakeries and confectioneries can create a wide range of one-of-a-kind treats. From personalized birthday cakes to artisanal chocolates with unique fillings, biotechnology enables confectioners to craft products that cater to individual tastes.

11.4.3 Gluten-Free Baking: Biotechnological Alternatives for Wheat-Free Delights

For individuals with gluten sensitivities, gluten-free baking has been revolutionized by biotechnology. Genetic engineering has led to the development of yeast strains and enzyme combinations that improve the texture and shelf life of gluten-free bread, muffins, and pastries. Biotechnological advancements also offer alternatives to traditional wheat-based flours. Yeast strains engineered to ferment alternative flours like rice, quinoa, or chickpea flour can produce gluten-free baked goods with comparable texture and taste.

11.4.3.1 Innovations in Gluten-Free Flour Substitutes

In response to the increasing demand for gluten-free products, there have been notable innovations in gluten-free flour substitutes. These innovations aim to replicate the texture and taste of traditional wheat-based products while catering to individuals with gluten sensitivities.

11.4.3.2 Microbes in Gluten-Free Fermentation

The role of microbes in gluten-free fermentation is essential for enhancing the quality and palatability of gluten-free baked goods. While gluten is absent, beneficial microbes, including lactic acid bacteria (LAB) and yeast, play vital roles. LAB ferments the sugar in the gluten-free flour, creating lactic acid. This acid preserves the food, adds a tangy flavor, and helps develop desirable textures. Whereas Yeast ferments sugars as well, but with a different outcome. Microbes produce carbon dioxide gas, which helps the dough rise and gives the final product a lighter texture. Gluten, found in wheat flour, helps dough become elastic and rise. In gluten-free flours, these microbes take over the job of gluten, creating a similar structure and improving the overall quality of the product. This can potentially boost the nutritional profile by increasing certain vitamins and making minerals more bioavailable (Kawamura-Konish et al. 2013).

11.4.4 Health-Enhanced Baked Goods

To meet the demand for healthier baked goods, there is a growing trend toward incorporating probiotics and prebiotics into recipes. These additions offer potential digestive health benefits and enhance the nutritional value of baked products:

11.4.4.1 Probiotics

Probiotics are beneficial live microorganisms, often strains of *Lactobacillus* and *Bifidobacterium*, known for their positive impact on gut health. The inclusion of probiotics in baked goods, such as bread, muffins, or yogurt-based products, can introduce live cultures that may support digestive health (Akamine et al. 2023).

The development of two distinct protocols for the administration of probiotics via bakery products are, probiotic pieces of bread made by adding microorganisms directly to the dough, and the addition of edible probiotic film based on sodium caseinate and chia mucilage for application on the bread surface, reports show the development of novel edible films containing probiotics that are used to coat bakery products like bread. The pullulant/starch films were developed and optimized to allow high viability and retention of encapsulated probiotic bacteria after storage. The edible films could slowly release probiotic cells like *Lactobacillus plantarum* over 72 h. When films were coated onto white bread and the bread was stored for a week, the counts of *Lactobacillus plantarum* probiotic cells were maintained at high viable levels (Kanmani and Lim 2013).

11.4.4.2 Prebiotics

Prebiotics are fibers that humans cannot digest, but certain gut bacteria can utilize them for growth. Prebiotics offer a double dose of delight. Firstly, they can be incorporated into baked goods to create a "gut-friendly" label, tapping into the growing consumer demand for functional foods. Secondly, prebiotics often boast technological benefits. They can improve dough handling, enhance moisture retention, and even extend shelf life. When prebiotics are included in foods like baked goods, they help stimulate the growth and activity of beneficial bacteria in the gut microbiome (Dimidi et al. 2019). Prebiotics like inulin and oligofructose add valuable fiber content to baked goods, appealing to health-conscious consumers. Inulin is a common prebiotic added to some types of bread, cookies, and other baked products. Inulin is extracted from chicory roots and composed of fructose chains that provide fuel for probiotic species like *Bifidobacterium* and *Lactobacillus*. As these bacteria digest the inulin, they produce beneficial by-products like short-chain fatty acids and certain B vitamins (Ansari et al. 2023).

11.4.4.2.1 Digestive Health and Functional Bakery Products

The integration of probiotics and prebiotics into bakery products aligns with the broader trend of functional foods. Such products are designed not only for taste and texture but also for their potential to promote well-being by enhancing gut health and aiding digestion. These are foods enriched with ingredients that offer potential health benefits beyond basic nutrition. They often target specific functions in the body, like gut health, immune function, or heart health. Consumers are increasingly interested in functional foods as they strive for a more proactive approach to wellness through their diet. Example: Probiotic Yogurt Muffins, These muffins incorporate live yogurt containing probiotic bacteria (*Bifidobacterium*) into the batter. The yogurt adds a tangy flavor and contributes to a moist texture. More importantly, the live bacteria in the yogurt potentially benefit gut health by maintaining a healthy balance of gut microbiota. Additionally, the muffins can be fortified with prebiotic fibers like inulin to further promote the growth of these beneficial bacteria Olson and Aryana (2022).

11.5 Conclusion

The use of microorganisms in baking provides a wonderful junction of science and culinary creation. Moreover, contemporary approaches have contributed to sustainable baking practices, such as reducing food waste and optimizing resource utilization. Biotechnological interventions have influenced the leavening process, resulting in an extensive range of bread varieties with improved quality and shelf life. The use of enzymes and genetically modified microbes has contributed to the enhanced texture and volume of pastries, such as croissants, Danish pastries, and puff pastries. It also leads to expanding the possibilities for creative and novelty cakes, providing unique products. From traditional bread-making to modern genetic engineering and the application of enzymes, biotechnology continues to drive innovation, leading to the production and availability of a wide variety of baked goods. Microorganisms, which include yeasts, bacteria, and the use of enzymes, play critical roles in molding the flavor, texture, and shelf life of baked goods. Microbe's adaptability enables the development of a wide range of products that appeal to a wide range of customer preferences, including gluten-free choices and health-enhanced baked goods. Gluten-free baked goods have proven to be advantageous for individuals sensitive to gluten consumption. Innovations in gluten-free flour substitutes have expanded the options for individuals with gluten sensitivities. Additionally, the integration of probiotics and prebiotics into baked goods reflects a growing emphasis on digestive health and functional bakery products, providing consumers with healthier choices. Encapsulating probiotic cultures allows nutritional fortification without excessive heat-killing microbes during baking.

References

Ahmad Z, Butt MS, Ahmed A, Riaz M, Sabir SM, Farooq U, Rehman FU (2014) Effect of *Aspergillus niger* xylanase on dough characteristics and bread quality attributes. J Food Sci Technol 51(10):2445–2453. https://doi.org/10.1007/s13197-012-0734-8

Akamine IT, Mansoldo FRP, Vermelho AB (2023) Probiotics in the sourdough bread fermentation: current status. Fermentation 9(2):90. https://doi.org/10.3390/fermentation9020090

Ansari F, Neshat M, Pourjafar H, Jafari SM, Samakkhah SA, Mirzakhani E (2023) The role of probiotics and prebiotics in modulating of the gut-brain axis. Front Nutr 10:1173660. https://doi.org/10.3389/fnut.2023.1173660

Arezoo F, Abedinzadeh S, Torbati MA, Damirchi SA, Savage GP (2024) Considering sourdough from a biochemical, organoleptic, and nutritional perspective. J Food Compos Anal 125:105853. https://doi.org/10.1016/j.jfca.2023.105853

Aslankoohi E, Rezaei MN, Vervoort Y, Courtin CM, Verstrepen KJ (2015) Glycerol production by fermenting yeast cells is essential for optimal bread dough fermentation. PLoS One 10(3):0119364. https://doi.org/10.1371/journal.pone.0119364

Ayivi RD, Ibrahim SA (2022) Lactic acid bacteria: an essential probiotic and starter culture for the production of yoghurt. Int J Food Sci Technol 57(11):7008–7025. https://doi.org/10.1111/ijfs.16076

Babu J, Kumar V, Ramteke PW (2019) Psychrophilic enzymes: potential biocatalysts for food processing. In: Enzymes in food biotechnology. Academic, pp 817–825. https://doi.org/10.1016/B978-0-12-813280-7.00047-5

Boukid F, Carini E, Curti E, Bardini G, Pizzigalli E, Vittadini E (2018) Effectiveness of vital gluten and transglutaminase in the improvement of physico-chemical properties of fresh bread. LWT 92:465–470. https://doi.org/10.1016/j.lwt.2018.02.059

Brochado RA, Matos C, Moller BL, Hansen J, Mortensen UH, Patil KR (2010) Improved vanillin production in baker's yeast through in silico design. Microb Cell Factories 9(1):84. https://doi.org/10.1186/1475-2859-9-84

Buche F, Davidou S, Verte F, Rouille J, Potus J, Nicolas J, Pommet M (2014) Influence of oxygen content of kneading atmosphere on oxygen uptake and relaxation index of bread dough with various additives. J Cereal Sci 60(2):282–288. https://doi.org/10.1016/j.jcs.2014.06.007

Burak A, Unal SS (2017) The effects of certain enzymes on the rheology of dough and the quality characteristics of bread prepared from wheat meal. J Food Sci Technol 54(6):1628–1637. https://doi.org/10.1007/s13197-017-2594-8

Celano G, Angelis MD, Minervini F, Gobbetti M (2016) Different flour microbial communities drive to sourdoughs characterized by diverse bacterial strains and free amino acid profiles. Front Microbiol 7:1770. https://doi.org/10.3389/fmicb.2016.01770

Cunha DQB, Cândida C, Gama AR, Cintra LC, Bataus LAM, Ulhoa CJ (2018) Improvement of bread making quality by supplementation with a recombinant xylanase produced by *Pichia pastoris*. PLoS One 13(2):e0192996. https://doi.org/10.1371/journal.pone.0192996

Curiel JA, Salvado Z, Tronchoni J, Morales P, Rodrigues AJ, Quiros M, Gonzalez R (2016) Identification of target genes to control acetate yield during aerobic fermentation with *Saccharomyces cerevisiae*. Microb Cell Factories 15(1):156. https://doi.org/10.1186/s12934-016-0555-y

Da Silva TC, Smith ML, Barnard AM, Kung L (2015) The effect of a chemical additive on the fermentation and aerobic stability of high-moisture corn. J Dairy Sci 98(12):8904–8912. https://doi.org/10.3168/jds.2015-9640

Dangi AK, Dubey KK, Shukla P (2017) Strategies to improve *Saccharomyces cerevisiae*: technological advancements and evolutionary engineering. Indian J Microbiol 57(4):378–386. https://doi.org/10.1007/s12088-017-0679-8

De Souza PM, Magalhães PDOE (2010) Application of microbial α-amylase in industry—a review. Braz J Microbiol 41(4):850–861. https://doi.org/10.1590/S1517-83822010000400004

Dimidi E, Cox SR, Rossi M, Whelan K (2019) Fermented foods: definitions and characteristics, impact on the gut microbiota and effects on gastrointestinal health and disease. Nutrients 11(8):1806. https://doi.org/10.3390/nu11081806

Fereidoon S, Hossain A (2022) Role of lipids in food flavor generation. Molecules 27(15):5014. https://doi.org/10.3390/molecules27155014

Fraberge V, Claudia A, Konrad JD (2020) Functional properties and sustainability improvement of sourdough bread by lactic acid bacteria. Microorganisms 8(12):1895. https://doi.org/10.3390/microorganisms8121895

Gobbetti M, Ganzle M (2013) Handbook on sourdough biotechnology. Springer, New York. https://doi.org/10.1007/978-1-4614-5425-0

Gottardi M, Knudsen JD, Prado L, Oreb M, Branduardi P, Boles E (2017) De novo biosynthesis of trans-cinnamic acid derivatives in *Saccharomyces cerevisiae*. Appl Microbiol Biotechnol 101(12):4883–4893. https://doi.org/10.1007/s00253-017-8220-x

Gupta R, Gigras P, Mohapatra H, Goswami VK, Chauhan B (2003) Microbial α-amylases: a biotechnological perspective. Process Biochem 38(11):1599–1616. https://doi.org/10.1016/S0032-9592(03)00053-0

Hansen EH, Moller BL, Kock DR, Bünner CM, Kristensen C, Jensen OR, Okkels FT, Olsen CE, Motawia MS, Hansen J (2009) De novo biosynthesis of vanillin in fission yeast (*Schizosaccharomyces pombe*) and baker's yeast (*Saccharomyces cerevisiae*). Appl Environ Microbiol 75(9):2765–2774. https://doi.org/10.1128/AEM.02681-08

Heredia-Sandoval N, Valencia-Tapia M, Barca ACDL, Islas-Rubio A (2016) Microbial proteases in baked goods: modification of gluten and effects on immunogenicity and product quality. Foods 5(4):59. https://doi.org/10.3390/foods5030059

Ignea C, Cvetkovic I, Loupassaki S, Kefalas P, Johnson CB, Kampranis SC, Makris AM (2011) Improving yeast strains using recyclable integration cassettes, for the production of plant terpenoids. Microb Cell Factories 10(1):4. https://doi.org/10.1186/1475-2859-10-4

Jayasekara S, Ratnayake R (2019) Microbial cellulases: an overview and applications. Cellulose 1(1). https://doi.org/10.5772/intechopen.84531

Jianzhong G, Jiang X, Liu W, Wang Y, Huang H, Bai Y, Su X, Yao B, Luo H (2020) Characterization, stability improvement, and bread baking applications of a novel cold-adapted glucose oxidase from Cladosporium neopsychrotolerans SL16. Food Chem 310:125970. https://doi.org/10.1016/j.foodchem.2019.125970

Kanmani P, Lim ST (2013) Development and characterization of novel probiotic-residing pullulan/starch edible films. Food Chem 141(2):1041–1049

Kawamura-Konish Y, Shoda K, Koga H, Honda Y (2013) Improvement in gluten-free rice bread quality by protease treatment. J Cereal Sci 58(1):45–50. https://doi.org/10.1016/j.jcs.2013.02.010

Keskin S, Sumnu G, Sahin S (2004) Usage of enzymes in a novel baking process. Nahrung 48(2):156–160. https://doi.org/10.1002/food.200300412

Konstantinos P, Zoumpopoulou G, Georgalaki M, Alexandraki V, Kazou M, Anastasiou R, Tsakalidou E (2019) Sourdough bread. In: Innovations in traditional foods. Woodhead Publishing, pp 127–158

Kumar VA, Kurup SC, Snishamol C, Prabhu GN (2018) Role of cellulases in food, feed, and beverage industries. In: Green bio-processes. Springer, pp 323–343

Lin X, Zhang CY, Xiao-Wen B, Hai-Yan S, Dong-Guang X (2014) Effects of MIG1, TUP1 and SSN6 deletion on maltose metabolism and leavening ability of baker's yeast in lean dough. Microb Cell Factories 13(1):93. https://doi.org/10.1186/s12934-014-0093-4

Lin Q, Hu Y, Qiu C, Li X, Sang S, McClements DJ, Chen L (2023a) Peanut protein-polysaccharide hydrogels based on semi-interpenetrating networks used for 3D/4D printing. Food Hydrocoll 137:108332. https://doi.org/10.1016/j.foodhyd.2022.108332

Lin Y, Yunzi F, Lin Z, Zhao M, Huang M (2023b) Improved protein production in yeast using cell engineering with genes related to a key factor in the unfolded protein response. Metab Eng 77:152–161. https://doi.org/10.1016/j.ymben.2023.04.004

Linko YY, Javanainen P, Linko S (1997) Biotechnology of bread baking. Trends Food Sci Technol 8(10):339–344. https://doi.org/10.1016/S0924-2244(97)01066-2

Lonner C (1988) Acidification properties of lactic acid bacteria in rye sour doughs. Food Microbiol 5(1):43–58. https://doi.org/10.1016/0740-0020(88)90007-X

Lorenzo JM, Munekata PE, Dominguez R, Pateiro M, Saraiva JA, Franco D (2018) Main groups of microorganisms of relevance for food safety and stability. In: Innovative technologies for food preservation. Academic, pp 53–107. https://doi.org/10.1016/B978-0-12-811031-7.00003-0

Mazorra-Manzano MA, Miguel AS, Robles-Porchas GR, González-Velázquez DA, Torres-Llanez MJ, Martínez-Porchas M, García-Sifuentes CO, Gonzalez-Cordova AF, Vallejo-Cordoba B (2020) Cheese whey fermentation by its native microbiota: proteolysis and bioactive peptides release with ACE-inhibitory activity. Fermentation 6(1):19. https://doi.org/10.3390/fermentation6010019

Mengesha Y, Tebeje A, Tilahun B (2022) A Review on Factors Influencing the Fermentation Process of Teff (Eragrostis teff) and Other Cereal-Based Ethiopian Injera. International Journal of Food Science 2022:4419955

Miguel ASM, Souza T, Erika Figueiredo VDC, Lobo BWP, Maria G (2013) Enzymes in bakery: current and future trends. Food Ind. https://doi.org/10.5772/53168

Mikumo D, Takaya M, Orikasa Y, Ohwada T (2015) Improved leavening ability of a wild yeast, *Saccharomyces cerevisiae* AK46 2-deoxyglucose resistant mutant. Food Sci Technol Res 21(4):623–630. https://doi.org/10.3136/fstr.21.623

Minervini F, Dinardo FR, Angelis MD, Gobbetti M (2019) Tap water is one of the drivers that establish and assembly the lactic acid bacterium biota during sourdough preparation. Sci Rep 9(1):570. https://doi.org/10.1038/s41598-018-36786-2

Molina MA, Cazzaniga A, Milde LB, Sgroppo SC, Zapata PD, Fonseca MI (2023) Purification and characterization of a fungal laccase expressed in *Kluyveromyces lactis* suitable for baking. J Food Sci 88(4):1365–1377. https://doi.org/10.1111/1750-3841.16497

Muhialdin BJ, Hassan Z, Saari N (2018) In vitro antifungal activity of lactic acid bacteria low molecular peptides against spoilage fungi of bakery products. Ann Microbiol 68(9):557–567. https://doi.org/10.1007/s13213-018-1363-x

Munialo CD, Stewart D, Campbell L, Euston SR (2022) Extraction, characterisation and functional applications of sustainable alternative protein sources for future foods: a review. Future Foods 6:100152. https://doi.org/10.1016/j.fufo.2022.100152

Nielsen JE, Borchert TV (2000) Protein engineering of bacterial α-amylases. Biochim Biophys Acta Protein Struct Mol Enzymol 1543(2):253–274. https://doi.org/10.1016/S0167-4838(00)00240-5

Okpara MO (2022) Microbial enzymes and their applications in food industry: a mini-review. Adv Enzyme Res 10(01):23–47. https://doi.org/10.4236/aer.2022.101002

Omedi JO, Huang W, Zhang B, Li Z, Zheng J (2019) Advances in present-day frozen dough technology and its improver and novel biotech ingredients development trends—a review. Cereal Chem 96(1):34–56. https://doi.org/10.1002/cche.10122

Olson DW, Aryana KJ (2022) Probiotic Incorporation into Yogurt and Various Novel Yogurt-Based Products. Applied Sciences 12(24): 12607. https://doi.org/10.3390/app122412607

Ozel B, Simsek O, Settanni L, Erten H (2020) The influence of backslopping on lactic acid bacteria diversity in Tarhana fermentation. Int J Food Microbiol 335:108886. https://doi.org/10.1016/j.ijfoodmicro.2020.108886

Parapouli M, Vasileiadi A, Afendra AS, Hatziloukas E (2020) *Saccharomyces cerevisiae* and its industrial applications. AIMS Microbiol 6(1):1–32. https://doi.org/10.3934/microbiol.2020001

Pasqualone A (2018) Traditional flat breads spread from the fertile crescent: production process and history of baking systems. J Ethnic Foods 5(1):10–19. https://doi.org/10.1016/j.jef.2018.02.002

Perez-Alvarado O, Zepeda-Hernández A, Garcia-Amezquita LE, Requena T, Vinderola G, García-Cayuela T (2022) Role of lactic acid bacteria and yeasts in sourdough fermentation during breadmaking: evaluation of postbiotic-like components and health benefits. Front Microbiol 8(13):969460. https://doi.org/10.3389/fmicb.2022.969460

Randez-Gil F, Sanz P, Prieto JA (1999) Engineering baker's yeast: room for improvement. Trends Biotechnol 17(6):237–244. https://doi.org/10.1016/S0167-7799(99)01318-9

Rasiah IA, Sutton KH, Low FL, Lin HM, Gerrard JA (2005) Crosslinking of wheat dough proteins by glucose oxidase and the resulting effects on bread and croissants. Food Chem 89(3):325–332. https://doi.org/10.1016/j.foodchem.2004.02.052

Raveendran S, Parameswaran B, Ummalyma SB, Abraham A, Mathew AK, Madhavan A, Rebello S, Pandey A (2018) Applications of microbial enzymes in food industry. Food Technol Biotechnol 56(1):16–30

Rebholz FG, Sebald K, Dirndorfer S, Dawid C, Hofmann T, Scherf KA (2021) Impact of exogenous α-amylases on sugar formation in straight dough wheat bread. Eur Food Res Technol 247(3):695–706. https://doi.org/10.1007/s00217-020-03657-y

Reyes R, Laura A, Barranco FV, Sandoval G (2022) Recent advances in lipases and their applications in the food and nutraceutical industry. Catalysts 12(9):960. https://doi.org/10.3390/catal12090960

Rivera SXC, Orias NE, Azapagic A (2014) Life cycle environmental impacts of convenience food: comparison of ready and home-made meals. J Clean Prod 73:294–309. https://doi.org/10.1016/j.jclepro.2014.01.008

Shevchenko A, Yang Y, Knaust A, Thomas H, Jiang H, Wang ELC, Shevchenko A (2014) Proteomics identifies the composition and manufacturing recipe of the 2500-year old sourdough bread from Subeixi cemetery in China. J Proteomics 105:363–371. https://doi.org/10.1016/j.jprot.2013.11.016

Singh S, Bajaj BK (2017) Potential application spectrum of microbial proteases for clean and green industrial production. Energy Ecol Environ 2(6):370–386. https://doi.org/10.1007/s40974-017-0076-5

Spencer JFT, Bizeau C, Reynolds N, Spencer DM (1985) The use of mitochondrial mutants in hybridization of industrial yeast strains: VI. Characterization of the hybrid, *Saccharomyces diastaticus* × *Saccharomyces rouxii*, obtained by protoplast fusion, and its behavior in simulated dough-raising tests. Curr Genet 9(8):649–652. https://doi.org/10.1007/BF00449817

Sridhar A, Ponnuchamy M, Ponnusamy SK, Kapoor A (2021) Food preservation techniques and nanotechnology for increased shelf life of fruits, vegetables, beverages and spices: a review. Environ Chem Lett 19(2):1715–1735. https://doi.org/10.1007/s10311-020-01126-2

Sridhar A, Ponnuchamy M, Ponnusamy SK, Kapoor A, Vo DVN, Rangasamy G (2023) Digitalization of the agro-food sector for achieving sustainable development goals: a review. Sustain Food Technol 1(6):783–802. https://doi.org/10.1039/D3FB00124E

Struyf N, Maelen EVD, Hemdane S, Verspreet J, Verstrepen KJ, Courtin CM (2017) Bread dough and baker's yeast: an uplifting synergy. Compr Rev Food Sci Food Saf 16(5):850–867. https://doi.org/10.1111/1541-4337.12282

Tian B, Zhou C, Li D, Pei J, Guo A, Liu S, Li H (2021) Monitoring the effects of hemicellulase on the different proofing stages of wheat aleurone-rich bread dough and bread quality. Foods 10(10):2427. https://doi.org/10.3390/foods10102427

Turfani V, Narducci V, Durazzo A, Galli V, Carcea M (2017) Technological, nutritional and functional properties of wheat bread enriched with lentil or carob flours. LWT 78:361–366. https://doi.org/10.1016/j.lwt.2016.12.030

Verma DK, Al-Sahlany STG, Niamah AK, Thakur M, Shah N, Singh S, Baranwal D, Patel AR, Utama GL, Aguilar CN (2022) Recent trends in microbial flavour compounds: a review on chemistry, synthesis mechanism and their application in food. Saudi J Biol Sci 29(3):1565–1576. https://doi.org/10.1016/j.sjbs.2021.11.010

Yang YL, Guan EQ, Li M, Li NQ, Wang KTJ, Lu CY, Chen MH, Xu F (2022) Effect of transglutaminase on the quality and protein characteristics of aleurone-riched fine dried noodles. LWT 154:112584. https://doi.org/10.1016/j.lwt.2021.112584

Zha M, Yin S, Sun B, Wang X, Wang C (2017) STR3 and CYS3 contribute to 2-furfurylthiol biosynthesis in Chinese sesame-flavored Baijiu yeast. J Agric Food Chem 65(27):5503–5511. https://doi.org/10.1021/acs.jafc.7b01359

Zhan FX, Wang QH, Jiang SJ, Zhou YL, Zhang GM, Ma YH (2014) Developing a xylanase XYNZG from *Plectosphaerella cucumerina* for baking by heterologously expressed in *Kluyveromyces lactis*. BMC Biotechnol 14(1):107. https://doi.org/10.1186/s12896-014-0107-7

Zhang CY, Lin X, Feng B, Liu XE, Bai XW, Xu J, Pi L, Xiao DG (2016) Enhanced leavening properties of baker's yeast by reducing sucrase activity in sweet dough. Appl Microbiol Biotechnol 100(14):6375–6383. https://doi.org/10.1007/s00253-016-7449-0

Zhang Y, Liu C, Yang M, Lin ZOY, Zhao F, Han S (2022) Characterization and application of a novel xylanase from *Halolactibacillus miurensis* in wholewheat bread making. Front Bioeng Biotechnol 10:1018476. https://doi.org/10.3389/fbioe.2022.1018476

Zhou N, Semumu T, Gamero A (2021) Non-conventional yeasts as alternatives in modern baking for improved performance and aroma enhancement. Fermentation 7(3):102. https://doi.org/10.3390/fermentation7030102

Chapter 12
Biopharmaceutical Production by Recombinant DNA Technology: Future Perspectives

Subramani Parasuraman, Lakshmi Narayanan Dinesh Kumar, Ganeson Thanapakiam, Abu Sadat Md. Sayem, Jia Jun Chuah, and Krishnamoorthy Venkateskumar

Abstract Biopharmaceuticals are complex therapeutic molecules derived by recombinant deoxyribonucleic acid (rDNA) technology and produced from microbial cells. The category of biopharmaceuticals includes cytokines, interferons, interleukins, hormones, hematopoiesis-stimulating factors, growth factors, clotting factors, vaccines, enzymes, monoclonal antibodies, cell therapy, antisense drugs, and peptide therapeutics. Insulin, the first biopharmaceutical, was developed in 1982, and in later years, over 300 biologics have been approved for the management of various clinical conditions. In 2020, five of the top ten selling medications are biologics. Biopharmaceuticals have revolutionized the way that many diseases, including diabetes, are treated since they can be customized to meet the unique medical conditions of individual patients. With biotechnology, the drug can be genetically modified using rDNA technologies to alter specificities for individual diseases. The discovery of rDNA technology helps in new drug development and has transformed the biopharmaceutical industry. The objective of this chapter is to discuss biopharmaceutical production by rDNA technology.

Keywords Recombinant DNA technology · Biopharmaceuticals · Interferons · Monoclonal antibodies · Vaccines

---

S. Parasuraman (✉)
Unit of Pharmacology, Toxicology and Basic Health Sciences, Faculty of Pharmacy, AIMST University, Bedong, Kedah Darul Aman, Malaysia

Centre for Global Health Research, Saveetha Medical College and Hospital, Chennai, Tamil Nadu, India
e-mail: parasuraman@aimst.edu.my

L. N. D. Kumar · G. Thanapakiam · K. Venkateskumar
Unit of Pharmaceutical Technology, Faculty of Pharmacy, AIMST University, Bedong, Kedah Darul Aman, Malaysia

A. S. M. Sayem · J. J. Chuah
Unit of Pharmacology, Toxicology and Basic Health Sciences, Faculty of Pharmacy, AIMST University, Bedong, Kedah Darul Aman, Malaysia

V. Kothari et al. (eds.), *Microbial Products for Health and Nutrition*,
https://doi.org/10.1007/978-981-97-4235-6_12

12.1 Introduction

Biopharmaceuticals are complex therapeutic molecules derived by recombinant deoxyribonucleic acid (rDNA) technology and produced from microbial cells (e.g., bacterial, yeast, insect, plant, or mammalian cell expression systems, and hybridomas). Insulin, the first biopharmaceutical, was developed by Genentech in 1982. In the last four decades, over 300 biologics have been approved for use in patients and are available on the market today, including treatments for multiple sclerosis, breast cancer, thyroid cancer, and acquired immunodeficiency syndrome (AIDS). In 2020, five of the top ten selling medications are biologics (Dasani et al. 2023). All over the world, innovations in biopharmaceutical production by rDNA technology gain attention. The PATENTSCOPE—World Intellectual Property Organization (WIPO) database presents 27 search results for patents related to "biopharmaceuticals by recombinant DNA" and 725 search results for patents related to "biopharmaceuticals by rDNA technology" on March 3, 2024.

The category of biopharmaceuticals includes cytokines, interferons (INFs), interleukins (ILs), hormones, hematopoiesis-stimulating factors, growth factors, clotting factors, vaccines, enzymes, monoclonal antibodies (mAbs), cell therapy, antisense drugs, and peptide therapeutics (Halpern and Hutto 2013). Biopharmaceuticals can be applied in many clinical settings, and it has various advantages in disease diagnosis, prevention, and treatment. Biologics and biosimilars are two types of biopharmaceuticals, which are used in recombinant therapy, antibody therapy, cell therapy, and gene therapy. Biologics are large protein-derived compounds produced in cells and biosimilars are compounds that closely resemble biologicals and are approved for clinical uses. In comparison with chemical drugs, biopharmaceuticals are complicated to produce, have many different routes of administration, and have different pharmacokinetic properties. As a result, the cost of producing them is high and they can induce antidrug antibodies which will eventually lead to a decrease in efficacy or deficiency in biosafety (Chen and Yeh 2018). The objective of this chapter is to discuss biopharmaceutical production by rDNA technology.

12.2 Biopharmaceuticals

Biopharmaceuticals are a type of medicinal product that is manufactured utilizing modern biological techniques such as rDNA technology and hybridoma technology. The term "biopharmaceutical" is associated with therapeutic proteins synthesized in engineered biological systems. However, direct extraction of therapeutic proteins from their naturally producing source is not considered biopharmaceuticals (Walsh 2002). Plants, animals, or microorganisms that have been genetically modified are known as transgenic organisms and are usually used to produce biopharmaceuticals (Chen and Yeh 2018). Biopharmaceuticals are classified as complex macromolecules with complex structures and specific biological activity requirements (Hu et al. 2006). They are distinguished by their high specificity and affinity for their target, as well as their decreased risk of adverse effects when compared to other types of therapeutics. Further, they also face challenges due to protein and peptide instability, complicating manufacturing, formulation, and storage, and potentially affecting the pharmacokinetic and pharmacodynamic properties of the drug. The therapeutic potential of biopharmaceuticals is also further affected due to their poor absorption and high susceptibility toward acid- and enzyme-catalyzed degradation in the gastrointestinal tract (Danielsen et al. 2022).

Unengineered murine mAbs and simple replacement proteins constitute a large number of first-generation biopharmaceuticals. Second-generation modern biopharmaceuticals have been engineered to suit their therapeutic properties. Examples of products in the second generation include altered amino acid sequence, altered glycocomponent, or a covalently attached chemical moiety (Walsh 2004). Advances in protein engineering have further expedited the production of novel "fusion" or hybrid pharmaceuticals. The characteristic of this product is that it consists of two or more polypeptides which occur naturally or more commonly, the fragments that are linked directly or by a short linker sequence. Generally, one section will functionally serve as a molecular recognition and the second will serve as an effector function. Examples of three biopharmaceutical fusion products that have been approved for marketing to the general public are "Ontak®," "Enbrel®," and "Amevive®" (Walsh 2004).

Biopharmaceutical classification system (BCS) assesses the contributions of three major factors such as dissolution, solubility, and intestinal permeability. These three factors will affect the absorption of oral drugs from the immediate release of their solid dosage forms. There are four different classes of drug substances according to the BCS. They are:

- Class I: high solubility-high permeability
- Class II: low solubility-high permeability
- Class III: high solubility-low permeability
- Class IV: low solubility-low permeability

12.3 Milestones and Revolution on Discoveries in Biopharmaceutical Products

Over the past few decades, the field of biopharmaceuticals has experienced remarkable advancements that have resulted in groundbreaking discoveries in patient care and medicine. Innovative research, state-of-the-art technologies, and a growing comprehension of biological processes have all contributed to the journey. The invention of rDNA technology was a groundbreaking achievement in the field of biopharmaceuticals, and by utilizing this approach, it is possible to modify DNA and introduce particular genes into their host organisms, which can produce therapeutic proteins. It commences with James Watson and Francis Crick's elucidation of DNA's double helix structure in 1953, which laid the foundation for understanding genetic information. Subsequently, restriction enzymes were invented in the 1960s which are capable of precisely cutting DNA at specific sequences, enabling the manipulation of genetic material (Roberts 2005). The discovery of DNA ligase in 1967 added a crucial piece to the puzzle, facilitating the joining of DNA fragments from different sources (Shuman 2009). Paul Berg successfully merged DNA from different organisms, birthing the first recombinant DNA molecule (Wright 1986). Herbert Boyer and Stanley Cohen followed suit in 1973 by creating the inaugural organism containing rDNA-a bacterium incorporating genes from a toad (Hughes 2001). The ensuing decades witnessed monumental breakthroughs, including Frederick Sanger's revolutionary DNA sequencing technique, the development of the first genetically engineered insulin in 1982, the invention of polymerase chain reaction (PCR) in 1983, the launch of the Human Genome Project, and the advent of genetically modified organisms (GMOs) (Mullis 1990). The year 2012 marked a significant milestone with the emergence of the CRISPR-Cas system, a transformative tool for genetic manipulation (Asmamaw and Zawdie 2021). A new era in biopharmaceuticals has begun with the development of RNA-based therapeutics, such as messenger RNA (mRNA) vaccines. As evidenced by the efficacy of coronavirus disease (COVID-19) vaccines, the development of mRNA vaccines shows how quickly this technology can address global health issues (Jackson et al. 2020).

Biopharmaceutical discoveries have revolutionized the treatment landscape by providing inventive solutions for a wide range of diseases, marking significant milestones in this regard. Every development in the field of medicine, from the early days of rDNA technology to the most recent developments in mRNA vaccines and CRISPR-Cas9 gene editing, has added to the expansion of available treatments. There is even greater potential for the creation of innovative biopharmaceutical products in the future as scientific research keeps pushing the boundaries of knowledge, which will influence medical practice and enhance patient outcomes.

12.4 Production of Biopharmaceuticals

The production of biopharmaceuticals is a multifaceted and tightly controlled procedure that emphasizes the preservation of product quality and uniformity, the assurance of safety, and the fulfillment of regulatory standards. The successful execution of this endeavor necessitates the integration of scientific proficiency, technical advancement, and strict adherence to meticulous quality control protocols. The advent of biopharmaceutical medications has had a profound and revolutionary effect on the management and therapy of various diseases, establishing themselves as indispensable components in nearly every branch of medicine. In contemporary times, the biopharmaceutical market has exhibited a notably accelerated development rate in comparison with the broader pharmaceutical sector, indicating considerable prospects for additional expansion owing to the considerable demand for these therapeutic agents (Kesik-Brodacka 2018). Therapeutic proteins, including mAbs, peptides, and recombinant proteins, are the predominant class of emerging products in the biopharmaceutical industry. The process entails a sequence of complex procedures, commencing with the identification of a suitable host cell line that possesses the capability to express the intended therapeutic protein. After the use of genetic engineering techniques, the process of cell line generation becomes imperative to ascertain the stability and productivity of the cell lines. The scope of upstream processing involves the cultivation of cells on a massive scale, whereas downstream processing mostly concentrates on employing purification procedures to separate and enhance the desired biopharmaceutical. A comprehensive range of analytical tests is conducted to ensure the quality of the product, encompassing evaluations of purity, identification, and potency. The formulation and fill-and-finish operations are implemented to prepare the biopharmaceutical product for distribution, ensuring compliance with rigorous aseptic conditions. Stringent quality control procedures are consistently upheld throughout the entirety of the process, and rigorous preparation of regulatory submissions is undertaken to seek permission from pertinent bodies (Dumont et al. 2016).

Most of the biopharmaceuticals are produced using rDNA technology. It often referred to as genetic engineering or gene cloning, stands as one of the most revolutionary advancements in the field of molecular biology and biotechnology. This groundbreaking technique has transformed our understanding of genetics, paved the way for numerous medical breakthroughs, and opened up a world of possibilities in agriculture, pharmaceuticals, and beyond. At its core, rDNA technology allows to manipulation and recombination of DNA from different sources, including different species, to create entirely new genetic sequences. This has profound implications for understanding the fundamental mechanisms of life, developing novel therapies, producing valuable bio-products, and addressing critical challenges in fields ranging from medicine and agriculture to environmental science (Khan et al. 2016). The genesis of rDNA technology can be traced back to the early 1970s when scientists first demonstrated the ability to cut and paste genes from one organism into another (Wright 1986). This pivotal moment marked the birth of genetic engineering and initiated a new era in biological research. Today, rDNA technology has evolved far beyond its initial stages, enabling the precise manipulation of genes, the creation of genetically modified organisms, and the production of essential proteins and enzymes through techniques such as gene cloning, gene therapy, and DNA sequencing. In the four decades since its inception, rDNA technology has continuously evolved, reshaping the landscape of modern biology and biotechnology. rDNA technology's history is a captivating journey through pivotal scientific discoveries (Fig. 12.1).

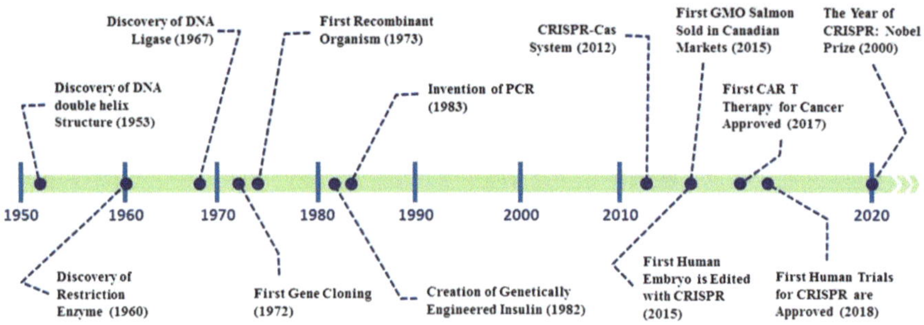

Fig. 12.1 Major advances of rDNA technology revolution

12.5 Biopharmaceuticals Products

12.5.1 Insulin and Hormones

Before the advent of rDNA technology, insulin for therapeutic purposes was extracted from the animal pancreas. The insulin obtained from such source however caused immune reactions and had the disadvantage of contamination and limited supply. Insulin made with rDNA technologies has been available since the 1980s. The molecular weight of human insulin is approximately 5800 kDa. It is a globular protein made up of 51 amino acid residues arranged into two polypeptide chains (A and B) connected by two disulfide linkages. The technique of rDNA has facilitated the industrial manufacture of human insulin through the application of genetic engineering (Gualandi-Signorini and Giorgi 2001). The most common microorganisms employed in insulin production by rDNA technology are *Escherichia coli*, *Saccharomyces cerevisiae*, *Pichia pastoris*, and a few cell lines (e.g., Chinese hamster ovary [CHO] cells). The selection of microorganisms depends on factors such as desired insulin characteristics, cost, and regulatory requirements. rDNA technology has revolutionized biopharmaceutical manufacturing, especially in humanized insulin production (Adrio and Demain 2010).

rDNA technology has played and playing a significant role in the production of several important hormones used as biopharmaceutical products. Human growth hormone (hGH), once sourced with difficulty from human cadavers, is now manufactured by rDNA technology for treating growth deficiencies (Miller and Eberhardt 1983). Similarly, fertility hormones like follicle-stimulating hormones and luteinizing hormones are manufactured with better consistency and purity, through rDNA technology (Lunenfeld 2004). Other examples include erythropoietin (EPO), essential for anemia treatment, and parathyroid hormone (PTH) for the management of hypoparathyroidism (Eschbach et al. 1987; Winer 1996). Additionally, calcitonin, used to treat bone disorders like osteoporosis, is also produced by rDNA technology (Reginster et al. 2000). The rDNA technology does ensure safer, more reliable, and accessible hormone therapies for patients.

12.5.2 Vaccines

The development of vaccines in the early stages was mostly based on empirical methods, relying on the attenuation or inactivation of pathogens. Such conventional method manufactured vaccines induced the immune response against the pathogens with low antigenic variability, via the B cell (antibody)-mediated immunity. Polio, tetanus, diphtheria, and measles vaccines were some examples of this conventional method of vaccines. However, these vaccines lack the ability/reduced ability to induce the T-cell (cell-mediated) immune response which will be effective for chronic infections and against intracellular pathogens. Also, conventional vaccines carry the disadvantage of virulence in susceptible host and potential reversal of attenuation. Recombinant vaccines are dependent on the ability of one or more specific antigens to elicit an immune response against the pathogen. This immune response can be achieved by administering the antigens alongside adjuvants or by expressing them by using plasmids or non-pathogenic bacterial/viral vectors. Recombinant protein vaccines offer a solution to various concerns associated with vaccines derived from purified macromolecules. These concerns include the potential co-purification of unwanted contaminants and the possibility of the toxoids reverting to their toxigenic forms, as observed in diphtheria or tetanus toxoid vaccines, for instance. This method addresses another crucial challenge, which pertains to the intricacy associated with acquiring adequate quantities of purified antigenic components. In comparison, recombinant vaccines have higher efficacy and specificity as compared to conventional vaccines (Nascimento and Leite 2012). Recombinant vaccines are generated through the integration of viral DNA or RNA genetic material into yeast cells or viruses, leading to the synthesis of specific constituents of the original pathogen. The three main technology directions at the forefront of recombinant vaccination research are (1) antigen targeting to dendritic cells, (2) combinations of several technologies in prime-boost vaccination administration regimes, and (3) antigen presentation in multiple forms (Le Grand et al. 2012). Examples of recombinant vaccines with their expression system are summarized in Table 12.1 (de Pinho Favaro et al. 2022). Types of recombinant vaccines include subunit vaccines, conjugate vaccines, DNA vaccines, T-cell receptor peptide vaccines and T-cell epitope conjugate vaccines, virus-like particles, and viral vector-based vaccines.

12.5.3 Monoclonal Antibodies

Monoclonal antibodies (mAbs) are laboratory-engineered proteins designed to mimic the immune response against targets like viruses and cancer cells. A therapeutic breakthrough since the 1990s, they have revolutionized treatments, such as rituximab targeting lymphoma (Liu 2014). The suffix -mab is used for mAbs or fragments. The mAbs can be generated by the conventional hybridoma method or

Table 12.1 Examples of recombinant vaccines with their expression system

Pathogen	Antigen for inducing immune response	Expression system	Vaccine name (Developer)
Hepatitis B virus	Hepatitis B Surface antigen (HBsAg)	*Saccharomyces cerevisiae*, *Hansenula polymorpha*	Recombivax HB® (Merck) Engerix-B® (GlaxoSmithKline) Twinrix® (GlaxoSmithKline) PreHevbrio® (VBI Vaccines)
Hepatitis E virus	ORF2 protein	*E. coli*	Hecolin® (Xiamen Innovax Biotech Co., Ltd) (Wu et al. 2016)
Influenza A & B virus	Hemagglutinin (HA)	Insect cells-*Baculo virus* expression system	Flublok® (Sanofi Pasteur) Flucelvax Quadrivalent® (Seqirus, Inc.)
Malaria	Hepatitis B surface antigen (HBsAg) + Repeat ("R") and T-cell epitope ("T") of the pre-erythrocytic circumsporozoite protein (CSP) of the *Plasmodium falciparum* malaria parasite (RTS chimera)	*Saccharomyces cerevisiae*	RTS, S/AS01 vaccine (GlaxoSmithKline) (Laurens 2020)
SARS-CoV-2	Spike (S) protein	Insect cells-*Baculo Virus* Expression System	Novavax (Novavax) (Dunkle et al. 2022)

by the recombinant method. The conventional method involves the mouse cells to produce mAbs whereas the recombinant method utilizes genetic engineering. The genes that code antibodies (proteins) of interest are inserted into the host cells such as CHO cells to produce the proteins. The conventional method may induce immunogenicity due to its production in mouse cells whereas the recombinant method provides more humanized antibodies. The efficacy of murine-derived immunoglobulin (conventional method utilizing mouse cells) preparations has the disadvantage of inducing anti-mouse immune responses. rDNA technology produces safer and more effective humanized mAbs compared to early mouse-derived versions (Hudson and Souriau 2003; Siegel 2002). Examples of mAb applications include targeting the influenza virus's hemagglutinin protein for prevention and treatment (Henry et al. 2018) and combating COVID-19 with antibodies against the severe acute respiratory syndrome coronavirus 2 (SARS-CoV-2) spike protein (Weinreich et al. 2021). Recombinant techniques enhance mAb development and expand their potential in transfusion medicine (Siegel 2002).

rDNA technology has also been employed in the manufacturing of mAbs for the treatment of non-communicable diseases, including adalimumab (Humira®) and infliximab (Remicade®) for rheumatoid arthritis (Weinblatt et al. 1999), ustekinumab, a selective inhibitor of interleukin (IL)-12 and IL-23 pathways (Stelara®) for psoriasis (Leonardi et al. 2008), vedolizumab (Entyvio®) for Crohn's disease and ulcerative colitis (Feagan et al. 2013), omalizumab (Xolair®) for asthma (Normansell et al. 2014), trastuzumab (Herceptin®) combat certain breast cancers (Slamon et al. 2001), and denosumab (Prolia®) for the treatment of osteoporosis (Cummings et al. 2009).

12.5.4 Human Growth Hormone

The pituitary gland in the brain produces human growth hormone (hGH) and a peptide hormone (somatotropin). It is essential to the human body's growth, development, and several metabolic activities. Growth hormone (GH) was first used for medicinal purposes in 1958. The availability of GH generated by rDNA technology expanded its prospective uses in 1985. Since then, GH has been approved by the US Food and Drug Administration (FDA) for three indications linked to disorders caused by GH insufficiency and nine indications unrelated to GH deficiency (Kemp and Frindik 2011). The process initiates with the incorporation of the plasmid containing the hGH gene into the microorganism such as *Bacillus* spp., *E. coli*, *S. cerevisiae*, *P. pastoris*. Once the successful genomic integration is achieved the strain with the desired gene is cultured either in a fed-batch bioreactor or shake flask based on the microorganism. The microorganisms secrete the hormone in the culture medium. The choice between the bacteria and yeast is based on many factors such as cost, final product outcome, and regulatory standards. When compared to yeast-mediated production, *E. coli* expression system has the advantage of cost-effective and fast growth (SECRETERS-European Union's Horizon 2020 Programme 2022).

12.5.5 Interferon

Interferons (INFs) are a group of signaling proteins produced and released by host cells in response to the presence of pathogens, such as viruses, bacteria, parasites, or tumor cells. These proteins play a crucial role in the innate immune system, serving as a first line of defense against infections and other threats to the body. There are three main types of INFs: type 1 (IFN-α, IFN-β), type 2 (IFN-γ), and type 3. INFs have been used therapeutically for various medical conditions, such as hepatitis B and C, multiple sclerosis, certain cancers, and viral infections. Recombinant INFs produced using biotechnology have been a significant advancement in medicine, providing a safe and standardized source of these important proteins for clinical use. Their unique properties, including their ability to inhibit cell proliferation, induce cell differentiation, and modulate the immune system, made them promising candidates for cancer treatment. The rDNA technology in the late 1970s facilitated the large-scale production of highly purified IFNs, enabling systematic clinical studies. Over several decades of research on INFs, there have been periods of optimism and skepticism regarding their therapeutic potential (Tossing 2001). The messenger RNA encoding the INF is extracted from cells, and a complementary DNA copy is synthesized. The gene of interest (e.g., IFN-α, IFN-β and IFN-γ) is isolated. Short nucleotides such as promoters, ribosome binding sites, and linker molecules are inserted for the gene's proper expression ("Interferon production by genetic engineering" 1981). Production involves four stages: (1) mRNA gene isolation, inserted into a plasmid, then introduced into microorganisms like *E. coli* (especially BL21 strain) or *P. pastoris*; (2) production stage with a suitable growth medium; (3) downstream purification using chromatographic techniques; (4) formulation with buffers, surface-active agents, etc. Recombinant INFs are widely used in the treatment of viral infections, cancer, and autoimmune diseases (Castro et al. 2021). Specific recombinant INF products have found significant applications in treating various diseases. IFN-α products, such as Intron A and Roferon A, are used for hepatitis B, C, and certain cancers (Zeuzem 2008). IFN-β products, like Betaseron and Avonex, are used for the treatment of multiple sclerosis (Dhib-Jalbut and Marks 2010). Actimmune®, an IFN-γ product, is used for chronic granulomatous disease (Aminizadeh et al. 2016).

Various clinical trials are in progress to investigate the use of INFs and the combination of INFs with other medications. The investigation is to assess the use of novel INF formulations, combinations with other drugs, and their potential for treating various cancers, viral infections, autoimmune diseases, and even less common conditions like specific types of hepatitis or immune deficiencies. These trials also cover the exploration of the potential of INF combined with resveratrol for hepatitis B, testing the effectiveness of nebulized IFN-β1b against COVID-19, and comparing pegylated IFN-α2b versus IFN-α therapy in adults with essential thrombocythemia. Additionally, researchers are investigating IFN-α for stage IV solid tumors, lymphoma, or myeloma (IFN-α trial), as well as evaluating IFN-α in INF-β neutralizing antibody positive multiple sclerosis patients (REPAIR trial). Furthermore, studies are assessing the efficacy and safety of IFN-α2β in the treatment of novel coronavirus (NCT01171209 2012).

12.5.6 Interleukin

Interleukins/inter-leukocyte communication (IL) is a type of cytokine that alters the behavior of cells and plays essential roles in the activation and differentiation of immune cells, as well as proliferation, maturation, migration, and adhesion. IL also has pro-inflammatory and anti-inflammatory properties (Justiz Vaillant and Qurie 2023). Currently, more than 40 ILs have been found so far, with distinct and overlapping roles (Melgrati et al. 2023). IL-2 and its receptor were the first cytokine and cytokine receptor to be cloned, in more than 20 years which plays a role in the regulation of protective immunity, as well as maintaining immune tolerance mediated by CD4+ regulatory T lymphocytes (Bachmann and Oxenius 2007; Dik and Heron 2020). The uses of recombinant human interleukin are summarized in Table 12.2.

Aldesleukin is a recombinant IL-2 produced by rDNA technology using a genetically engineered *E. coli* strain and has potent immunomodulatory and antitumor activity. In 1992, Aldesleukin was approved for the treatment of malignant melanoma (National Institute of Diabetes and Digestive and Kidney Diseases 2012). In 2001, the FDA approved the first biological agent "Anakinra" an IL-1 receptor antagonist for the treatment of rheumatoid arthritis. Anakinra inhibits the pro-inflammatory effects of IL-1 in patients with rheumatoid arthritis (Cvetkovic and Keating 2002). Anakinra is also indicated in the treatment (off-label) of Schnitzler's syndrome and systemic juvenile idiopathic arthritis (Giancane et al. 2022; Vandenhende et al. 2011). In July 2021, the European Medicines Agency approved Anakinra for the treatment of COVID-19 in adults with pneumonia (EMA 2021). In phase 1/2 clinical study, recombinant IL-21 (10–40 mcg/kg) + Sorafenib exhibited antitumor activity against metastatic renal cell carcinoma by enhancing anticancer immunity by modulating both innate and adaptive immune responses (Bhatia et al. 2014). Akram et al. cloned IL-29 using *Chlamydomonas reinhardtii* and demonstrated in vitro anticancer activity using HepG2 cells (Akram et al. 2023). Clinical use of recombinant IL-15 and IL-29 is also under investigation. In a clinical trial recombinant Human NIZ985, a recombinant heterodimer of IL-15/IL-15 receptor alpha in combination with spartalizumab exhibits preliminary antitumor activity in patients with melanoma, gastric cancer, and pancreatic cancer (Conlon et al. 2021). In an in vitro study, exosomes encapsulated with recombinant IL-29 exhibited significant cytotoxic activity (Fujimura et al. 2023). Recombinant IL including anti-interleukin-17A, anti-interleukin-23, IL-30, IL-34, IL-37, IL-38, IL-39, and IL-40 is important potential targets for future therapeutics (Catalan-Dibene et al. 2018).

Table 12.2 Recombinant human interleukin

Recombinant human interleukin	Genetic name	Indications/use
Recombinant interleukin-2	Aldesleukin	Approved for the treatment of metastatic renal cell carcinoma and malignant melanoma (Jetanalin et al. 2013)
Recombinant interleukin-11	Oprelvekin	Approved for the prevention of severe thrombocytopenia (Kuter 2007)
Recombinant human interleukin-12	HemaMax™	Exhibited radioprotective properties (Basile et al. 2012)
Recombinant human IL-1 receptor antagonist	Anakinra	Approved for the treatment of rheumatoid arthritis (Cvetkovic and Keating 2002)

12.5.7 *Erythropoietin*

Erythropoietin (EPO) is a glycoprotein hormone that regulates the synthesis of red blood cells, a process known as erythropoiesis. EPO is largely produced by the kidneys in response to low blood oxygen levels (hypoxia). It enters the bloodstream and stimulates the generation of red blood cells from hematopoietic stem cells in the bone marrow. EPO is largely generated by the kidney and liver, and its tissue-specific expression is controlled by distinct gene sequences. A 3′ enhancer regulates the activation of EPO gene expression under low oxygen circumstances (hypoxia). This enhancer is bound by hypoxia-inducible factor 1 (HIF-1) alpha and beta proteins. Hypoxia-driven gene regulation is a common event in many cell types and impacts many genes in addition to EPO (Lacombe and Mayeux 1998). Recombinant human erythropoietin (rHuEPO) is the first EPO. In 1989, the US FDA approved the first rHuEPO product for clinical use and used in the treatment of anemia associated with conditions like chronic kidney disease and cancer chemotherapy (Jelkmann 2007).

12.6 Regulations on Manufacturing of Biopharmaceutical Products

As biopharmaceuticals continue to revolutionize modern medicine, their production using rDNA technology introduces a unique set of challenges and considerations. Hence, it is important to delve into the intricate web of regulatory requirements governing the development, manufacturing, and approval processes for rDNA-based biopharmaceuticals. The major challenges that have emerged since the production size and the number of goods in corporate pipelines expanded. This has led to more obstacles in locating and resolving issues that surfaced throughout manufacturing processes, resulting in batch failures and product waste (FDA 2004). The regulatory considerations and approval process for biopharmaceuticals produced using rDNA technology involve several key steps. Patient safety and access to effective rDNA-based biopharmaceuticals are safeguarded by a rigorous, multi-step regulatory process. While regionally adapted, the process adheres to core principles. International collaboration, led by organizations like the International Council for Harmonisation (ICH), seeks to harmonize these standards globally. This collaborative approach, along with regulatory flexibility, was crucial in the rapid development and authorization of COVID-19 vaccines, ensuring both safety and timely access. From risk assessments and compliance with Good Manufacturing Practice (GMP) guidelines to the submission of Investigational New Drug (IND) applications and the final New Drug Application (NDA) or Marketing Authorization Application (MAA), there is a pivotal role played by regulatory agencies, the meticulous scrutiny of clinical trial data, and the evolving standards shaping the approval pathways for these cutting-edge therapeutic agents (FDA 2023). Prior to initiating clinical trials, manufacturers conduct comprehensive risk assessments to identify and mitigate potential hazards associated with rDNA technology. This risk assessment is critical in assuring the safety of the final product and directing the creation of effective risk management techniques. As the global standard to the regulatory framework for rDNA-based biopharmaceuticals the World Health Organization's (WHO) Good Manufacturing Practice (GMP) guidelines serve in assuring the quality and consistency of manufacturing processes (WHO 2015). Furthermore, the ICH produces guidelines, such as Q5A and Q5D, that provide recommendations for the quality of biotechnological goods (ICH n.d.). Ensuring the safety and effectiveness of these innovative medicinal agents is greatly dependent on the demanding and complex approval process for rDNA-based biopharmaceuticals. The procedure moves on to the filing of IND applications after starting with preclinical development, when comprehensive studies evaluate the product's viability and its dangers. These applications, which are under regulatory review, indicate the shift to human trials, which are carried out in meticulously planned clinical investigations to determine safety and efficacy profiles. The completion of these trials results in the filing of New Drug Applications (NDAs) or Marketing Authorization Applications (MAAs), which are essentially data dossiers that include risk assessments, manufacturing information, and clinical trial results (FDA 1995).

The comprehensive assessment conducted by regulatory bodies is pivotal in ascertaining the product's appropriateness for market authorization. The effort to maintain strict guidelines and guarantee that rDNA-based biopharmaceuticals fulfil the highest standards of quality and safety prior to being administered to patients is reflected in the intricate approval process. Regulatory assessment continues after approval. Post-marketing surveillance provides ongoing monitoring to ensure that any unanticipated adverse events or changes in efficacy are quickly identified and treated. Sustaining the confidence of patients and regulatory bodies alike depends on this continuous dedication to safety (Lee et al. 2020). Regulatory considerations change with the biopharmaceutical industries. The future of approvals for rDNA-based biopharmaceuticals is being shaped by advances in regulatory science and global harmonization initiatives, which are also encouraging innovation and streamlining procedures (Kesik-Brodacka 2018).

12.7 Future Perspectives

rDNA technology advances have made life considerably simpler for humans in recent years. rDNA technology is expected to expand its application horizons such as the development of new therapeutics, genetically modified crop development, veterinary product development, gene therapy, and biopesticides and biofuel production, and many more niche areas to fulfil the market needs in future (Khan et al. 2016). It was also postulated that rDNA technology could be utilized to help clean up the environment (phytoremediation and microbial remediation) and improve plant tolerance to various stresses (drought, pests, and salt) (Khan et al. 2016).

Market research reports indicate that the Global rDNA technology market is poised to have a great leap and is expected to reach 1395.50 billion by 2032 and is poised to grow at a compound annual growth rate (CAGR) of 6.40%. The availability of good infrastructure for research and development purposes in this niche research area could make this expected growth possible in the coming years. The major market players are focusing on investing in developing nations to expand their applications window in various sectors not only in human applications but also in plants and microbes ("Recombinant DNA technology market" 2023).

The development of new drugs and vaccines through rDNA technology is significantly improving health conditions (Khan et al. 2016). The successful emergence of genetically engineered human insulin in diabetes treatment and recombinant hGH to treat children lacking the ability to produce hGH triggered the development of several recombinant therapeutics, human proteins, as well as drugs.

The desired qualities or characteristics by using a desirable sequence of the gene could be possible by utilizing rDNA technology. The external alteration of genetic material of a particular organism to achieve the desired characteristics has been well reported and the increased use of this technology in improving the quality of the crops has been increasing and has revolutionized its application in agriculture segments (Ullah et al. 2015).

The medical segment in RNA technology is poised for a huge positive growth in the coming years. These developments also provided a platform that could lead to massive success for the rDNA technology, which could potentially prospect for developing new moieties that could address the limitations in disorders and their effective treatment. RNA technology has also made great strides in the development of new, safe, and effective vaccines which was evident during the COVID-19 pandemic (Thomas 2022). Under the non-medical category, RNA technology can harness biological processes and could lead to the development of new therapeutics, and in production of transportation fuels and chemicals. The use of this technology in the production of genetically modified products such as animals could provide a great opportunity for carrying out preclinical research using such models.

Though this technology tends to carry a huge potential in diversified areas, its pitfalls and limitations should not be ignored. For example, utilization of this technique could negatively hamper the existence of the original species in the environment since the genetics of the organisms are manipulated and introduced into new habitats. Such manipulation in therapeutics development could lead to new moieties with higher adverse effects rather than beneficial effects. Further, crops developed using this technology could result in cross-contamination and may grow uncontrollably and extremely hard to control its impact on the environment. It is imminent to devise strategies to address the limitations of this technology and implement various measures to close the gaps in the exploration of this technology and unveil its vast potential (Šlesingerová 2021).

12.8 Conclusion

Biopharmaceuticals offer targeted therapies rather than symptomatic treatment and biopharmaceuticals are gaining major clinical attention for disease control and prevention. The efficacy and safety of biopharmaceuticals and the ability to treat previously incurable diseases allow the pharmaceutical industry to invest more in this field. Nemours biopharmaceuticals are in the development stage, though some regulatory issues are still unsolved. The increased need for biopharmaceuticals also helps in the development of bioseparation and chromatographic methods. Biopharmaceuticals are also part of the emerging field of gene and cell therapy. Biopharmaceuticals market size is expanding especially during the COVID-19 pandemic era.

References

Adrio J-L, Demain AL (2010) Recombinant organisms for production of industrial products. Bioeng Bugs 1:116–131. https://doi.org/10.4161/bbug.1.2.10484

Akram M, Khan MA, Ahmed N, Bhatti R, Pervaiz R, Malik K, Tahir S, Abbas R, Ashraf F, Ali Q (2023) Cloning and expression of an anti-cancerous cytokine: human IL-29 gene in Chlamydomonas reinhardtii. AMB Expr 13:23. https://doi.org/10.1186/s13568-023-01530-1

Aminizadeh E, Alavian SM, Akbari Sari A, Ebrahimi Daryani N, Behnava B (2016) Safety and efficacy of adding ribavirin to interferon or peginterferon in treatment of hepatitis C infection in patients with thalassemia: a systematic review on randomized controlled trials. Hepat Mon 16:e28537. https://doi.org/10.5812/hepatmon.28537

Asmamaw M, Zawdie B (2021) Mechanism and applications of CRISPR/Cas-9-mediated genome editing. Biologics 15:353–361. https://doi.org/10.2147/BTT.S326422

Bachmann MF, Oxenius A (2007) Interleukin 2: from immunostimulation to immunoregulation and back again. EMBO Rep 8:1142–1148. https://doi.org/10.1038/sj.embor.7401099

Basile LA, Ellefson D, Gluzman-Poltorak Z, Junes-Gill K, Mar V, Mendonca S, Miller JD, Tom J, Trinh A, Gallaher TK (2012) HemaMax™, a recombinant human interleukin-12, is a potent mitigator of acute radiation injury in mice and non-human primates. PLoS One 7:e30434. https://doi.org/10.1371/journal.pone.0030434

Bhatia S, Curti B, Ernstoff MS, Gordon M, Heath EI, Miller WH, Puzanov I, Quinn DI, Flaig TW, Van Veldhuizen P, Byrnes-Blake K, Freeman JA, Bittner R, Hunder N, Souza S, Thompson JA (2014) Recombinant interleukin-21 plus sorafenib for metastatic renal cell carcinoma: a phase 1/2 study. J Immunother Cancer 2:2. https://doi.org/10.1186/2051-1426-2-2

Castro LS, Lobo GS, Pereira P, Freire MG, Neves MC, Pedro AQ (2021) Interferon-based biopharmaceuticals: overview on the production, purification, and formulation. Vaccines (Basel) 9:328. https://doi.org/10.3390/vaccines9040328

Catalan-Dibene J, McIntyre LL, Zlotnik A (2018) Interleukin 30 to interleukin 40. J Interf Cytokine Res 38:423–439. https://doi.org/10.1089/jir.2018.0089

Chen Y-C, Yeh M-K (2018) Introductory chapter: Biopharmaceuticals. In: Yeh M-K, Chen Y-C (eds) Biopharmaceuticals. InTech. https://doi.org/10.5772/intechopen.79194

Conlon K, Watson DC, Waldmann TA, Valentin A, Bergamaschi C, Felber BK, Peer CJ, Figg WD, Potter EL, Roederer M, McNeel DG, Thompson JA, Gupta S, Leidner R, Wang-Gillam A, Parikh NS, Long D, Kurtulus S, Ho Lee L, Chowdhury NR, Bender F, Pavlakis GN (2021) Phase I study of single agent NIZ985, a recombinant heterodimeric IL-15 agonist, in adult patients with metastatic or unresectable solid tumors. J Immunother Cancer 9:e003388. https://doi.org/10.1136/jitc-2021-003388

Cummings SR, San Martin J, McClung MR, Siris ES, Eastell R, Reid IR, Delmas P, Zoog HB, Austin M, Wang A, Kutilek S, Adami S, Zanchetta J, Libanati C, Siddhanti S, Christiansen C, Freedom Trial (2009) Denosumab for prevention of fractures in postmenopausal women with osteoporosis. N Engl J Med 361:756–765. https://doi.org/10.1056/NEJMoa0809493

Cvetkovic RS, Keating G (2002) Anakinra. BioDrugs 16:303–311; discussion 313–314. https://doi.org/10.2165/00063030-200216040-00005

Danielsen M, Hempel C, Andresen TL, Urquhart AJ (2022) Biopharmaceutical nanoclusters: towards the self-delivery of protein and peptide therapeutics. J Control Release 347:282–307. https://doi.org/10.1016/j.jconrel.2022.04.050

Dasani S, Palanki R, Menon P, Bose SK (2023) Biopharmaceuticals. In: Translational surgery. Elsevier, pp 535–538. https://doi.org/10.1016/B978-0-323-90300-4.00056-2

de Pinho Favaro MT, Atienza-Garriga J, Martínez-Torró C, Parladé E, Vázquez E, Corchero JL, Ferrer-Miralles N, Villaverde A (2022) Recombinant vaccines in 2022: a perspective from the cell factory. Microb Cell Factories 21:203. https://doi.org/10.1186/s12934-022-01929-8

Dhib-Jalbut S, Marks S (2010) Interferon-beta mechanisms of action in multiple sclerosis. Neurology 74(Suppl 1):S17–S24. https://doi.org/10.1212/WNL.0b013e3181c97d99

Dik WA, Heron M (2020) Clinical significance of soluble interleukin-2 receptor measurement in immune-mediated diseases. Neth J Med 78:220–231

Dumont J, Euwart D, Mei B, Estes S, Kshirsagar R (2016) Human cell lines for biopharmaceutical manufacturing: history, status, and future perspectives. Crit Rev Biotechnol 36:1110–1122. https://doi.org/10.3109/07388551.2015.1084266

Dunkle LM, Kotloff KL, Gay CL, Áñez G, Adelglass JM, Barrat Hernández AQ, Harper WL, Duncanson DM, McArthur MA, Florescu DF, McClelland RS, Garcia-Fragoso V, Riesenberg RA, Musante DB, Fried DL, Safirstein BE, McKenzie M, Jeanfreau RJ, Kingsley JK, Henderson JA, Lane DC, Ruíz-Palacios GM, Corey L, Neuzil KM, Coombs RW, Greninger AL, Hutter J, Ake JA, Smith K, Woo W, Cho I, Glenn GM, Dubovsky F (2022) Efficacy and safety of NVX-CoV2373 in adults in the United States and Mexico. N Engl J Med 386:531–543. https://doi.org/10.1056/NEJMoa2116185

EMA (2021) EMA starts evaluating the use of Kineret in adult COVID-19 patients at increased risk of severe respiratory failure. https://www.ema.europa.eu/en/news/ema-starts-evaluating-use-kineret-adult-covid-19-patients-increased-risk-severe-respiratory-failure

Eschbach JW, Egrie JC, Downing MR, Browne JK, Adamson JW (1987) Correction of the anemia of end-stage renal disease with recombinant human erythropoietin. N Engl J Med 316:73–78. https://doi.org/10.1056/NEJM198701083160203

FDA (1995) Guidance for industry: content and format of investigational new drug (IND) applications for phase 1 studies of drugs, including well-characterized, therapeutic, biotechnology-derived products. FDA

FDA (2004) Guidance for industry PAT—a framework for innovative pharmaceutical development, manufacturing, and quality assurance. FDA

FDA (2023) Biotechnology products, including therapeutic biotechnology products. FDA

Feagan BG, Rutgeerts P, Sands BE, Hanauer S, Colombel J-F, Sandborn WJ, Van Assche G, Axler J, Kim H-J, Danese S, Fox I, Milch C, Sankoh S, Wyant T, Xu J, Parikh A, GEMINI 1 Study Group (2013) Vedolizumab as induction and maintenance therapy for ulcerative colitis. N Engl J Med 369:699–710. https://doi.org/10.1056/NEJMoa1215734

Fujimura NA, Fatima SE, Ahmed N, Akram M, Tahir S, Khan MA, Amirzada I, Nadeem T, Bashir H, Malik K (2023) Evaluation of exosomes encapsulated recombinant Interleukin-29 for its in vitro anticancer studies. J Biotechnol 373:24–33. https://doi.org/10.1016/j.jbiotec.2023.06.008

Giancane G, Papa R, Vastert S, Bagnasco F, Swart JF, Quartier P, Antón J, Kamphuis S, Sanner H, Glerup M, De Benedetti F, Tsitsami E, Remesal A, Moreno E, De Inocencio J, Myrup C, Pallotti C, Koné-Paut I, Franck-Larsson K, Malmström H, Cederholm S, Pistorio A, Wulffraat

N, Ruperto N, Paediatric Rheumatology International Trials Organisation (PRINTO) (2022) Anakinra in patients with systemic juvenile idiopathic arthritis: long-term safety from the Pharmachild Registry. J Rheumatol 49:398–407. https://doi.org/10.3899/jrheum.210563

Gualandi-Signorini AM, Giorgi G (2001) Insulin formulations—a review. Eur Rev Med Pharmacol Sci 5:73–83

Halpern W, Hutto D (2013) Biopharmaceuticals. In: Haschek and Rousseaux's handbook of toxicologic pathology. Elsevier, pp 751–782. https://doi.org/10.1016/B978-0-12-415759-0.00025-X

Henry C, Palm A-KE, Krammer F, Wilson PC (2018) From original antigenic sin to the universal influenza virus vaccine. Trends Immunol 39:70–79. https://doi.org/10.1016/j.it.2017.08.003

Hu X, Ma Q, Zhang S (2006) Biopharmaceuticals in China. Biotechnol J 1:1215–1224. https://doi.org/10.1002/biot.200600083

Hudson PJ, Souriau C (2003) Engineered antibodies. Nat Med 9:129–134. https://doi.org/10.1038/nm0103-129

Hughes SS (2001) Making dollars out of DNA: the first major patent in biotechnology and the commercialization of molecular biology, 1974–1980. Isis 92:541–575

ICH (n.d.) ICH guidelines. ICH

Interferon production by genetic engineering (1981) Br Med J (Clin Res Ed) 282:674–675. https://doi.org/10.1136/bmj.282.6265.674-a

Jackson NAC, Kester KE, Casimiro D, Gurunathan S, DeRosa F (2020) The promise of mRNA vaccines: a biotech and industrial perspective. npj Vaccines 5:11. https://doi.org/10.1038/s41541-020-0159-8

Jelkmann W (2007) Erythropoietin after a century of research: younger than ever. Eur J Haematol 78:183–205. https://doi.org/10.1111/j.1600-0609.2007.00818.x

Jetanalin P, Lee SJ, Kavanaugh A (2013) Biologic modifiers of inflammatory diseases. In: Clinical immunology. Elsevier, pp 1131–1142. https://doi.org/10.1016/B978-0-7234-3691-1.00105-7

Justiz Vaillant AA, Qurie A (2023) Interleukin. StatPearls Publishing, Treasure Island, FL

Kemp SF, Frindik JP (2011) Emerging options in growth hormone therapy: an update. Drug Des Devel Ther 5:411–419. https://doi.org/10.2147/DDDT.S23140

Kesik-Brodacka M (2018) Progress in biopharmaceutical development. Biotechnol Appl Biochem 65:306–322. https://doi.org/10.1002/bab.1617

Khan S, Ullah MW, Siddique R, Nabi G, Manan S, Yousaf M, Hou H (2016) Role of recombinant DNA technology to improve life. Int J Genomics 2016:2405954. https://doi.org/10.1155/2016/2405954

Kuter DJ (2007) Platelet growth factors. In: Platelets. Elsevier, pp 1211–1235. https://doi.org/10.1016/B978-012369367-9/50828-4

Lacombe C, Mayeux P (1998) Biology of erythropoietin. Haematologica 83:724–732

Laurens MB (2020) RTS,S/AS01 vaccine (Mosquirix™): an overview. Hum Vaccin Immunother 16:480–489. https://doi.org/10.1080/21645515.2019.1669415

Le Grand, L.R., White, M., Siegel, E.B., Barnard, R.T., 2012. Recombinant vaccines: development, production, and application, in: Kayser, O., Warzecha, H. (Eds.), Pharmaceutical biotechnology. Wiley, pp. 423–449. https://doi.org/10.1002/9783527632909.ch17

Lee I, Lee TA, Crawford SY, Kilpatrick RD, Calip GS, Jokinen JD (2020) Impact of adverse event reports from marketing authorization holder-sponsored patient support programs on the performance of signal detection in pharmacovigilance. Expert Opin Drug Saf 19:1357–1366. https://doi.org/10.1080/14740338.2020.1792883

Leonardi CL, Kimball AB, Papp KA, Yeilding N, Guzzo C, Wang Y, Li S, Dooley LT, Gordon KB, PHOENIX 1 Study Investigators (2008) Efficacy and safety of ustekinumab, a human interleukin-12/23 monoclonal antibody, in patients with psoriasis: 76-week results from a randomised, double-blind, placebo-controlled trial (PHOENIX 1). Lancet 371:1665–1674. https://doi.org/10.1016/S0140-6736(08)60725-4

Liu JKH (2014) The history of monoclonal antibody development—progress, remaining challenges and future innovations. Ann Med Surg (Lond) 3:113–116. https://doi.org/10.1016/j.amsu.2014.09.001

Lunenfeld B (2004) Historical perspectives in gonadotrophin therapy. Hum Reprod Update 10:453–467. https://doi.org/10.1093/humupd/dmh044
Melgrati S, Sozzani S, Thelen M (2023) Editorial: Insights in cytokines and soluble mediators in immunity: 2022. Front Immunol 14:1194553. https://doi.org/10.3389/fimmu.2023.1194553
Miller WL, Eberhardt NL (1983) Structure and evolution of the growth hormone gene family. Endocr Rev 4:97–130. https://doi.org/10.1210/edrv-4-2-97
Mullis KB (1990) The unusual origin of the polymerase chain reaction. Sci Am 262:56–61, 64–65. https://doi.org/10.1038/scientificamerican0490-56
Nascimento IP, Leite LCC (2012) Recombinant vaccines and the development of new vaccine strategies. Braz J Med Biol Res 45:1102–1111. https://doi.org/10.1590/s0100-879x2012007500142
National Institute of Diabetes and Digestive and Kidney Diseases (2012) Aldesleukin. In: LiverTox: clinical and research information on drug-induced liver injury. National Institute of Diabetes and Digestive and Kidney Diseases, Bethesda, MD
NCT01171209 (2012) REsPonse to interferon-alpha in InterfeRon-β neutralizing antibody positive multiple sclerosis patients [WWW Document]. https://classic.clinicaltrials.gov/ct2/show/results/NCT01171209?view=results. Accessed 1 Mar 2024
Normansell R, Walker S, Milan SJ, Walters EH, Nair P (2014) Omalizumab for asthma in adults and children. Cochrane Database Syst Rev 2014:CD003559. https://doi.org/10.1002/14651858.CD003559.pub4
Recombinant DNA technology market [WWW Document] (2023). https://www.precedenceresearch.com/recombinant-dna-technology-market
Reginster J-Y, Minne HW, Sorensen OH, Hooper M, Roux C, Brandi ML, Lund B, Ethgen D, Pack S, Roumagnac I, Eastell R (2000) Randomized trial of the effects of risedronate on vertebral fractures in women with established postmenopausal osteoporosis. Osteoporos Int 11:83–91. https://doi.org/10.1007/s001980050010
Roberts RJ (2005) How restriction enzymes became the workhorses of molecular biology. Proc Natl Acad Sci USA 102:5905–5908. https://doi.org/10.1073/pnas.0500923102
SECRETERS—European Union's Horizon 2020 Programme (2022) Microbial protein cell factories fight back? Trends Biotechnol 40:576–590. https://doi.org/10.1016/j.tibtech.2021.10.003
Shuman S (2009) DNA ligases: progress and prospects. J Biol Chem 284:17365–17369. https://doi.org/10.1074/jbc.R900017200
Siegel DL (2002) Recombinant monoclonal antibody technology. Transfus Clin Biol 9:15–22. https://doi.org/10.1016/s1246-7820(01)00210-5
Slamon DJ, Leyland-Jones B, Shak S, Fuchs H, Paton V, Bajamonde A, Fleming T, Eiermann W, Wolter J, Pegram M, Baselga J, Norton L (2001) Use of chemotherapy plus a monoclonal antibody against HER2 for metastatic breast cancer that overexpresses HER2. N Engl J Med 344:783–792. https://doi.org/10.1056/NEJM200103153441101
Šlesingerová E (2021) Recombinant DNA and genome-editing technologies: embodied utopias and heterotopias. Body Soc 27:32–57. https://doi.org/10.1177/1357034X21998449
Thomas L (2022) Current and future perspectives of mRNA technology [WWW Document]. https://www.news-medical.net/news/20220308/Current-and-future-perspectives-of-mRNA-technology.aspx
Tossing G (2001) New developments in interferon therapy. Eur J Med Res 6:47–65
Ullah MW, Khattak WA, Ul-Islam M, Khan S, Park JK (2015) Encapsulated yeast cell-free system: a strategy for cost-effective and sustainable production of bio-ethanol in consecutive batches. Biotechnol Bioproc E 20:561–575. https://doi.org/10.1007/s12257-014-0855-1
Vandenhende M-A, Bentaberry F, Morlat P, Bonnet F (2011) Anakinra: an effective treatment in the Schnitzler syndrome. Joint Bone Spine 78:636–637. https://doi.org/10.1016/j.jbspin.2011.03.021
Walsh G (2002) Biopharmaceuticals and biotechnology medicines: an issue of nomenclature. Eur J Pharm Sci 15:135–138. https://doi.org/10.1016/s0928-0987(01)00222-6
Walsh G (2004) Second-generation biopharmaceuticals. Eur J Pharm Biopharm 58:185–196. https://doi.org/10.1016/j.ejpb.2004.03.012

Weinblatt ME, Kremer JM, Bankhurst AD, Bulpitt KJ, Fleischmann RM, Fox RI, Jackson CG, Lange M, Burge DJ (1999) A trial of etanercept, a recombinant tumor necrosis factor receptor: Fc fusion protein, in patients with rheumatoid arthritis receiving methotrexate. N Engl J Med 340:253–259. https://doi.org/10.1056/NEJM199901283400401

Weinreich DM, Sivapalasingam S, Norton T, Ali S, Gao H, Bhore R, Xiao J, Hooper AT, Hamilton JD, Musser BJ, Rofail D, Hussein M, Im J, Atmodjo DY, Perry C, Pan C, Mahmood A, Hosain R, Davis JD, Turner KC, Baum A, Kyratsous CA, Kim Y, Cook A, Kampman W, Roque-Guerrero L, Acloque G, Aazami H, Cannon K, Simón-Campos JA, Bocchini JA, Kowal B, DiCioccio AT, Soo Y, Geba GP, Stahl N, Lipsich L, Braunstein N, Herman G, Yancopoulos GD, Trial Investigators (2021) REGEN-COV antibody combination and outcomes in outpatients with Covid-19. N Engl J Med 385:e81. https://doi.org/10.1056/NEJMoa2108163

WHO (2015) Good manufacturing practices. WHO

Winer KK (1996) Synthetic human parathyroid hormone 1-34 vs calcitriol and calcium in the treatment of hypoparathyroidism: results of a short-term randomized crossover trial. JAMA 276:631. https://doi.org/10.1001/jama.1996.03540080053029

Wright S (1986) Recombinant DNA technology and its social transformation, 1972–1982. Osiris 2:303–360. https://doi.org/10.1086/368659

Wu X, Chen P, Lin H, Hao X, Liang Z (2016) Hepatitis E virus: current epidemiology and vaccine. Hum Vaccin Immunother 12:2603–2610. https://doi.org/10.1080/21645515.2016.1184806

Zeuzem S (2008) Interferon-based therapy for chronic hepatitis C: current and future perspectives. Nat Clin Pract Gastroenterol Hepatol 5:610–622. https://doi.org/10.1038/ncpgasthep1274

Chapter 13
Microbial Production of Amino Acids and Their Applications in Health and Nutrition Sectors

Ankita Agrawal, Jitesh Kumar Maharana, Amisha Mohanty, and Amiya Kumar Patel

Abstract Amino acids are attractive biomolecules that have multibillion-dollar market demand due to their widespread applications in pharmaceuticals, food, cosmetics, animal feed, and agricultural sector. Amino acids derived either from natural resources or chemically synthesized have limitations in terms of cost, sustainability, and environmental impact. Increase in microbial way of amino acid production has revolutionized the biotechnology sector, offering more economical, cost-effective, and sustainable alternative for commercial production. In particular, fermentation stands as a primary cornerstone in the realm of commercial amino acid production, contributing to 80% of the global biosynthesis of amino acids. This chapter underscores recent progress and crucial insights into the microbial-mediated synthesis of amino acids while highlighting the advantages and disadvantages of each approach. Moreover, genetic engineering presents distinct benefits compared to traditional methods, encompassing enhanced yield, reduced production costs, specificity, a diminished carbon footprint, and targeted synthesis. With the recent advancements in omics, throughput, and computational tools, the paradigm of strain improvement has transitioned from random mutagenesis to system metabolic engineering (target-specific). The system metabolic engineering integrates synthetic and systems biology with evolutionary engineering, offering an efficient method to unlock the maximum potential of commercial microbial strains, notably *Escherichia coli* and *Corynebacterium glutamicum*. This chapter delves into the application of genetic engineering alongside novel and innovative strategies implemented in microbial strains to augment amino acid production in a more cost-effective and eco-friendly manner.

Keywords Amino acids · Fermentation · Genetic engineering · Microbes

A. Agrawal · J. K. Maharana · A. Mohanty · A. K. Patel (✉)
Department of Biotechnology and Bioinformatics, Sambalpur University, Burla, Odisha, India
e-mail: amiya.patel@suniv.ac.in

V. Kothari et al. (eds.), *Microbial Products for Health and Nutrition*,
https://doi.org/10.1007/978-981-97-4235-6_13

13.1 Introduction

Amino acids are molecules encoded by DNA that influence a variety of physiological processes (Zhang et al. 2021). It comprises an amino group, a carboxyl group, a hydrogen atom, and a unique side chain (R group) all bonded to a central carbon atom. Amino acids play a crucial role as metabolic intermediates and fundamental building blocks of proteins and peptides that regulate key metabolic pathways and activities vital for the development and functioning of life forms (Wu 2009). Consequently, amino acids contribute significantly to human and animal nutrition, serving a vital part in maintaining health (Duan et al. 2023). The 22 amino acids based on dietary necessity fall into essential and non-essential categories (Choi and Coloff 2019). Essential amino acids, referred to as indispensable amino acids, cannot be synthesized and require to obtained through an exogenous diet, while non-essential amino acids can be generated by the body itself. Their significance is linked with nutritional value, especially for nine amino acids (valine, phenylalanine, tryptophan, threonine, isoleucine, methionine, lysine, histidine, and leucine) essential for humans and animals (Kramer et al. 2004; Choi and Coloff 2019). Except for glycine, all amino acids have optically active isomers (D- and L-forms) and only L-amino acids exist in natural proteins (Okafor and Okeke 2017).

Amino acids are the primary products in white biotechnology, holding significance both in regard to volume and value within the global sector (Wendisch et al. 2006). To meet the escalating demand, amino acids are produced on a multibillion-ton scale. In 2021, the global demand for various amino acids reached 10.3 million tons, generating a gross sale of $28 billion, with a growth rate per year of 6.76% (Tuo et al. 2023). Further, the annual amino acid market is projected to grow at a CAGR of 8.7% from 2023 to 2033. Several predominant manufactures of microbial amino acids globally comprise Ajinomoto group, Meihua group, Evonik, CJ CheilJedang, ADM, and Kyowa Hakko Bio (Mahmood 2015). Additionally, in 2022, India arose as the second-biggest feed amino acid market in the Asia-Pacific region, representing 8.4% of the overall market esteem. As of 2024, the assessed market size for India Feed amino acids stands at 207.04 million USD. In India, major plays in the amino acid manufacturing industry includes Ajinomoto India Limited, PharmaZell, Globion India, Aumgene Biosciences, Maple Biotech, and Chaitanya Groups, which plays crucial role in meeting the rising demand of amino acids across different sectors within the country.

The rising market demand for amino acids underscores their diverse applications across various industries, including food and feed industry, cosmetics, polymer materials, agriculture, chemical industries, medicines, and pharmaceutical industries (Pandey et al. 2020). The animal feed sector drives about 56% of the entire amino acid market (Sanchez et al. 2018), making it the world's largest and most rapidly expanding market for amino acids (Wendisch 2020). Amino acids utilized in food industries as food additives, artificial sweetener, and flavoring agents consume 40% of amino acid production (Kramer et al. 2004). Further, certain amino acids have been used in the biofuels and antibiotic production as crucial intermediary precursors (Park and Lee 2008). Amino acids also find applications in health improvement, regulation of muscle protein metabolism, reduction of adiposity, and growth control, showcasing prominent therapeutic potential (Mahmood 2015; Tonouchi and Ito 2016). Amino acids, when added to food, are recommended for purposes such as sleep aid, smoking cessation, addressing attention-deficit hyperactivity disorder, and managing depression (Ivanov et al. 2013). Besides, amino acids have garnered ample attention for their potential to alleviate mental exhaustion. Amino acids are also employed in various medical applications like ammonia detoxification in blood, heart failure treatment, transfusions, peptic ulcer therapy, and male infertility (Okafor and Okeke 2017). Further, the pandemic has heightened consumer demand for immunity-boosting products, sustaining the rising need for amino acids. The broad industrial applications of various amino acids are depicted (Table 13.1). Increasing demand for amino acids driven by their nutritional value, metabolic activity, and medicinal significance, research is rapidly progressing to explore and enhance techniques for commercial-scale production. This entails innovations of new routes and the improvement of strain efficacy to achieve the highest possible levels of production (Lee and Kim 2015).

Amino acid production was started in 1908 by Prof. K. Ikeda in Japan after the discovery of kelp or monosodium glutamate (MSG), a unique flavoring component of seaweed (Konbu) through a protein extraction process (Nakamori 2017; Zhang et al. 2021). Subsequently, MSG was industrially produced from acid hydrolysates of wheat and soybean protein (Hermann 2003), the first commercialized production of amino acids but with significant drawbacks (Nakamori 2017). Further, in 1935, amino acid production for pharmaceutical applications began (Tonouchi and Ito 2016). Glutamic acid became the first commercially produced amino acid using the gram-positive bacterium *Corynebacterium glutamicum* through microbial fermentation, setting a benchmark for industrialized amino acid synthesis (Kumagai 2000). In the 1960s, glutamic acid was also chemically synthesized using the Strecker cyanohydrin reaction by Ajinomoto Co. (Nakamori 2017). Following the success of *C. glutamicum*, bacterial strains such as *E. coli* were implemented for large-scale production (Dong et al. 2011). Greater need for amino acids necessitates the development of inventive strategies for their commercial production, focusing on higher efficiency and productivity through the utilization of high-yielding mutant strains.

Amino acids are promptly produced on an industrial basis through three primary processes: extraction from natural resources, chemical synthesis, and microbial-mediated processes involving enzymatic and fermentation approaches (Pandey et al. 2020; Tuo et al. 2023). Yet, the most beneficial strategy for commercial production of specific amino acids has been picked from an economic standpoint (Ikeda 2003). The biotechnologically mediated microbial amino acid synthesis is considered a green manufacturing approach with greater advantages and reliability (Zhang et al. 2021). About 80% of global amino acid synthesis is attributed by microbial fermentation (Tuo et al. 2023). L-Amino acids comprising L-threonine, L-glutamic acid, L-cysteine, L-alanine, L-aspartic acid, L-dihydroxyphenylalanine, L-lysine, and hydroxy-L-proline, are produced industrially through microbial-mediated processes (Kumagai 2000; Ikeda 2003). Notably, about 3.3 million tons of L-glutamate/

Table 13.1 Industrial application of different amino acids across diverse sectors

Amino acid(s)	Use(s) and application(s)	Reference(s)
Glycine	Used as a sweetener in juices, manufactured herbicide (glyphosate), flavor enhancer, buffering agents in antacids, antiperspirants, and cosmetics	Ivanov et al. (2013)
L-Alanine	Used as flavor and taste enhancer, food supplements for body building, food preservative	Zhang et al. (2007)
L-Arginine	Used in dental products and cosmetics, an ingredient of infusion fluids to treat liver diseases	Wu (2009), Park et al. (2014)
L-Asparagine	Used as a nutritional supplement to treat dietary deficiency or imbalance	Ivanov et al. (2013)
L-Aspartic acid	Used in artificial sweeteners (aspartame), agricultural industry, drug industry, as food supplements, and buffering agents in cosmetics	Kumagai (2000), Derbikov et al. (2017)
L-Cysteine	Used as dietary supplements for muscle building and growth, serving as an antioxidant, as a precursor in the medicinal products, cosmetics and livestock industries, and employed as an ingredient in infusion fluids	Ikeda and Takeno (2013), Tonouchi and Ito (2016)
L-Glutamic acid	Use to manufacture synthetic leather, to enhance flavor, to increase nutritional value in animal feeds, surfactants in soaps and shampoo, fertilizer and dye industries, in pharmaceuticals as neurotransmitter	Nishio et al. (2013), Tonouchi and Ito (2016)
L-Glutamine	Used for healing gastric ulcers, as a component in nutritional products to promote muscle growth, utilized for treating injuries and neuropathic diseases, and effective in reducing healing time after surgical operations	Hayashi and Tabata (2013)
L-Histidine	Used as antioxidants to preserve milk powder, food supplements to minimize premenstrual pain, anti-inflammatory, used to treat peptide ulcers	Ivanov et al. (2013)
L-Isoleucine	Constituents for dietary supplements for muscle builder, a component in infusion fluids, used as motor tonic	Sanchez et al. (2018)
L-Leucine	Used as flavoring products, ingredients in cosmetic and infusion fluids, used in animal feed additives, fitness supplements, and tablet lubricant	Tonouchi and Ito (2016), Wang et al. (2019)
L-Lysine	Used as nutritional supplements and flavor enhancer in food, feed, and beverage industries; used in the agricultural industry, nervous system drugs, and nutritional therapy; cosmetics and personal care products (eye makeup, baby products); treatment of cold sores and genital herpes, aids in calcium absorption and bone health, used in infusions	Hermann (2003), Mahmood (2015), Pandey et al. (2020)
L-Methionine	To improve nutritional value in animal food and feed additives, raw materials in medical supplies, as antioxidant	Mahmood (2015), Li et al. (2016)
L-Phenylalanine	Used in artificial sweetener (aspartame), nutritional supplements for antidepressant effect	Ikeda and Takeno (2013)

L-Proline	Used in supplements for muscle strength and body building, osmoprotectant in pharmaceutical industries	Ivanov et al. (2013)
L-Serine	Used as a moisturizing agent in cosmetics, utilized in food supplements to cure chronic fatigue syndrome and enhance mental health, as nutritional supplement in health drinks and also used in medicines	Zhang et al. (2018)
L-Threonine	Used to manufacture herbicide (aztreonam), used in animal feed industry, flavor enhancer in food such as cheese, fish, and meat, used in cosmetics and pharmaceutical products	Du et al. (2020), Duan et al. (2023)
L-Tryptophan	Used as antioxidants to preserve milk powder, food supplements to cure sleep disorder, premenstrual syndrome, depression and phenylketonuria, used in agricultural application, used as flavoring agent in beverage and food industry	Ikeda (2006), Patnaik et al. (2008), Cordente et al. (2019)
L-Tyrosine	Used as ingredients of infusion fluids, protein supplements to treat phenylketonuria, used in low-calorie sweetener aspartame	Ikeda (2006), Cordente et al. (2019)
L-Valine	Used in food supplements for body building and muscle strength, an ingredient in infusion solutions, as moisturizing agent in cosmetics, as a precursor for antibiotics, antiviral, and herbicides	Cheng and Zhang (2014), Westbrook et al. (2018)

MSG, 3.3 lakh tons of L-threonine, 3000 tons of L-glutamine, 1500 tons of L-arginine, and others are primarily produced through fermentation (Sanchez et al. 2018). Enzymatic production includes 17,000 tons of L-aspartate, 4000 tons of cysteine, and 600 tons of L-alanine. Additionally, about 14,000 tons of L-tryptophan are manufactured jointly by fermentation and enzymatic methods (Sanchez et al. 2018). Chemical processes contribute 8.5 lakh tons of DL-methionine per year. Researchers are currently directing on producing specialty amino acids and increasing amino acid synthesis utilizing commercial strains (*C. glutamicum* and *E. coli*). Moreover, each commercialization strategy must undergo continuous improvement through technological innovations, optimizing for sustainable and cost-effective amino acid production (Zhang et al. 2021).

This study delves into various methodologies employed for commercial synthesis of diverse amino acids emphasizing the utilization of high-yielding mutant strains, reactor design and operation, process optimization, and scaling-up of operation. It also summarizes the pros and cons associated with each production strategy. Traditional approaches to meet the escalating demand for amino acids including production strategies, commercial strain development, and the exploration of microbial strains are discussed below. Moreover, the study explores the novel and emerging metabolic engineering strategies (CRISPRi, biosensors) not only employed for microbial strain improvement but also for the rapid and precise screening of amino acid overproducing microbial strains. These innovative strategies aim to ensure the creation of potent overproducers and high-yielding microbial strains to maximize amino acid productivity in a cost-effective, sustainable, and environmentally friendly manner.

13.2 Amino Acid Production Strategies

Amino acids have been synthesized through three primary strategies: (1) extraction, (2) chemical synthesis, and (3) microbial processes. Each tactic comes with its own set of rewards as well as drawbacks. The efficacy of different strategies hinges on a multitude of factors such as economic viability, types of amino acids, accessibility of raw components, market demand, and ecological factors (Ikeda 2003). Out of these, the fermentation stands out as more promising for mass production of amino acid since it involves the utilization of appropriate microbial strains specific to each amino acid and low-cost carbon sources (Paloyan et al. 2022). The details of each production strategy are outlined below.

13.2.1 Extraction Process

The primary method of amino acid production relies on the extraction of distinct amino acids from protein hydrolysates (Ivanov et al. 2013). This extraction method finds optimal use for commercial synthesis of limited amino acids only (Ikeda

2003). It is a simple and reliable approach with an established technological protocol for amino acid production using industrial wastes or byproducts like plant hydrolysis products, meat extracts, and hair (Breuer et al. 2004). The reagents required for production, like hydrochloric acid and sodium hydroxide, are readily available. However, the major drawback includes its exploitation of different physiochemical properties that make the downstream processing challenging (Zhang et al. 2016). Additionally, elevated pH not only causes protein degradation, leading to low productivity, but also leads to formation of undesired toxic compounds lowering product quality (Sereewatthanawut et al. 2008). Further, the process of washing residual alkali solution generates a huge amount of wastewater.

For an ecofriendly approach, efficient extraction requires the implementation of sub-critical and supercritical water hydrolysis to restore proteins and amino acids (Sereewatthanawut et al. 2008). Scientific analysis reveals that reaction time and temperature during hydrolysis play a significant role in maximizing amino acid production (D'Este et al. 2017). As the extraction process relies on the accessibility of natural compounds with high protein content like hair, keratin, feather, blood meal, or soyabean, satisfying the rising demands is challenging. Consequently, this technique is utilized in conjunction with other strategies for mass production of amino acids (Pandey et al. 2020).

13.2.2 Chemical Synthesis

Chemical synthesis (Strecker synthesis) is being employed successfully in the production of glycine, methionine as well as the racemic mixture of alanine and methionine (D'Este et al. 2017). This process involves converting aldehyde/ketone and ammonia/amine into alpha amino acids using an acid catalyst, cyanide (as a nitrogen source), and water (Deng et al. 2018). This typically consists of three steps: (1) the condensation of aldehyde or ketone with amine or ammonia to synthesize imine, (2) the reaction of the imine with cyanide to produce amino nitrile, and (3) ultimately, hydrolyzing amino nitrile to yield the corresponding alpha amino acids (Deng et al. 2018). The most prevalent example is the ammonolysis of trichloroethylene to produce glycine (Pandey et al. 2020). In addition, chemical synthesis has also been utilized for the biosynthesis of alanine, phenylalanine, threonine, methionine, tryptophan, and valine. However, the use of complex and expensive catalysts and the requirement of toxic cyanide sources have limited the utility of this approach (Zuend et al. 2009).

In industrial setting, the Bucherer-Bergs method (a modified Strecker synthesis) is employed to generate racemic mixtures of amino acids. These racemic mixtures are particularly valuable, as they can be readily converted into enantiomerically pure compounds using diverse biocatalytic methods (Breuer et al. 2004). The Bucherer-Bergs method entails reacting an aldehyde or ketone with ammonium carbonate and sodium cyanide to form hydantoins, then hydrolyzed to generate a racemic mixture of amino acids (Breuer et al. 2004). Subsequently, the desired L-forms of amino acids are obtained through the treatment of racemic mixture with the immobilized enzyme (amino acylase) produced from *Aspergillus niger* (Ivanov et al. 2013). However, the involvement of costly enzymes to obtain pure compounds, as well as the extended reaction time and high temperature requirements, this approach is deemed suitable solely for the commercial biosynthesis of glycine (achiral) and methionine (non-natural enantiomer) (Breuer et al. 2004). Presently, an altered version of the chemical process known as the "catalytic asymmetric method" has been introduced, which utilizes a robust catalyst (chiral amido-thiourea) and non-toxic, easy-to-handle aqueous cyanide salts to enable the direct formation of enantiomerically pure amino acids (Zuend et al. 2009). However, studies revealed that the chemical process has not achieved commercial viability for biosynthesis of optically pure isomers of amino acids.

13.2.3 Microbial-Mediated Amino Acid Production

The microbiological method stands out as the predominant green production approach for amino acid biosynthesis owing to its potential to generate high-quality amino acids that is economically viable (Zhang et al. 2021). A diverse array of microbes encompassing prokaryotic and eukaryotic strains is harnessed for the purpose. The production process empathizes with the breeding of amino acids producing microbes, strategies employed based on the type of amino acid and genetic engineering aspects. Microbial method produces stereospecific L-amino acids making the process more advantageous compared to other synthetic processes (Zhang et al. 2021). Microbial-mediated amino acid production employed can be categorized as an enzymatic method, semi-fermentation and direct fermentation approach (Ivanov et al. 2013).

13.2.3.1 Enzymatic Process

Enzymatic catalysis method is extensively employed in the chemical industry, where purified enzymes sourced from microbes are applied for amino acid biosynthesis (Ivanov et al. 2013). In this method, the amino acid precursor molecule is turned into the target molecule by utilizing one or more enzymes, eliminating the process of microbial growth (D'Este et al. 2017). Ammonia lyases, NAD^+-dependent L-amino acid dehydrogenase, and hydrolytic enzymes isolated from *Pseudomonas dacunhae*, *Saccharomyces cerevisiae*, *Escherichia coli*, and *Cryptococcus lurendii* have all been employed (Pollegioni et al. 2008). The primary benefit relies on its capacity to yield optically pure isomers of amino acids at higher concentrations while minimizing the generation of byproducts (Ikeda 2003; D'Este et al. 2017).

For the very first time, enzymatic catalysis has been implemented for the synthesis of *N*-acetyl D, L-amino acids using acylase in an immobilized form (Leuchtenberger et al. 2005). Presently, this method is successfully used in the production of methionine, aspartic acid, alanine, tryptophan, cystine, phenylalanine, serine, cysteine, and valine (Ivanov et al. 2013). For instance, L-methionine and L-valine synthesis utilize acylase from *Aspergillus oryzae* in an enzymatic membrane reactor (Leuchtenberger et al. 2005) and L-alanine from aspartate using aspartate β-decarboxylase from *P. dacunhae* (Sanchez et al. 2018). Further, three distinct enzymes derived from *Pseudomonas thiazolinophilum* are used in the large scale of L-cysteine, which catalyze the breakdown of DL-2 amino-2-thiazoline-4-carboxylate into specific amino acids (Sanchez et al. 2018). Moreover, L-aspartic acid is biosynthesized from ammonia and fumarate using an immobilized aspartate enzyme isolated from *E. coli* (Kumagai 2000).

Conversely, the involvement of expensive enzymes with limited stability is the major drawback of the process (Ikeda 2003). Economic feasibility of the approach is predominantly influenced by the expense of producing the substrate. While enzyme catalysis stands as specific and sophisticated method for the biosynthesis of D-amino acids and non-proteogenic L-amino acids (Leuchtenberger et al. 2005), it is not the optimal strategy for large-scale L-amino acid biosynthesis. This method is also limited to synthesizing specific amino acids only.

13.2.3.2 Semi-Fermentation Technique

Metabolic intermediates in amino acid biosynthesis or their precursors are introduced to a growth medium supplemented with various nutrient resources, which undergoes fermentation to produce amino acids (Ivanov et al. 2013; Okafor and Okeke 2017). Biosynthesis of L-serine using methane-utilizing bacterium *Hyphomicrobium* or *Pseudomonas* sp. is a prime example of this strategy. Further, *Brevibacterium* sp. is used for the conversion of inexpensive intermediates like DL-α-aminobutyric acid and ethanol into valuable L-isoleucine (Okafor and Okeke 2017).

13.2.3.3 Fermentative Production of Amino Acids

Microbial fermentation stands out as the most cost-effective, potent, and ecofriendly method employed for amino acid biosynthesis accounting for 80% of global amino acid biosynthesis (Tuo et al. 2023). Amino acid fermentation plays a crucial component in commercial biotechnology, facilitated by the advancement of commercial strains through genetic engineering strategies (Okafor and Okeke 2017; Wendisch 2020). Nearly all amino acids are exclusively generated through the fermentation process, yielding only the biologically active L-enantiomers and eliminating the need for additional downstream processing (Becker and Wittmann 2012). Besides, the fermentation process can be optimized at moderate conditions preventing product deterioration and low maintenance expenses (D'Este et al. 2017). Fermentation is a conversion process that involves three steps such as (1) cultivation of microbial strains in a suitable medium, (2) exploitation of microbial strain in optimized culture conditions, and (3) downstream processing involving the extraction and purification of amino acids with desired degree of purity with minimal cost (Ikeda and

Takeno 2013; Ivanov et al. 2013). Design and operation of bioreactor and optimization of culture conditions are prerequisites leading to high capital investment, which are some of the issues related to the strategy. However, the cost of nutrient supplements, productivity and yield, purification cost, and overall productivity largely influence the reliability of the technique (Ikeda 2003).

13.3 Reactor Design and Process Optimization

13.3.1 Bioreactors Used for Amino Acid Production

Fed-batch reactor is the preferred reactor for industrial amino fermentation compared to a continuous culture reactor (D'Este et al. 2017). Main benefits include high reproducibility and higher concentration with reduced risk of inhibition. Besides, the reactor supplemented with low substrate concentration ensures adequate oxygen capacity and thereby prevents oxygen limitation and undesired byproduct formation (Hermann 2003). Compared to batch, the fed-batch reactor is mostly employed for large-scale production. Furthermore, the fed-batch fermentation extends its duration by periodically withdrawing broth and replenishing it with nutrient input to enhance overall productivity (Ikeda and Takeno 2013). In contrast, continuous production processes exhibit a higher degree of substrate conversion but entail an elevated risk of contamination due to the open system, where substrate feeding and product removal occur simultaneously (D'Este et al. 2017). Further, continuous culture reactor generally yields relatively lower productivity as compared to fed-batch reactors, particularly in certain fermentation processes with distinct growth and production phases, such as fermentation of L-threonine by *E. coli* and alanine by *Arthrobacter oxydans* (Ikeda 2003). Yet, continuous culture is regarded as a valuable method to examine microbial physiology in response to environment and biosynthesis that batch culture is unable to address (Ikeda and Takeno 2013).

13.3.2 Process Optimization

With increasing demand and applications of amino acids, microbial fermentation process has been enhanced through rector design and process optimization. Key parameters influencing productivity include pH, temperature, inflow rate, aeration capacity, inoculum size, and quality to ensure substrate conversion rate to obtain the desired product in maximum quantity (Ikeda 2003). Inoculum quality and quantity are of utmost importance, influencing productivity and yield, and thus require rigorous monitoring to ensure stability (Ikeda 2003). Besides, mass transfer and gaseous exchange are critical factors affecting productivity. Maintenance of optimum pH and temperature are pre-requisite specific for target compound. Optimum temperature for amino acid production using *C. glutamicum* is around 30–34 °C (Hermann 2003). Relatively higher production of L-glutamate is achieved at high temperatures (41 °C) using *C. glutamicum*, whereas thermotolerant bacteria, i.e., *Bacillus methanolicus*, yield more L-lysine and L-glutamate at 50 °C (Brautaset et al. 2007). The refinement of culture conditions is critical for massive amino acid biosynthesis.

Scaling processes (scale-up and scale-down) are key factors in enhancing productivity and maintenance of quality in amino acid fermentation (Takors et al. 2007), which is influenced by culture condition, substrate concentration, metabolic precision, oxygen transfer, inoculums size, and phase of microbial growth (Takors et al. 2007). High glucose concentration with lower oxygen leads to overproduction of byproducts resulting in increased acidity, which inhibits microbial growth. Several approaches, including computational fluid dynamics (CFD) and scale-down approaches, are being developed, which not only elucidate the hydrodynamics by offering spatial and temporal information for estimating process efficiency and product titers but also enable the visualization and control of parameters that are challenging to access in the fermentation process (Nadal-Rey et al. 2021). Besides, scale-down approaches are gaining attention for comprehending and optimizing the fermentation process. Integrating results from CFD models with data obtained from scale-down apparatus, process optimization, is feasible in the initial phases of development (D'Este et al. 2017). Additionally, downstream processing also plays a crucial role that adds a substantial capital investment for manufacturers and significant loss of product in purification (Ecker et al. 2012). To improve productivity, membrane-based purification processes are increasingly employed for amino acid biosynthesis. A recent study implies nanofiltration as highly convenient technology for recovering high-purity amino acids, where pressure-driven membrane filtration technology is integrated with fermentation reactors that allow the production as well as purification simultaneously, thereby reducing cost and improving efficiency (Ecker et al. 2012).

13.4 Microbial Strains Used in Amino Acid Production

Microbial strains such as bacterial, yeast, and fungal species are being involved with the commercial processing of amino acids. Choosing high-yielding microbial strains is essential to precise amino acid biosynthesis, a process primarily governed by various criteria. These criteria include the strain's non-pathogenic nature, optimal growth with readily available substrates, a relatively high degree of substrate utilization, tolerance to a broad spectrum of temperatures and pH, bacteriophage resistance, and minimal substrate utilization for byproduct formation. Exploitation of both prokaryotic and eukaryotic cells for commercial production has been a hallmark of amino acid biosynthesis (Patnaik et al. 2008). There are four categories of microbial strains employed for the biosynthesis of amino acids: wild-type strains, auxotrophic mutants, regulatory mutants, and auxotrophic regulatory mutants.

13.4.1 Prokaryotic Microbial Strains

Traditionally, wild-type prokaryotic strains, exemplified by *C. glutamicum,* have been the go-to choose for commercial amino acid fermentation due to their high production efficiency. Genetically engineered prokaryotic species, like *E. coli*, is used for commercial amino acid synthesis. Both *C. glutamicum* and *E. coli* remain the most preferred bacteria, undergoing continual refinement through metabolic engineering tactics (Liu et al. 2022). Specifically, *C. glutamicum* stands out as a potent and efficient strain utilized for the manufacture of a diverse spectrum of amino acids, including L-glutamate, L-lysine (Wendisch et al. 2016), L-phenylalanine (Hermann 2003), L-threonine, L-tryptophan, L-serine, L-glutamine, L-proline, L-isoleucine, and L-arginine (D'Este et al. 2017). *C. glutamicum* offers advantages as an industrial strain as it lacks endotoxins and utilizes a diverse array of carbon sources including glucose, mannose, sucrose, ribose, fructose, and maltose (Wendisch et al. 2016). Additionally, auxotrophic strains of *C. glutamicum*, lacking homoserine dehydrogenase activity, are harnessed to boost L-lysine output in industrial settings. Various strains of *Brevibacterium* sp. (*B. flavum*, *B. divaricatum*, *B. lactofermentum*, *B. ammoniagenes, B. alanium*, and *B. roseum*) have been widely utilized for the synthesis of valine, proline, isoleucine, glutamate, and lysine on a massive scale (Ivanov et al. 2013; Zhang et al. 2021).

Moreover, *Escherichia coli* is well recognized as a key microbial strain for commercial biosynthesis of various amino acids, often subjected to metabolic engineering to enhance its efficiency. Both wild and mutant strains of *E. coli* are being harnessed for amino acid biosynthesis, encompassing L-threonine (Dong et al. 2011), L-lysine (Hayashi and Tabata 2013), L-tyrosine, L-phenylalanine, L-alanine, L-tryptophan, (Zhang et al. 2007), L-methionine (Ikeda and Takeno 2013), L-aspartic acid (Derbikov et al. 2017), L-glutamine (Hayashi and Tabata 2013), L-valine, L-leucine, and L-isoleucine, all of these have significant industrial applications (D'Este et al. 2017). Scientific literature supports the utilization of both wild and mutant bacterial strains, such as *Serratia marcescens* (Dong et al. 2011), *Bacillus methanolicus* (Brautaset et al. 2007), *Bacillus subtilis*, *Lactobacillus plantarum* (Zhang et al. 2021), *Corynebacterium crenatum, Halomonas bluephagenesis* (Du et al. 2020), *Arthrobacter aminofaciens*, and *Zymomonas mobilis* (Ikeda 2003), for large-scale amino acid production in industrial settings (Table 13.2).

Table 13.2 Various microbes employed for the biosynthesis of all 22 amino acids through microbial fermentation

Amino acid	Titer (g/L)	Commercial strains used in microbial fermentation	Reference(s)
L-Lysine	120	*C. glutamicum*	Sanchez et al. (2018)
	33.92	*B. flavum*	Irshad et al. (2015)
	134.9	*E. coli*	Pandey et al. (2020)
	–	*B. lactofermentum*	Kumagai (2000)

(continued)

Table 13.2 (continued)

Amino acid	Titer (g/L)	Commercial strains used in microbial fermentation	Reference(s)
L-Glutamic acid	>100	*C. glutamicum*	Mahmood (2015)
	120	*Escherichia coli*	Zhang et al. (2016)
	101	*Bacillus subtilis*	Huang et al. (2011)
L-Aspartic acid	193.3	*E. coli*	Derbikov et al. (2017)
L-Methionine	35	*E. coli*	Shim et al. (2017)
	16	*C. glutamicum*	Shim et al. (2017)
L-Threonine	111	*E. coli*	Sanchez et al. (2018)
	52	*C. glutamicum*	Zhang et al. (2021)
	58	*B. flavum*	Kumagai (2000)
	100	*Serratia marcescens*	Sanchez et al. (2018)
	–	*Proteus rettgeri*	Kumagai (2000)
	33	*Halomonas bluephagenesis*	Du et al. (2020)
L-Glutamine	73.1	*E. coli*	Hayashi and Tabata (2013)
	40	*C. glutamicum*	Wendisch et al. (2006)
L-Tryptophan	58	*C. glutamicum*	Zhang et al. (2021)
	29	*Escherichia coli*	Ivanov et al. (2013)
L-Tyrosine	21.6	*C. glutamicum*	Ikeda et al. (2005)
	2.6	*E. coli*	Juminaga et al. (2012)
L-Proline	27	*E. coli*	Csonka and Leisinger (2007)
	12.7	*C. glutamicum*	Wendisch et al. (2016)
L-Asparagine	–	*C. glutamicum*	Ivanov et al. (2013)
	–	*B. flavum*	Ivanov et al. (2013)
	–	*E. coli*	Ivanov et al. (2013)
L-Phenylalanine	28	*C. glutamicum*	Ikeda (2006)
	50	*E. coli*	Zhang et al. (2021)
L-Alanine	113.9	*E. coli*	Zhang et al. (2007)
	216	*C. glutamicum*	Yamamoto et al. (2012)
	7.5	*Zymomonas mobilis*	Ikeda (2003)
L-Glycine	168	*C. glutamicum*	Ikeda (2003)
L-Cysteine	31.4	*P. thiazolinophilum*	Ivanov et al. (2013)
		E. coli	Ivanov et al. (2013)
L-Histidine	22.5	*C. glutamicum*	Kulis-Horn et al. (2014)
	42.6	*S. marcescens*	Kulis-Horn et al. (2014)
	66.5	*E. coli*	Kulis-Horn et al. (2014)
L-Arginine	96	*C. crenatum*	Zhang et al. (2017)
	92.5	*C. glutamicum*	Park et al. (2014)
L-Valine	227	*C. glutamicum*	Oldiges et al. (2014)
	325	*B. flavum*	Cheng and Zhang (2014)
	60	*E. coli*	Oldiges et al. (2014)
	–	*S. cinnamonensis*	Wendisch et al. (2016)

(continued)

Table 13.2 (continued)

Amino acid	Titer (g/L)	Commercial strains used in microbial fermentation	Reference(s)
L-Serine	36.26	*C. glutamicum*	Stolz et al. (2007)
	266.3	*E. coli*	Zhang et al. (2018)
	–	*Methylobacterium* spp.	Ivanov et al. (2013)
L-Isoleucine	40	*C. glutamicum*	Sanchez et al. (2018)
	2.4	*P. aureofaciens*	Yamamoto et al. (2016)
	3.0	*Erwinia carotovora*	Yamamoto et al. (2016)
	1.3	*S. marcescens*	Yamamoto et al. (2016)
	25	*S. typhimurium*	Yamamoto et al. (2016)
L-Leucine	38	*C. glutamicum*	Wang et al. (2019)
	30	*B. lactofermentum*	Yamamoto et al. (2016)
	11.4	*E. coli*	Wang et al. (2019)
L-Selenocysteine	–	*Streptococcus thermophilus*	Castañeda-Ovando et al. (2019)
	–	*Bacillus subtilis*	Yin et al. (2023)
L-Pyrrolysine	Trace	*E. coli*	Gaston et al. (2011)
		Methanosarcina barkeri	Krzycki (2013)

13.4.2 Eukaryotic Microbial Strains

In addition to prokaryotic microbial strains, eukaryotic cells such as yeast and molds are also employed for industrial-scale amino acid fermentation (Patnaik et al. 2008; Qin et al. 2015; Zhang et al. 2021). Various commercial mutant eukaryotic strains have been developed not only to enhance substrate utilization but also to augment amino acid biosynthesis. For instance, *Aspergillus niger* produces TCA cycle intermediates, which have been employed as precursor molecules for amino acid biosynthesis, especially L-lysine (Karaffa and Kubicek 2019). *Saccharomyces cerevisiae*, through fermentation, generates aromatic amino acids and their derivatives that supplement the aroma to wine (Cordente et al. 2019). Further, genetically engineered *S. cerevisiae* has been employed for L-glutathione production that serves as a crucial supplement in pharmaceuticals and food industry as well as for the synthesis of L-ornithine, an intermediate of L-arginine (Qin et al. 2015). Additionally, *Pichia pastoris* is renowned for its role in L-arginine production (Karbalaei et al. 2020).

Mold strains such *as Rhizopus oryzae* and *Rhizopus delemar* have been utilized in food industry and also employed in industrial fermentation for a wide variety amino acid (Odoni et al. 2017). Being acid-tolerant and using a variety of carbon sources (maize straw, sugarcane bagasse, and wheat stem) are beneficial features that accompany molds as an efficient commercial strain used for bulk amino acid production (Kamilari et al. 2023). Previous research has proven that mold strains are beneficial in fermenting amino acids and their derivatives. As an instance, a mutant variant of *Aspergillus oryzae* has been employed to yield a valuable product, L-DOPA (3,4-dihydroxy phenyl L-alanine) utilizing L-tyrosine (Ul-Haq and Ali

2002). Similarly, a pin mold strain (*Mucor mucedo* KP736529) was implemented to manufacture lysine and glutamic acid on a large scale (El-Hersh et al. 2016).

13.5 Metabolic Pathway Involved in Amino Acid Biosynthesis

Carbon metabolism pathway is a conserved and crucial step for amino acid biosynthesis (Ma et al. 2017). Metabolic pathway involved in microbial-mediated production of amino acids is well-regulated with a series of intermediates to produce specific amino acids (Fig. 13.1). Analyzing the metabolic pathway in microbial species is a way to reset natural pathways or engineer novel routes for enhancing commercial amino acid production (Shimizu and Matsuoka 2022). Microbial synthesis of nearly all L-amino acids encompasses the Embden-Meyerhof-Parnas (EMP) pathway, pentose phosphate pathway (PPP), and tricarboxylic acid (TCA) cycle. The metabolic pathway central to amino acid biosynthesis is depicted in Fig. 13.1.

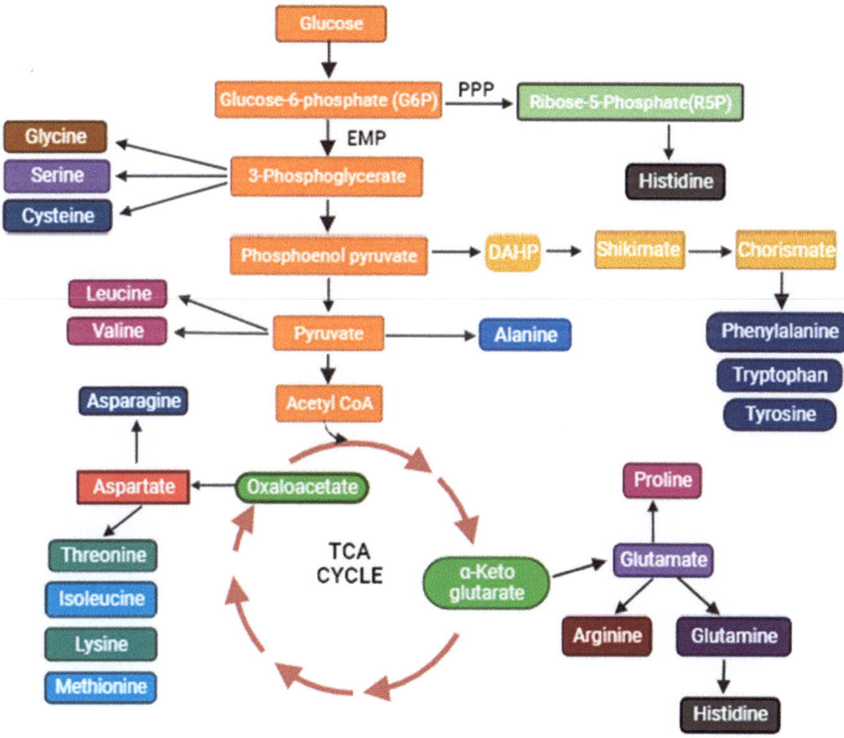

Fig. 13.1 Microbial-mediated metabolic pathways involved in amino acid biosynthesis

13.6 Strategies for Commercial Strain Development

The worldwide demand for amino acids is perpetually increasing, posing a great challenge to enhance the potency of microbial strains for higher productivity using various biotechnological approaches (Mahmood 2015; Wendisch 2020) including the discovery of robust strains, rational genetic engineering to enhance specific amino acid production, reactor design, and process optimization to achieve higher yield. Further, efforts are directed toward exploring novel pathways for specific amino acids with minimal cost. The pathways involved in amino acid biosynthesis are highly complex and auto-regulated. Natural overproducing microbial strains are very rare, and certain strategies are adopted to produce the desired quantity of amino acids (Park and Lee 2008). Genetic modification has been instrumental in developing high-yielding strains, involving amplification of key rate-limiting enzymes, creation of thermotolerant microbial strains through site-directed mutagenesis, modifications in cellular transport system creation of resistant mutant to minimize the feedback inhibition, knockout of repressor genes regulating gene expression, and fine-tuning of metabolic pathways by blocking side chains to achieve higher productivity (Ikeda 2006; Shimizu and Matsuoka 2022). The resulting improved strain, with enhanced efficiency, marks the initial step toward establishing an optimized cellular system for amino acid overproduction. However, further advancements are still possible in the fabrication of highly potent microbial strains for certain amino acids (Bott and Eggeling 2017).

13.6.1 Random Mutagenesis to Systems Metabolic Engineering

Advances in strain enhancement for synthesizing amino acid encompass a spectrum of technologies, commencing with classic mutagenesis, genetic engineering involving addition, deletion, or alteration of genetic material, systems biology, synthetic biology, and systems metabolic engineering (Lee and Wendisch 2017; Ma et al. 2017). The advent of multiple omics techniques, bioinformatics, and high-throughput screening has led to a transition in microbial improvement strategies, moving away from random mutagenesis toward systems metabolic engineering, that melds synthetic, system, and evolutionary engineering based on entire metabolic and regulatory pathways. Besides, genome breeding and biosensor screening are also prominent in the strain development paradigm (Ikeda and Takeno 2020). The altered landscape of strain development tactics for improved amino acid production is illustrated in Fig. 13.2.

Previously, random mutagenesis and selection processes were employed to enhance amino acid production by microbial strains. However, these techniques usually result in undesired genomic changes, growth defeats, and genetic instability (D'Este et al. 2017). Additionally, the unwanted modifications due to random

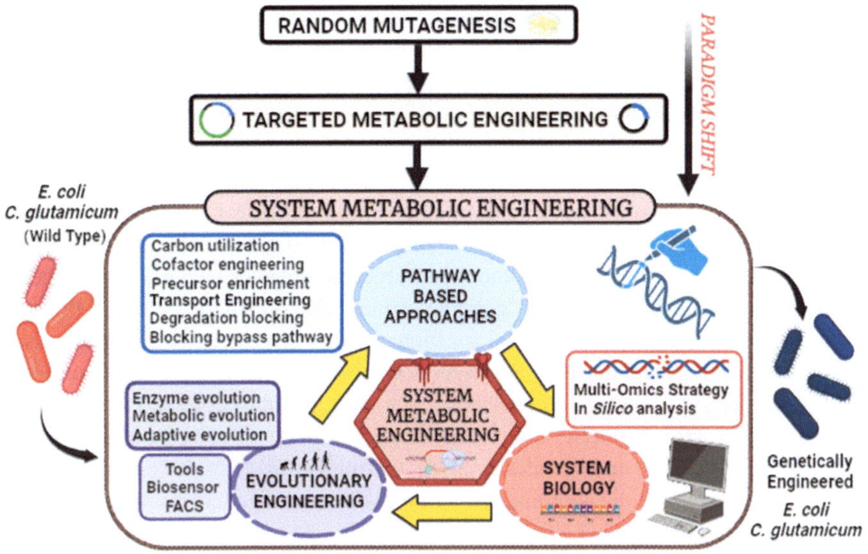

Fig. 13.2 Paradigm shift in microbial strain development from random mutagenesis to system metabolic engineering to enhance amino acid biosynthesis

mutation create challenges in further improvement of cellular performances (Park and Lee 2008). For instance, *C. glutamicum* strain resistant to aminoethyl-cysteine has been developed by subjecting the wild strain to mutagen (*N*-methyl-*N*′-nitro-nitrosoguanidine, NTG) that can produce 44 g/L L-lysine (Zhang et al. 2021). However, mutated strains underwent a series of selection and screening tests, which were usually time-consuming and labor-intensive. In contrast, targeted engineering techniques that allow for precise modifications to desired genes and pathways associated with the biosynthesis of desired amino acids have been implemented, resulting in enhanced productivity (Zhang et al. 2021). However, site-specific mutagenesis has certain limitations in the extent of improvement as it is based on local rather than genome-wide. Advent in system and synthetic biology has successfully overcome such limitations through the implementation of the combined approach to obtain rationally designed microbial strain (Ma et al. 2017).

Systems metabolic engineering is a booming interdisciplinary area, merging principles of systems biology, synthetic biology, and evolutionary engineering (Lee and Kim 2015). It involves the use of omics technology, computational methods (Flux profiling and in silico modeling), and high-throughput screening approaches to predict the combination of genetic alternations useful for strain development as well as the improvement in their efficiencies toward amino acid biosynthesis (Ma et al. 2017). The techniques used for systems metabolic engineering involve pathway-based, systems biology, and evolutionary approaches.

Pathway-focused methods typically target to amplify the production capacity for specific products by employing local metabolic engineering tactics. These include improving carbon source usage (alteration of regulatory gene *csrA* to promote gluconeogenesis and decrease glycolysis for phenylalanine synthesis in *C. glutamicum*) and optimizing key enzyme expression. Other strategies include eliminating feedback inhibition, attenuating transcription, and blocking bypass routes (Ma et al. 2017). Both systems biology and evolutionary techniques provide complete perspectives, implementing this could readily target critical genes for future metabolic engineering and remove hidden barriers to strain enhancement. Systems biology utilizes omics-based approaches and in silico simulation for target-specific alteration in amino acid overproducers. For instance, in silico gene knockout simulations identified three essential genes as knockout targets to enhance L-valine biosynthesis (Park et al. 2007). Subsequently, genome reduction serves as a synthetic biology strategy to build chassis hosts suited for biotech purposes by reducing unnecessary genomic elements. The initial genome-reduced *C. glutamicum* was originated from wild strain (ATCC 13032) and L-lysine generating strain DM1933. This strain exhibited a growth phenotype like the wild type but with enhanced transformation efficacy (Pérez-García and Wendisch 2018). Currently, the genome-reduced chassis strain C1* boasts a 13.4% genome reduction (412 deleted genes) compared to wild type but retains growth behavior akin to wild-type and shows robustness against stress conditions (Baumgart et al. 2018). Evolutionary approaches include transcriptional regulator-based biosensors, allowing quick, accurate, and high-throughput screening of potent mutant strains for industrial processes. With the advancement in biotechnology, systems metabolic engineering is poised to produce commercial microbial strains with higher efficiency toward amino acid biosynthesis.

13.6.2 Feedstock Flexibility Concept

The concept of flexible feedstock emphasizes providing accessibility to diverse carbon sources (Wendisch et al. 2016). Microbial strains used in production, like *B. subtilis*, *C. glutamicum*, *E. coli*, *Pseudomonads*, and *S. cerevisiae*, were altered to efficiently exploit these alternative carbon sources (Wendisch et al. 2016). Wendisch (2020) highlights a diverse range of biomass resources, including starch, lipids, glycerol, xylose, arabinose, corn straw, and cellulose, which can be converted into bioproducts. As an illustration, glycerol, a major byproduct of the biodiesel industry, has been employed as an alternative carbon source in biotechnological processes through the feedstock flexibility concept (Wendisch 2020).

Litsanov et al. (2013) demonstrated the modification of *C. glutamicum* to express *E. coli* glycerol usage genes (glpF, glpK, and glpD) to overproduce L-glutamate, L-lysine, and L-arginine utilizing glycerol as a carbon source. Besides, engineering efforts enabled *C. glutamicum* to utilize polymers like starch or glucans (Wendisch et al. 2016). Further, both *C. glutamicum* and *E. coli* strains have undergone genetic modifications to produce methanol-based L-lysine (Leßmeier et al. 2015). Metabolic engineering of *C. glutamicum* and *E. coli* facilitates the utilization of non-native carbon substrates, fulfilling the flexible feedstock concept as discussed (Wendisch 2020).

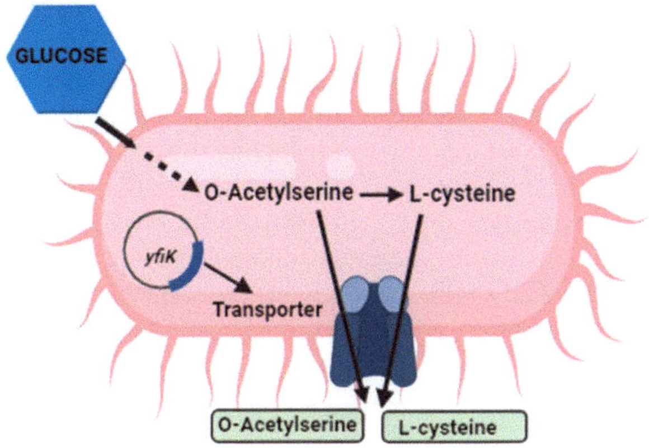

Fig. 13.3 Enhanced production of L-cysteine by upregulation of *yfiK* gene (encoding a transporter) in *E. coli*

13.6.3 Transport Engineering

Transport engineering plays a vital role in advancing the innovation of high-yielding microbial strains, improving nutrient uptake for their growth and proliferation, efficient substrate utilization, and increasing product export to mitigate feedback product-induced inhibition or avoid intermediates loss (Pérez-García and Wendisch 2018). Some microbial strains, with modified transport systems, have proven to be potent amino acid producers (Ikeda 2003).

For instance, elevating the activity of the *yfiK* gene, encoding for multidrug efflux, in *E. coli* induces a significant multiplex increase in L-cysteine synthesis (Fig. 13.3) (Okafor and Okeke 2017). Further, integrating *MurP* and *Crr* genes from *E. coli* into *C. glutamicum* facilitated enhanced uptake of *N*-acetyl-D-muramic acid (MurNAc), resulting in accelerated growth when utilizing MurNAc as a substrate (Sgobba et al. 2018). Furthermore, transport engineering techniques have proven effective in modifying *C. glutamicum* to efficiently utilize xylose and arabinose found in lignocellulosic hydrolysates (Wendisch et al. 2016).

13.6.4 New Cultivation Approach

Co-cultures and co-production approaches along with metabolic engineering and process control strategies have been employed to maximize productivity (Wendisch 2020). In co-culture, synthetic microbial consortia (two strains) are developed to divide the labor to produce the desired product, for instance, the development of a synthetic mutualistic consortium of α-amylase negative, L-lysine overproducing *C. glutamicum* strain collaborating with L-lysine auxotrophic, α-amylase secreting *E. coli* for L-lysine biosynthesis using starch (Sgobba et al. 2018). Stable synthetic co-cultures known as CoNoS (Communities of Niche-optimized Strains) utilizing *C. glutamicum* were built to enhance amino acid output (Zuchowski et al. 2023). Subsequently, the co-production approach explores the possibility of simultaneously producing amino acids and a second value-added molecule (like cell-bound product like lipid). For instance, Henke et al. (2018) designed L-lysine overproducing *C. glutamicum* strain (with reduced genome) capable of accumulating carotenoid astaxanthin alongside L-lysine. Furthermore, new and advanced genetic engineering is being developed to enhance the efficacy of microbial strains, reducing costs and efforts while increasing specificity and productivity.

13.7 Emergence of Novel and Innovative Technologies

Metabolic engineering of microbial strains employed in amino acid biosynthesis has embraced innovative techniques, including CRISPR tools for gene repression and genome editing, flux enforcement, and genetically encoded biosensors (Wendisch 2020), all of which have been elaborated upon.

13.7.1 CRISPRi/Cas9 Technology

CRISPR-Cas technology promises benefits like engineering recalcitrant strains, offering an efficient counterstain system, facilitating gene regulation, and reducing genome manipulation time and multiplex genome editing and regulation (Bott and Eggeling 2017). Initially, the CRISPR interference technique with deactivated Cas9 (dCas9) protein was utilized in *C. glutamicum* to explore the impact of target gene repression on the synthesis of L-lysine and L-glutamate (Wendisch et al. 2016). Wendisch (2020) implemented multiplex CRISPRi/Cas9-based target gene repressions (pgi, pck, and pyk), aiming to enhance L-lysine and L-glutamate biosynthesis in *C. glutamicum*. The study revealed that single-guide RNAs/dCas9 to reduce the expression of *pgi* and *pck* genes (up to 98%) and *Pyk* (up to 97%) resulted in increased titer ratio of L-lysine and L-glutamate compared to one obtained by gene elimination (Cleto et al. 2016).

Subsequently, CRISPR-Cas system is being utilized in *E. coli* to induce gene expression (Bikard et al. 2013). Furthermore, CRISPR tools due to its design simplicity have been used for genome editing to enhance L-valine biosynthesis in *Bacillus subtilis* (Westbrook et al. 2018) as well as L-ornithine production in *S. cerevisiae* (Wendisch 2020). All of these indicate that CRISPR-Cas serves as a rapid and efficient approach for metabolic pathway remodeling that requires no gene deletions or mutations (Cleto et al. 2016).

13.7.2 Genetically Encoded Biosensor

Genetically encoded biosensors relying on transcriptional regulators are being used for ultra-high-throughput screening of mutant overproducing amino acids (Bott and Eggeling 2017). Evidently, in *C. glutamicum*, the regulator Lrp was utilized to create biosensors for screening L-valine, L-methionine, L-isoleucine, and L-leucine overproducing mutant strains (Wang et al. 2019; Wendisch 2020). Lrp stimulates the activity of the brnFE operon at elevated intracellular concentrations of these amino acids (Wendisch et al. 2016). Further, employing fluorescence-activated cell sorting, the transcriptional factor LysG was employed as a biosensor to isolate mutants of *C. glutamicum* overproducing L-lysine (Binder et al. 2012). Similarly, ArgP of *E. coli* (regulating argO transcription, encoding for an arginine exporter) was found as an excellent pick involved in the overproduction of L-arginine (Ginesy et al. 2015). Moreover, biosensors utilizing precursors (shikimate), core metabolite (cAMP), and reduced redox factor (NADPH) were designed to boost amino acid output (Wendisch 2020).

13.7.3 Flux Enforcement

Metabolic engineering by flux enforcement strategy integrates the biosynthetic pathway into the metabolic pathway essential for growth. The first flux enforcement was achieved in *E. coli* by coupling the production of 4-hydroxy-L-isoleucine via 2-oxoglutarate-dependent hydroxylase to growth by deleting gene sucA (Wendisch 2020). Further, Kind et al. (2013) employed flux coupling to enhance L-lysine biosynthesis by a succinyl-CoA synthetase-negative (ΔsucCD) strain of *C. glutamicum*, through specific disruption of the TCA cycle. Similarly, flux enforcement was employed to enhance *C. glutamicum* glutarate synthesis by coupling transamination processes catalyzed by PutA and GabT during glutarate biosynthesis with growth condition deficient of L-glutamate dehydrogenase (Gdh) (Pérez-García et al. 2018). Further, the efficacy of flux enforcement was higher when integrating gdh deletion at a single coupling site, rather than coupling it with two transaminases (PutA and GabT) for glutarate biosynthesis (Haupka et al. 2020). This emphasizes the significance of pathway-specific flux enforcement stoichiometry for microbial strain development to enhance amino acid biosynthesis.

13.8 Conclusion and Future Prospective

Amino acids are vital biomolecules that have applications in food and feed supplements, cosmetics, and medicines, along with serving as precursors for antibiotics and biofuels. Microbial-mediated commercial production of amino acids is expanding enormously to fulfill the growing needs and diverse applications. Natural overproducing microbial strains are scarce in nature and rarely produce the desired level of amino acids. Thus, with recent developments in metabolic and genetic engineering, potent tools for microbial strain improvement and fermentation processes have evolved tremendously. These innovative strategies integrate whole-genome sequencing, systems metabolic engineering, and bioinformatics techniques that undoubtedly develop high-yielding strains with enhanced amino acid synthesis capability. Further, the development of high-throughput, precise, time-efficient, and universal screening methods to identify overproducing strains within a diverse mutant library has guaranteed improvement in amino acid production. Moreover, delving into potential targets indirectly linked to amino acid production and utilization can deepen our comprehension of the underlying mechanisms driving amino acid overproduction. Incorporating these novel targets with existing microbial strains has the potential to revolutionize amino acid fermentation, reducing costs, enhancing production efficiency, improving quality, and accelerating industrial growth.

Funding Ms. Ankita Agrawal adores the funding of the DST, Govt. of India, New Delhi, through DST/INSPIRE FELLOWSHIP/2020/IF200208.

Conflict of Interest None.

Acknowledgement The authors wish to express gratitude to the Department of BT & BI, Sambalpur University, for fostering a conducive scientific environment.

References

Baumgart M, Unthan S, Kloß R, Radek A, Polen T, Tenhaef N, Müller MF, Küberl A, Siebert D, Brühl N, Marin K (2018) *Corynebacterium glutamicum* chassis C1*: building and testing a novel platform host for synthetic biology and industrial biotechnology. ACS Synth Biol 7(1):132–144

Becker J, Wittmann C (2012) Systems and synthetic metabolic engineering for amino acid production–the heartbeat of industrial strain development. Curr Opin Biotechnol 23(5):718–726

Bikard D, Jiang W, Samai P, Hochschild A, Zhang F, Marraffini LA (2013) Programmable repression and activation of bacterial gene expression using an engineered CRISPR-Cas system. Nucleic Acids Res 41(15):7429–7437

Binder S, Schendzielorz G, Stäbler N, Krumbach K, Hoffmann K, Bott M, Eggeling L (2012) A high-throughput approach to identify genomic variants of bacterial metabolite producers at the single-cell level. Genome Biol 13(5):1–12

Bott M, Eggeling L (2017) Novel technologies for optimal strain breeding. Amino acid fermentation. Adv Biochem Eng Biotechnol 159:227–254

Brautaset T, Jakobsen ØM, Josefsen KD, Flickinger MC, Ellingsen TE (2007) Bacillus methanolicus: a candidate for industrial production of amino acids from methanol at 50°C. Appl Microbiol Biotechnol 74:22–34

Breuer M, Ditrich K, Habicher T, Hauer B, Keßeler M, Stürmer R, Zelinski T (2004) Industrial methods for the production of optically active intermediates. Angew Chem Int Ed Engl 43(7):788–824

Castañeda-Ovando A, Segovia-Cruz JA, Flores-Aguilar JF, Rodríguez-Serrano GM, Salazar-Pereda V, Ramírez-Godínez J, Contreras-López E, Jaimez-Ordaz J, González-Olivares LG (2019) Serine-enriched minimal medium enhances conversion of selenium into selenocysteine by Streptococcus thermophilus. J Dairy Sci 102(8):6781–6789

Cheng K, Zhang C (2014) Screening, mutagenesis of *Brevibacterium flavum* for the enhancement of L-valine production. In: Proceedings of the 2012 international conference on applied biotechnology (ICAB 2012), vol 1(249), pp 285–289

Choi BH, Coloff JL (2019) The diverse functions of non-essential amino acids in cancer. Cancers 11(5):675–691

Cleto S, Jensen JV, Wendisch VF, Lu TK (2016) *Corynebacterium glutamicum* metabolic engineering with CRISPR interference (CRISPRi). ACS Synth Biol 5(5):375–385

Cordente AG, Schmidt S, Beltran G, Torija MJ, Curtin CD (2019) Harnessing yeast metabolism of aromatic amino acids for fermented beverage bioflavouring and bioproduction. Appl Microbiol Biotechnol 103(11):4325–4336

Csonka LN, Leisinger T (2007) Biosynthesis of proline. EcoSal Plus 2(2):10–1128

D'Este M, Alverado-Morales M, Angelidaki I (2017) Amino acids production focusing on fermentation technologies—a review. Biotechnol Adv 36(1):14–25

Deng W, Wang Y, Zhang S, Gupta KM, Hülsey MJ, Asakura H, Liu L, Han Y, Karp EM, Beckham GT, Dyson PJ (2018) Catalytic amino acid production from biomass-derived intermediates. PNAS 115(20):5093–5098

Derbikov DD, Novikov AD, Gubanova TA, Tarutina MG, Gvilava IT, Bubnov DM, Yanenko AS (2017) Aspartic acid synthesis by *Escherichia coli* strains with deleted fumarase genes as biocatalysts. Appl Biochem Microbiol 53(9):859–866

Dong X, Quinn PJ, Wang X (2011) Metabolic engineering of *Escherichia coli* and *Corynebacterium glutamicum* for the production of L-threonine. Biotechnol Adv 29(1):11–23

Du H, Zhao Y, Wu F, Ouyang P, Chen J, Jiang X, Ye J, Chen GQ (2020) Engineering *Halomonas bluephagenesis* for L-threonine production. Metab Eng 60:119–127

Duan M, Chen S, Liu X, Liu J, Zhu D (2023) The application of *Corynebacterium glutamicum* in L-threonine biosynthesis. Fermentation 9(9):822–830

Ecker J, Raab T, Harasek M (2012) Nanofiltration as key technology for the separation of LA and AA. J Membr Sci 389:389–398

El-Hersh MS, Saber WI, El-Fadaly HA, Mahmoud MK (2016) Lysine and glutamic acids as the end products of multi-response of optimized fermented medium by *Mucor mucedo* KP736529. PJBS 19(7):279–288

Gaston MA, Zhang L, Green-Church KB, Krzycki JA (2011) The complete biosynthesis of the genetically encoded amino acid pyrrolysine from lysine. Nature 471(7340):647–650

Ginesy M, Belotserkovsky J, Enman J, Isaksson L, Rova U (2015) Metabolic engineering of *Escherichia coli* for enhanced arginine biosynthesis. Microb Cell Factories 14(29):1–11

Haupka C, Delépine B, Irla M, Heux S, Wendisch VF (2020) Flux enforcement for fermentative production of 5-aminovalerate and glutarate by *Corynebacterium glutamicum*. Catalysts 10(9):1065–1079

Hayashi M, Tabata K (2013) Metabolic engineering for L-glutamine overproduction by using DNA gyrase mutations in *Escherichia coli*. Appl Environ Microbiol 79(9):3033–3039

Henke NA, Wiebe D, Pérez-García F, Peters-Wendisch P, Wendisch VF (2018) Coproduction of cell-bound and secreted value-added compounds: simultaneous production of carotenoids and amino acids by *Corynebacterium glutamicum*. Bioresour Technol 247:744–752

Hermann T (2003) Industrial production of amino acids by *coryneform* bacteria. J Biotechnol 104(1–3):155–172

Huang B, Qin P, Xu Z, Zhu R, Meng Y (2011) Effects of $CaCl_2$ on viscosity of culture broth, and on activities of enzymes around the 2-oxoglutarate branch, in *Bacillus subtilis* CGMCC 2108 producing poly-(γ-glutamic acid). Bioresour Technol 102(3):3595–3598

Ikeda M (2003) Amino acid production processes. In: Faurie R et al (eds) Microbial production of l-amino acid, Advances in biochemical engineering/biotechnology, vol 79. Springer, Berlin, pp 1–35

Ikeda M (2006) Towards bacterial strains overproducing L-tryptophan and other aromatics by metabolic engineering. Appl Microbiol Biotechnol 69(6):615–626

Ikeda M, Takeno S (2013) Amino acid production by *Corynebacterium glutamicum*. In: Yukawa H, Inui M (eds) *Corynebacterium glutamicum*, Microbiology monographs, vol 23. Springer, Berlin, pp 107–147

Ikeda M, Takeno S (2020) Recent advances in amino acid production. In: Inui M, Toyoda K (eds) *Corynebacterium glutamicum*, Microbiology monographs, vol 23. Springer, Cham, pp 175–226

Ikeda M, Ohnishi J, Mitsuhashi S (2005) Genome breeding of an amino acid-producing *Corynebacterium glutamicum* mutant. In: Barredo JL (ed) Microbial processes and products, Methods in biotechnology, vol 18. Humana Press, pp 179–190

Irshad S, Hashmi AS, Javed MM, Babar ME, Awan AR, Anjum AA (2015) Optimization of physico-chemical parameters for hyper-production of lysine by mutated strain of *Brevibacterium flavum*. JAPS 25(3):784–791

Ivanov K, Stoimenova A, Obreshkova D, Saso L (2013) Biotechnology in the production of pharmaceutical industry ingredients: amino acids. Biotechnol Biotechnol Equip 27(2):3620–3626

Juminaga D, Baidoo EE, Redding-Johanson AM, Batth TS, Burd H, Mukhopadhyay A, Petzold CJ, Keasling JD (2012) Modular engineering of L-tyrosine production in *Escherichia coli*. Appl Environ Microbiol 78(1):89–98

Kamilari E, Stanton C, Reen FJ, Ross RP (2023) Uncovering the biotechnological importance of *Geotrichum candidum*. Foods 12(6):1124–1148

Karaffa L, Kubicek CP (2019) Citric acid and itaconic acid accumulation: variations of the same story? Appl Microbiol Biotechnol 103(7):2889–2902

Karbalaei M, Rezaee SA, Farsiani H (2020) *Pichia pastoris*: a highly successful expression system for optimal synthesis of heterologous proteins. J Cell Physiol 235(9):5867–5881

Kind S, Becker J, Wittmann C (2013) Increased lysine production by flux coupling of the tricarboxylic acid cycle and the lysine biosynthetic pathway-metabolic engineering of the availability of succinyl-CoA in *Corynebacterium glutamicum*. Metab Eng 15:184–195

Kramer RM, Li C, Carter DC, Stone MO, Naik RR (2004) Engineered protein cages for nanomaterial synthesis. J Am Chem Soc 126(41):13282–13286

Krzycki JA (2013) The path of lysine to pyrrolysine. Curr Opin Chem Biol 17(4):619–625

Kulis-Horn RK, Persicke M, Kalinowski J (2014) Histidine biosynthesis, its regulation and biotechnological application in *Corynebacterium glutamicum*. Microb Biotechnol 7(1):5–25

Kumagai H (2000) Microbial production of amino acids in Japan. Adv Biochem Eng Biotechnol 69:71–85

Lee SY, Kim HU (2015) Systems strategies for developing industrial microbial strains. Nat Biotechnol 33(10):1061–1072

Lee JH, Wendisch VF (2017) Production of amino acids–genetic and metabolic engineering approaches. Bioresour Technol 245:1575–1587

Leßmeier L, Pfeifenschneider J, Carnicer M, Heux S, Portais JC, Wendisch VF (2015) Production of carbon-13-labeled cadaverine by engineered *Corynebacterium glutamicum* using carbon-13-labeled methanol as co-substrate. Appl Microbiol Biotechnol 99:10163–10176

Leuchtenberger W, Huthmacher K, Drauz K (2005) Biotechnological production of amino acids and derivatives: current status and prospects. Appl Microbiol Biotechnol 69:1–8

Li Y, Cong H, Liu B, Song J, Sun X, Zhang J, Yang Q (2016) Metabolic engineering of *Corynebacterium glutamicum* for methionine production by removing feedback inhibition and increasing NADPH level. Antonie Van Leeuwenhoek 109:1185–1197

Litsanov B, Brocker M, Bott M (2013) Glycerol as a substrate for aerobic succinate production in minimal medium with *Corynebacterium glutamicum*. Microb Biotechnol 6(2):189–195

Liu J, Xu JZ, Rao ZM, Zhang WG (2022) Industrial production of L-lysine in *Corynebacterium glutamicum*: progress and prospects. Microbiol Res 262:127101

Ma Q, Zhang Q, Xu Q, Zhang C, Li Y, Fan X, Xie X, Chen N (2017) Systems metabolic engineering strategies for the production of amino acids. Synth Syst Biotechnol 2(2):87–96

Mahmood ZA (2015) Microbial amino acids production. In: Microbial biotechnology. CRC Press, pp 202–227

Nadal-Rey G, McClure DD, Kavanagh JM, Cornelissen S, Fletcher DF, Gernaey KV (2021) Understanding gradients in industrial bioreactors. Biotechnol Adv 46:107660

Nakamori S (2017) Early history of the breeding of amino acid-producing strains. In: Yokoto A, Ikeda M (eds) Amino acid fermentation, Advances in biochemical engineering/biotechnology, vol 159. Springer, Tokyo, pp 35–53

Nishio Y, Ogishima S, Ichikawa M, Yamada Y, Usuda Y, Masuda T, Tanaka H (2013) Analysis of L-glutamic acid fermentation by using a dynamic metabolic simulation model of *Escherichia coli*. BMC Syst Biol 7(92):1–12

Odoni DI, Tamayo-Ramos JA, Sloothaak J, van Heck RG, Dos Santos VA, de Graaff LH, Suarez-Diez M, Schaap PJ (2017) Comparative proteomics of *Rhizopus delemar* ATCC 20344 unravels the role of amino acid catabolism in fumarate accumulation. PeerJ 5:e3133

Okafor N, Okeke BC (2017) Modern industrial microbiology and biotechnology, 2nd edn. CRC Press, pp 1–488

Oldiges M, Eikmanns BJ, Blombach B (2014) Application of metabolic engineering for the biotechnological production of L-valine. Appl Microb Biotechnol 98:5859–5870

Paloyan AM, Melkonyan LH, Avetisova GY (2022) Microbial approaches for amino acids production. In: Microbial syntrophy-mediated eco-enterprising. Academic Press, pp 177–208

Pandey AK, Pandey K, Singh LK (2020) Microbial production and applications of L-lysine. In: Mishra P, Mishra RR, Adetunji CO (eds) Innovations in food technology. Springer, Singapore, pp 211–229

Park JH, Lee SY (2008) Towards systems metabolic engineering of microorganisms for amino acid production. Curr Opin Biotechnol 19(5):454–460

Park JH, Lee KH, Kim TY, Lee SY (2007) Metabolic engineering of *Escherichia coli* for the production of L-valine based on transcriptome analysis and in silico gene knockout simulation. Proc Natl Acad Sci 104(19):7797–7802

Park SH, Kim HU, Kim TY, Park JS, Kim SS, Lee SY (2014) Metabolic engineering of *Corynebacterium glutamicum* for L-arginine production. Nat Commun 5(1):4618–4626

Patnaik R, Zolandz RR, Green DA, Kraynie DF (2008) L-tyrosine production by recombinant *Escherichia coli*: fermentation optimization and recovery. Biotechnol Bioeng 99(4):741–752

Pérez-García F, Wendisch VF (2018) Transport and metabolic engineering of the cell factory*C orynebacterium glutamicum*. FEMS Microbiol Lett 365(16):fny166

Pérez-García F, Jorge JM, Dreyszas A, Risse JM, Wendisch VF (2018) Efficient production of the dicarboxylic acid glutarate by *Corynebacterium glutamicum via* a novel synthetic pathway. Front Microbiol 9(2589):1–13

Pollegioni L, Molla G, Sacchi S, Rosini E, Verga R, Pilone MS (2008) Properties and applications of microbial D-amino acid oxidases: current state and perspectives. Appl Microbiol Biotechnol 78:1–6

Qin J, Zhou YJ, Krivoruchko A, Huang M, Liu L, Khoomrung S, Siewers V, Jiang B, Nielsen J (2015) Modular pathway rewiring of *Saccharomyces cerevisiae* enables high-level production of L-ornithine. Nat Commun 6(8224):1–11

Sanchez S, Rodríguez-Sanoja R, Ramos A, Demain AL (2018) Our microbes not only produce antibiotics, they also overproduce amino acids. J Antibiot 71(1):26–36

Sereewatthanawut I, Prapintip S, Watchiraruji K, Goto M, Sasaki M, Shotipruk A (2008) Extraction of protein and amino acids from deoiled rice bran by subcritical water hydrolysis. Bioresour Technol 99(3):555–561

Sgobba E, Stumpf AK, Vortmann M, Jagmann N, Krehenbrink M, Dirks-Hofmeister ME, Moerschbacher B, Philipp B, Wendisch VF (2018) Synthetic *Escherichia coli-Corynebacterium glutamicum* consortia for l-lysine production from starch and sucrose. Bioresour Technol 260:302–310

Shim J, Shin Y, Lee I, Kim SY (2017) L-methionine production. In: Amino acid fermentation, Advances in biochemical engineering/biotechnology, vol 159. Springer, Tokyo, pp 153–177

Shimizu K, Matsuoka Y (2022) Feedback regulation and coordination of the main metabolism for bacterial growth and metabolic engineering for amino acid fermentation. Biotechnol Adv 55:107887

Stolz M, Peters-Wendisch P, Etterich H, Gerharz T, Faurie R, Sahm H, Fersterra H, Eggeling L (2007) Reduced folate supply as a key to enhanced L-serine production by *Corynebacterium glutamicum*. Appl Environ Microbiol 73(3):750–755

Takors R, Bathe B, Rieping M, Hans S, Kelle R, Huthmacher K (2007) Systems biology for industrial strains and fermentation processes-example: amino acids. J Biotechnol 129(2):181–190

Tonouchi N, Ito H (2016) Present global situation of amino acids in industry. In: Amino acid fermentation, Advances in biochemical engineeering/biotechnology, vol 159. Springer, Tokyo, pp 3–14

Tuo J, Nawab S, Ma X, Huo YX (2023) Recent advances in screening amino acid overproducers. Eng Microbiol 3(1):100066

Ul-Haq I, Ali S (2002) Microbiological transformation of L-tyrosine to 3, 4-dihydroxyphenyl L-alanine (L-dopa) by a mutant strain of *Aspergillus oryzae* UV-7. Curr Microbiol 45(2):88–93

Wang YY, Xu JZ, Zhang WG (2019) Metabolic engineering of l-leucine production in *Escherichia coli* and *Corynebacterium glutamicum*: a review. Crit Rev Biotechnol 39(5):633–647

Wendisch VF (2020) Metabolic engineering advances and prospects for amino acid production. Metab Eng 58:17–34

Wendisch VF, Bott M, Eikmanns BJ (2006) Metabolic engineering of *Escherichia coli* and *Corynebacterium glutamicum* for biotechnological production of organic acids and amino acids. Curr Opin Microbiol 9(3):268–274

Wendisch VF, Jorge JM, Pérez-García F, Sgobba E (2016) Updates on industrial production of amino acids using *Corynebacterium glutamicum*. World J Microbiol Biotechnol 32(6):105–116

Westbrook AW, Ren X, Moo-Young M, Chou CP (2018) Metabolic engineering of *Bacillus subtilis* for l-valine overproduction. Biotechnol Bioeng 115(11):2778–2792

Wu G (2009) Amino acids: metabolism, functions, and nutrition. Amino Acids 37(1):1–17

Yamamoto S, Gunji W, Suzuki H, Toda H, Suda M, Jojima T, Inui M, Yukawa H (2012) Overexpression of genes encoding glycolytic enzymes in *Corynebacterium glutamicum* enhances glucose metabolism and alanine production under oxygen deprivation conditions. Appl Environ Microbiol 78(12):4447–4457

Yamamoto K, Tsuchisaka A, Yukawa H (2016) Branched-chain amino acids. In: Amino acid fermentation, Advances in biocehmical engineering/biotechnology, vol 159. Springer, Tokyo, pp 103–128

Yin X, Zhou Y, Yang H, Liao Y, Ma T, Wang F (2023) Enhanced selenocysteine biosynthesis for seleno-methylselenocysteine production in *Bacillus subtilis*. Appl Microbiol Biotechnol 107(9):2843–2854

Zhang X, Jantama K, Moore JC, Shanmugam KT, Ingram LO (2007) Production of L-alanine by metabolically engineered *Escherichia coli*. Appl Microbiol Biotechnol 77:355–366

Zhang F, Zhao S, Yan W, Xia Y, Chen X, Wang W, Zhang J, Gao C, Peng C, Yan F, Zhao H (2016) Branched chain amino acids cause liver injury in obese/diabetic mice by promoting adipocyte lipolysis and inhibiting hepatic autophagy. EBioMedicine 13:157–167

Zhang B, Ren L, Yu M, Zhou Y, Ye B (2017) Enhanced l-ornithine production by systematic manipulation of lornithine metabolism in engineered *Corynebacterium glutamicum* S9114. Bioresour Technol 250:60–68

Zhang X, Xu G, Shi J, Koffas MA, Xu Z (2018) Microbial production of l-serine from renewable feedstocks. Trends Biotechnol 36(7):700–712

Zhang G, Ren X, Liang X, Wang Y, Feng D, Zhang Y, Xian M, Zou H (2021) Improving the microbial production of amino acids: from conventional approaches to recent trends. Biotechnol Bioprocess Eng 6:708–727

Zuchowski R, Schito S, Neuheuser F, Menke P, Berger D, Hollmann N, Gujar S, Sundermeyer L, Mack C, Wirtz A, Weiergräber OH (2023) Discovery of novel amino acid production traits by evolution of synthetic co-cultures. Microb Cell Factories 22(71):1–22

Zuend SJ, Coughlin MP, Lalonde MP, Jacobsen EN (2009) Scaleable catalytic asymmetric Strecker syntheses of unnatural α-amino acids. Nature 461(7266):968–970

Chapter 14
Microbial Production of Water-Soluble Vitamins

V. Barghavi, Lopita Swain, Siddhi Gurudeo, and Soumi Sadhu

Abstract Water-soluble vitamins (vitamin B complex and C) are essential coenzymes for various metabolic activities that are necessary for the body to function properly. Our body cannot store these vitamins; thus, frequent dietary intake is required to restore them. A deficiency in these vitamins can result in several illnesses that can severely impair daily functioning or even lead to fatal diseases. Most vitamins can be obtained in sufficient amounts through a diet comprising whole grains, vegetables, fruits, and lean protein-rich foods. People with specific medical conditions, dietary limitations, or poor absorption might need to take supplements. Besides, certain microorganisms can prevent the absorption of vitamins produced by the gut microflora. Genetic engineering, a valuable tool, is utilized to fight against malnutrition-related deficiency illnesses. Large-scale production of vitamins is achieved through genetic modification of microbes. Supplementing conventional chemical synthesis techniques, microbial generation of vitamins, has shown great promise, especially, for those vitamins that are costly or difficult to produce chemically. The food and pharmaceutical industries stand to gain a great deal from the ability to synthesize vitamins in an economical, sustainable, and highly specialized manner. They address micronutrient deficiencies and enhance general health and well-being by producing targeted supplements and fortified foods.

Keywords Water-soluble vitamins · Microbial production · Genetic engineering · Deficiency illnesses · Gut bacteria · Industrial applications

V. Barghavi · L. Swain · S. Gurudeo · S. Sadhu (✉)
Department of Life Sciences, Sharda School of Basic Sciences and Research, Sharda University, Greater Noida, UP, India
e-mail: soumi.sadhu@sharda.ac.in

V. Kothari et al. (eds.), *Microbial Products for Health and Nutrition*,
https://doi.org/10.1007/978-981-97-4235-6_14

14.1 Introduction

Nine water-soluble vitamins (vitamin B complex and vitamin C) are essential coenzymes for various metabolic activities that are necessary for the body to function properly (Fig. 14.1). The history of vitamins dates to the early 1800s, when researchers started to discover that some foods might either prevent or treat specific diseases. It was found that sailors who consumed citrus fruits had a lower risk of scurvy, a condition marked by skin sores and bleeding gums. The word "vitamin" was first used to refer to these vital nutrients in the early 1900s by Polish scientist Casimir Funk. In 1912, Umetaro Suzuki, a Japanese chemist, made the first isolation of vitamin B1, marking a significant discovery in the field of vitamins. Our body does not store vitamins B and C, and hence, it needs to be replenished regularly through dietary intake. These B complex vitamins and vitamin C play crucial role in several metabolic functions in the human body as compiled in Fig. 14.2. Deficiency in these vitamins can result in several illnesses and can prove fatal. Most vitamins can be obtained in sufficient amounts by eating a healthy balanced diet comprising whole grains, fruits, vegetables, and lean protein sources. However, individuals with certain health conditions, dietary restrictions, or poor absorption may require supplementation to meet their daily vitamin requirements. This chapter on vitamins is primarily focused on the different types of vitamin-producing microorganisms and the genetic modifications of these microbes for large-scale production. Current market demands are driving the commercial production of these vitamins using biotechnology or genetic engineering as a tool to combat malnutrition-related deficiency illnesses. The sources of water-soluble vitamins, microbial sources, and large-scale genetic modifications of microbes for increased yield of vitamins are listed in Table 14.1.

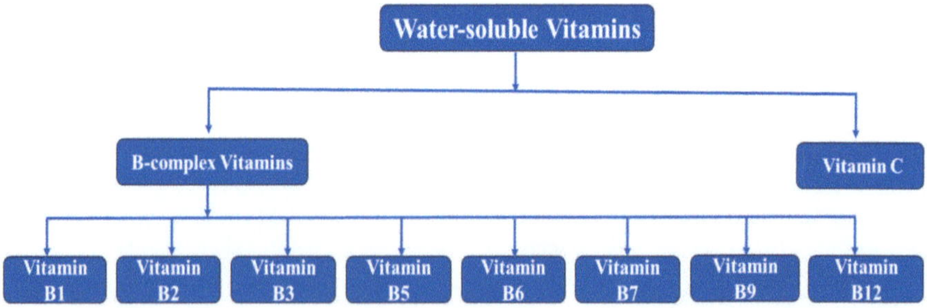

Fig. 14.1 Various water-soluble vitamins. Seven B vitamins and vitamin C represented in a tabular manner

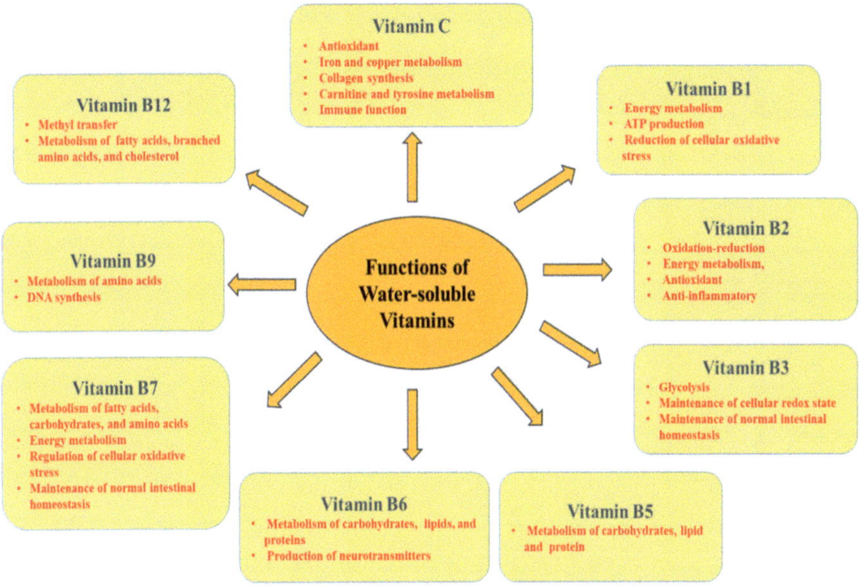

Fig. 14.2 Various functions of water-soluble vitamins. Water-soluble vitamins are involved in various metabolic activities that are necessary for the body to function properly viz., metabolism of carbohydrates, amino acids, and fats

14.2 Vitamin B1

Vitamin B1 or thiamine chloride has five phosphate derivatives namely thiamine monophosphate (ThMP), thiamine diphosphate (ThDP), thiamine triphosphate (ThTP), adenosine thiamine diphosphate (AThDP), and adenosine thiamine triphosphate (AThTP). Thiamine plays a role of coenzyme in several metabolic functions, chiefly those that involve the synthesis of energy from carbohydrates. The main source of energy is obtained by converting glucose into ATP (adenosine triphosphate). To keep nervous system functioning properly, thiamine is necessary. It plays a role in the creation of myelin, which is essential for the healthy operation of neurons and neurotransmitters. It also possesses antioxidant qualities, which aid in defending cells against oxidative stress brought on by free radicals. These mechanisms that produce energy would be disturbed in the absence of sufficient amounts of vitamin B1, which might result in health issues. Wernicke-Korsakoff Syndrome (WKS), Beriberi, and Polyneuritis are the illnesses caused by low vitamin B1 levels (Ledesma-Amaro et al. 2013). Nuts, legumes, whole grains, lean pork, and cereals are sources of thiamine. Although found in many foods, its concentration is frequently low because it is degraded during cooking.

Table 14.1 Water-soluble vitamins, their natural food sources, and microbial production of these vitamins

S. no.	Water-soluble vitamins and their structure	Natural sources of vitamins	Microbes producing vitamins	Genetically modified microbes producing vitamins	References
1.	Vitamin B1 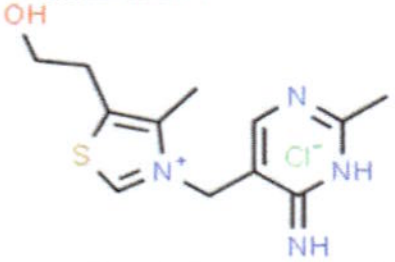	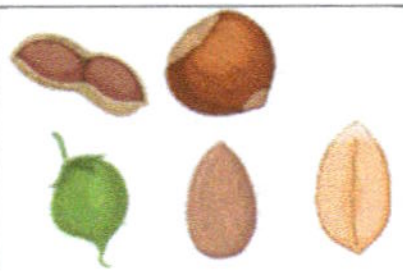	*Prevotella*, *Actinobacteria*, *Fusobacterium varium*, *Bacteroides fragilis*, and *Clostridium*	Overexpression of thiA, nmtA, and thiP in *Aspergillus oryzae*	Wan et al. (2022), Wang et al. (2021)
2.	Vitamin B2 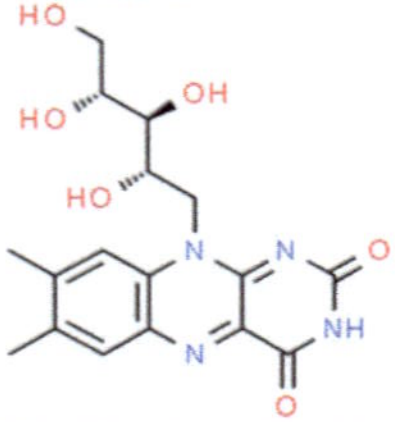		*Eremothecium ashbyii*, *Ashbya gossypii*, *Candida famata*, *Candida flaeri*, *Bacillus subtilis*, and *Corynebacterium ammoniagenes*	De-regulation of ribA gene and purR gene in *B. subtilis*	Ledesma-Amaro et al. (2013), Shi et al. (2014)
3.	Vitamin B3 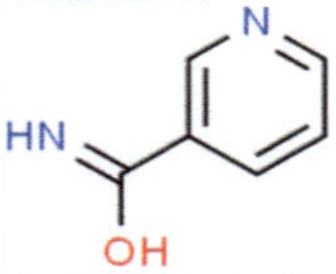		*Prevotella copri*, *Bifidobacterium infantis*, *Clostridium difficile*, *F. varium*, *H. pylori*, and *B. fragilis*	Overexpression of nitrilase in *Rhodococcus rhodochrous*	Deguchi et al. (1985), Nagasawa and Yamada (1989)

4.	Vitamin B5 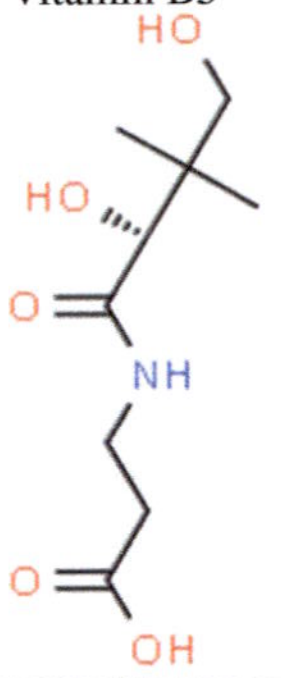		*Salmonella typhimurium* and *E. coli*	Overexpression of glyA, panBCDE, ilvBHCD, and serA in *B. subtilis*	Wan et al. (2022), Guo et al. (2023)
5.	Vitamin B6 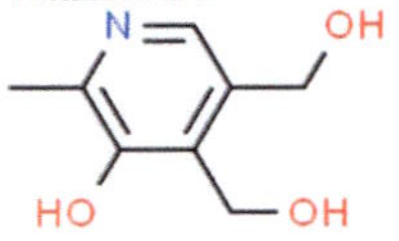		*Bifidobacterium longum* and *Collinsella aerofaciens*, *H. pylori*, and *B. fragilis* and *P. copri*	Overexpression of pdxA genes in *E. coli* and pdxJ genes in *S. meliloti*	Wan et al. (2022), Commichau et al. (2015)
6.	Vitamin B7 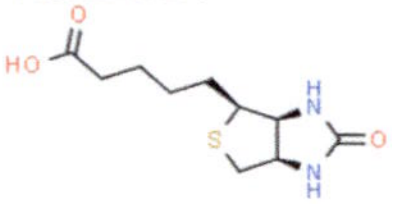		*Bacteroides fragilis*, *Prevotella copri*, *Fusobacterium varium*, and *Campylobacter coli*	Recombinants of *B. sphaericus*, *Serratia marcescens*, and *Sphingomonas* sp.	Hossain et al. (2022), Chandel et al. (2022)

(continued)

Table 14.1 (continued)

S. no.	Water-soluble vitamins and their structure	Natural sources of vitamins	Microbes producing vitamins	Genetically modified microbes producing vitamins	References
7.	Vitamin B9 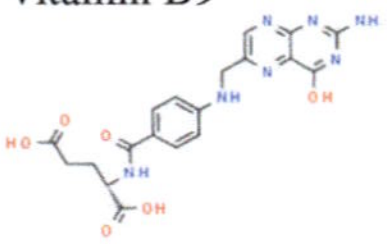		*Bacteroides fragilis*, *Prevotella copri*, *Clostridium difficile*, *Lactobacillus plantarum*, *L. reuteri*, *L. delbrueckii* ssp. *bulgaricus*, *Streptococcus thermophilus*, *Bifidobacterium* spp., *Salmonella enterica*, and *Fusobacterium varium*	Overexpression of the folate operon in *Ketogulonicigenium vulgare*	Hossain et al. (2022), Cai et al. (2012)
8.	Vitamin B12 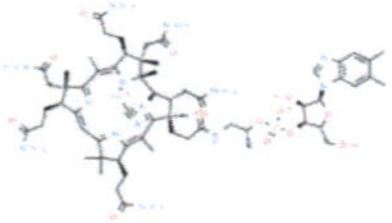		*Bacteroides fragilis*, *Faecalibacterium prausnitzii*, *L. reuteri*, *Clostridium difficile*, *Pseudomonas denitrificans*, and *Propionibacterium freudenreichii*, *Prevotella copri*, *Ruminococcus lactaris*, *Bifidobacterium animalis*, *B. infantis*, *B.longum*, *Fusobacterium varium*, *Bacillus megaterium*, and *Enterococcus faecium*	Co-culturing *P. freudenreichii* and *Ralstonia eutropha*	Wan et al. (2022), Miyano et al. (2000)
9.	Vitamin C 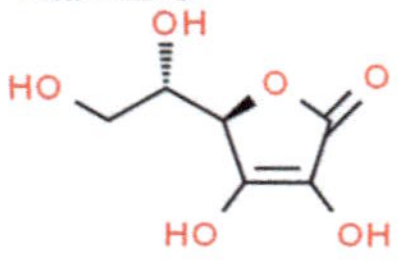		*Gluconobacter oxydans*	Overexpression of L-galactose dehydrogenase and D-arabinono-1,4-lactone oxidase in budding yeast cells	Ledesma-Amaro et al. (2013), Sauer et al. (2004)

Structure, natural sources, microbial sources, and genetic modification of microbes for scaling up industrial production of water-soluble vitamins are listed in the table

14.2.1 Microbes Producing Vitamin B1

The large intestine largely absorbs vitamin B1 obtained from diet, whereas the small intestine absorbs vitamin B1 generated by intestinal microorganisms. Free thiamine cannot be stored by humans, and phosphorylated thiamine is found in very minute amounts in cells. The two most prevalent bacterial phyla that can synthesize TPP (thiamine pyrophosphate) are *Bacteroidetes* and *Fusobacteria.* Certain gut microbiota, including *Prevotella*, *Actinobacteria*, *Fusobacterium varium*, *Bacteroides fragilis*, and *Clostridium*, can produce vitamin B1. However, there is gut microbiota that lacks the mechanism for synthesizing vitamin B1 like *Ruminococcaceae* (Wan et al. 2022). Adding thiamine to diet of cows may boost synthesis of acetic acid in the rumen and increases bacteria including *Succinivibrio*, *Ruminococcus 1*, *Pyramidobacter*, and *Ruminobacter* (Pan et al. 2017). LAB such as *Enterococcus*, *Lactobacillus*, *Streptococcus*, and *Bifidobacterium* produce acids within gastrointestinal system. It is subsequently inferred that the expansion of these microbiota enhances thiamine absorption through control of gut pH (Wan et al. 2022).

14.2.2 Genetically Modified Microbes Producing Vitamin B1

Vitamin B1 is synthesized using chemical, genetic, and metabolic engineering techniques to meet the nutritional requirements of humans and animals. It can be chemically synthesized by condensation of thiazole and pyrimidine rings (Ledesma-Amaro et al. 2013). Vitamin B1-producing microbes are *Escherichia coli*, *Salmonella typhimurium*, and *Bacillus subtilis*. Thiamine pyrimidine moiety production in *S. typhimurium* can occur via de novo purine biosynthesis or alternate pyrimidine biosynthesis (APB) route, which operates independently of the *purF* gene (Wang et al. 2021). *Escherichia coli* has become the ideal microorganism for producing thiamine pyrophosphate with the overexpression of *thiFSGHCE* and *thiD* genes linked to Fe-S metabolism (Cardinale et al. 2017). Mutations in thiamine-related transport protein (*YuaJ* and *YkoD*) genes and TPP kinase activity (*thiN*) were created in *B. subtilis* which are components of riboswitches regulating thiamine biosynthesis pathway in bacteria (Hwang et al. 2017). In a study, *nmtA*, *thiP*, and *thiA* overexpression of *Aspergillus oryzae* quadrupled thiamine output than the yield obtained from wild-type strain (Tokui et al. 2011). A patent covering use of *B. subtilis* in metabolic engineering to accumulate high levels of vitamin B1 in media was recently developed. The microorganisms with the mutation that causes thiamine production to be deregulated were detailed in the patent with reference to *Brevibacteriaceae*, *Lactobacillaceae*, *Streptococcaceae*, and *Corynebacteriaceae* species (Ledesma-Amaro et al. 2013).

14.2.3 Industrial Applications

The industrial applications of vitamin B1 (thiamine) lie in its use in various sectors and applications, primarily in the food and pharmaceuticals. The following are some significant facets of vitamin of the industrial applications of the vitamin:

Pharmaceutical industry: Thiamine has a key function in energy generation, and thus, thiamine deficiency (TD) has been related to heart failure, neurological problems, and oxidative stress (lactic acidosis and sepsis). In RCTs (randomized controlled trials), the incidence of ICU (intensive care unit) delirium was found to be lower in patients receiving thiamine supplementation (Sedhai et al. 2021). In a quasi-experimental trial, Aminotrofic® sachets comprising vitamins B1 (thiamine) and B6 (pyridoxine), primarily consisting of amino acids, were given for patients with metastatic cancer (Johal et al. 2022).

Nutraceutical industry: As a fortification measure, vitamin B1 is frequently added to a variety of food items. This is carried out to guarantee that individuals have enough thiamine, particularly in areas where thiamine deficiency poses a risk. Foods like rice, wheat, and cereals are frequently fortified (Muthayya et al. 2012). In India, idli, a fermented breakfast dish primarily composed of black gram and rice, is extremely popular. It was demonstrated that folate, thiamine, and riboflavin contents improved after fermentation of idli batter using yeast, *Leuconostoc mesenteroides* and *Lactobacillus* ssp. (Ghosh and Chattopadhyay 2011).

Animal feed industry: For poultry and cattle, thiamine is a necessary nutrient. It is added to animal feed to support the growth and well-being of the animals. Thiamine is used by the livestock and agricultural sectors to meet the nutritional demands of animals (Pan et al. 2018).

Brewing and fermentation industry: During the fermentation process in winemaking and beer, thiamine is essential for yeast metabolism (Labuschagne and Divol 2021).

14.3 Vitamin B2

The word "riboflavin," which refers to vitamin B2, comes from its yellow hue (flavus). Riboflavin is primary constituent of the FAD (flavin adenine dinucleotide) and FMN (flavin mononucleotide) cofactors, and it is necessary for the correct operation of all flavoproteins. Riboflavin, one among B complex vitamins, is essential for sustaining general health and well-being. Protein, fat, and carbohydrate metabolism all depend on riboflavin. Riboflavin is essential for both tissue healing and cell proliferation. It helps the body's capacity to keep the neurological system, skin, and eyes healthy. Free radicals are dangerous chemicals that can damage cells and riboflavin functions as an antioxidant by aiding in their neutralization. The deficiency of this vitamin leads to skin dermatitis, retarded growth, and fatigue (Ledesma-Amaro et al. 2013). Riboflavin is abundant in milk, yogurt, and cheese. They are found in lean meats including lamb, hog, and cattle, while liver is among the organ meats that are particularly high in nutrients. Fish, especially mackerel and salmon, are good sources of riboflavin. They are also present in eggs, particularly in the yolks. Certain types of mushrooms, like shiitake mushrooms, contain riboflavin. Although they do not contain as much riboflavin as animal products, vegetables like spinach, kale, and collard greens do contain riboflavin. A little amount of riboflavin may be found in almonds and sunflower seeds. Whole grains are rich in vitamin B2 including brown rice, whole wheat bread, and fortified cereals. The addition of riboflavin to a lot of processed goods, such energy bars and cereal for breakfast, helps consumers achieve their daily needs.

14.3.1 Microbes Producing Vitamin B2

All *Bacteroidetes*, *Fusobacteria*, and *Proteobacteria* genomes were found to follow the de novo production route of vitamin B2. A thorough genome study of the production of B vitamins indicates that 256 genomes, or 65% of bacterial genomes, are thought to be capable of producing vitamin B2 (Magnúsdóttir et al. 2015). Numerous microorganisms, including Ascomycete fungus such as *Ashbya gossypii* and *Eremothecium ashbyii* and yeasts including *Candida famata* and *Candida flaeri*, and bacteria in particular, *Corynebacterium ammoniagenes* and *Bacillus subtilis* naturally produce riboflavin (Ledesma-Amaro et al. 2013). Riboflavin produced by the microorganisms of the gut is primarily absorbed in the colon. In *Lactococcus lactis*, overexpression of genes responsible for riboflavin biosynthesis may cause the bacteria to become producers of vitamin B2 rather than only consumers. It has been studied that the reduced pace of emptying gastric contents enhances the absorption of riboflavin. The pace of emptying of gastric contents was slowed down by *Lactobacillus gasseri* and *Lactobacillus reuteri* led to the improvement of vitamin B2 absorption (Indrio et al. 2011). Additionally, a different study found a negative correlation between vitamin B2 intake and the relative abundance of *Streptococcus*. Many *Streptococcus* species, including *Streptococcus pyogenes* and *Streptococcus pneumoniae*, are believed to be associated with strep infections. Getting enough vitamin B2 each day may help shield the body from strep infections (Maitiniyazi et al. 2022).

14.3.2 Genetically Modified Microbes Producing Vitamin B2

Vitamin B2 is made chemically from D-glucose using three distinct methods. A prime example of how biotechnology can use genetic and metabolic bioengineering to transform a chemical synthesis into a bioprocess with appreciable cost savings is industrial riboflavin production (Ledesma-Amaro et al. 2013).

The two most significant industrial producers of riboflavin are *B. subtilis* and *Ashbya gossypii.* The *ribABCGH* genes can be overexpressed to boost riboflavin synthesis. Following deletions and nucleotide substitutions in *rib* operon, riboflavin production was significantly increased in *B. subtilis* (Burgess et al. 2004). Riboflavin production was greater than 826.52 mg/L after de-regulation of *ribA* gene and *purR* gene in *B. subtilis* (Shi et al. 2014). A strain of *B. subtilis* employed by a Chinese pharmaceutical company for riboflavin synthesis showing resistance to a proline analog led to an elevated vitamin B2 production of 26 g/L (Schwechheimer et al. 2016). In *Ashbya gossypii*, overexpression of malate synthase gene involved in glyoxylate cycle led to 70% increased yield (Sugimoto et al. 2009). *sef1* and *imh3* overexpression in *Candida famata* was used to create a strain AF-4 which yielded around 1026 mg/L of vitamin B2 (Dmytruk et al. 2014).

14.3.3 Industrial Applications

Pharmaceutical industry: Riboflavin is employed as a precursor in manufacture of several drugs and diagnostic tests, among other pharmaceutical uses (Hobby and Deuschle 1959). Multivitamins and B complex supplements are among the dietary supplements that frequently include riboflavin (Shabbir et al. 2020).

Nutraceutical industry: To prevent vitamin B2 deficiency, riboflavin fortification of basic foods such as flour, cereals, and dairy products is beneficial (Darnton-Hill 2019). About 10 different *Lactobacillus* strains were assessed for features of growth and viability in soymilk enriched with B vitamins in a study. The results of the study demonstrated that B vitamin supplementation of soymilk increased the proliferation of all lactobacilli and accelerated production of riboflavin and folic acid. Strains, such as ATCC314 of *L. acidophilus* or FTDC 8833 as starter cultures, raised the concentration of riboflavin in soymilk (657% and 243%) in comparison with starting levels before a 24-h fermentation period (Ewe et al. 2010). In order to fortify bread, *L. fermentum* PBCC11.5 was employed, the best strain for riboflavin overproduction (Russo et al. 2014).

Medical research: It is noteworthy that vitamin-producing strains are used as probiotics, with their ability to produce vitamins being associated with certain therapeutic benefits. Reports have suggested that deficiencies in folate may be linked to IBD (irritable bowel syndrome). Vitamin-producing strains of LAB (riboflavin and folate) were tested in various IBD models due to anti-inflammatory properties of these vitamins. *Lactobacillus plantarum* CRL2130, a LAB that produces riboflavin, was able to reduce inflammation of the gut in a mouse model of colitis by inhibiting an uncontrollably high rise in pro-inflammatory cytokines. LAB strains which produce folate and riboflavin may be used as adjuvant therapy for patients with inflammatory disorders (LeBlanc et al. 2020). It has been shown in another study that vitamin B2 can be encapsulated in chitosan nanoparticles. According to the release profiles, vitamin B2 may be retained in this system for at least 5 months by using alginate/chitosan nanoparticles. Therefore, they may be added to various food and beverages to improve the stability of vitamins (Azevedo et al. 2014).

14.4 Vitamin B3

Vitamin B3 often known as niacin is a water-soluble substance. NAD abbreviated as nicotinamide adenine dinucleotide and NADP as nicotinamide adenine dinucleotide phosphate (NADP), the two vital coenzymes involved in cellular metabolism, are precursors to niacin. By taking part in redox reactions, niacin supports DNA repair processes and energy generation. Pellagra is a disorder characterized by dermatitis, diarrhea, and neurological symptoms that can be brought on by niacin deficiency. Niacin, like all B vitamins, is essential for healthy eyes, hair, skin, and nerves. Additionally, vitamin B3 supports sex and stress-associated hormones (Ledesma-Amaro et al. 2013). Tryptophan is the endogenous source, and dietary NA (nicotinic acid) and NAM (nicotinamide) are the external source for vitamin B3. NA is primarily found in plant-based meals like beans, whereas NAM is primarily found in animal-based foods. Meat sources include, hog, beef, and chicken, and fishes like Salmon and tuna are best sources of this vitamin. Almonds, peanuts, and sunflower seeds, beans and lentils, whole grains such as rice, barley, and wheat, green leafy vegetables and mushrooms contain vitamin B3. Niacin is also present in milk and eggs.

14.4.1 Microbes Producing Vitamin B3

Vitamin B3 can be produced from tryptophan by several gut bacteria species. *Prevotella copri*, *Bifidobacterium infantis*, *Clostridium difficile*, *F. varium*, *H. pylori*, and *B. fragilis* are among the bacteria that have a route for production of vitamin B3 (Deguchi et al. 1985). LAB, including *Bifidobacterium*, *Lactobacillus*, *Enterococcus*, and *Streptococcus*, alter gut pH and vitamin B3 absorption (Wan et al. 2022).

14.4.2 Genetically Modified Microbes Producing Vitamin B3

Chemical synthesis of vitamin B3 involves either the complete hydrolysis of 3-cyanopyridine or the oxidation of 5-ethyl-2-methylpyridine. Nitrilase (which produces nicotinic acid) or nitrile hydratase (which produces nicotinamide) can also be used in a bioprocess to carry out this latter transformation. *Rhodococcus rhodochrous* is a strain that can fully convert 3-cyanopyridine to nicotinic acid when nitrilase is overexpressed (Nagasawa and Yamada 1989).

14.4.3 Industrial Applications

Niacin, or vitamin B3, is important for industry because it is used in so many different fields. The food and beverage, pharmaceutical, are among those that use niacin and its derivatives.

Pharmaceutical industry: The pharmaceutical industry uses niacin because of its ability to treat diseases like hyperlipidemia and niacin deficiency. Niacin and niacinamide-containing medications are used to treat cardiovascular problems as well as other medical conditions (Ganji et al. 2003). Supplementing with nicotinamide (vitamin B3) led to the improvement of function of retina in patients with glaucoma, according to a randomized clinical trial (RCT) (Hui et al. 2020).

Nutraceutical industry: To improve the nutritional value of processed foods like cereals and flour, niacin is added (Darnton-Hill 2019).

Cosmetic and personal care industry: Several dermatological applications of nicotinamide have been reported, including the prophylaxis of non-melanoma skin cancer (NMSC), blistering diseases, acne vulgaris, and anti-aging cosmetic uses. Oral and topical nicotinamide have been utilized as preventative measures in NMSC. It inhibits UV radiation's immunosuppressive effects as well as the oxidative stress that results from it (Elgharably et al. 2023).

14.5 Vitamin B5

Pantothenic acid, or vitamin B5, is a water-soluble substance that is necessary for a variety of metabolic functions in living things. It plays a crucial role in energy metabolism as a precursor to coenzyme A (CoA) involved in synthesis of fats, cholesterol, and amino acids. Conversion of pantothenic acid to pantetheine and its further conversion into acetyl CoA is essential in metabolism of fats and carbohydrates. Vitamin B5 deficiency although rare can cause insomnia, depression, stomach, and respiratory infections (Wan et al. 2022). Numerous foods contain pantothenic acid including pork, turkey, and chicken. Whole grains like oats, wheat, and brown rice, as well as dairy products like cheese, yogurt, and milk are best sources. Legumes like split peas and lentils, broccoli, sweet potatoes, and avocado contain vitamin B5.

14.5.1 Microbes Producing Vitamin B5

Bacteria that produce vitamin B5 include *Salmonella typhimurium* and *E. coli.* Aspartate and intermediate intermediates of valine biosynthesis are used as substrates for *Escherichia coli.* Using the enzymes dihydroxy acid dehydratase and acetohydroxy acid synthase isozyme I, *Salmonella typhimurium* converts alpha-ketoisovalerate to pantothenate. Certain bacteria such as *Lactobacillus helveticus*, *E. faecalis*, and *Streptococcus* require vitamin B5 for their growth, and hence, they compete with the host. There is copious evidence suggesting that gut bacteria played a crucial role for vitamin B5 utilization, particularly during vitamin B5 shortage. The gut microbiota is altered by chemicals containing sulfonamide functional group, such as sulfathiazole and succinyl sulfathiazole (Wan et al. 2022).

14.5.2 Genetically Microbes Producing Vitamin B5

Isobutyraldehyde, formaldehyde, and cyanide are the chemical sources of pantolactone. D-Pantolactone and beta-alanine are condensed to create pantothenic acid (Ledesma-Amaro et al. 2013). Following several genetic bioengineering procedures for production of pantothenic acid, *E. coli* is a suggested and patented effective vitamin producer reaching 66 g/L output (Moriya et al. 1999). In *Corynebacterium glutamicum*, 1.75 g/L of pantothenate was produced by deleting *ilvA* gene, suppressing *ilvE* gene, and upregulating *ilvBNCD* gene (Hüser et al. 2005). *B. subtilis* is frequently used to make pantothenic acid. Overexpression of *glyA*, *panBCDE*, *ilvBHCD*, and *serA* led to maximum output of 82–86 g/L of pantothenic acid (Hohmann et al. 2017). DPA171 produced 4.1 g/L of D-pantothenic acid which is the greatest titer of *S. cerevisiae* to date by perfecting fed-batch fermentation (Guo et al. 2023).

14.5.3 Industrial Applications

Cosmetics and personal care industry: Because of its moisturizing qualities, pantothenic acid and its derivatives, such as panthenol, are frequently used in skincare and cosmetic products. Our skin gets benefited from pantothenic acid, which is derived from panthenol, to stay hydrated (Camargo et al. 2011).

Pharmaceutical industry: Vitamin B5 is useful in synthesizing coenzyme A (CoA), which is indispensable for cellular metabolism. It is employed in the pharmaceutical sector for a range of therapeutic uses (Sampedro et al. 2015). While vitamins play a crucial role in adjuvant therapy for infections, their full spectrum of immunoregulatory effects is not well understood. Over the course of its extensive research, B5 is emerging as a pioneer not only in the field of rheumatic disorders and immunological modulation, but also in the healing of wounds, potentially having significant therapeutic implications for clinical management (Gheita et al. 2020).

Animal feed industry: A vital component of animal nutrition is pantothenic acid. It is a component of animal feed that promotes livestock growth, reproduction, and general health (Alagawany et al. 2021).

14.6 Vitamin B6

Vitamin B6, also known as pyridoxine, is a coenzyme that facilitates metabolism of proteins, lipids, and carbohydrates as well as aids in producing neurotransmitters. There are three natural forms of vitamin B6: pyridoxal (PL), pyridoxine (PN), and pyridoxamine (PM). It is necessary for healthy immune and nervous system function as well as normal brain development (Muhamad et al. 2023). The phosphorylated derivatives of vitamin B6 such as pyridoxamine 5′-phosphate (PMP), pyridoxine 5′-phosphate (PNP), and pyridoxal 5′-phosphate (PLP) can be obtained from food sources. Upon consumption, they are dephosphorylated by phosphatases and converted into active form in the liver. Insufficient levels of pyridoxine have been linked to anemia and seizures (Alsaeedi et al. 2023). Vitamin B6 is found naturally in milk, eggs, turkey, pork, tuna, salmon, sunflower seeds, flax seeds, brown rice, oats, lentils, banana, avocado, potatoes, and spinach.

14.6.1 Microbes Producing Vitamin B6

Vitamin B6 is primarily obtained by humans from two sources: their diet and their normal gut microbiota. Some bacteria that can produce vitamin B6 are *Bifidobacterium longum* and *Collinsella aerofaciens*, *H. pylori*, and *B. fragilis* and *P. copri*. PLP production of gut flora utilizes pathway reliant on deoxyxylulose 5-phosphate (DXP), and PNP production uses pathway free of DXP. In addition, the pH of the intestinal lumen is directly linked to breakdown of pyridoxine, an important process for its absorption. Variations in the quantity of certain lactic acid producing bacteria such as *Enterococcus*, *Bifidobacterium*, *Lactobacillus*, and *Streptococcus* can lower pH thereby affect assimilation of this vitamin. Certain microbiota species, including *Veillonella*, *Faecalibacterium*, *Ruminococcus*, and *Lactobacillus* spp., are incapable of biosynthesizing vitamin B6, and therefore, they obtain exogenous vitamin B6 from the gastrointestinal tract (Wan et al. 2022).

14.6.2 Genetically Modified Microbes Producing Vitamin B6

The Diels-Alder reaction of oxazoles with maleic acid is one of the ways used to create pyridoxine hydrochloride in large-scale (Ledesma-Amaro et al. 2013). Besides, the best biological source for genetically engineered microbes producing this vitamin is the Gram-negative bacterium *Sinorhizobium meliloti*, which can produce 103 mg/L of vitamin B6 in 168 h, according to extensive screening investigations of several strains (Hoshino et al. 2006). Currently, there are two pathways for PLP production naturally. Increased production of vitamin B6 has resulted from genetic modification of the bacteria in both these routes.

B. subtilis have been employed for converting 4-hydroxy L-threonine (4HT) to PN. Both 4-phosphohydroxy-L-threonin (4HTP) and 4HT affect biosynthesis of threonine and isoleucine, and hence, the bacteria need to adapt its genome for 4HT conversion to PN. Thus, the altered *B. subtilis* had developed mutations in amino acid transporters leading to very less uptake of 4HT from the medium and the *hom* promoter region, thereby increasing resistance to 4HT by dysregulation of biosynthesis of threonine. Moreover, overexpression of genes such as *pdxA* (*E. coli*) and *pdxJ* (*S. meliloti*) yielded 65 mg/L from 120 mg/L 4HT (Commichau et al. 2015).

14.6.3 Industrial Applications

Vitamin B6 is a necessary substance that is involved in various biological functions. The key aspects of its industrial applications are its manufacture, applications, and usage in diverse industries. The following are important business sources regarding vitamin B6.

Pharmaceutical industry: A variety of medications and dietary supplements are produced using vitamin B6 as an active component. It is frequently found in vitamin and mineral supplements and prescribed drugs for illnesses where a vitamin B6 deficiency could be problematic. Pyridoxine hydrochloride (vitamin B6) was measured in pharmaceutical formulations using an affordable and sensitive method involving SDS, a surfactant and chlorazol black, an analytical reagent (Khudhair et al. 2019). In a case study, vitamin B6 was also administered as a replacement therapy at 150 mg/day as a treatment for neutropenia in patients for squamous cell carcinoma (Rauf et al. 2011).

Animal feed industry: To guarantee that pets, poultry, and animals receive the right nourishment, vitamin B6 is added to animal feed formulas. It supports the general health and growth of animals. Pyridoxine added to flaxseed cake in chicken feed increases the number of omega-3-enriched eggs produced (Khan 2019).

Cosmetics and personal care industry: As vitamin B6 may be beneficial to skin, some cosmetics and personal care items, like skin creams and lotions, contain this vitamin. Vitamin B6 has an intriguing therapeutic application in doxorubicin-induced alopecia. According to a study by Agostini et al., vitamin B6 shields mice from the harmful effects of the chemotherapeutic drug doxorubicin. Doses ranging from 800 to 1600 mg/kg/day were given to mice; however, more research is required to determine the ideal dosage for humans. Doxorubicin produces reactive oxygen species which can cause damage to hair follicles resulting in alopecia. Vitamin B6 is important for shielding cells from the oxidative damage caused by doxorubicin. Hence, vitamin B6 supplementation may be beneficial in addition to doxorubicin-based chemotherapy regimens for patients (D'Agostini et al. 2013).

Medical research: Vitamin B6 is used extensively in medical research to study how it affects human metabolism and health. On promonocytic lymphoma cell lines, they discovered that the vitamins B2, B6, and B9 had anticancer activity. The anti-inflammatory effect of vitamins could be attributed to various mechanisms such as nitric oxide synthesis, oxidative stress, varied PD-L1 expression, altered cytokine secretion, and angiogenesis. Compared to control groups, monoblastic leukemia cell lines secreted more IL-8 and IL-10 in response to vitamins B2, B6, and B9. Besides, important enzymes in the pathways leading to methyl metabolism require PLP as a cofactor (Stach et al. 2021).

14.7 Vitamin B7

Vitamin B7, also called as biotin, acts as a CO_2 carrier in carboxylation reactions. These carboxylation reactions are crucial for the metabolism of amino acids, lipids, and sugars (Scott 2020). It is common knowledge that biotin supports strong nails, skin, and hair. It is a common ingredient in a lot of hair and beauty products, and some people take supplements to look better. Biotin deficiency is comparatively uncommon and can be brought on by several illnesses, inadequate nutrition, or prolonged antibiotic use. Hair loss, skin rashes, fatigue, and neurological issues are possible symptoms. A wide range of foods naturally contain biotin. One of the best natural sources of biotin is found in egg yolks. Liver, especially beef liver, is an organ meat that has a high biotin content. Besides, sunflower seeds and almonds, peanuts, and walnuts are among the nuts and seeds that are good providers of this vitamin. Peas, beans, soybeans, and lentils are examples of legumes that contain biotin.

14.7.1 Microbes Producing Vitamin B7

Since microbiota, which includes bacteria and yeast, and plants alone can synthesize vitamin B7, the microbiota of human large intestine is essential for daily supplementation of this vitamin in addition to diet. *Prevotella copri*, *Bacteroides fragilis*, *Campylobacter coli*, and *Fusobacterium varium* are capable of producing vitamin B7 (Hossain et al. 2022). Bacteria that consume vitamin B7 are deficient in the pathways necessary for the biosynthesis of vitamin B7; thus, they must obtain it from the environment to sustain their metabolism. Genes involved in acquiring biotin from surroundings, for example, are present in *Lactobacillus*. Moreover, vitamin B7 may affect gut microbiome of the host. *Prevotella*, *Bifidobacteria*, *Ruminococcus*, and *Lactobacillus* are among the bacteria that require vitamin B7 to sustain their microbiological processes. These bacteria also have free biotin transporters. Therefore, low levels of vitamin B7 may affect their abundance in the gut (Wan et al. 2022).

14.7.2 Genetically Modified Microbes Producing Vitamin B7

Biotin is chemically synthesized from two precursors pimeloyl-coA and alanine by three enzymes (Ledesma-Amaro et al. 2013). In *Bacillus sphaericus* and *Serratia marcescens*, there are several overproducing mutants that are resistant to biotin analogs. Among other species, *B. subtilis*, *B. sphaericus*, and *E. coli* have been reported to use the synthetic biotin pathway. There are several strategies that has been explored to boost biotin synthesis. A recombinant plasmid of *Sphingomonas* sp. containing a portion of biotin operon and an extra copy of the biotin operon from *S. marcescens* yielded an output of 600 mg/L (Chandel et al. 2022). Some microbes, viz., *Mesorhizobium loti*, *C. glutamicum*, and *S. meliloti*, can produce excessive amounts of biotin. A yield of 110 mg/L was achieved in *Rhizobium* HK40 and *Agrobacterium by* overexpressing biotin operon of *Escherichia coli* using potent artificial *tac* promoter. Also, an altered RBS (ribosome binding site) was introduced before *bioB* to achieve this yield (Streit and Entcheva 2003). In *B. subtilis*, biotin levels of 21 mg/L were obtained through overexpression of the native biotin operon (Van Arsdell et al. 2005).

14.7.3 Industrial Applications

Pharmaceutical industry: Biotin is used in the pharmaceutical industry to produce medications and supplements. It is often included in formulations designed to address specific medical conditions, particularly those related to biotin deficiency. Four weeks of probiotic and biotin supplementation demonstrated a positive overall clinical therapy benefit in patients diagnosed with major depression (Reininghaus et al. 2020). Vitamin B7 is widely used for healthy skin, hair, and nails. Supplementing with vitamin B7 appears to be beneficial for patients suffering from seborrheic dermatitis, brittle nail syndrome, and uncombable hair syndrome. While some studies have found a neutral effect, others have also indicated benefits in healthy persons (Coerdt et al. 2021).

Animal feed industry: Biotin has been found to have improved the efficiency and physiological traits of broiler chicken exposed to oxidative stress caused by adding H_2O_2 to drinking water (Hasan Kadhim et al. 2022).

14.8 Vitamin B9

Folate or vitamin B9 is required for DNA replication and is involved in metabolism of amino acids and vitamins. To avoid B9 deficiency, folic acid is chiefly employed in dietary supplements and fortified foods. A few abnormalities in the mother such as anemia or peripheral neuropathy and the baby like congenital malformations have been related to folate deficiency. It has been demonstrated that taking folic acid supplements during pregnancy lowers the chance of neural tube abnormalities in the fetus. The maintenance of genomic integrity and normal bodily functions depends on folate absorption (Shams 2022).

Neural tube anomalies account for around 10% of the 7.9 million infants born each year with congenital disability. Alzheimer's disease, coronary heart disease, stroke, and several malignancies have been associated with folate deficiency (Zheng et al. 2022). World Health Organization recommends daily oral supplementation with folic acid and iron to mitigate the possibility of very low birth weight, anemia during pregnancy, and iron deficit. For all pregnant adult women and adolescents, folic acid supplementation should start as soon as possible (McNulty et al., 2011).

The sources of vitamin B9 include fruits such as cantaloupe, grapefruit juice, banana, raspberry, strawberry, papaya, grapefruit, honeydew melon, vegetables like avocado, peas, gram, spinach, cabbage, cauliflower, kidney beans, mustard greens, sunflower seeds, avocados, peanuts, corn, wheat germ, fish, and liver (Singh 2021).

14.8.1 Microbes Producing Vitamin B9

Bacteroides fragilis, *Prevotella copri*, *Clostridium difficile*, *Lactobacillus plantarum*, *L. reuteri*, *Streptococcus thermophilus*, *L. delbrueckii* ssp. *bulgaricus*, *Bifidobacterium* spp., *Salmonella enterica*, and *F. varium* are folate producers in the gut. Producers of high folate content are *Bifidobacterium bifidum* and *Bifidum* spp. *infantis*; producers of low folate content are *Bifidobacterium breve*, *Bifidobacterium adolescentis*, and *Bifidobacterium longum* subsp. Additionally, common Lactobacilli that produce folate include *L. sakei*, *L. plantarum*, *L. helveticus L. reuteri*, and *L. delbrueckii* (Hossain et al. 2022). Human gut microbes including *Actinobacteria* (26%), *Proteobacteria* (71%), *Fusobacteria* (79%), and *Firmicutes* (15%) have the capacity to synthesize vitamin B9 de novo (Magnúsdóttir et al. 2015). *Lactobacillus sakei* LZ217, an excellent generator of vitamin B9, has been identified from raw milk by Liu et al. (2022). *Propionibacterium* sp., widely recognized for their capacity to generate vitamin B12, can also produce rather large quantities of folate (Hugenholtz and Smid 2002).

14.8.2 Genetically Modified Microbes Producing Vitamin B9

Folic acid is chemically produced via several methodologies (Ledesma-Amaro et al. 2013). Certain vitamins, including B9 vitamins like folate, have been produced through fermentation processes involving bacteria, fungi, and other microorganisms. It is possible to genetically modify microorganisms to overproduce folates. An eightfold rise in folate levels was seen when *B. subtilis* was subjected to computer-aided flux analysis (Zhu et al. 2005). In *Lactococcus lactis*, metabolic engineering has resulted in an increase of almost three times in the levels of this vitamin (Sybesma et al. 2003). With the recent engineering of *Ketogulonicigenium vulgare* leading to the overexpression of the folate operon, greater production of folate was achieved (Cai et al. 2012). When subjected to metabolic engineering, an increase of folic acid synthesis to 6.59 mg/L from 0.04 mg/L was observed in *A. gossypii* (Serrano-Amatriain et al. 2016). *S. stipitis* was utilized in folate production in which the bioreactor yielded 3.7 ± 0.07 mg/L of folate, the greatest amount reported in literature to date when considering wild-type bacteria (Mastella et al. 2022).

14.8.3 Industrial Applications

Pharmaceutical industry: Folic acid supplementation is advised to meet the increased requirements as well as to avoid unfavorable pregnancy outcomes such as neural tube abnormalities (NTDs). Although folic acid fortification of food has increased the intake of folate, it still falls short of the recommended level for reducing the risk of non-transformed diseases (NTDs) (Lamers 2011).

Medical research: Drugs that are hydrophilic or hydrophobic can be encapsulated using folic acid (FA) hydrogel (Panja et al. 2018).

Nutraceutical industry: Adding folate to flour and other baking ingredients is a common practice. Fortification of rice with vitamins and minerals increases the nutrition among people in rice-consuming countries who suffer from micronutrient deficiencies. Combining a non-traditional starter culture (*L. amylovorus* CRL887) with chosen strains of *L. bulgaricus* CRL871 and *S. thermophilus* was found effective to create a novel yogurt producing high concentrations of folate (260 μg/L), which made the above new bio-enriched product an interesting substitute for fortification with chemical folic acid (Laiño et al. 2014). Angadharan and Nampoothiri also showed that adding *Lactobacillus lactis* to cucumber and watermelon juice was a great way to boost their folate levels (Gangadharan and Nampoothiri 2011).

14.9 Vitamin B12

Cobalamin, another name for vitamin B12, is necessary for preserving overall health. Among the B vitamins, vitamin B12 is distinct because it is the only one that contains cobalt, hence its term cobalamin. It helps convert food into energy and is essential for proper cellular function. Red blood cell formation requires vitamin B12. Megaloblastic anemia is a disorder where red blood cells grow abnormally large and become unable to function normally due to lack of vitamin B12 (Rizzo and Laganà 2020).

Vegans typically suffer from inadequate dietary intake when it comes to vitamins, and B12 malabsorption is linked to digestive disorders. Vitamin B12 binds to haptocorrin (HC) in the stomach after being liberated from food carrier proteins. The conversion of B12 to IF in the duodenum is brought about by the breakdown of HC by pancreatic proteases (Guéant et al. 2022).

Vitamin B12 is majorly acquired from animal-based foods, which makes it challenging for individuals following a strict vegetarian or vegan diet to obtain enough naturally. Meat sources include Pork, lamb, organ livers, chicken, turkey, salmon, tuna, and dairy sources such as cheese and milk contain cobalamin. Clams and oysters are abundant in vitamin B12.

14.9.1 Microbes Producing Vitamin B12

In the human gut, certain bacteria are capable of producing vitamin B12, and many require a precursor named precorrin-2. Approximately, 20% of gut bacteria are capable of producing vitamin B12, and over 80% of gut bacteria depend on B12 for metabolic processes. The microbes that produce vitamin B12 are *Bacteroides fragilis, Faecalibacterium prausnitzii, L. reuteri, Clostridium difficile, Pseudomonas denitrificans*, and *Propionibacterium freudenreichii, Prevotella copri, Ruminococcus lactaris, Bifidobacterium animalis, B. infantis, B. longum, Fusobacterium varium, Bacillus megaterium*, and *Enterococcus faecium*. Almost 30 genes are involved in cobalamin synthesis by bacteria, which can use either aerobic or anaerobic routes. *Pseudomonas denitrificans* has been studied for the aerobic pathway, and *Salmonella typhimurium, Bacillus megaterium*, and *P. shermanii* have been studied for the anaerobic pathway of vitamin B12 biosynthesis (Hossain et al. 2022).

14.9.2 Genetically Modified Microbes Producing Vitamin B12

Early cobalt insertion pathway (anerobic) was adopted by *S. enterica, Propionibacterium freudenreichii*, and *B. megaterium*, and late cobalt insertion pathway (aerobic) was adopted by *Sinorhizobium meliloti, P. denitrificans*, and *Ensifer casida* for cobalamin biosynthesis. Vitamin B12 is produced by the Gram-negative bacterium *P. denitrificans* using the aerobic biosynthesis pathway. Currently, industrial manufacturers like Sanofi and Huarong Pharmacy Corporation employ *P. denitrificans* as their primary vitamin B12 producer, even though it is not generally recognized as safe (GRAS). While official information regarding the volumetric production of the Sanofi strain is lacking, one could assume that, considering other aerobic strains, it produces approximately 200–300 mg/L. The strain developed by Sanofi was initially created by combining molecular biology methods with random mutagenesis. Rhône-Poulenc, a French corporation, amplified multiple cob genes essential in vitamin B12 production, leading to creation of high-yield strains, SBL27 and subsequently SC510. The primary producer of vitamin B12 in Europe is currently Sanofi, formerly known as Rhône-Poulenc.

It is to be noted that *P. freudenreichii* has been granted both GRAS status by the FDA (Food and Drug Administration) and QPS (qualified presumption of safety) status by the EFSA (European Food Safety Authority). In a study, it was possible to produce folate and vitamin B12 simultaneously by co-cultivating *P. freudenreichii* and *Lactiplantibacillus plantarum*. Additionally, a recent patent for the simultaneous production of vitamins D and B1 was obtained by co-fermenting a strain of *Basidiomycota* with *P. freudenreichii* (Calvillo et al. 2022). The production of cobalamin increased to 19 mg/L from 6.73 mg/L after co-culturing *P. freudenreichii* and *Ralstonia eutropha* (Miyano et al. 2000).

14.9.3 Industrial Applications

Pharmaceutical industry: B12, a highly soluble, non-toxic vitamin with an incredibly efficient absorption mechanism, is a desirable oral delivery system (Clardy et al. 2011). Drinking demineralized water treated by a RO (reverse osmosis) system has been linked to negative health impacts, according to numerous research. The consumption of RO processed water was positively correlated with vitamin B12 insufficiency, according to a cross-sectional study from western India. Positive trends were noted even though there was no statistically significant correlation found between the male gender and drinking fewer than 100 mL of milk/day and the development of vitamin B12 deficiency (Gupta et al. 2016). A case-control study determined that a vegetarian diet, dark complexion, being from a lower socioeconomic background as well as drinking RO-treated water were all independently linked to vitamin B12 deficiency in adults undergoing treatment at a tertiary care hospital in Vadodara (Patel et al. 2022). A clinical trial was conducted in the primary healthcare context to compare the effects of oral against intramuscular B12 delivery in the individuals with vitamin B12 deficit (Project OB12). The findings of this investigation should contribute to the determination of whether vitamin B12 taken orally is a viable substitute for intramuscular injection, with consideration given to quality of life (Sanz-Cuesta et al. 2012). In an RCT, two quasi-experimental trials, and an observational research, vitamin B9 (folate) was given in addition to vitamin B12 and a chemotherapeutic drug pemetrexed (Ohe et al. 2008).

Nutraceutical industry: A promising alternative therapy to avoid vitamin deficiency is a soybean-based functional food that has been biofortified with vitamins using a lactobacilli strain that can produce compounds with B12 activity (Molina et al. 2012). So, naturally (i.e., without supplementation) fortifying fermented dairy products with B vitamins is possible. *Propionibacterium* sp., a food-grade bacteria, have additional uses outside large-scale vitamin production (Hugenholtz and Smid 2002).

14.10 Vitamin C

Humans require vitamin C or ascorbic acid as a micronutrient because of its pleiotropic properties, which are associated with its electron-donating capacity. It functions as a cofactor for a range of gene-regulating and biosynthetic enzymes and is a strong antioxidant. Vitamin C can offer protection against oxidative stress by boosting capacity of the skin to eliminate oxidants while preserving its capacity to fend off infections. Vitamin C deficiency can lead to scurvy, a condition marked by skin patches, swelling in the joints, and bleeding gums (Carr and Maggini 2017).

Vitamin C is present in several foods, primarily fruits and vegetables. Citrus fruits such as oranges, strawberries, lemons, kiwi, and cantaloupes are excellent sources, and vegetables including broccoli, cauliflower, cabbage, cooked kale, and brussels sprouts contain ascorbic acid.

14.10.1 Microbes Producing Vitamin C

D-Sorbitol is selectively oxidized into L-sorbitol by *Gluconobacter oxydans*. Oxidation of this compound yields 2-keto-L-gluconic acid (2KLGA). After that, it is lactonized and enolized to produce 50% final yield of L-ascorbic acid (Ledesma-Amaro et al. 2013).

14.10.2 Genetically Modified Microbes Producing Vitamin C

Starch from wheat or corn is broken down into glucose, which then undergoes a chemical transformation to become sorbitol. Vitamin C is made from sorbitol through a sequence of chemical, biological processing followed by purification procedures. Prior to the last two decades, Reichstein method dominated the market. However, several bioconversions have made this procedure simpler (Ledesma-Amaro et al. 2013). *G. oxydans* was genetically engineered to express genes from other strains producing 130 g/L 2-KLGA (2-keto-L-gulonic acid) (Saito et al. 1997). *Erwinia herbicola* was bioengineered to express heterologous genes from *Corynebacterium*, reaching a production of 120 g/L of 2-KLGA (Chotani et al. 2000). Under static culture conditions, *Ketogulonicigenium vulgare* produced 1.37 g/L of L-AA (ascorbic acid) (Sugisawa et al. 2005). *G. oxydans*, *B. endophyticus*, and *K. vulgare* yielded 73.7 g/L of 2-KGA within 30 h (Ma et al. 2019). *Prototheca moriformis* and *Chlorella pyrenoidosa* are the two microalgae that have been shown to yield 2 g/L of AA (Running et al. 1994). Overexpression of L-galactose dehydrogenase and D-arabinono-1,4-lactone oxidase in budding yeast cells led to production of 100 mg/L of AA (Sauer et al. 2004).

14.10.3 Industrial Applications

Pharmaceutical industry: Collagen is a structural protein that is vital to connective tissues, skin, and blood vessels. Collagen synthesis requires vitamin C. Vitamin C-containing pharmaceutical formulations may be used to support the formation of collagen during wound healing and tissue repair (Boo 2022). The use of vitamin C as a treatment for COVID-19 is being explored, given its track record in combating respiratory illnesses (Lordan et al. 2021).

Cosmetics and personal care industry: Biopolymeric films made of a 1:1 mix of gelatin and cassava starch were used to evaluate the possible applications of environmentally friendly cosmetics. Natural antioxidants like vitamin C and catuaba extract were added to these films to further improve them. Films made with all vitamin C formulations and 1.5 g of catuaba extract per 100 g of polymer show promise as sustainable cosmetic product substitutes (Pereira et al. 2022).

14.11 Conclusion

The human gut bacteria can generate specific amounts of vitamins through metabolic cycles. Numerous factors, including genetics, lifestyle factors, diet, and exposure to free radicals and antibiotics, affect the synthesis of vitamins in the gut. The vitamins produced by bacteria are insufficient to meet daily requisites of the host and gut flora. If food supply of vitamins is not optimal, there is a risk of vitamin shortfall in the gut due to competition between the microorganisms in the gut and those of the host cells.

Food-grade lactic acid bacteria (LAB), which is a significant class of microorganisms employed in industry, presents special opportunities to boost the nutritional content of food items. These bacteria even led to the creation of novel functional foods with increased vitamin content, which could expand the market for these goods. LAB are utilized globally for fermentation of a vast range of products, including dairy, breads, and wine. The number of vitamins produced by LAB varies greatly, depending on the strain or species. This characteristic is typically linked to a partial or whole disruption of the genetic information necessary to produce vitamins (Bove et al. 2012). Therefore, to choose appropriate vitamin-producing LAB or to develop techniques to improve vitamin production in food, a thorough understanding of the genes and/or operons involved in vitamin biosynthesis is needed. The development of new functional cereal-based foods with an increased vitamin content through fermentation technique would help in gradual replacement of chemically synthesized vitamins for fortification. This will subsequently lead to enhanced nutritional and commercial worth. Additionally, they could reduce vitamin inadequacies, still widespread in many developing nations. Genetic engineering methods can be used on strains that are crucial for industry but are incapable of producing vitamins, or on strains that generate vitamins at low levels (Capozzi et al.

2012). Even though B vitamins are abundant in cereal grains, processing methods such as grounding or heating leads to reduced availability. The fermentation of cereals and LAB pre-treatment boost synthesis of these vitamins. Probiotics like *Lactobacillus reuteri*, *L. fermentum*, *L. subtilis*, and *L. rossiae* have been shown to synthesize vitamin B12 and may be a good substitute for industrial production due to the discovery of genes that code for the enzymes required to produce cobalamin (vitamin B12). Additionally, *B. animalis* subsp. *lactis* HNO19 supplementation resulted in a rise in levels of vitamins B6 and B12 in the blood of pregnant women (Thorakkattu et al. 2022). The microbes that are employed in fermentation process of "tape ketan" (an indigenous cuisine of Indonesia) can synthesize thiamine. These bacteria are also responsible for restored thiamine content of unpolished rice (Nout and Kiers 2005). Consequently, this is beneficial for rice-eaters. Soyabeans are soaked, dehulled, and partially cooked using *Rhizopus oligosporus* or related molds to create "Tempe" another cuisine with strong Indonesian roots. Riboflavin content is doubled, and niacin is increased sevenfold in this process (Nout and Kiers 2005). Therefore, the microbiologically derived vitamins used in food supplements and fortified foods like yogurt, cheese, milk, bread, or as probiotics can find a niche in the nutraceutical market and contribute toward mitigating malnutrition.

There are various steps involved in removing vitamins from bacteria, viz., (1) cell disruption, (2) filtration and clarification, (3) purification, (4) concentration and drying, and (5) quality control. Six of the thirteen vitamins required by humans are produced by microorganisms. Production of vitamin B12 is done by strains of *Propionibacterium freudenreichii* or *Pseudomonas denitrificans*, while vitamin B2

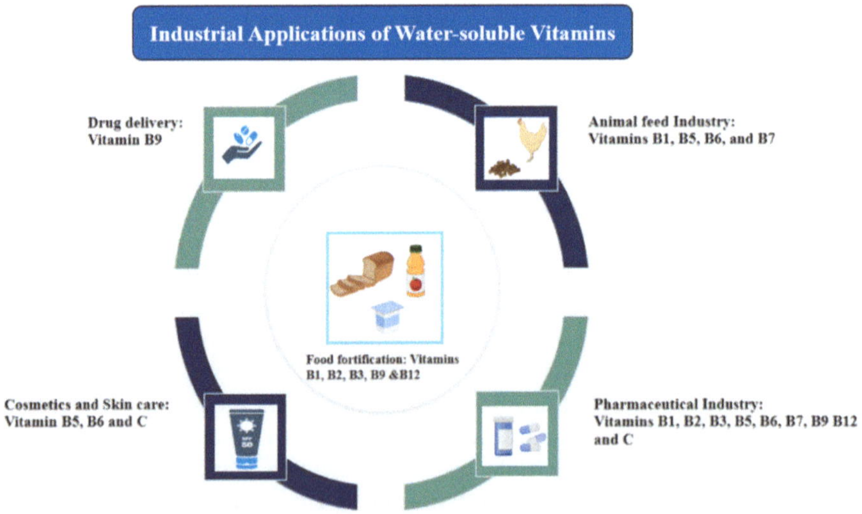

Fig. 14.3 Vitamins B and C and their industrial applications. Vitamin production is scaled up industrially for variety of uses as in pharmaceutical industry, as nutritional supplements, animal feed industry, drug delivery, cosmetic and skin care industry, and food fortification for eliminating diseases associated with nutritional deficiencies

(riboflavin) is produced by the bacterium *Bacillus subtilis* or the fungus *Ashbya gossypii.* Other potential producers include *Sinorhizobium meliloti*, various *Lactobacillus* strains, and even *Escherichia coli*. Microbial fermentation of bacteria and yeast can be used to produce vitamin C or ascorbic acid. *Gluconobacter oxydans* and *Ketogulonicigenium vulgare* are the most often used microorganisms for large-scale manufacturing. Apart from vitamin B12, *P. freudenreichii* may also synthesize other chemicals such as propionic acid, and trehalose, which can increase its industrial attractiveness. Moreover, microbe strains, growth conditions, and fermentation methods may change based on required industry standards and the targeted vitamin (Rusu et al. 2023). There are numerous benefits to using bacteria, yeasts, or microalgae for the fermentative manufacture of vitamins as opposed to conventional chemical synthesis techniques. When it comes to safety, biological activity, absorption rate, and other factors, vitamins produced via biological processes may be more appropriate for use both externally and internally (Yuan et al. 2020). While large-scale commercial production of vitamin B2 and vitamin B12 has already reached a technically advanced stage, the remaining B complex vitamins still require major yield improvement. On the other hand, the manufacture of vitamin C using single-bacteria fermentation replaces accompanying bacteria with related active agents, hence eliminating the dependence of the auxiliary bacteria (Vandamme and Revuelta 2016).

Despite the significant challenges of bioengineered production of vitamins due to poorly designed enzymes and highly complex biosynthetic pathways, several other factors suggest that bio-based processes will eventually replace conventional ones. Environmental restrictions are one of the factors already driving up the cost of using pollution-causing chemical processes. Life-cycle assessment (LCA) for riboflavin found that fermentation method outperformed the chemical synthesis method by showing a 30% reduction in the potential for global warming (CO_2), a 50% acidification (SO_4), a 60% creation of ozone (NO_3), and a 40% reduction in cumulative energy consumption. However, the eutrophication potential increased by 40% (PO_4). Besides, our understanding of basic processes involved in vitamin biosynthesis and our capability to modify microbes is rapidly improving. Lastly, we are seeing an increased consumer demand for naturally derived biochemicals in certain markets, such as food supplements and cosmetics (Acevedo-Rocha et al. 2019).

The obvious conclusion drawn from this chapter is that microbial production is replacing chemical synthesis in large-scale industrial vitamin production. Recent successful industrial production and research papers both demonstrate that biotechnology is the main force behind this transition. It is particularly important for certain vitamins that are difficult or expensive to produce chemically. The ability to produce vitamins sustainably, with high specificity and in a cost-effective manner, has significant implications for the food and pharmaceutical industries (Fig. 14.3). It enables the development of innovative products, such as fortified foods and targeted supplements, that can address micronutrient deficiencies and promote overall health and well-being.

14.12 Future Perspectives

Microbes are the exclusive source of some vitamins, including B2 and B12. While some vitamins, like D, K, B3, and B5, are made through at least one microbial enzyme process, vitamins C and A are made both chemically and microbiologically (Ledesma-Amaro et al. 2013). Currently, the synthesis of D-PA (D-pantothenic acid) in industry mainly relies on chemical processes and enzymatic catalysis. Due to market demand, microbial fermentation using renewable resources must take the place of chemical-based D-PA (vitamin B5) synthesis (Zhao et al. 2023).

Microbial production of vitamins has emerged as a promising alternative to traditional chemical synthesis methods. Between 1999 and 2012, the percentage of the global fermented vitamin market grew from 5% to 75% (Acevedo-Rocha et al. 2019). The fact that vitamin B12 is being produced and consumed in a wide range of markets should come as no surprise, given its diverse applications. However, the lack of accurate global market data makes it challenging to determine precise global market values. In 1989, the annual production was approximately 3 tons; however, by 2005, it had increased to 10 tons, with a nearly EUR 77 million market value. In 2020, 31.41 tons of vitamin B12 were produced in China with a market worth of USD 339.48 million. Besides, the market for vitamin B12 is expected to achieve USD 410 million soon (Calvillo et al. 2022). By 2033, size of world market for fermented ingredients is projected to reach a market valuation of US$ 93.9 billion. It is anticipated that the market would expand at a notable CAGR (compound annual growth rate) of 6.7%. Growing public knowledge of the health advantages of organic acids, B vitamins, and other fermented components is one of the things driving the industry's expansion (Fermented Ingredients Market: Report 2023).

The compatibility of flexible biological systems with engineering systems and the universality of biological system reconstruction are just two of the many obstacles that lie ahead for the area of synthetic biotechnology. Developing newer microbial cell factories to ferment vitamins in industrial scale will require multidisciplinary research, advancing current technology, and combining it with new approaches. As a concluding remark, we can say that as synthetic biotechnology and metabolic engineering continue to progress, microbial production of vitamins in industrial scale will lead to a more widespread, secure, and environmental-friendly manufacturing process.

Acknowledgements The authors wish to thank Sharda University for sanction of the SEED fund (Grant No. #SU/SF/2023/11) to Dr. Soumi Sadhu for support of studies which contributed to this chapter.

References

Acevedo-Rocha CG, Gronenberg LS, Mack M, Commichau FM, Genee HJ (2019) Microbial cell factories for the sustainable manufacturing of B vitamins. Curr Opin Biotechnol 56:18–29

Alagawany M, Elnesr SS, Farag MR, Tiwari R, Yatoo MI, Karthik K, Michalak I, Dhama K (2021) Nutritional significance of amino acids, vitamins and minerals as nutraceuticals in poultry production and health—a comprehensive review. Vet Q 41(1):1–29
Alsaeedi A, Welham S, Rose P (2023) Impact of lifestyle factors on dietary vitamin B6 intake and plasma pyridoxal 5′-phosphate level in UK adults: National Diet and Nutrition Survey Rolling Programme (NDNS) (2008–2017). Br J Nutr 130:1–13
Azevedo MA, Bourbon AI, Vicente AA, Cerqueira MA (2014) Alginate/chitosan nanoparticles for encapsulation and controlled release of vitamin B2. Int J Biol Macromol 71:141–146
Boo YC (2022) Ascorbic acid (vitamin C) as a cosmeceutical to increase dermal collagen for skin antiaging purposes: emerging combination therapies. Antioxidants 11(9):1663
Bove P, Gallone A, Russo P, Capozzi V, Albenzio M, Spano G, Fiocco D (2012) Probiotic features of *Lactobacillus plantarum* mutant strains. Appl Microbiol Biotechnol 96:431–441
Burgess C, O'Connell-Motherway M, Sybesma W, Hugenholtz J, Van Sinderen D (2004) Riboflavin production in *Lactococcus lactis*: potential for in situ production of vitamin-enriched foods. Appl Environ Microbiol 70(10):5769–5777
Cai L, Yuan MQ, Li ZJ, Chen JC, Chen GQ (2012) Genetic engineering of *Ketogulonigenium vulgare* for enhanced production of 2-keto-l-gulonic acid. J Biotechnol 157(2):320–325
Calvillo Á, Pellicer T, Carnicer M, Planas A (2022) Bioprocess strategies for vitamin B12 production by microbial fermentation and its market applications. Bioengineering 9(8):365
Camargo FB Jr, Gaspar LR, Maia Campos PM (2011) Skin moisturizing effects of panthenol-based formulations. J Cosmet Sci 62(4):361
Capozzi V, Russo P, Dueñas MT, López P, Spano G (2012) Lactic acid bacteria producing B-group vitamins: a great potential for functional cereals products. Appl Microbiol Biotechnol 96:1383–1394
Cardinale S, Tueros FG, Sommer MOA (2017) Genetic-metabolic coupling for targeted metabolic engineering. Cell Rep 20(5):1029–1037
Carr AC, Maggini S (2017) Vitamin C and immune function. Nutrients 9(11):1211
Chandel V, Andhare P, Upadhyay D (2022) A review on microbial production of vitamins. Int J Innov Res Comput Sci Technol 10(2):519–525
Chotani G, Dodge T, Hsu A, Kumar M, LaDuca R, Trimbur D, Weyler W, Sanford K (2000) The commercial production of chemicals using pathway engineering. Biochim Biophys Acta 1543(2):434–455
Clardy SM, Allis DG, Fairchild TJ, Doyle RP (2011) Vitamin B12 in drug delivery: breaking through the barriers to a B12 bioconjugate pharmaceutical. Expert Opin Drug Deliv 8(1):127–140
Coerdt KM, Goggins CA, Khachemoune A (2021) Vitamins A, B, C, and D: A short review for the dermatologist. Altern Ther Health Med 27(4):41–48
Commichau FM, Alzinger A, Sande R, Bretzel W, Reuss DR, Dormeyer M et al (2015) Engineering *Bacillus subtilis* for the conversion of the antimetabolite 4-hydroxy-l-threonine to pyridoxine. Metab Eng 29:196–207
D'Agostini F, Fiallo P, Ghio M, De Flora S (2013) Chemoprevention of doxorubicin-induced alopecia in mice by dietary administration of L-cystine and vitamin B6. Arch Dermatol Res 305:25–34
Darnton-Hill I (2019) Public health aspects in the prevention and control of vitamin deficiencies. Curr Dev Nutr 3(9):075
Deguchi Y, Morishita T, Mutai M (1985) Comparative studies on synthesis of water-soluble vitamins among human species of *Bifidobacteria*. Agric Biol Chem 49(1):13–19
Dmytruk K, Lyzak O, Yatsyshyn V, Kluz M, Sibirny V, Puchalski C, Sibirny A (2014) Construction and fed-batch cultivation of *Candida famata* with enhanced riboflavin production. J Biotechnol 172:11–17
Elgharably N, Al Abadie M, Al Abadie M, Ball PA, Morrissey H (2023) Vitamin B group levels and supplementations in dermatology. Dermatol Rep 15(1):9511

Ewe JA, Wan-Abdullah WN, Liong MT (2010) Viability and growth characteristics of *Lactobacillus* in soymilk supplemented with B-vitamins. Int J Food Sci Nutr 61(1):87–107

Fermented Ingredients Market: Report (2023) Survey on fermented ingredients market covering sales outlook, up-to-date key trends, market size and forecast, production capacity, export and import analysis, regulations, trade analysis, brand analysis and pricing analysis, p 333

Gangadharan D, Nampoothiri KM (2011) Folate production using *Lactococcus lactis ssp cremoris* with implications for fortification of skim milk and fruit juices. LWT Food Sci Technol 44(9):1859–1864

Ganji SH, Kamanna VS, Kashyap ML (2003) Niacin and cholesterol: role in cardiovascular disease. J Nutr Biochem 14(6):298–305

Gheita AA, Gheita TA, Kenawy SA (2020) The potential role of B5: a stitch in time and switch in cytokine. Phytother Res 34(2):306–314

Ghosh D, Chattopadhyay P (2011) Preparation of idli batter, its properties and nutritional improvement during fermentation. J Food Sci Technol 48:610–615

Guéant JL, Guéant-Rodriguez RM, Alpers DH (2022) Vitamin B12 absorption and malabsorption. In: Vitamins and hormones, vol 119. Academic Press, pp 241–274

Guo J, Sun X, Yuan Y, Chen Q, Ou Z, Deng Z, Ma T, Liu T (2023) Metabolic engineering of *Saccharomyces cerevisiae* for vitamin B5 production. J Agric Food Chem 71(19):7408–7417

Gupta ES, Sheth SP, Ganjiwale JD (2016) Association of vitamin B12 deficiency and use of reverse osmosis processed water for drinking: a cross-sectional study from western India. J Clin Diagn Res 10(5):37

Hasan Kadhim A, Shamkhi Noor A, Amer Ali M (2022) The effectiveness of biotin (vitamin b7) added to the diet in improving the efficiency of productivity, and some physiological traits for broiler chickens (ross-308) exposed to oxidative stress. Arch Razi Inst 77(5):1805–1811

Hobby GL, Deuschle KW (1959) The use of riboflavin as an indicator of isoniazid ingestion in self-medicated patients. Am Rev Respir Dis 80(3):415–423

Hohmann HP, van Dijl JM, Krishnappa L, Prágai Z (2017) Host organisms: *Bacillus subtilis*. In: Industrial biotechnology: microorganisms, vol 1. Wiley, pp 221–297

Hoshino T, Ichikawa K, Nagahashi Y, Tazoe M, inventors, DSM IP Assets BV, assignee (2006) Microorganism and process for preparing vitamin B6. U.S. patent application US 10/528,881

Hossain KS, Amarasena S, Mayengbam S (2022) B vitamins and their roles in gut health. Microorganisms 10(6):1168

Hugenholtz J, Smid EJ (2002) Nutraceutical production with food-grade microorganisms. Curr Opin Biotechnol 13(5):497–507

Hui F, Tang J, Williams PA, McGuinness MB, Hadoux X, Casson RJ, Coote M, Trounce IA, Martin KR, van Wijngaarden P, Crowston JG (2020) Improvement in inner retinal function in glaucoma with nicotinamide (vitamin B3) supplementation: a crossover randomized clinical trial. Clin Experiment Ophthalmol 48(7):903–914

Hüser AT, Chassagnole C, Lindley ND, Merkamm M, Guyonvarch A, Elišáková V, Pátek M, Kalinowski J, Brune I, Pühler A, Tauch A (2005) Rational design of a *Corynebacterium glutamicum* pantothenate production strain and its characterization by metabolic flux analysis and genome-wide transcriptional profiling. Appl Environ Microbiol 71(6):3255–3268

Hwang S, Cordova B, Abdo M, Pfeiffer F, Maupin-Furlow JA (2017) *ThiN* as a versatile domain of transcriptional repressors and catalytic enzymes of thiamine biosynthesis. J Bacteriol 199(7):10–1128

Indrio F, Riezzo G, Raimondi F, Bisceglia M, Filannino A, Cavallo L, Francavilla R (2011) *Lactobacillus reuteri* accelerates gastric emptying and improves regurgitation in infants. Eur J Clin Investig 41(4):417–422

Johal J, Han CY, Joseph R, Munn Z, Agbejule OA, Crawford-Williams F, Wallen MP, Chan RJ, Hart NH (2022) Dietary supplements in people with metastatic cancer who are experiencing malnutrition, cachexia, sarcopenia, and frailty: a scoping review. Nutrients 14(13):2642

Khan SA (2019) Inclusion of pyridoxine to flaxseed cake in poultry feed improves productivity of omega-3 enriched eggs. Bioinformation 15(5):333

Khudhair AF, Saeed SI, Marhoon AA, Alesary HF (2019) A new spectrophotometric method to determine vitamin B6 in pharmaceutical formation samples using a micelle form. J Phys Conf Ser 1234(1):012087

Labuschagne PWJ, Divol B (2021) Thiamine: a key nutrient for yeasts during wine alcoholic fermentation. Appl Microbiol Biotechnol 105:953–973

Laiño JE, del Valle MJ, de Giori GS, LeBlanc JGJ (2014) Applicability of a *Lactobacillus amylovorus* strain as co-culture for natural folate bio-enrichment of fermented milk. Int J Food Microbiol 191:10–16

Lamers Y (2011) Folate recommendations for pregnancy, lactation, and infancy. Ann Nutr Metab 59(1):32–37

LeBlanc JG, Levit R, Savoy de Giori G, de Moreno de LeBlanc A (2020) Application of vitamin-producing lactic acid bacteria to treat intestinal inflammatory diseases. Appl Microbiol Biotechnol 104:3331–3337

Ledesma-Amaro R, Santos MA, Jiménez A, Revuelta JL (2013) Microbial production of vitamins. In: Microbial production of food ingredients, enzymes and nutraceuticals. Woodhead Publishing Limited, pp 571–594

Liu M, Chen Q, Sun Y, Zeng L, Wu H, Gu Q, Li P (2022) Probiotic potential of a folate-producing strain *Latilactobacillus sakei* LZ217 and its modulation effects on human gut microbiota. Foods 11(2):234

Lordan R, Rando HM, Greene CS (2021) Dietary supplements and nutraceuticals under investigation for COVID-19 prevention and treatment. Msystems 6(3):10–1128

Ma Q, Bi YH, Wang EX, Zhai BB, Dong XT, Qiao B, Ding MZ, Yuan YJ (2019) Integrated proteomic and metabolomic analysis of a reconstructed three-species microbial consortium for one-step fermentation of 2-keto-L-gulonic acid, the precursor of vitamin C. J Ind Microbiol 46(1):21–31

Magnúsdóttir S, Ravcheev D, de Crécy-Lagard V, Thiele I (2015) Systematic genome assessment of B-vitamin biosynthesis suggests co-operation among gut microbes. Front Genet 6:148

Maitiniyazi G, Cao X, Chen Y, Zhang R, Liu Y, Li Z, Gu D, Li T, Xia S (2022) Impact of gut microbiota on the association between diet and depressive symptoms in breast cancer. Nutrients 14(6):1186

Mastella L, Senatore VG, Guzzetti L, Coppolino M, Campone L, Labra M, Beltrani T, Branduardi P (2022) First report on vitamin B9 production including quantitative analysis of its vitamers in the yeast *Scheffersomyces stipitis*. Biotechnol Biofuels Bioprod 15(1):98

McNulty B, Pentieva K, Marshall B, Ward M, Molloy AM, Scott JM, McNulty H (2011) Women's compliance with current folic acid recommendations and achievement of optimal vitamin status for preventing neural tube defects. Hum Reprod 26(6):1530–1536

Miyano KI, Ye K, Shimizu K (2000) Improvement of vitamin B12 fermentation by reducing the inhibitory metabolites by cell recycle system and a mixed culture. Biochem Eng J 6(3):207–214

Molina V, Médici M, de Valdez GF, Taranto MP (2012) Soybean-based functional food with vitamin B12-producing lactic acid bacteria. J Funct Foods 4(4):831–836

Moriya, T., Hikichi, Y., Moriya, Y. and Yamaguchi, T., Takeda Chemical Industries Ltd., 1999. Process for producing D-pantoic acid and D-pantothenic acid or salts thereof. U.S. patent 5,932,457

Muhamad R, Akrivaki A, Papagiannopoulou G, Zavridis P, Zis P (2023) The role of vitamin B6 in peripheral neuropathy: a systematic review. Nutrients 15(13):2823

Muthayya S, Hall J, Bagriansky J, Sugimoto J, Gundry D, Matthias D, Prigge S, Hindle P, Moench-Pfanner R, Maberly G (2012) Rice fortification: an emerging opportunity to contribute to the elimination of vitamin and mineral deficiency worldwide. Food Nutr Bull 33(4):296–307

Nagasawa T, Yamada H (1989) Microbial transformations of nitriles. Trends Biotechnol 7(6):153–158

Nout MR, Kiers JL (2005) Tempe fermentation, innovation and functionality: update into the third millenium. J Appl Microbiol 98(4):789–805

Ohe Y, Ichinose Y, Nakagawa K, Tamura T, Kubota K, Yamamoto N, Adachi S, Nambu Y, Fujimoto T, Nishiwaki Y, Saijo N (2008) Efficacy and safety of two doses of pemetrexed supplemented with folic acid and vitamin B12 in previously treated patients with non-small cell lung cancer. Clin Cancer Res 14(13):4206–4212

Pan X, Xue F, Nan X, Tang Z, Wang K, Beckers Y, Jiang L, Xiong B (2017) Illumina sequencing approach to characterize thiamine metabolism related bacteria and the impacts of thiamine supplementation on ruminal microbiota in dairy cows fed high-grain diets. Front Microbiol 8:1818

Pan X, Nan X, Yang L, Jiang L, Xiong B (2018) Thiamine status, metabolism and application in dairy cows: a review. Br J Nutr 120(5):491–499

Panja A, Das S, Chakraborty A, Chakraborty P, Pal S, Nandi AK (2018) Injectable hydrogel of vitamin B9 for the controlled release of both hydrophilic and hydrophobic anticancer drugs. ChemMedChem 13(22):2427–2436

Patel SV, Makwana AB, Gandhi AU, Tarani G, Patel J, Bhavsar V (2022) Factors associated with vitamin B12 deficiency in adults attending tertiary care Hospital in Vadodara: a case control study. Egypt J Intern 34(1):11

Pereira JF, Lonni AAG, Mali S (2022) Development of biopolymeric films with addition of vitamin C and catuaba extract as natural antioxidants. Prep Biochem Biotechnol 52(1):1–10

Rauf M, Gleason C, Nooka AK, Husman A, Waller EK (2011) Treatment of severe neutropenia with high-dose pyridoxine in a patient with chronic graft versus host disease and squamous cell carcinoma: a case report. J Med Case Rep 5(1):1–3

Reininghaus EZ, Platzer M, Kohlhammer-Dohr A, Hamm C, Mörkl S, Bengesser SA, Fellendorf FT, Lahousen-Luxenberger T, Leitner-Afschar B, Schöggl H, Amberger-Otti D (2020) PROVIT: supplementary probiotic treatment and vitamin B7 in depression—a randomized controlled trial. Nutrients 12(11):3422

Rizzo G, Laganà AS (2020) A review of vitamin B12. In: Molecular nutrition. Academic Press, pp 105–129

Running JA, Huss RJ, Olson PT (1994) Heterotrophic production of ascorbic acid by microalgae. J Appl Phycol 6:99–104

Russo P, Capozzi V, Arena MP, Spadaccino G, Dueñas MT, López P, Fiocco D, Spano G (2014) Riboflavin-overproducing strains of *Lactobacillus fermentum* for riboflavin-enriched bread. Appl Microbiol Biotechnol 98:3691–3700

Rusu AV, Trif M, Rocha JM (2023) Microbial secondary metabolites via fermentation approaches for dietary supplementation formulations. Molecules 28(16):6020

Saito Y, Ishii Y, Hayashi H, Imao Y, Akashi T, Yoshikawa K, Noguchi Y, Soeda S, Yoshida M, Niwa M, Hosoda J (1997) Cloning of genes coding for L-sorbose and L-sorbosone dehydrogenases from *Gluconobacter oxydans* and microbial production of 2-keto-L-gulonate, a precursor of L-ascorbic acid, in a recombinant *G. oxydans* strain. Appl Environ Microbiol 63(2):454–460

Sampedro A, Rodriguez-Granger J, Ceballos J, Aliaga L (2015) Pantothenic acid: an overview focused on medical aspects. Eur Sci J 11(21):1857

Sanz-Cuesta T, Gonzalez-Escobar P, Riesgo-Fuertes R, Garrido-Elustondo S, del Cura-Gonzalez I, Martin-Fernandez J, Escortell-Mayor E, Rodriguez-Salvanes F, Garcia-Solano M, Gonzalez-Gonzalez R, Martín-de la Sierra-San Agustín MA (2012) Oral versus intramuscular administration of vitamin B12 for the treatment of patients with vitamin B12 deficiency: a pragmatic, randomised, multicentre, non-inferiority clinical trial undertaken in the primary healthcare setting (Project OB12). BMC Public Health 12:1–11

Sauer M, Branduardi P, Valli M, Porro D (2004) Production of L-ascorbic acid by metabolically engineered *Saccharomyces cerevisiae* and *Zygosaccharomyces bailii*. Appl Environ Microbiol 70(10):6086–6091

Schwechheimer SK, Park EY, Revuelta JL, Becker J, Wittmann C (2016) Biotechnology of riboflavin. Appl Microbiol Biotechnol 100:2107–2119

Scott WC (2020) Literature review of both classic and novel roles of biotin (vitamin B7) in cellular processes. J Nat Sci 1(1):45–51

Sedhai YR, Shrestha DB, Budhathoki P, Jha V, Mandal SK, Karki S, Baniya R, Cable CA, Kashiouris MG (2021) Effect of thiamine supplementation in critically ill patients: a systematic review and meta-analysis. J Crit Care 65:104–115

Serrano-Amatriain C, Ledesma-Amaro R, López-Nicolás R, Ros G, Jiménez A, Revuelta JL (2016) Folic acid production by engineered *Ashbya gossypii*. Metabol Eng 38:473–482

Shabbir H, Shabbir I, Aslam M, Sarwar MF, Sarwar MH, Sarwar M (2020) Fundamental aspects of vitamin B complex in human nourishment and fitness. Am J Food Sci Health 6(4):109–118

Shams A (2022) Folates: an introduction. In: B-complex vitamins-sources, intakes and novel applications. IntechOpen

Shi Z, Zhen S, Wittert GA, Yuan B, Zuo H, Taylor AW (2014) Inadequate riboflavin intake and anemia risk in a Chinese population: five-year follow up of the Jiangsu Nutrition Study. PLoS One 9(2):88862

Singh J (2021) Vitamin B9 in dark green vegetables: deficiency disorders, bio-availability, and fortification issues. In: B-complex vitamins-sources, intakes and novel applications. IntechOpen

Stach K, Stach W, Augoff K (2021) Vitamin B6 in health and disease. Nutrients 13(9):3229

Streit WR, Entcheva P (2003) Biotin in microbes, the genes involved in its biosynthesis, its biochemical role and perspectives for biotechnological production. Appl Microbiol Biotechnol 61:21–31

Sugimoto T, Kanamasa S, Kato T, Park EY (2009) Importance of malate synthase in the glyoxylate cycle of *Ashbya gossypii* for the efficient production of riboflavin. Appl Microbiol Biotechnol 83:529–539

Sugisawa T, Miyazaki T, Hoshino T (2005) Microbial production of L-ascorbic acid from D-sorbitol, L-sorbose, L-gulose, and L-sorbosone by *Ketogulonicigenium vulgare* DSM 4025. Biosci Biotechnol Biochem 69(3):659–662

Sybesma W, Starrenburg M, Kleerebezem M, Mierau I, de Vos WM, Hugenholtz J (2003) Increased production of folate by metabolic engineering of *Lactococcus lactis*. Appl Environ Microbiol 69(6):3069–3076

Thorakkattu P, Khanashyam AC, Shah K, Babu KS, Mundanat AS, Deliephan A, Deokar GS, Santivarangkna C, Nirmal NP (2022) Postbiotics: current trends in food and pharmaceutical industry. Foods 11(19):3094

Tokui M, Kubodera T, Gomi K, Yamashita N, Nishimura A (2011) Construction of a thiamine pyrophosphate high-producing strain of *Aspergillus oryzae* by overexpression of three genes involved in thiamine biosynthesis. J Biosci Bioeng 111(4):388–390

Van Arsdell SW, Perkins JB, Yocum RR, Luan L, Howitt CL, Prasad Chatterjee N, Pero JG (2005) Removing a bottleneck in the *Bacillus subtilis* biotin pathway: bioA utilizes lysine rather than S-adenosylmethionine as the amino donor in the KAPA-to-DAPA reaction. Biotechnol Bioeng 91(1):75–83

Vandamme EJ, Revuelta JL (2016) Industrial fermentation of vitamin C. In: Vandamme EJ, Revuelta JL (eds) Industrial biotechnology of vitamins, biopigments, and antioxidants. Wiley, pp 161–192

Wan Z, Zheng J, Zhu Z, Sang L, Zhu J, Luo S, Zhao Y, Wang R, Zhang Y, Hao K, Chen L (2022) Intermediate role of gut microbiota in vitamin B nutrition and its influences on human health. Front Nutr 9:1031502

Wang Y, Liu L, Jin Z, Zhang D (2021) Microbial cell factories for green production of vitamins. Front Bioeng Biotechnol 9:661562

Yuan P, Cui S, Liu Y, Li J, Liu L (2020) Metabolic engineering for the production of fat-soluble vitamins: advances and perspectives. Appl Microbiol Biotechnol 104:935–951

Zhao K, Tang H, Zhang B, Zou S, Liu Z, Zheng Y (2023) Microbial production of vitamin B5: current status and prospects. Crit Rev Biotechnol 43(8):1172–1192

Zheng J, Wang X, Wu B, Qiao L, Zhao J, Pourkheirandish M, Wang J, Zheng X (2022) Folate (vitamin B9) content analysis in bread wheat (*Triticum aestivum L.*). Front Nutr 9:933358

Zhu T, Pan Z, Domagalski N, Koepsel R, Ataai MM, Domach MM (2005) Engineering of *Bacillus subtilis* for enhanced total synthesis of folic acid. Appl Environ Microbiol 71(11):7122–7129

Chapter 15
Microbial Production of Polyketides and Non-ribosomal Peptides and Their Applications

Mahima S. Mohan, Simi Asma Salim, Paramita Pakhira, and Siddhardha Busi

Abstract There is an increasing demand of novel bioactive molecules as the drug resistance is emerging worldwide. Microorganisms are adapted to already existing antimicrobials used for various microbial infections. Researchers have to focus more for the identification and development of secondary metabolites from the environment. Bioactive secondary metabolites synthesized by microorganisms are widely explored for the development of drugs. The two main groups of secondary metabolites are polyketides and non-ribosomal peptides synthesized by key enzymes such as polyketide synthases and non-ribosomal peptide synthetases respectively. These metabolites exhibit wide range of applications, which includes clinically relevant metabolites for the treatment of microbial infections and cancer. The major hurdle in the identification of bioactive molecules from the environmental samples is the deficit in microbial cultivation. Nevertheless, the advancement in metagenomic studies and genome mining makes feasible to predict biosynthetic gene clusters of bioactive molecules from the environment. Biosynthesis, evolution, production or bioengineering of the bioactive molecules are discussed.

Keywords Polyketides · Non-ribosomal peptides · Biosynthetic gene clusters (BGCs) · Polyketide synthase · Non-ribosomal synthetases

15.1 Introduction

The prevalence of multidrug-resistant (MDR) microbial infection poses serious threat worldwide due to limited therapeutic options. Since the discovery of antibiotics, the emergence of multidrug resistance in microorganisms is increasing rapidly. There is an urgent demand of novel therapeutic agents to tackle MDR pathogens.

M. S. Mohan · S. A. Salim · P. Pakhira · S. Busi (✉)
Department of Microbiology, School of Life Sciences, Pondicherry University, Puducherry, India

V. Kothari et al. (eds.), *Microbial Products for Health and Nutrition*,
https://doi.org/10.1007/978-981-97-4235-6_15

Bioactive molecules from bacteria, fungi and plants are widely employed as therapeutic molecules that include antibiotics, immunosuppressive agents and antitumour agents. The major classes of bioactive secondary metabolites are polyketides and non-ribosomal peptides, which are synthesized by the enzymes, polyketide synthase (PKS) and non-ribosomal peptide synthetases (NRPS), respectively. They belong to two distinct families having broad spectrum of biological and pharmaceutical activities that include antibiotics, pigments, siderophores, cytotoxicity, immunomodulatory effects and toxins (Masschelein et al. 2015; Wang et al. 2014). Actinomycetes are one of the foremost producers of bioactive molecules, especially antibiotics that include β-lactams, macrolides, tetracyclines and aminoglycosides (Robertsen and Musiol-Kroll 2019). The diverse structure and biological properties of PKs and NRPs can be explored to produce novel therapeutics. Moreover, engineering of biosynthetic gene clusters (BGCs) of bioactive metabolites yields complex products with diverse chemical framework and biological activities. Detection of gene clusters responsible for the synthesis of such bioactive molecules helps to understand more about the biosynthetic pathways. Combination of bioinformatic tools such as genome mining and metagenomics along with culturing techniques aids in drug discovery (Boddy 2014; Wang et al. 2014). This chapter discusses about polyketide and non-ribosomal peptide bioactive molecules, biosynthesis, evolution, production, identification using bioinformatic platforms and their applications.

15.2 Polyketide Antibiotics

Polyketides are distinct group of secondary metabolites having great pharmacological potential and ecological functions. In the worldwide market of antibiotics, polyketides occupy around 20% of the market. Yearwise revenue of this polyketide-based antibiotics is over 18 billion USD (Baerson and Rimando 2007). In market, these polyketide antibiotics are present as different brand name such as ilosone, sigmamycin, rovamycin, biaxin, Zithromax, ketek, solithera, CEM-101, T-4288 and WCk 4873, which are derived from erythromycin. Tylocine, tylan, pulmotil, micotil, timovet, zuprevo, draxxin and zactran are tylosin derivatives. Coban, rumensin and monensin are obtained from monensin A. Dificid is derived from tiacumicin B. Aemcolo, relafalk, rifadin, rimactane, mycobutin, priftin, normix, rifacol and xifaxan are derived from rifamycin. Tetracycline, aureomycin, vibramycin, minocin, tygacil, nuzyra, zerava, seysara and ty-271 are tetracycline derivatives. Pyostacine, synercid and Nxl-103 are derivatives of pristinamycin. Mycostatin and mystop are derived from nystalin (Robertsen and Musiol-Kroll 2019). They have remarkably distinct carbon backbones such as polyphenols, polyether's, macrolides, polyenes and enediynes. They have various pharmacological effects and some of the examples like antimicrobial, anticancer, anti cholesterolemic, anti-parasitic and immunomodulatory activities. Among 10,000 PKs reported, more than 1% of them were predicted as a drug molecule. Polyketides are having high level of structural diversity, complexity and activity against biological substrate. PKs are generally constructed with sugar moieties. PKs were widely produced by microorganisms, plants and fungi; among them, microorganism, for instance, actinomycetes, was reported as the major antibiotic producer (Xu et al. 2020; Rohr 2000; Helfrich et al. 2014). Polyketide synthases are known as PKSs; these enzyme groups are accountable for different types of PK antibiotics synthesis (Kennedy et al. 1999). Some of the PKs-based pharmaceutical drugs recently developed are FK506 and rapamycin (immunosuppressant), lovastatin (cholesterol lowering agent), doxorubicin (anticancer agent), and tetracycline and erythromycin (antimicrobial compound) (Xu et al. 2020).

The three main classes of PKs are as follows:

Type I polyketides are bioactive molecules which is synthesized by a distinct class of enzyme called type I polyketide synthases (PKSs). Type I PKS class consists of diverse molecules, and around one-third of them are known for their pharmaceutical properties. Type I PKSs were grouped into two: iterative and non-iterative PKSs. Former PKS class was identified in fungi, while latter was seen in bacteria. The difference between iterative and non-iterative classes is based on the substrate number for the elongation process. One of the major examples of non-iterative PKSs is giant multi-modular megasynthases which includes two types of enzymes: one is archetypal erythromycin PKS and the other one is 6-deoxyerythronolide B (6-dEB) synthase (DEBS) (Xu et al. 2020). Some of the bacterial iterative PKS-derived secondary metabolites are avilamycin, myxochromide and neocarzinostatin (Jenke-Kodama and Dittmann 2009). Examples of fungal iterative PKs are austinol, fumagillin, altertoxin, arthrosporol, rugulosin A and skyrin (Stroe et al. 2023).

Type II polyketide classes are synthesized from a class of enzyme called type II polyketide synthases. These enzymes are made up of multiple enzymes that consist of proteins with mono functions. These molecules are predominant in bacteria and synthesize varied aromatic polyketide biomolecules. Aromatic polyketides exhibit multiple roles such as anticancer, antiviral and antibacterial properties, for instance, tetracycline and anthracyclines (Wang et al. 2020). Actinomycetes are the predominant group synthesizes type II polyketides. Based on the aromatic nature and biosynthesis, bioactive molecules produced through type II PKSs were categorized into tetracyclines, aureolic acids, angucyclines, anthracyclines, pradimicin-type polyphenols, tetracenomycins and benzoisochromanequinones. Anthracyclines belong to the topmost category of anticancer molecules. Rhodomycin and cinerubicin are the first anthracyclines identified and isolated during 1950s; later, daunomycin and doxorubicin were discovered. Doxorubicin is employed for the treatment of breast cancer, sarcoma of soft tissue and lymphomas, while daunorubicin has activities against acute lymphoblastic or myeloblastic leukaemia. Tetracyclines, protein synthesis inhibitors, represent the antibiotic class produced by type II PKSs. First-generation tetracyclines were identified during 1950–1960s, which include oxytetracycline, tetracycline, chlortetracycline and demeclocycline. Doxycycline and minocycline belong to the second-generation tetracycline class. Benzoisochromanequinone class consists of antimicrobial drug, actinorhodin isolated from *Streptomyces coelicolor*, medermycin (*Streptomyces* sp. AM7161), griseusin A (*Streptomyces griseus*) and so on (Hertweck et al. 2007).

Type III polyketides are bioactive secondary metabolites that were synthesized by type III polyketide synthases. They are homodimeric in nature and have simpler and distinct structure unlike type I and II PKSs. PKS enzyme was initially considered as plant-specific, but later, reports confirm its presence in various species of bacteria and in some of the fungi (Yu et al. 2012; Lim et al. 2016). These bioactive molecule synthesized through type III PKS is present as a precursor of antibiotics, lipids, UV protective pigments, antibiotic production regulators and so on (Shimizu et al. 2017).

First time, type III PKS was discovered in the year of 1999; furthermore, structures were predicted and characterized. RppA is the first type III PKS which was isolated as well as identified from *Streptomyces griseus* (Yu et al. 2012). Phloroglucinol, a predecessor molecule of antifungal compound 2,4-diacetylphloroglucinol, is synthesized from phloroglucinol synthase of *Pseudomonas fluorescens*. This enzyme showed broad-spectrum activity and has 40% similarity with RppA. 3,5-Dihydrophenylacetic acid synthase obtained from *Amycolatopsis mediterranei* catalyses the synthesis of 3,5-dihydrophenylacetic acid, which serves as a predecessor molecule of (*S*)-3,5-dihydroxyphenylglycine, an amino acid residue involved in the synthesis of a vancomycin derivative, balhimycin (Lim et al. 2016).

15.3 Non-ribosomal Peptides (NRPs)

These molecules are produced by enzyme complex with multiple modules and are non-ribosomal in nature, as suggested by their name. The gene clusters encode for the multi-modular enzyme complex which is called as non-ribosomal peptide synthetases (NRPSs) that spans more than 10 kb in the genome. Nevertheless, NRPSs are ribosomal in origin and are activated by phosphopantetheinyl transferase (PPTase) and involved in the catalysis of NRPS into active NRPS via phosphopantetheine addition. Each multiple modules present in the NRPS take part in the addition of one peptide monomers into the protein (Vassaux et al. 2019). NRPs are widely reported to produce medically relevant bioactive peptides like antibiotics, immunosuppressive agents, anticancer agents and toxins. Most of the NRPs are nonlinear in nature and made up of non-standard amino acid residues in the range of 20 to more than 100 (Mohimani et al. 2014). Examples of non-ribosomal peptide antibiotics include gramicidin S, vancomycin and tyrocidine A. Other NRP molecules reported to have wide application that includes cyclosporin A (immunosuppressive agent); bleomycin A2 and epothilone having cytotoxic activity; and bacillibactin, enterobactin and vibriobactin are siderophores (Schwarzer et al. 2003). Some bioactive molecules were heterologously produced in *Escherichia coli* either partially or entirely, which include echinomycin, triostin A, ecolimycin C, valinomycin and lyngbyatoxin (Li and Neubauer 2014).

15.4 Biosynthesis of PKs and NRPs

15.4.1 Polyketide Biosynthesis

Polyketides are synthesized through Claisen condensation reactions between a starter unit such as acetyl/propionyl-CoA and an extender unit, mostly, malonyl/methylmalonyl-CoA which results in the formation of a β-keto functional group. Subsequent growth of the polyketide chain attached to enzyme will occur by the stepwise incorporation of several extender units (Ma et al. 2020).

The complex family of enzymes termed polyketide synthases (PKSs) are involved in the manufacturing of polyketides. These enzymes are homologous to Fatty acid synthase (FAS), and various types of PKSs have been identified. Broadly, they are classified into three (type I, II and III) on the basis of their domain organization and catalytic mechanisms (Gokulan et al. 2014). Detailed structure and mechanism of biosynthesis is described below.

15.4.1.1 Type I Polyketide Synthases

Type I PKSs (also called modular or non-iterative PKSs) are responsible for the synthesis of PKs like macrolides, and polyether polyketides are organized into several modules. Each module is assigned for the non-iteratively acting functions to carry out one round of polyketide chain elongation (Gokulan et al. 2014).

Generally, every module of canonical type I PKS is comprised a minimum of three domains such as acyltransferase (AT), ketosynthase (KS) and acyl carrier protein (ACP), each of which has defined functions. In several cases, it may contain additional domains including dehydratase (DH), ketoreductase (KR) and enoylreductase (ER). The main catalytic domain AT catalyses the stimulation and transfer of acyl group from its donor to an acceptant. Usually, the initial AT domain has the capability to identify different kinds of acyl-CoA starter units. The KS domain carries out the chain extension through condensation and decarboxylation reactions. The developing chain of polyketide is tethered to the ACP domain (which contains a phosphopantetheine moiety) through thioester bond (Gokulan et al. 2014; Singh et al. 2019).

Here, the starting precursors are small organic acids like acetic and propionic acids, which will be converted into their corresponding CoA ester (such as acetyl-CoA, propionyl-CoA, malonyl-CoA and methylmalonyl-CoA) forms. Following ester formation, the specified acyltransferase will attach them to the PKS complex. The starter acyl-CoA attaches to the KS active site, whereas the extender acyl-CoA

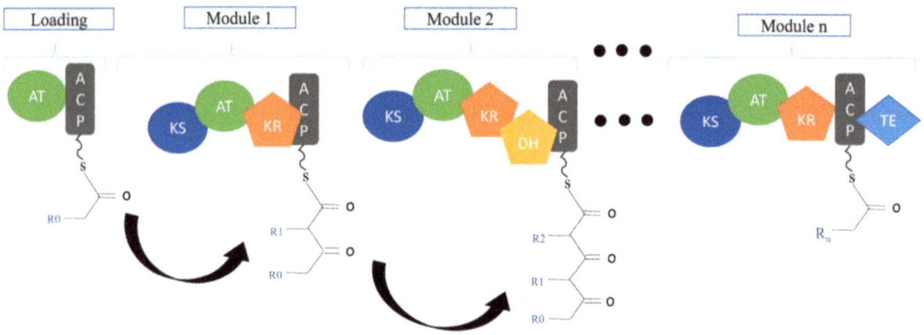

Fig. 15.1 Type I polyketide synthesis: PKS domains and subdomains are arranged into different modules

attaches to the structural component acyl carrier protein (ACP). The enzyme ketosynthase initiates condensation reaction. Then, first AT enzyme of the activation module begins the reaction via the formation of thioester bond between first precursor and ACP. The activated substrate is then shifted to ACP of next module called ping-pong mechanism. A four-carbon chain tethered to ACP will be formed as the result of condensation reaction. CO_2 will be eliminated during the reaction. Further elongation of the ACP-tethered chain will be carried out by the next KS where condensation of the extender units takes place (Gokulan et al. 2014; Singh et al. 2019).

Other domains such as dehydratase (DE) catalyse double bond formation following elimination of water; ketoreductase (KR) catalyses hydroxylation reaction. The last module of PKS consists of a TE (thioesterase) domain at the end which helps in the release of nascent polyketide by hydrolysis. The detached nascent polyketides are later on modified by numerous tailoring enzymes including oxidoreductases, group transferases, halogenases and cyclases to transform into different types of bioactive polyketides (Fig. 15.1) (Olano et al. 2010; Gokulan et al. 2014; Singh et al. 2019). The structural diversities of polyketides are introduced at various stages of synthesis. This includes priming of different starter units and number of elongation step by choosing various types of acyl-CoA as extender units, chirality and number of modules. And as said earlier, the finishing modifications are presented by tailoring enzymes. Duplication of genes, modules/domain losses, recombination and horizontal gene transfer between various bacterial species also cause diversification of polyketides (Gokulan et al. 2014; Singh et al. 2019).

15.4.1.2 Type II Polyketide Synthase

Contrasting to type I PKSs, type II PKSs are monofunctional multienzyme complexes, taking part in the production of aromatic polyketides including tetracyclines and anthracyclines. While the synthesis mechanism of type I PKS is non-iterative, type II PKSs are iterative, i.e. continuous activity of same enzyme takes place for

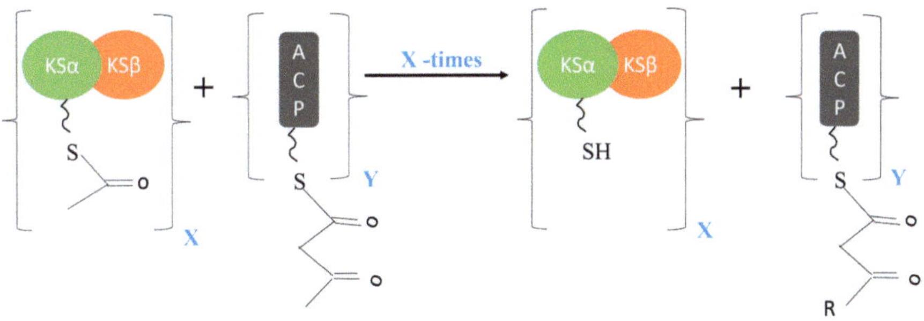

Fig. 15.2 Type II polyketide synthesis. Type II PKS composed of two ketosynthase subunits—KSα and KSβ and one ACP domain

the synthesis of polyketides. It is composed of two ketosynthase subunits like KSα and KSβ and one ACP. The primary precursor (e.g. acetyl-CoA) initially tethers to the ACP via thioester bond and acts as the starter unit of polyketide biosynthesis. Following that, condensation reactions are initiated by the KSα, forming Claisen type C–C bond between activated starter unit and decarboxylated malonyl-CoA. On the other hand, KSβ domain (chain length factor/CLF) regulates the elongation of carbon chain during the iterative cycle. Type II PKS uses only malonyl-CoA as the substrate for expansion of chain. Two additional domains like an AT domain and FabH-like KS domain are present in few type II PKS in order to choose a starter unit and to begin the chemical reaction (Fig. 15.2) (Gokulan et al. 2014; Singh et al. 2019).

15.4.1.3 Type III Polyketide Synthase

Type III PKSs also termed chalcone-like synthases are comparatively simpler enzymes, and unlike the other two types of PKSs, they lack the ACP structural domain. The mode of polyketide synthesis is iterative like type II PKS. All the substrate priming, decarboxylation, extension and other reactions like cyclization occur at single active site. Starter acyl-CoA differs among various type III PKSs, while malonyl-CoA or methylmalonyl-CoA is the universal extender units.

Structurally, two cavities are present in type III PKSs, one of which will occupy the starter unit (acyl-CoA), whereas the other one will be occupied by extender unit. At first, enzyme selects the starter acyl-CoA and it get binds to the substrate-binding pocket. A covalent thioester bond will form due to nucleophilic attack on starter unit by the cysteine residue there. Extender units bind to CoA binding pocket, and the active site amino acids carry out decarboxylation and handover these two carbon units to the developing chain. Condensation reaction differs for every type III PKS. The redox state of catalytic site and unsteadiness of various chemical groups present in growing chain encourage spontaneous chain cyclization (Fig. 15.3) (Gokulan et al. 2014; Singh et al. 2019).

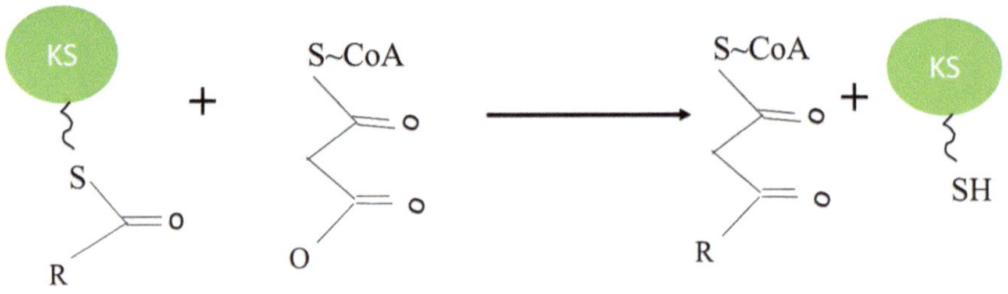

Fig. 15.3 Type III polyketide synthesis. Unlike type I and II PKSs, type III does not have the ACP structural domain

15.4.2 Biosynthesis of Non-ribosomal Peptides

Non-ribosomal peptides of varying structural and functional diversities are produced by complex multi-modular enzymes called non-ribosomal peptide synthetases (NRPSs). The building blocks are selected, recruited and added in a gradual manner to construct these peptides. The building blocks, i.e. amino acids, are not only proteinogenic, it can also be non-proteinogenic amino acid monomers such as L-, D- and α-, β- or δ-amino acids. Unusual amino acids such as 2-aminoisobutyric acid found in peptaibols and hydroxyphenylglycine found in glycopeptide antibiotics and dihydroxybenzoate of various siderophores are examples. Like ribosomal peptide synthesis, the peptide synthesis will always commence from N- to C-terminal (Duban et al. 2022).

Similar to type I PKSs, these enzymes are organized into different modules (Fig. 15.4). Each module comprises adenylation (A), thiolation (T) and condensation (C) domains. These are the core or essential domains. Apart from that, a thioesterase terminal core domain positioned at the end module will carry out the release of peptide chain from the enzyme. Sometimes, modules may contain secondary domains carrying out catalytic functions like epimerization and methylation. These domains are called supplementary domains. Adenylation or A-domain, which possess a substrate-binding pocket, carries out recognition of related building units and transferring of these building blocks to the thiolation domain of same module. Thiolation (T) or peptide carrier protein (PCP) domain contains a phosphopantetheine group; likewise, we explained about ACP domain of PKSs. Transferring of the substrate to PCP by A-domain leads to creation of a thioester linkage between phosphopantetheine group and the substrate. Condensation or C-domain carries out the peptide (C–N) bond formation between the amino acid monomers. Different types of C-domains have been recognized based on the condensation they perform. LCL subtypes carry out the peptide bond formation between two L-monomers, whereas DCL subtypes catalyse peptide bond formation between the D-monomer at the C-terminus of growing chain with an L-monomer added onto the succeeding module (Gokulan et al. 2014; Duban et al. 2022).

The terminal thioesterase domain releases the NRP from the enzyme complex via hydrolysis. Linear compounds undergo intramolecular reaction to produce

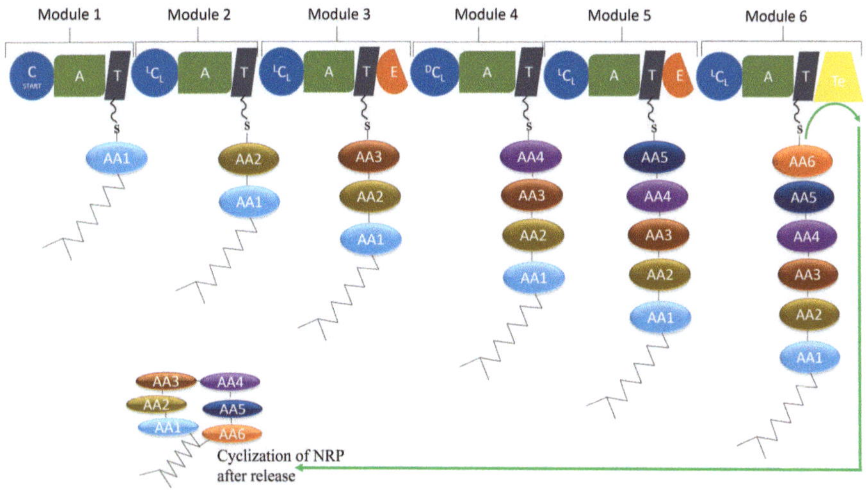

Fig. 15.4 Non-ribosomal peptide synthesis. C is the condensation domain; based on peptide bond they form, two types of C-domains either LCL or DCL are present in every module. A is the adenylation domain which activates the monomers by adenylation. T is the thiolation domain which contains the phosphopantetheine arm. E is the most common epimerase secondary domain. Te is the terminal core domain which helps the release of peptide chain from enzyme complex and decides the mode of NRP synthesis. AA1 to AA6 represents various amino acid residues

various cyclic peptides. Different cyclization strategies such as head-to-tail and side chain-to-tail will produce macrocycles of various ring sizes (Duban et al. 2022).

Secondary or auxiliary/supplementary domains catalyse the modification of incorporated monomers during the peptide synthesis. Modifications of amino acids can be by methylation, formylation, epimerization, etc. Epimerization (E) domain is the most recurrent secondary domain. It converts the L-amino acids to D-amino acids by isomerization of its α-carbon. Generally, E-domain will be located following the thiolation domain. Methylation (MT) domain, another secondary domain located generally upstream to thioesterase domain, catalyses the methylation of peptide backbone by transferring a $-CH_3$ group from *S*-adenosylmethionine (SAM) to the C or N atom of the amide bond. Based on their sequence, *N*-methyltransferase and *C*-methyltransferase domains get differentiated. Formylation (F) domains are rarely present secondary domain found at the initiation module. It formylates the N-terminus of the amino acid (Duban et al. 2022).

There are different modes of synthesis of non-ribosomal peptide. They are type A or linear, type B or iterative, and type C or nonlinear. In the type A mode of biosynthesis, monomers are added as per the order of module assembly line of NRPS. The final peptide product can either be linear (e.g.: penicillin precursor ACV) or macrocyclized forms as a result of head-to-tail condensation (e.g.: gramicidin S) and side chain-to-tail condensation (e.g.: fengicin). In case of type B biosynthesis, the terminal core domain thioesterase allows the multienzyme complex to repeat twice the synthesis process hitherto. The cyclic decapeptide gramicidin S and the NRP siderophores including enterobactin are synthesized through this mode of synthesis. The third mode of biosynthesis, type C, includes non-strict linear and iterative modes, depending on the order in the use of the modules and domains. An uncommon organization of the core domains is observed in the type C NRPSs which results in rare internal cyclization or branch-point syntheses. So, the product from this NRPS cannot be predicted easily from the enzyme architecture (Duban et al. 2022).

15.5 Evolution of Polyketide Synthase and Non-ribosomal Peptide Synthetases

In the previous section, we discussed that the type I polyketide synthases and non-ribosomal peptide synthetases are huge enzymes consisting of different modules. Because of their multi-modular architecture, these megasynthases manufacture their products in assembly line logic, i.e. repeated chemical condensation steps cause the incorporation of small monomeric units to produce a linear oligomer (Fischbach and Walsh 2006; Amoutzias et al. 2008). This modular organization is definitely creating a great energy burden to the cells. The fact that why these microbial cells rely on this high energy-dependent mechanisms to produce these metabolites even though similar products can be synthesized by processes like ribosomal peptide synthesis is still elusive (Nikolouli and Mossialos 2012). The gene-cluster-level phylogenetic analysis of PKS and NRPS is difficult due to their varied genetic organization. Mostly, evolutionary history is obtained by analysis of phylogenetic trees of individual domains. Current studies are focussed on a family of same or metabolites alike, which share similar or related pathway of biosynthesis and supposed to have either a common ancestor or evolved via horizontal gene transfer (HGT). This inference will be made by determining the congruence of gene and species tree. Phylum or domain level analysis suggested possibilities of the combined gene transfer events (vertical and horizontal). PKSs and NRPS sporadic distribution were also been explained by HGT between bacteria and fungi (Wang et al. 2015). Apparently, a series of events like mutation, duplication, insertion/deletion, recombination and reshuffling made the evolution of these complex enzyme systems. Various studies clearly evidenced and confirmed these hypotheses (Jenke-Kodama et al. 2005; Gallo et al. 2013; Medema et al. 2014).

A comprehensive phylogenetic analysis based on KS domains of multi-modular and iterative PKS of bacteria and fungi and of the various FAS from various organisms shown that the PKS in bacteria display joint evolution with FAS of animals. High influence of gene duplications and gene losses has been detected in the phylogenomic analysis of type I PKSs (Jenke-Kodama et al. 2005; Jenke-Kodama and Dittmann 2009). Phylogenomic analysis based on activation domains points out that, during evolution, PKS-I genes of bacteria were separated in two different clades. Precisely, first clade includes all acyltransferase domains that activate malonyl-CoA and few not having any anticipated substrates. On the other hand, second clade comprises those activating methylmalonyl-CoA or other uncommon substrates (Jenke-Kodama and Dittmann 2009).

A systematic analysis of various enzymatic types may unravel the evolutionary origins of the PKSs and NRPSs. The speculated convergent evolution of modular PKSs and NRPSs can be assessed by phylogenetic study of the hybrid gene cluster, particularly hybrid proteins with chimeric PKS and NRPS domains (Wang et al. 2015).

15.6 Production and Bioengineering of PKs and NRPs

Biomanufacturing, i.e. use of biological systems such as microbes, animal and plant cells or tissues, enzymes, or in vitro synthetic (enzymatic) systems for the production of commercially significant biomolecules, aimed at usage in the food, agricultural and biomedical industries (Zhang et al. 2017). Low yield and excessive loss during the downstream processing were always been the hurdles for the production of industrially important microbial secondary metabolites like polyketides and non-ribosomal peptides. Undoubtedly, introduction of bioengineering especially metabolic engineering, synthetic and systems biology, and adaptive laboratory evolution contributed to the industrial production of desired products in high quality and quantity, low cost and gained more consumer acceptance. These techniques that aim straight to the genome cause redirecting metabolic network flux of carbon and nitrogen towards the target product, circumvent the inhibition and suppression mechanisms which restricts the maximum productivity, and even rise the substrate concentration of the cell for production of desired product thereby helping in optimization and improvement of microbial strains for the mass production of these bioactive secondary metabolites.

An important tool in the genomic era of discovery and developing natural products is heterologous expression. Next-generation sequencing technologies confirmed the presence of several 'cryptic' putative biosynthetic gene clusters (BGCs) which codes for non-ribosomal peptides and polyketides. Since these organisms are uncultivable or do not produce these bioactive metabolites in the tested laboratory culture conditions, engineering these genes to other organisms, i.e. heterologous expression, can help (Cook and Pfleger 2019). Recently, Park et al. isolated a giant BGC encoding an antifungal polyene polyketide called neotetrafibricin from a strain of *Streptomyces rubrisoli* and heterologous expression of this BGC boosted its yield level by more than 2.5-fold compared to wild type, and the I-NTF aglycone-producing heterologous host strain yielded more than 30 times higher than mutant strain of *S. rubrisoli* (Park et al. 2022). In another study, *Saccharomyces cerevisiae* was engineered for the production of a bacterial NRP, indigoidine. Indigoidine is a blue pigment which can be considered as an excellent environment-friendly alternative in dye industry. The *Streptomyces lavendulae* NRPS-BpsA was introduced to the yeast *S. cerevisiae*. The active holo-BpsA can convert L-glutamine to indigoidine. They succeeded in a high-titre production of the indigoidine using the fungal host (Wehrs et al. 2018). Similarly, they also engineered fungus *Rhodosporidium toruloides* by integrating codon-optimized sequences of the *B. subtilis* PPTase *sfp* and the NRPS gene *bpsA* via *A. tumefaciens*-mediated transformation. The indigoidine titre reached 2.9 ± 0.8 g/L, the highest reported titre of indigoidine till date (Wehrs et al. 2019). Heterologous expression is also applicable for the production of polyketide-non-ribosomal peptide hybrids. *Thm* genes, which are responsible for the production of nematocidal polyketide-non-ribosomal peptide hybrid-macrolactone from the thermophilic fungi *Talaromyces thermophilus*, were heterologously expressed and bioengineered in the type strains of *Aspergillus nidulans* and *E. coli*. A drastic increase in yield was reported (Zhang et al. 2020).

Ahmadi et al. used metabolic engineering and induction variable optimization for extending the production of a polyketide-non-ribosomal peptide-yersiniabactin. Yersiniabactin is a siderophore which can be used to selectively eliminate metals from different samples. Here, they altered various supporting pathways of substrate precursor via gene expression and supplementation of substrate precursors exogenously in order to enhance the synthesis of target compound. They also optimized temperature and inducer concentration through gene expression induction and succeeded in extending yersiniabactin production to a concentration of ~175 mg/L (Ahmadi and Pfeifer 2016).

With the advent of synthetic biology, in vitro, cell-free systems (CFS) have emerged as potent platforms for the biomanufacturing of various bioactive metabolites. Exactly, CFS allows the study of enzyme biosynthesis within a test tube (Moore et al. 2023). A recent example is the in vitro synthesis of NRP antibiotic valinomycin by Zhuang et al. For the study, initially they used a cell-free protein synthesis (CFPS) system for the expression of entire valinomycin BGC in a single-pot reaction. After optimization, the valinomycin titre was ~37 μg/L. After that, two enzyme-enriched cell lysates were mixed for coupling CFPS with cell-free metabolic engineering system and thereby performed a two-stage biosynthesis. This approach enriched valinomycin yield up to 5000-fold (Zhuang et al. 2020). Synthesis of ikarugamycin, a bioactive compound from *Streptomyces*, was achieved via in vitro total one-pot synthesis using three recombinant enzymes—IkaA, IkaB and IkaC (Greunke et al. 2017).

Another study by Goering et al. validated the use of CFPS for the expression of two NRPS (GrsA and GrsB1) responsible for the synthesis of cyclic dipeptide D-Phe-L-Pro diketopiperazine. An *E. coli*-based CFPS system was used to express these NRPS proteins, where DNA is given as the direct input. Inactive cell-free expressed form of GrsA and GrsB1 was then transformed to their active form by supplementation of post-CFPS reaction mixture with purified PPTase Sfp. Later on, single-pot reactions performed for the co-expression of these NRPSs and thereby diketopiperazine were synthesized. LC-MS analysis confirmed high-level production of the product diketopiperazine (~12 mg/L), and the production level was more than previous cell-based recombinant expression reports (Goering et al. 2017).

15.7 Identification of Bioactive Compounds Through Genome Mining

For several decades, researchers used to identify and characterize bioactive secondary metabolites through conventional methods such as isolation, purification, bioactive molecule identification through biochemical assays and structural elucidation. Such way of identification requires labour as well as time. Later, the advancement in bioinformatics and the lower in cost of whole genome sequencing hasten the identification process. Through whole genome sequencing, the genomic sequences of the secondary metabolite producer strains can be identified earlier or in similar with conventional isolation methods. Highly compact gene clusters responsible for the bioactive molecule production facilitate such kind of analysis (Blin et al. 2013).

Combination of computational studies with culturing techniques aids in drug discovery. Genome scanning became an influential tool for the exploration of secondary metabolites that includes polyketides (PKs), non-ribosomal peptides (NRPs) and terpenoids. Computational methods include multiple sequence alignment, and crystallographic studies allow to explicate the role and characteristics of vital enzymes encompassed in the synthesis of bioactive molecules such as non-ribosomal peptide synthetases (NRPSs) and polyketide synthases (PKSs) (Li et al. 2009).

Numerous challenges have to be endured for the identification of PK and NRP through mining approaches. Initial step is to recognize the position of PK and NRP biosynthetic pathway in the genome. One such conventional approach is to identify with an ortholog sequences of known/predicted protein in the pathway. These conserved catalytic domains include ketosynthases (KS) from PK pathways, and adenylation (A) or condensation (C) domains from NRP pathways are frequently used. One such query sequence from *Bacillus subtilis* is PksJ, which comprised both type 1 PKS and NRPS domains, assisting in the detection of both pathways from the genome (Boddy 2014).

Identification of known as well as novel bioactive clusters from genomes and characterization of such biosynthetic gene clusters (BGC) are highly complex processes. In order to ease the analysis, machine learning algorithms are used to predict BCGs and novel secondary metabolite structures from the sequences. With computational tools and known BCG data, secondary metabolites can be identified with their biosynthetic pathways. antiSMASH is an example of such type of tool, which uses profile hidden Markov models (pHMMs) to detect gene clusters. A novel class of cryptic BCG, lanthipeptides, was categorized with ML and deep learning (DL) approaches. In some cases, BCGs may not show expression of any bioactive molecules; however, the use of elicitors such as small molecules or co-culturing can induce the expression and feasible to predict BGCs using ML tools (Sahayasheela et al. 2022). Mining of fungal endophyte, *Ascomycete* sp. F53 from *Taxus yunnanensis*, a Chinese medicinal plant discovered 35 plausible BGCs. An in silico tool includes antibiotics, and Secondary Metabolite Analysis Shell (antiSMASH) detects the additional clusters from separated fragments or the clusters that were too fragmented. F53 contains a tandem iterative type 1 PKS pathway, and *lij* encodes for the novel polyketide compound, azaphilone (Cain et al. 2020).

In a study, strains were subjected to genome mining for cluster identification contained in the secondary metabolite production using in silico tools, namely antiSMASH, NapDos, Np.search and BAGEL3. The existence of biosynthetic clusters in Gram-negative marine bacteria was found, and some strains possess eight distinct PKS and NRPS gene clusters. These clusters encoding bioactive molecules were spread among alpha and gamma proteobacteria, where the latter encodes more NRPS and PKS clusters (Machado et al. 2015).

Researchers found that exploring the genome of *Streptomyces avermitilis*, an industrially important actinomycete, possesses 38 bioactive gene clusters and 46 insertion sequence (IS) like sequences. Before the availability of genome mining, the structure and synthesis of avermectin and polyketide macrocyclic lactone, oligomycins, have established. From genome sequence, 11 polyketide synthase (PKS) gene clusters were distinguished, which leads to the detection of filipins, a third group of macrocyclic lactones. Moreover, *S. avermitilis* genome harbours bioactive gene clusters for 8 NRPs, 6 terpenoids assorted pigments and so on (Ikeda et al. 2014).

In a study, the abundance of NRPs and PKS in sphagnum bog ecosystem was identified using genome mining and metagenomics. Around 279 NRPSs, 346 PKSs and 40 PKS-NRPS hybrid gene sequences were identified. NRPS is abundant in *Burkholderia* genus of *Proteobacteria* phylum, whereas PKS gene clusters were predominant in Actinobacteria. Using PCR screening, 13 novel NRPS-associated sequences were distinguished, which were encoded in the phyla *Proteobacteria*, *Actinobacteria* and *Cyanobacteria* (Müller et al. 2015).

Several bioinformatic algorithms were developed in recent years in order to identify BCGs based on genomic sequences. ClustScan, SMURF, CLUSEAN and SBSPKS are the tools for the detection of enzymatic domains in polyketide synthases and non-ribosomal peptide synthetases. NRPSpredictor, NRPSSP and PKS/NRPS predictive BLAST server are used for the substrate identification of the biosynthetic pathways by predicting the structure of compounds based on genomic sequences (Blin et al. 2013).

In a study, *Sorangium cellulosum* genome was used to detect the PK-NRP gene clusters using antiSMASH, NP.searcher and NaPDoS. NP.searcher detects 2 polyketides, 1 non-ribosomal peptide and 2 PK-NRP hybrid gene clusters; antiSMASH identified 5 polyketides, 2 non-ribosomal peptides and 4 PK-NRP hybrids; NaPDoS discovered 50 ketosynthases (KS) and 20 condensation (C) domains from the NRP and PK-NRP hybrid gene clusters. Authors manually annotated the genome and identified 3 polyketides, 2 non-ribosomal peptides and 4 PK-NRP hybrids. Software such as antiSMASH provides insights for pathway identification, and NaPDoS offers KS and C-domain identification. NP.searcher and NaPDoS analyse the data and provide results rapidly, while antiSMASH analyses the results in more than 4 h based on the load (Boddy 2014).

Some other tools for the exploring the bioactive compounds and biosynthetic pathways include Basic Local Alignment Search Tool (BLAST) and the profile hidden Markov model suite of tools (HMMER). BLAST allows to identify the protein with common short region with known NRPS or PKS, whereas HMMER offers to explore the proteins that contain characteristic domains such as C-domains (a characteristic of NRPS). GOLD (Genomes Online) database contains genome and metagenomic data of more than 2700 bacterial genomes and more than 150 archaeal entries (http://www.genomesonline.org). Another database called NORINE focussed on the biological activity, structure and reference of the non-ribosomal peptides and provides access to more than 1000 peptides (http://bioinfo.lifl.fr/norine/) (Nikolouli and Mossialos 2012). ClusterBLAST, a sequence alignment tool in BLAST, provides comparison of BCGs, where already known and characterized BCGs were eliminated. PRISM and MetaBCG are two BCG mining tools, where former detects and predicts the biological activity via machine learning, while latter detects gene clusters from gene sequences. The Integrated Microbial Genomes Atlas of Biosynthetic Gene Clusters (IMG-ABC) offers access to >4 lakhs BGCs identified from fungi, bacteria, archaea and metagenomes. Another database called Minimum Information about a Biosynthetic Gene Cluster (MIBiG) offers facts about all the known BCGs and the metabolites synthesized. This database consists of 1923 entries of bioactive compounds (Malit et al. 2022). Advancement in mining with artificial intelligence (AI) improves the discovery of putative BCGs through algorithms. AdenylPred, ClusterFinder and DeepBGC are some of the notable AI-assisted BCG mining tools. Mining of bioactive metabolites from the unexplored resources leads to the detection of unique molecules from the biosynthetic gene clusters (Li et al. 2023). Table 15.1 enlists the bioactive secondary metabolites identified through genome mining analysis. Hence, genome mining provides an insight for selecting potential bioactive metabolite producers, which needed to be tested in vitro. Using the bioinformatic tools, rapid screening, identification and development of novel bacteria and their metabolites would be easier.

Table 15.1 Different bioactive molecules identified through genome mining and metagenomic methods with their source and properties

Compounds	Source	Activity/mode of action	References
Genome mining			
Gramibactin	Rhizosphere bacteria, *Paraburkholderia graminis*	Siderophore	Hermenau et al. (2018)
Malleilactone	*Burkholderia pseudomallei*	Cytotoxic siderophore	Biggins et al. (2012)
Varicidin A	*Penicillium variabile*	Antifungal	Tan et al. (2019)
Lobatamide A	*Gynuella sunshinyii*	Cytotoxin	Ueoka et al. (2020)
Brevicidine and laterocidine	*Brevibacillus laterosporus* DSM 25	Antibacterial activity against *P. aeruginosa* and colistin-resistant *E. coli*	Li et al. (2018)
Pyxidicyclines	*Pyxidicoccus fallax* An d48	Bacterial and human topoisomerase	Panter et al. (2018)
Kanglemycins	*Amycolatopsis mediterranei*	Activity against rifampicin resistant strains such as *Mycobacterium tuberculosis*	Peek et al. (2018)
Formicamycins	*Streptomyces formicae*	Active against methicillin-resistant *Staphylococcus aureus* (MRSA) and vancomycin-resistant e*nterococci* (VRE)	Qin et al. (2017)
Metagenomic approach			
Misakinolide Theonellamide	*Theonella* sp.	Cytotoxic	El Samak et al. (2023)
Lasonolide A	Marine sponge, *Forcepia* sp.	Anticancer	Uppal et al. (2022)
Malacidin	Soil	Against MDR Gram-positive organisms	Hover et al. (2018)
Clarepoxcins A–E and landepoxcins A and B	Soil microbiomes	Proteasome inhibitors	Owen et al. (2015)
Mycalamides, pateamines	Sponge, *Mycale hentscheli*	Cytotoxin	Rust et al. (2020), Dyshlovoy et al. (2012), Northcote et al. (1991)
Pelorusides	Sponge, *Mycale hentscheli*	Anticancer	Rust et al. (2020)
Curacin A	Tropical marine cyanobacterium, *Lyngbya majuscula*	Antiproliferative and cytotoxicity	Chang et al. (2004)
Turbomycin A and B	*Legionella pneumophilia*	Wide-spectrum inhibition towards Gram-positive and negative bacteria	Gillespie et al. (2002)
Metathramycin	*Streptomyces albus*	Colon cancer	Stevenson et al. (2021)

15.8 Metagenomics Approaches in Exploring PKs and NRPs

Plenitude of bioactive metabolites was identified from the cultured microorganisms; however, some organisms remain uncultured. The exploration of diverse environment through metagenomic studies facilitates the identification of microbial consortium or association and their interaction within the community. The advancement in genome sequencing techniques aids in the generation of huge data. Similarly, progression in genome mining extends the study in depth. Metagenomic approaches include sample collection, DNA extraction, library construction for sequencing and taxonomic and functional analysis (Puig-Castellví et al. 2023).

Metagenomics offers two discrete approaches, comparative and functional metagenomics. In comparative metagenomics, extraction of DNA from crude samples followed by high-throughput sequencing leads to the detection of BGCs based on sequence similarity. Sequence-based approaches are widely used in metagenomic studies as they identify the gene without gene expression in the host (Nikolouli and Mossialos 2012). Identification of BGCs is via shotgun sequencing or targeted sequence amplification using polymerase chain reaction (PCR). Second-generation massively parallel short-read sequencing has improved PCR-based analysis. However, shotgun approaches faced difficulties in the identification of entire BGCs because of large enough assemblies of complex metagenomic data. Third-generation long-read sequencing empowers shotgun sequencing and provides information about the BGC diversity within the microbiomes (Rosenzweig et al. 2023).

In functional metagenomics, isolated DNAs were used to construct libraries, which are further screened for the identification of bioactive compounds. Once the specific bioactive molecule is identified, the clone containing biosynthetic gene is subjected to sequencing and compared with the sequences of other organisms, through which known and unknown clusters can be identified (Nikolouli and Mossialos 2012). *E. coli* is widely used as the host, and other bacteria such as *Streptomyces*, *Pseudomonas* and *Bacillus* were also used. Heterologous expression of BCGs was studied in most cases; however, such expression could be hindered due to large size of BCGs, variation in codon usage, lack of controlling factors, protein folding alterations, gene product toxicity, impaired protein secretions and so on (Amoutzias et al. 2016).

Wei et al. identified PKs and NRPS producers from the 16 s rRNA fragments of yellow sea sediments. Around 62% of the bacterial species are dominated by *Proteobacteria* and *Bacteroidetes*, while *Actinobacteria*, commonly considered as bioactive molecule producers, accounts for only 0.82% of the total bacteria identified. Based on the distribution of ketosynthase domain, adenylation domain, dTDP-glucose-4,6-dehydratase (dTGD) and the diverseness of PKs and NRPs were assessed. They showed the richness of adenylation and dTGD genes, where some of these genes have less than 50% similarities with the known genes. However, the abundance of KS genes was lesser and showed more than 60% similarities with the known ones. Furthermore, fosmid library clones were constructed to screen bioactive gene clusters of the sediment. Out of 9 fosmid clones having PK and NRP clusters, three were allotted to proteobacteria, five were assigned to actinobacteria and one for unknown bacteria. Assembly of these 9 clones showed 11 contigs consisting of NRPS, PKS and PK-NRPS hybrid BGCs. Moreover, these clusters have less than 60% similarities with the already known candidates, which might be contained in novel bioactive secondary metabolite synthesis (Wei et al. 2018).

Hover et al. 2018 identified a new antibiotic, malacidin, through sequence-based screening of BGC coding for the conserved Asp-X-Asp-Gly calcium binding motifs from the soil samples. The incidence of PKs and NRPs in the microorganism of Krubera-Voronja caves, an oligotrophic deepest cave in the world, was studied. In this study, samples such as sediments, soils, clay from walls, water, speleothems and coloured spots from the cave wall were analysed. The predominant phyla obtained are *Actinobacteria* and *Proteobacteria*; however, the presence of *Firmicutes*, *Acidobacteria*, *Cyanobacteria*, other phyla and unclassified bacteria was detected from the sequences obtained from the cave. Abundance of KSα and KSβ sequences was identified from the *Streptomyces* sp. High diverseness of NRPS was also reported, and the adenylation domain showed presumptive NRPS sequences in the GenPept database. In this study, they predicted novel metabolites such as an angucycline group PK, a massetolide A-like cyclic lipopeptide and a surfactin-like lipopeptide (Lukoseviciute et al. 2021).

Rhizosphere is a complex habitat, where plant roots, soil and microorganisms interact each other. This area is plentiful in organic matter and nutrients, which stimulates the root microbiome. These organisms enhance the plant growth and prevent the soil-borne pathogens by producing metabolites or induce resistance to plants. Most of these metabolites are NRPs or PKs. Dror et al. investigated the NRPSs and PKS clusters in the root samples of tomato and lettuce and adjacent bulk soil samples. BGCs of NRPSs and PKSs of soil and root ecosystem are highly abundant and diverse and have fewer identities (50–70%) with that of MIBiG repository. These clusters were screened with known cosmid libraries, and five putative whole BCGs were identified; three were assigned to *Actinobacteria*, while remaining belongs to unknown ones. One BCG assigned to *Streptomyces* was predicted to synthesize an antifungal molecule, polyene (Dror et al. 2020). In another study, the occurrence of NRPS- and PKS-encoding BCGs from soil and root microbiome was studied from the public metagenome data. antiSMASH-BD tool annotated the abundance of BCGs in both soil- and root-associated microbiome is almost similar. The identified sequences were mostly novel clusters with low similarity with known sequences. This study confirms the presence of unexplored diversity and BCGs of soil and root microbiota, which are involved in the plant-root-soil interaction (Dror et al. 2023).

Marine sponges harbour diverse range of unique bioactive molecules. Researchers are identifying such unexplored microorganisms through metagenomic studies. In a study, the heterogeneity of NRP and PK gene clusters was studied by exploring the Australian marine sponges using deep sequencing tag-encoded FLX amplicon pyrosequencing (TEFAP) method. They found sequences of 100 unique ketosynthase domains, and 400 condensation domains were obtained from six sponges. Both pathways were identified in *Scopalina* sp., *Cymbastela coralliophila* and *Cymbastela concentrica*, while PKS pathways were detected in *Stylissa* sp. and *Rhopaloeides odorabile.* This study concluded that diverse niches can be explored by this approach and gains knowledge about the biosynthetic pathways involved in the bioactive metabolite production (Woodhouse et al. 2013). EI Samak et al. reported gene clusters in the microbiome of Egyptian red sea sponge, *Theonella* sp. Out of 38 phyla identified, Poribacteria represented as the plentiful one with an average of 27.5%. The gene clusters detected were involved in the synthesis of terpenes, NRPs, PKs, ribosomally synthesized and post-translationally modified peptides (RiPPs), hybrid PKS-NRPS and others. The predominant cluster involved is terpene pathway, followed by NRPS, PKS and RiPPs biosynthetic clusters. Some of the identified clusters were included in the synthesis of molecules already reported for the anticancer activity that includes theonellamide, misakinolide, microsclerodermin, cryptophycin, hapalosin and ecotine. Similarly, some of the clusters take part in the production of antimicrobial compounds such as hygromycin A, friulimicin A–D, atratumycin, anikasin and icosalide A and B (El Samak et al. 2023).

A study reported the occurrence of bioactive gene clusters in the gut microbiome of mud crab, *Scylla serrata*. Out of 36 bacterial isolates, 3 (FKP2-4, FKP4-1 and FKP2-16) possess NRPS and 2 (FKP2-4 and FKP4-1) have PKS genes. After sequencing and BLAST analysis, the isolates FKP2-4, FKP4-1 and FKP2-16 were identified as *Acinetobacter variabilis*, *Vagococcus fluvialis* and *Staphylococcus arlettae*. These isolates showed antagonistic activities against *E. coli* and *S. aureus* (Banu et al. 2023). Some of the bioactive molecules reported through metagenomic analysis are enlisted in Table 15.1. Metagenomics provides information about those uncultured microorganisms and their diversity in the environmental samples. Combined action of traditional culture methods with metagenomics approaches can unravel the microbial diversities and metabolites from diverse niches.

15.9 Applications of PKs and NRP Molecules

Bioactive secondary metabolites such as PKs and NRPs harbour wide range of applications from health to industries. Several medically relevant polyketide antibiotics and non-ribosomal peptide antibiotics were identified through various bioinformatic tools. These molecules also showed wide array of environmental applications. Some of these molecules are enlisted in Table 15.2. NRP and PK molecules are greatly applied in health fields. Several antimicrobials and anti-tumour molecules from microorganisms and plants are NRPS-PKS origin (Martínez-Núñez and López 2016).

Table 15.2 Bioactive metabolites produced through PK/NRPS biosynthetic pathways and their applications

PK/ NRPS	Compound class	Compounds	Application/mode of action	Organisms	References
NRPS	β-Lactams	Penicillin Cephalosporin	Antibacterial; cell wall synthesis	Fungal genera, *Penicillium*, *Cephalosporium*, *Aspergillus*; Bacteria, *Streptomyces*, *Nocardia*, *Flavobacterium*, *Lysobacter* sp.	Felnagle et al. (2008)
	Cyclic lipopeptide	Daptomycin	Skin infection by Gram-positive bacteria-methicillin- and vancomycin-resistant *S. aureus*	*Streptomyces roseosporus*	Miao et al. (2005)
	Undecapeptides	Cyclosporin A	Fungicide, anti-inflammatory, immunosuppressant	*Tolypocladium inflatum*	Felnagle et al. (2008)
	Glycopeptides	Vancomycin Teicoplanin	MRSA infections, endocarditis Vancomycin-resistant enterococci	*Amycolatopsis orientalis* *Actinoplanes teichomyceticus*	Durand et al. (2019)
	Quinoxalines	Thiocoraline Echinomycin Luzopeptins Quinoxapeptins	DNA polymerase inhibition; anti-tumour activity RNA polymerase disruption; anti-HIV	*Micromonospora marina*	Felnagle et al. (2008)
	Aminoglycosides	Capreomycin	Antituberculosis	*Saccharothrix mutabilis* subsp. *capreolus*	Heifets et al. (2005)
	Beta-lactams	Carbapenem	Antibacterial	*Streptomyces cattleya*	Birnbaum et al. (1985)
	Beta-lactam	Monobactam	Antibacterial	*Chromobacterium violaceum*	Durand et al. (2019)
	Lincosamides	Lincomycin	Active against Gram-positive bacteria	*Streptomyces lincolnensis*	Wang et al. (2018)
	Polypeptides	Polymixin	Active against Gram-negative bacteria	*Paenibacillus polymyxa*	Tambadou et al. (2015)
	Streptogramins	Streptogramin B Pristinamycin	Bacteriostatic activity by protein synthesis inhibition Wide-spectrum activity towards Gram-positive and Gram-negative bacteria	*Streptomyces graminofaciens* *Streptomyces pristinaespiralis*	Mast and Wohlleben (2014)

NRPS/ hybrid	Glycosylated peptides	Bleomycins	Oesophageal cancer, lymphomas, squamous cell carcinoma, recalcitrant warts, testicular cancer	*Streptomyces* sp.	Galm et al. (2005), Hecht (2000)
	Rifamycin	Rifampicin	Broad-spectrum activity towards Gram-positive and Gram-negative bacteria	*Streptomyces mediterranei*	Floss and Yu (2005)
PKS	Tetracyclines	Chlortetracycline Oxytetracycline	Active against Gram-positive and Gram-negative bacteria, mycoplasma, etc.	*Streptomyces aureofaciens*, *Streptomyces rimosus*	Petković et al. (2017)
	Macrolides	Erythromycin	Broad-spectrum activity towards Gram-positive bacteria, pulmonary infections	*Saccharopolyspora erythraea*	Staunton and Wilkinson (1997)
		Josamycin	Used for respiratory tract infections and sexually transmitted diseases	*Streptomyces narbonensis* var. *josamyceticus*	Arsic et al. (2018)
		Spiramycin	Used for *Toxoplasma gondii* infections and soft tissue infections; active against Gram-positive and Gram-negative bacterial infections	*Streptomyces ambofaciens*	Arsic et al. (2018)
	Macrolides	Fidaxomicin	Used against *Clostridium difficile* infections; RNA polymerase inhibition	*Dactylosporangium aurantiacum*	Venugopal and Johnson (2012)
	Polyketides	Mupirocin	Used against MRSA	*Pseudomonas fluorescens*	Gurney and Thomas (2011)

15.10 Conclusion

Pharmaceutical drugs used for the treatment of disease were discovered from the BGCs of several microorganisms. Polyketides and non-ribosomal peptides were the two important classes of bioactive molecules, having wide array of applications in health and industries. Diverse antibiotics, anti-tumour, immunosuppressive drugs, biosurfactants, etc. were emerged from these classes. Manipulation of bioactive biosynthetic machinery can enhance the enormous production of these molecules. Due to the advancement in bioinformatic tools such as sequencing, metagenomics and mining, identification of bioactive metabolite producers from the environment become easier. Machine learning and algorithms paved direction for the detection and characterization of novel as well as known clusters. Although bioactive properties of the microbial metabolites were established, transformation of such molecules into pharmaceutical products is challenging and needs to be addressed.

References

Ahmadi MK, Pfeifer BA (2016) Improved heterologous production of the nonribosomal peptide polyketide siderophore yersiniabactin through metabolic engineering and induction optimization. Biotechnol Prog 32:1412–1417

Amoutzias GD, Chaliotis A, Mossialos D (2016) Discovery strategies of bioactive compounds synthesized by nonribosomal peptide synthetases and type-I polyketide synthases derived from marine microbiomes. Mar Drugs 14:80

Amoutzias GD, Van de Peer Y, Mossialos D (2008) Evolution and taxonomic distribution of nonribosomal peptide and polyketide synthases. Future Microbiol 3:361–370

Arsic B, Barber J, Čikoš A et al (2018) 16-membered macrolide antibiotics: a review. Int J Antimicrob Agents 51:283–298

Baerson SR, Rimando AM (2007) A plethora of polyketides: structures, biological activities, and enzymes. In: Polyketides ACS symposium series. American Chemical Society, pp 2–14

Banu S, Alva S, Prabhu PJ et al (2023) Detection of non-ribosomal and polyketide biosynthetic genes in bacteria from green mud crab Scylla serrata gut microbiome and their antagonistic activities. Fish Shellfish Immunol Rep 24:100104

Biggins JB, Ternei MA, Brady SF (2012) Malleilactone, a polyketide synthase-derived virulence factor encoded by the cryptic secondary metabolome of *Burkholderia pseudomallei* group pathogens. J Am Chem Soc 134:13192–13195

Birnbaum J, Kahan FM, Kropp H et al (1985) Carbapenems, a new class of beta-lactam antibiotics: discovery and development of imipenem/cilastatin. Am J Med 78:3–21

Blin K, Medema MH, Kazempour D et al (2013) antiSMASH 2.0—a versatile platform for genome mining of secondary metabolite producers. Nucleic Acids Res 41:W204–W212

Boddy CN (2014) Bioinformatics tools for genome mining of polyketide and non-ribosomal peptides. J Ind Microbiol Biotechnol 41:443–450

Cain JW, Miller KI, Kalaitzis JA et al (2020) Genome mining of a fungal endophyte of Taxus yunnanensis (Chinese yew) leads to the discovery of a novel azaphilone polyketide, lijiquinone. Microb Biotechnol 13:1415–1427

Chang Z, Sitachitta N, Rossi JV et al (2004) Biosynthetic pathway and gene cluster analysis of Curacin A, an Antitubulin natural product from the tropical Marine Cyanobacterium Lyngbya m ajuscula. J Nat Prod 67:1356–1367

Cook TB, Pfleger BF (2019) Leveraging synthetic biology for producing bioactive polyketides and non-ribosomal peptides in bacterial heterologous hosts. Medchemcomm 10:668–681

Dror B, Jurkevitch E, Cytryn E (2023) Global-scale analysis reveals distinct patterns of non-ribosomal peptide and polyketide synthase encoding genes in root and soil bacterial communities. Soil Ecol Lett 5:38–45

Dror B, Wang Z, Brady SF et al (2020) Elucidating the diversity and potential function of nonribosomal peptide and polyketide biosynthetic gene clusters in the root microbiome. mSystems 5:e00866-20

Duban M, Cociancich S, Leclère V (2022) Nonribosomal peptide synthesis definitely working out of the rules. Microorganisms 10:577

Durand GA, Raoult D, Dubourg G et al (2019) Antibiotic discovery: history, methods and perspectives. Int J Antimicrob Agents 53:371–382

Dyshlovoy SA, Fedorov SN, Kalinovsky AI et al (2012) Mycalamide A shows cytotoxic properties and prevents EGF-induced neoplastic transformation through inhibition of nuclear factors. Mar Drugs 10:1212–1224

El Samak M, Zakeer S, Hanora A et al (2023) Metagenomic and metatranscriptomic exploration of the Egyptian Red Sea sponge *Theonella* sp. associated microbial community. Mar Genomics 70:101032

Felnagle EA, Jackson EE, Chan YA et al (2008) Nonribosomal peptide synthetases involved in the production of medically relevant natural products. Mol Pharm 5:191–211

Fischbach MA, Walsh CT (2006) Assembly-line enzymology for polyketide and nonribosomal peptide antibiotics: logic, machinery, and mechanisms. Chem Rev 106:3468–3496

Floss HG, Yu TW (2005) Rifamycin mode of action, resistance, and biosynthesis. Chem Rev 105:621–632

Gallo A, Ferrara M, Perrone G (2013) Phylogenetic study of polyketide synthases and nonribosomal peptide synthetases involved in the biosynthesis of mycotoxins. Toxins 5:717–742

Galm U, Hager MH, Van Lanen SG et al (2005) Antitumor antibiotics: bleomycin, enediynes, and mitomycin. Chem Rev 105:739–758

Gillespie DE, Brady SF, Bettermann AD et al (2002) Isolation of antibiotics turbomycin A and B from a metagenomic library of soil microbial DNA. Appl Environ Microbiol 68:4301–4306

Goering AW, Li J, McClure RA et al (2017) In vitro reconstruction of nonribosomal peptide biosynthesis directly from DNA using cell-free protein synthesis. ACS Synth Biol 6:39–44

Gokulan K, Khare S, Cerniglia C (2014) Production of secondary metabolites of bacteria. In: Encyclopedia of food microbiology. Academic Press, pp 561–569

Greunke C, Glöckle A, Antosch J et al (2017) Biocatalytic total synthesis of ikarugamycin. Angew Chem Int Ed 56:4351–4355

Gurney R, Thomas CM (2011) Mupirocin: biosynthesis, special features and applications of an antibiotic from a Gram-negative bacterium. Appl Microbiol Biotechnol 90:11–21

Hecht SM (2000) Bleomycin: new perspectives on the mechanism of action. J Nat Prod 63:158–168

Heifets L, Simon J, Pham V (2005) Capreomycin is active against non-replicating M. tuberculosis. Ann Clin Microbiol Antimicrob 4:1–7

Helfrich EJ, Reiter S, Piel J (2014) Recent advances in genome-based polyketide discovery. Curr Opin Biotechnol 29:107–115

Hermenau R, Ishida K, Gama S et al (2018) Gramibactin is a bacterial siderophore with a diazeniumdiolate ligand system. Nat Chem Biol 14:841–843

Hertweck C, Luzhetskyy A, Rebets Y et al (2007) Type II polyketide synthases: gaining a deeper insight into enzymatic teamwork. Nat Prod Rep 24:162–190

Hover BM, Kim SH, Katz M et al (2018) Culture-independent discovery of the malacidins as calcium-dependent antibiotics with activity against multidrug-resistant Gram-positive pathogens. Nat Microbiol 3:415–422

Ikeda H, Kazuo SY, Omura S (2014) Genome mining of the *Streptomyces avermitilis* genome and development of genome-minimized hosts for heterologous expression of biosynthetic gene clusters. J Ind Microbiol Biotechnol 41:233–250

Jenke-Kodama H, Dittmann E (2009) Evolution of metabolic diversity: insights from microbial polyketide synthases. Phytochemistry 70:1858–1866
Jenke-Kodama H, Sandmann A, Müller R et al (2005) Evolutionary implications of bacterial polyketide synthases. Mol Biol Evol 22:2027–2039
Kennedy J, Auclair K, Kendrew SG et al (1999) Modulation of polyketide synthase activity by accessory proteins during lovastatin biosynthesis. Science 284:1368–1372
Li J, Neubauer P (2014) Escherichia coli as a cell factory for heterologous production of nonribosomal peptides and polyketides. New Biotechnol 31:579–585
Li MH, Ung PM, Zajkowski J et al (2009) Automated genome mining for natural products. BMC Bioinform 10:1–10
Li X, Gadar-Lopez AE, Chen L et al (2023) Mining natural products for advanced biofuels and sustainable bioproducts. Curr Opin Biotechnol 84:103003
Li YX, Zhong Z, Zhang WP et al (2018) Discovery of cationic nonribosomal peptides as Gram-negative antibiotics through global genome mining. Nat Commun 9:3273
Lim YP, Go MK, Yew WS (2016) Exploiting the biosynthetic potential of type III polyketide synthases. Molecules 21:806
Lukoseviciute L, Lebedeva J, Kuisiene N (2021) Diversity of polyketide synthases and nonribosomal peptide synthetases revealed through metagenomic analysis of a deep oligotrophic cave. Microb Ecol 81:110–121
Ma L, Zhang Q, Yang C et al (2020) 1.06-Assembly line and post-PKS modifications in the biosynthesis of marine polyketide natural products. In: Liu H-W, Begley TP (eds) Comprehensive natural products III. Elsevier, Oxford, pp 139–197
Machado H, Sonnenschein EC, Melchiorsen J et al (2015) Genome mining reveals unlocked bioactive potential of marine Gram-negative bacteria. BMC Genomics 16:1–12
Malit JJ, Leung HY, Qian PY (2022) Targeted large-scale genome mining and candidate prioritization for natural product discovery. Mar Drugs 20:398
Martínez-Núñez MA, López VE (2016) Nonribosomal peptides synthetases and their applications in industry. Sustain Chem Process 4:1–8
Masschelein J, Clauwers C, Awodi UR et al (2015) A combination of polyunsaturated fatty acid, nonribosomal peptide and polyketide biosynthetic machinery is used to assemble the zeamine antibiotics. Chem Sci 6:923–929
Mast Y, Wohlleben W (2014) Streptogramins—two are better than one! Int J Med Microbiol 304:44–50
Medema MH, Cimermancic P, Sali A et al (2014) A systematic computational analysis of biosynthetic gene cluster evolution: lessons for engineering biosynthesis. PLoS Comput Biol 10:e1004016
Miao V, Coeffet-LeGal MF, Brian P et al (2005) Daptomycin biosynthesis in Streptomyces roseosporus: cloning and analysis of the gene cluster and revision of peptide stereochemistry. Microbiology 151:1507–1523
Mohimani H, Liu WT, Kersten RD et al (2014) NRPquest: coupling mass spectrometry and genome mining for nonribosomal peptide discovery. J Nat Prod 77:1902–1909
Moore SJ, Lai HE, Li J et al (2023) Streptomyces cell-free systems for natural product discovery and engineering. Nat Prod Rep 40:228–236
Müller CA, Oberauner-Wappis L, Peyman A et al (2015) Mining for nonribosomal peptide synthetase and polyketide synthase genes revealed a high level of diversity in the Sphagnum bog metagenome. Appl Environ Microbiol 81:5064–5072
Nikolouli K, Mossialos D (2012) Bioactive compounds synthesized by non-ribosomal peptide synthetases and type-I polyketide synthases discovered through genome-mining and metagenomics. Biotechnol Lett 34:1393–1403
Northcote PT, Blunt JW, Munro MH (1991) Pateamine: a potent cytotoxin from the New Zealand marine sponge, Mycale sp. Tetrahedron Lett 44:6411–6414
Olano C, Méndez C, Salas JA (2010) Post-PKS tailoring steps in natural product-producing actinomycetes from the perspective of combinatorial biosynthesis. Nat Prod Rep 27:571

Owen JG, Charlop-Powers Z, Smith AG et al (2015) Multiplexed metagenome mining using short DNA sequence tags facilitates targeted discovery of epoxyketone proteasome inhibitors. Proc Natl Acad Sci 112:4221–4226
Panter F, Krug D, Baumann S et al (2018) Self-resistance guided genome mining uncovers new topoisomerase inhibitors from myxobacteria. Chem Sci 9:4898–4908
Park HS, Park JH, Kim HJ et al (2022) BAC cloning and heterologous expression of a giant biosynthetic gene cluster encoding antifungal neotetrafibricin in streptomyces rubrisoli. Front Bioeng Biotechnol 10:964765
Peek J, Lilic M, Montiel D et al (2018) Rifamycin congeners kanglemycins are active against rifampicin-resistant bacteria via a distinct mechanism. Nat Commun 9:4147
Petković H, Lukežič T, Šušković J (2017) Biosynthesis of oxytetracycline by Streptomyces rimosus: past, present and future directions in the development of tetracycline antibiotics. Food Technol Biotechnol 55:3
Puig-Castellví F, Pacheco-Tapia R, Deslande M et al (2023) Advances in the integration of metabolomics and metagenomics for human gut microbiome and their clinical applications. TrAC Trends Anal Chem 167:117248
Qin Z, Munnoch JT, Devine R et al (2017) Formicamycins, antibacterial polyketides produced by *Streptomyces formicae* isolated from African Tetraponera plant-ants. Chem Sci 8:3218–3227
Robertsen HL, Musiol-Kroll EM (2019) Actinomycete-derived polyketides as a source of antibiotics and lead structures for the development of new antimicrobial drugs. Antibiotics 8:157
Rohr J (2000) A new role for polyketides. Angew Chem 39:2847–2849
Rosenzweig AF, Burian J, Brady SF (2023) Present and future outlooks on environmental DNA-based methods for antibiotic discovery. Curr Opin Microbiol 75:102335
Rust M, Helfrich EJ, Freeman MF et al (2020) A multiproducer microbiome generates chemical diversity in the marine sponge *Mycale hentscheli*. Proc Natl Acad Sci 117:9508–9518
Sahayasheela VJ, Lankadasari MB, Dan VM et al (2022) Artificial intelligence in microbial natural product drug discovery: current and emerging role. Nat Prod Rep 39:2215–2230
Schwarzer D, Finking R, Marahiel MA (2003) Nonribosomal peptides: from genes to products. Nat Prod Rep 20:275–287
Shimizu Y, Ogata H, Goto S (2017) Type III polyketide synthases: functional classification and phylogenomics. Chembiochem 18:50–65
Singh A, Chaudhary S, Shankar A et al (2019) Polyketide and its derivatives. In: New and future developments in microbial biotechnology and bioengineering. Elsevier, pp 219–227. https://doi.org/10.1016/b978-0-444-63503-7.00013-9
Staunton J, Wilkinson B (1997) Biosynthesis of erythromycin and rapamycin. Chem Rev 97:2611–2630
Stevenson LJ, Bracegirdle J, Liu L et al (2021) Metathramycin, a new bioactive aureolic acid discovered by heterologous expression of a metagenome derived biosynthetic pathway. RSC Chem Biol 2:556–567
Stroe MC, Gao J, Pitz M et al (2023) Complexity of fungal polyketide biosynthesis and function. Mol Microbiol 121:18
Tambadou F, Caradec T, Gagez AL et al (2015) Characterization of the colistin (polymyxin E1 and E2) biosynthetic gene cluster. Arch Microbiol 197:521–532
Tan D, Jamieson CS, Ohashi M et al (2019) Genome-mined Diels–Alderase catalyzes formation of the cis-octahydrodecalins of varicidin A and B. J Am Chem Soc 141:769–773
Ueoka R, Meoded RA, Gran-Scheuch A et al (2020) Genome mining of oxidation modules in trans-acyltransferase polyketide synthases reveals a culturable source for lobatamides. Angew Chem Int Ed 59:7761–7765
Uppal S, Metz JL, Xavier RK et al (2022) Uncovering Lasonolide A biosynthesis using genome-resolved metagenomics. MBio 13:e01524-22
Vassaux A, Meunier L, Vandenbol M et al (2019) Nonribosomal peptides in fungal cell factories: from genome mining to optimized heterologous production. Biotechnol Adv 37:107449

Venugopal AA, Johnson S (2012) Fidaxomicin: a novel macrocyclic antibiotic approved for treatment of *Clostridium difficile* infection. Clin Infect Dis 54:568–574

Wang H, Fewer DP, Holm L et al (2014) Atlas of nonribosomal peptide and polyketide biosynthetic pathways reveals common occurrence of nonmodular enzymes. Proc Natl Acad Sci 111:9259–9264

Wang H, Sivonen K, Fewer DP (2015) Genomic insights into the distribution, genetic diversity and evolution of polyketide synthases and nonribosomal peptide synthetases. Curr Opin Genet Dev 35:79–85

Wang J, Zhang R, Chen X et al (2020) Biosynthesis of aromatic polyketides in microorganisms using type II polyketide synthases. Microb Cell Factories 19:1

Wang M, Chen D, Zhao Q et al (2018) Isolation, structure elucidation, and biosynthesis of a cysteate-containing nonribosomal peptide in *Streptomyces lincolnensis*. J Org Chem 83:7102–7108

Wehrs M, Gladden JM, Liu Y et al (2019) Sustainable bioproduction of the blue pigment indigoidine: expanding the range of heterologous products in R. toruloides to include non-ribosomal peptides. Green Chem 21:3394–3406

Wehrs M, Prahl JP, Moon J et al (2018) Production efficiency of the bacterial non-ribosomal peptide indigoidine relies on the respiratory metabolic state in *S. cerevisiae*. Microb Cell Factories 17:1–11

Wei Y, Zhang L, Zhou Z et al (2018) Diversity of gene clusters for polyketide and nonribosomal peptide biosynthesis revealed by metagenomic analysis of the yellow sea sediment. Front Microbiol 9:295

Woodhouse JN, Fan L, Brown MV et al (2013) Deep sequencing of non-ribosomal peptide synthetases and polyketide synthases from the microbiomes of Australian marine sponges. ISME J 7:1842–1851

Xu X, Qu R, Wu W et al (2020) Applications of microbial co-cultures in polyketides production. J Appl Microbiol 130:1023–1034

Yu D, Xu F, Zeng J et al (2012) Type III polyketide synthases in natural product biosynthesis. IUBMB Life 64:285–295

Zhang JM, Wang HH, Liu X et al (2020) Heterologous and engineered biosynthesis of nematocidal polyketide–nonribosomal peptide hybrid macrolactone from extreme thermophilic fungi. J Am Chem Soc 142:1957–1965

Zhang YH, Sun J, Ma Y (2017) Biomanufacturing: history and perspective. J Ind Microbiol Biotechnol 44:773–784

Zhuang L, Huang S, Liu WQ et al (2020) Total in vitro biosynthesis of the nonribosomal macrolactone peptide valinomycin. Metab Eng 60:37–44

Chapter 16
Microorganisms in the Dairy Industry

Delower Hossain, Anas Bin Harun, Md Jisan Ahmed, Abdullah Al Bayazid, Sabiha Zarin Tasnim Bristi, Md Robiul Karim, Amina Khatun, Tahmina Sikder, and Nasir Uddin

Abstract The dairy industry is crucial for human nutrition, producing a variety of products essential for health and well-being. However, the quality and safety of milk and dairy products are influenced by microorganisms like coliforms, psychrotrophic, thermoduric, and thermophilic bacteria, fungi, and yeasts. Lactic acid bacteria and probiotics play a vital role in the transformation and preservation of dairy products, shaping their sensory attributes and shelf life. However, spoilage

D. Hossain (✉)
Department of Medicine and Public Health, Faculty of Animal Science and Veterinary Medicine, Sher-e-Bangla Agricultural University (SAU), Dhaka, Bangladesh

Department of Veterinary Medicine and Animal Sciences (DIVAS), Università degli Studi di Milano (UNIMI), Lodi, Italy

Udder Health Bangladesh (UHB), Chattogram, Bangladesh
e-mail: delowervet@sau.edu.bd; delower.hossain@unimi.it; https://www.uhb.org.bd

A. B. Harun · A. Al Bayazid
Department of Medicine, Faculty of Veterinary Medicine and Animal Science, Bangabandhu Sheikh Mujibur Rahman Agricultural University, Gazipur, Bangladesh

M. J. Ahmed · A. Khatun · T. Sikder
Department of Pathology, Faculty of Animal Science and Veterinary Medicine, Sher-e-Bangla Agricultural University (SAU), Dhaka, Bangladesh

S. Z. T. Bristi
Department of Veterinary Medicine and Animal Sciences (DIVAS), Università degli Studi di Milano (UNIMI), Lodi, Italy

Department of Medicine and Surgery, Faculty of Veterinary Medicine, Chattogram Veterinary and Animal Sciences University, Chattogram, Bangladesh

M. R. Karim
Department of Medicine, Faculty of Veterinary Medicine and Animal Science, Bangabandhu Sheikh Mujibur Rahman Agricultural University, Gazipur, Bangladesh

N. Uddin
Centre for Integrative Conservation, Xishuangbanna Tropical Botanical Garden, Chinese Academy of Sciences, Xishuangbanna, Yunnan, People's Republic of China

V. Kothari et al. (eds.), *Microbial Products for Health and Nutrition*, https://doi.org/10.1007/978-981-97-4235-6_16

and pathogenic microorganisms pose significant challenges to the industry, undermining product quality and consumer health. These challenges include biological, chemical, allergenic, and physical hazards. This chapter provides a comprehensive analysis of microorganisms in milk and dairy products, including their sources, occurrence, significance in public health, and the global risk of antimicrobial resistance linked to the dairy supply chain. A detailed discussion on probiotics, yeasts, fungi, and mycotoxins was also included in the chapter. A significant portion is dedicated to elucidating the methods and technologies employed for the detection of the quality of raw milk and the identification of pathogens. It also explores the issue of biofilm formation in dairy products and its control and importance in dairy industry and food safety. Addressing these challenges and implementing strategies for control measures, food safety, and public awareness is essential.

Keywords Milk · Dairy products · Probiotics · Pathogenic microorganisms · Biofilm formation · Mycotoxins · Antimicrobial resistance

Abbreviations

AFM1	Aflatoxin M1
AMP	Ampicillin
AMR	Antimicrobial resistance
AUG	Amoxicillin/clavulanate
CFU	Colony-forming unit
CIP	Ciprofloxacin
CIP	Clean-in-place
CN	Gentamycin
CTX	Ceftriaxone
E	Erythromycin
ELISA	Enzyme-linked immunosorbent assay
EPS	Extracellular polymeric substances
ESBL	Extended-spectrum beta-lactamases
GABA	Gamma-aminobutyric acid
GOS	Galactooligosaccharide
IMI	Intramammary infections
LAB	Lactic acid bacteria
LAMP	Loop-mediated isothermal amplification
mNGS	Metagenomic next-generation sequencing
mPCR	Multiplex polymerase chain reaction
MRSA	Methicillin-resistant *Staphylococcus aureus*
NA	Nalidixic acid
NAS	Non-aureus staphylococci

NASBA	Nucleic acid sequence-based amplification
P	Penicillin
PCR	Polymerase chain reaction
PDO	Protected designation of origin
PLC	Plate loop count
qPCR	Quantitative polymerase chain reaction
S	Streptomycin
SCC	Somatic cell counts
STEC	Shiga toxin-producing *Escherichia coli*
STX	Sulfamethoxazole-trimethoprim
TE	Tetracycline
TPC	Total plate count
UHT	Ultra-heat treatment or ultra-high temperature
VTEC	Verocytotoxin-producing *Escherichia coli*

16.1 Introduction

Milk, as a fundamental component of human nutrition and culinary heritage, plays a central role in global agriculture and has had a significant influence on our dietary habits over time. Dairy products, including cheese, butter, yogurt, sweets, cream, and ice cream, represent a significant portion of human diets, providing a diverse array of essential proteins, vitamins, and minerals (Górska-Warsewicz et al. 2019). However, these foods harbor a complex community of microorganisms that exert significant influence over their safety and quality. Freshly and aseptically collected milk from healthy cows has a bacterial count of a few hundred to thousands per milliliter (Tegegne and Tesfaye 2017). The quantity and types of these bacteria are influenced by many factors such as seasonal variations, farm sanitation, animal feed, and cooling process efficiency (Muehlhoff et al. 2013). Key participants in this microbial orchestra include lactic acid bacteria (LAB), coliforms, psychrotrophic, thermoduric, and thermophilic bacteria, which perform multifaceted roles that extend beyond contamination. The LAB-induced fermentation process transforms raw milk into a variety of delectable products and sculpts their sensory attributes with distinctive flavors and textures inclusive of influencing the release of beneficial compounds with anti-allergenic, anti-carcinogenic, and immunomodulatory effects (García-Burgos et al. 2020). Probiotics, predominantly a subset of LAB, have emerged as potential health-promoting agents with the ability to improve gut health, reduce blood pressure, alleviate specific digestive ailments, and boost the immune system by producing vitamins and other bioactive compounds (Abdul Hakim et al. 2023).

While most microorganisms found in dairy products are harmless or even beneficial, there exists a sinister aspect to this microbial realm. If left unchecked, spoilage organisms can compromise product quality and shelf life, while pathogenic bacteria pose a significant threat to public health. The primary concern when assessing microorganisms in the context of dairy is their potential to pose health hazards to consumers. These hazards encompass an array of biological, chemical, allergenic, and physical risks associated with milk and dairy products (Van Asselt et al. 2016). The presence of antibiotic residues and antimicrobial-resistant bacteria in milk adds another layer of concern, necessitating constant surveillance (Kumar et al. 2021; Rahman et al. 2021). Understanding and mitigating these risks is crucial for ensuring dairy safety, as these products are integral to human diets worldwide. To accomplish this, methods for determining microbial composition and identifying pathogens are essential. A range of microbiological analyses, nucleic acid amplification-based technologies, metagenomic approaches, and emerging biosensor-based techniques provide us with the tools necessary to identify specific bacteria, viruses, and parasites found in dairy products, thereby aiding in preventing foodborne disease outbreaks. By analyzing their microbial composition, we can assess the overall quality of dairy products and thereby ensure their compliance with the necessary safety standards.

This chapter underscores the significance of understanding the roles and consequences of the complex world of microorganisms in the dairy sector, both in terms of human health and in terms of the economy. It elucidates the challenges related to their survival and prevalence throughout the various stages of the dairy chain and acknowledges the growing concern about antimicrobial resistance (AMR), emphasizing pathogen identification methodologies and highlighting areas requiring immediate attention.

16.2 Health Risks of Milk and Dairy Products

Despite the numerous health benefits of milk and dairy products (Yan et al. 2022), they can occasionally cause a variety of health issues due to the presence of potential hazards. Therefore, it is necessary to maintain a contamination-free source by practicing good hygiene, better farm management, and maintaining strict hygienic protocols during the transportation, processing, and storage of milk and dairy products, to reduce the possible contamination by various hazards. The dairy industry maintains strict safety protocols from farm to table of consumer to provide safe and reliable milk and dairy products (Pierezan et al. 2022). However, despite all the advantages of the dairy industry, dairy production faces the challenge of contamination by microbial (bacteria, yeast, fungi, parasites, and viruses), chemical (antibiotics, metals, pesticides, insecticides, mycotoxins, hormones, feed additives, and allergenic hazards), and physical hazards (metal fragment, bone fragment, glass pieces, wood splinters, stones, insect fragment, dirt, dust, or hair) from farm to consumption (Ahmedsham et al. 2018; Mittal and Bharadwaj 2021; Calahorrano-Moreno et al. 2022).

16.3 Sources of Microorganisms in Milk and Dairy Products

Like other food industries, the dairy industry confronts a multitude of challenges to provide safe and high-quality food for consumers. One such challenge pertains to the potential for milk and dairy products to be contaminated with diverse microbial and chemical hazards throughout production, transportation, and storage process. Contamination pathways are complicated due to the presence of many sources of contamination. These sources and their hazards make it complicated to understand these pathways. Milk and dairy products can become contaminated in different ways (Fig. 16.1 and Table 16.1). It is essential to identify and analyze all the sources of contamination to develop strategies to minimize the food safety risks associated with milk and dairy products (Owusu-Kwarteng et al. 2020; Calahorrano-Moreno et al. 2022).

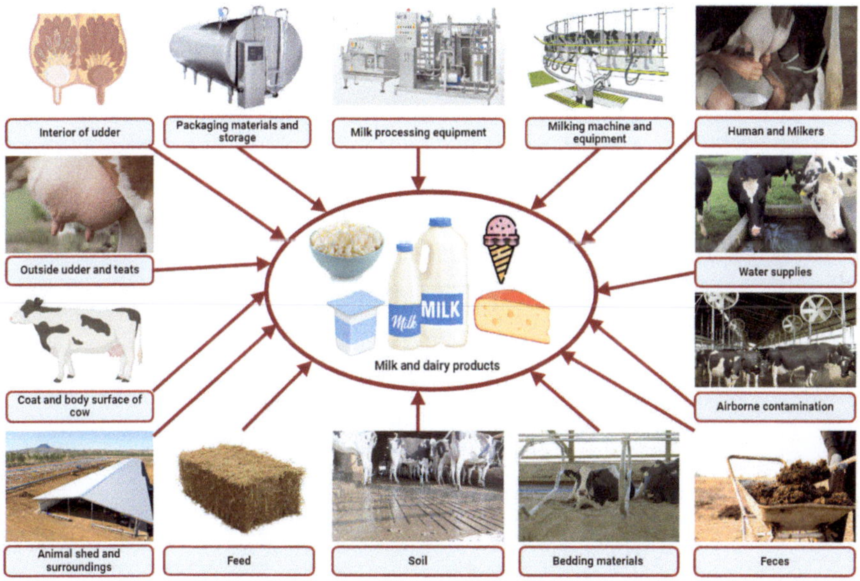

Fig. 16.1 Routes of microbial contamination in milk and dairy products at dairy farms and processing plants

Table 16.1 Major sources of microbial contaminants in milk and dairy products at dairy farms and processing plants

Major sources	Microbial contaminants	Sources and factors of microbial contaminants	Control measures	References
Interior of udder	*Streptococcus* (*St.*) spp., *St. agalactiae*, ***Micrococcus*** spp., and *Corynebacterium* spp.	Microorganisms enter the mammary gland through the teat tips	• Proper management of dairy sheds • Follow strict biosecurity • Care of dairy cows • Promote clean udders	Ledenbach and Marshall (2009), Kable et al. (2016), Kumar et al. (2016), Artursson et al. (2018), Velázquez-Ordoñez et al. (2019), Weber et al. (2019), Owusu-Kwarteng et al. (2020) and Calahorrano-Moreno et al. (2022)
Outside udder and teats	**Major**: *Micrococcus* spp., non-aureus Staphylococcus (NAS), *Enterococcus* spp., *Bacillus* spp. (*B. licheniformis*, *B. subtilis*, *B. pumilus*, *B. cereus*, *B. circulans* and *B. firmus*), *Clostridium* spp., and *Streptococcus* spp. **Others**: *Pseudomonas* (*P.*) *fluorescens*, *Alcaligenes* spp., *Flavobacterium* spp., and coliforms	• Environmental conditions of farm such as dirty floor, soil, manure, mud, feed, and bedding materials • Udder and teat tips become soiled with manure, dirty floor bedding material (sawdust, straw, etc.)	• Cleaning the body of the lactating cows to remove loose dirt • Cleaning of dirty and soiled udder before milking • Discarding of first few strips of milk that will help to lower the bacterial counts	
Coat and body surface of cow	*Bacillus* spp., and *Streptococcus* spp.	Air and stagnant water pools		
Animal shed and surroundings	Coliforms, *Staphylococcus* spp., *Streptococcus* spp., spore formers, and other Gram-negative bacteria	• Poor hygiene practices • Poor bedding materials	• Maintain proper biosecurity in the farm • Regular change of bedding materials	
Feed	***Clostridium*** spp., *Listeria* spp., *Bacillus* sp., and LAB	–	Avoid feeding during milking	

Soil	*Clostridium* spp., *Bacillus* spp., *Pseudomonas* spp., *Mycobacterium* spp., yeast, and molds	Contaminated and dirty floor	• Proper cleaning of teats and udder before milking • Proper cleaning of milking parlor • Quietness of cows during milking	
Bedding materials	Coliforms, *Clostridium* spp., *Bacillus* spp., *Staphylococcus* spp., *Streptococcus* spp., *Klebsiella* spp., and other Gram-negative bacteria	Poor quality bedding materials	• Proper cleaning of milking parlor • Regular change of bedding materials	
Feces	*Escherichia* (*E.*) *coli*, *Staphylococcus* spp., *Listeria* spp., *Mycobacterium* spp., and *Salmonella* spp.	–	• Proper management of manure • Restfulness of the cows during milking	
Airborne contamination	*Micrococcus* spp. (>50%), *Streptococcus* spp., *Coryneforms*, *Bacillus* spp., Gram-negative rods, yeast, and molds	• Sweeping of floors right before milking • Handling hay and animal feed right before milking • Brushing of animals before milking • Having dusty bedding materials • Allowing dust and dirt to accumulate on the walls, ceiling of sheds, and milking machine	• Free from suspended dirt and dusts • Maintain proper ventilation in the farm • Closed milking system	
Water supplies	Coliforms, *Streptococcus* spp., *Bacillus* spp., *Listeria* spp., *Pseudomonas* spp., and *Alcaligenes* spp.	• Poorly cleaned water storage tanks • Untreated water supplies from rivers, tanks, and deep wells	• Use treated good quality water for drinking and different farm operations • Regular screening of microbiological quality of water quality • Maintain sanitation and disinfection in water supply	
Human and Milkers	Coliforms, *Salmonella* spp., *Enterococcus* spp., and *Staphylococcus* spp.	• Milker's hand contacts • Hands with dust and dirt particles • Soiled clothes and hands of milkers • Milker with infected wounds	• Thoroughly cleaning and sanitizing the hands of milker • Good personal hygienic practice by milers • Clean clothing for milking time	

(continued)

Table 16.1 (continued)

Major sources	Microbial contaminants	Sources and factors of microbial contaminants	Control measures	References
Milking machine and equipment	*Micrococcus* spp., *Streptococcus* spp., *Bacillus* spp., coliforms, and *Microbacterium* spp.	• Improperly cleaned milking and cooling equipment • Improper cleaning of bulk milk tanker and collecting pipes and left milk residue • Dirty or poorly cleaned milk cans and lids	Proper cleaning and sanitation of milking machine and equipment	
Milk processing equipment	*Micrococcus* spp., *Streptococcus* spp., *Bacillus* spp., and coliforms	Milk processing equipment	Equipment used in milk processing should be cleaned and sanitized properly	
Packaging materials and storage	*Micrococcus* spp., *Streptococcus* spp., *Bacillus* spp., coliforms, yeast, and molds	Milk storage tank and packaging materials	• Packing machine and packing materials should be cleaned and sanitized properly • Packed milk products should be stored correctly	

16.4 Beneficial Microorganisms of Milk and Dairy Products and Its Significances

16.4.1 The Microbes and Fermented Dairy Products

Numerous dairy products that have been fermented in association with a wide range of microorganisms are part of our diet. Yogurt, cheese, and cultured buttermilk are the most popular fermented dairy products worldwide, but kefir, koumiss, leben, shrikhand, and lacchi are also popular in different countries. Milk fermentation is mostly carried out by LAB, which results in lactic acid from lactose, the main milk sugar. This process increases the acidity of the milk, which makes it more difficult for other microorganisms to proliferate. A wide range of bacteria families are included in the LAB used in dairy product fermentation (Table 16.2). The *Lactobacillus*, *Streptococcus*, *Enterococcus*, *Lactococcus*, and *Leuconostoc* genera account for the vast majority of LAB (Quigley et al. 2011).

Along with diversified groups of bacteria, there are also a few yeast and mold species that affect the physicochemical and organoleptic characteristics of dairy products. The most used yeast genera in the dairy industry are *Candida* spp., *Kluyveromyces lactis*, *Yarrowia lipolytica*, *Geotrichum candidum*, and *Debaryomyces hansenii*, while the most used mold genera in the dairy industry are *Aspergillus* spp., *Penicillium* spp., *Geotrichum* spp., *Fusarium* spp., and mucor (Lavoie et al. 2012).

16.4.2 Events During Fermentation

Lactose is the main source of carbohydrates used by LAB for energy and growth. The enzyme permease transports lactose into the cell, where lactase hydrolyzes it into glucose and galactose (Burgess et al. 2010). Above 95% of the fermentation by-product is lactic acid, which lowers the milk pH to preserve the product. Proteolysis enhances the peptide and amino acid content of fermented products, but it does not significantly boost gross proteolysis by LAB (Amer and Lammerding 1983). Only a small amount of LAB lipases hydrolyzes the triglycerides, whereas the lipases from *L. delbrueckii* subspp. *bulgaricus* and *St. thermophilus* are very potent to triglycerides that are lower in molecular weight compared to higher ones.

Furthermore, as low-fat milk or skim milk is increasingly used in the production of fermented milk products, lipid breakdown by LAB lipases contributes little to the product's qualities. Minerals are also required for growth because they catalyze enzymatic processes. However, their demand is low and would not have a substantial impact on the product's overall mineral content. B vitamins and minerals are also vital because they influence enzymatic reactions, and the LAB uses most B vitamins, except for folic acid.

Table 16.2 Fermented milk products

Items	Type of milk	Microorganisms involved	Where produced	Texture	References
Yogurt	Cow	*Lactobacillus* (*Lb.*) *acidophilus*, *St. thermophilus*, *Lb. bulgaricus*	Worldwide	Semi-solid	Hamann and Marth (1984)
Cheese	Cow, buffalo, sheep, goat	*Lactococcus* (*L.*) *lactis*, *Lb. bulgaricus*, *St. thermophilus*, *Penicillium roqueforti*	Worldwide	Soft, semi-soft	Desmazeaud and Le Bars (1977)
Lacchi	Cow, buffalo	*Lb. bulgaricus*	India	Creamy smooth texture	Krishna et al. (2019)
Cultured buttermilk		*L. lactis* subspp. *cremoris*, *L. lactis* subspp. *lactis biovar diacetylactis*	Worldwide	Thicker than plain milk	Barman et al. (2018)
Shrikhand		*Lb. bulgaricus*, *Streptococcus* (*St.*) *thermophiles*	India	Semi-soft	Nagaraja Roopashri et al. (2023)
Leben	Goat, sheep	*Lb. bulgaricus*, *St. thermophilus*, *L. lactis*, Lactose fermenting yeast	Arabia and North Africa	Liquid	Bensalah et al. (2009)
Kefir	Cow, buffalo, sheep, goat	*L. lactis*, *Leuconostoc* spp., *Saccharomyces kefir*, *Torula kefir*, *Micrococcus*	Southwestern Asia	Like buttermilk	Zeng et al. (2022)
Kumiss	Mare, camel, ass	*Lb. acidophilus*, *Lb. bulgaricus*, *Micrococcus*, *Saccharomyces*	Russia	Creamy	Ishii et al. (2014)
Acidophilus milk	Cow, buffalo, sheep, goat	*Lb. acidophilus*	Many countries	Thicker than regular milk	Jay (1998)

16.4.3 Beneficial and Toxic Compounds Released by LAB, Yeasts, and Molds During Fermentation

During the metabolic activity of bacteria or fungi in milk or dairy products, some worthy bioactive compounds have been produced and released that are renowned for health-promoting characteristics, such as conjugated linoleic acid, exopolysaccharides, bioactive peptides, vitamins, gamma-aminobutyric acid (GABA), and oligosaccharides (Table 16.3) (Bernardo et al. 2012; Hidalgo-Cantabrana et al. 2012).

However, there are occasional instances of releasing hazardous substances during dairy fermentation. Mycotoxins, which are harmful substances derived from some pathogenic fungi, and biogenic amines, which are primarily derived from some LAB, have both been found in fermented milk and other dairy products (Table 16.3).

Aspergillus spp., *Penicillium* spp., and *Fusarium* spp. are the three most common genera of filamentous fungi responsible for mycotoxin production. Although fungi can collectively produce hundreds of mycotoxins, the intake of such contaminated foods could cause serious health problems (detail in Sect. 16.7) (Tadesse et al. 2020).

16.4.4 Probiotics in Dairy Products

Presently, the widest range of probiotics is being used largely in the dairy industry to enhance the product's quality. Incorporating probiotics into dairy products has seen two significant advancements in recent years. These probiotics can either be utilized as starter cultures alone or in combination with conventional starters, or they can be added to these products after microbial fermentation, where they give the finished product with a lot of beneficial attributes such as enhanced flavor, smell, or texture (Gao et al. 2021).

The biological mood of action of probiotics includes enhancing host's immunity by anti-inflammatory activity, production and response of immunoglobulins, performing phagocytosis of foreign particles, synthesis of valuable organic acids, preventing pathological bacterial colonization through competition for both space and nutrients, enhancing barrier activities and synthesis of beneficial molecules which cause health-promoting effects, the synthesis of bile salt hydrolase and lactase enzymes, and promoting the gut microflora by substrate transformation and cross-feeding, production of antimicrobial residues, and promoting the consistency of other beneficial microbes (Sanders et al. 2019). It is thought that 10^6–10^8 colony-forming unit (CFU)/g is adequate to provide probiotic advantages (Champagne et al. 2011).

Table 16.3 Bacteria-derived bioactive compounds and biogenic amines produced in milk and dairy products

Compounds	Types	Production mechanism	Functions/comments
Bioactive compounds	Conjugated linoleic acid	Through conversion of unsaturated fatty acids (e.g., linoleic and linolenic acids) by LAB (Abd El-Salam et al. 2010)	Antiatherogenic, anti-inflammatory (Penedo et al. 2013) and antioxidant (Chinnadurai et al. 2013)
	Bioactive peptides	Mostly from caseins and from whey proteins by proteolytic action	Antimicrobial, antihypertensive, immunostimulant, antioxidative (Parvez et al. 2006). For example, serine and threonine are two amino acids that are popular for their immunomodulatory characteristics that have been produced as major extracellular proteins by *Lactiplantibacillus plantarum* (Bernardo et al. 2012)
	Bacteriocins	Synthesize by LAB during fermentation	Inhibit the growth of pathogenic and unwanted microbes (Inoue et al. 2003)
	Gamma-aminobutyric acid (GABA)	Through bioconversion of glutamic acid from milk proteins by *L. lactis* and *Lb. casei* Shirota	Prevent bradycardia (Inoue et al. 2003)
	Galactooligosaccharide (GOS)	Through transgalactosylation reaction of galactosidase enzyme on lactose by LAB	Prebiotic impact on the gut microbiota enhances selective growth and multiplication of bifidobacteria bacteria (Padilla et al. 2012). For example, some microbes produce carbohydrates which are extracellular polymers and complex in structure. They defend the starter organisms from harmful environmental elements and enhance host's immune system (Hidalgo-Cantabrana et al. 2012)
	Vitamins	LAB fermentation can result in the fortification of specific types of these vitamins, e.g., 522.7 μg/mL Vit B3 by *L. acidophilus*	Propionibacteria produces vitamin B3, B7, B9, and B12 (Hamzehlou et al. 2018)

(continued)

Table 16.3 (continued)

Compounds	Types	Production mechanism	Functions/comments
Biogenic amines	Histamine (*Lb. buchner* and *St. thermophiles*)	Decarboxylation of histidine	Regulate brain functions, such as movement, behavior, emotions, temperature, blood pressure, and endocrine secretion
	Tyramine (*Enterococcus* spp., *Lb. curvatus*, and *Lb. brevis*)	Decarboxylation of tyrosine	
	Putrescine (*Enterococcus* spp., *Lb. curvatus*, *Lb. brevis*, and *L. Lactis*)	Decarboxylation of ornithine and deamination of agmatine	
	Cadaverine (*Enterobacteriaceae*)	Decarboxylation of lysine	

There are three types of probiotics, such as LAB probiotics, non-LAB probiotics, and next-generation probiotics. The sources, characteristics, mode of action, and antimicrobial targets of probiotics are presented in Table 16.4.

16.4.5 Strategies to Improve the Viability and Functionality of Probiotics in Fermented Dairy Products and Gastrointestinal Tract

The ability of probiotics to survive in food must ensure the availability of the daily minimum dosage required to deliver health benefits. Nevertheless, probiotics frequently exhibit decreased longevity in food due to a variety of factors such as low pH, temperature, oxygen content, and the presence of other microbes. Probiotics must also maintain adequate sustainability throughout the intestinal journey to reach the site where they will work. They must overcome numerous biological challenges and harsh conditions, notably by the acidic pH in the stomach, bile salt in the duodenum, and proteolytic enzymes in the GI tract.

The optimal pH range for the growth of *L. acidophilus* is 5.5–6.0 while it is 6.0–7.0 for *Bifidobacteria* spp. (De Vuyst 2000). Hydrogen ions harm probiotic cells by interfering with transport across the cell membrane (Tamime et al. 2005). Lactobacilli are either anaerobic or aerotolerant and purely fermentative, whereas *bifidobacteria* are saccharoclastic and anaerobic (De Vuyst 2000). As a result, oxygen is harmful to probiotic growth and survival.

It is crucial for probiotics used in foodstuffs to be resistant to stress factors like heating, freezing, drying, salt, acid, bile, oxygen, and others. They also have to fight for restricted fermentable ingredients with the existing microflora of the intestine. Additionally, maintaining the metabolic process and the beneficial characteristics of strains are equally as crucial as their viability.

A number of methods are employed to increase the survivability of the microbes in fermented dairy products. Selection and integration of the proper LAB strains are one of the important factors. Selecting the probiotics that produce exopolysaccharides could be a good strategy for selecting strains with sufficient viability because

Table 16.4 Various probiotics with their sources, characteristics, and antimicrobial activity

Probiotics	Source	Characteristics	Mode of action	Antimicrobial targets	References
Pediococcus acidilactici SMVDUDB2	Fermented cheese (kalarei)	Strong hydrophobicity affinity with ethyl acetate, 77.6% autoaggregation, and 80% survivability at 0.3% bile salt concentration and pH between 2.0 to 3.0	Produce pediocins	*Staphylococcus (S.) aureus*, *Bacillus (B.) subtilis*, *E. coli*, *Mycobacterium smegmatis*, *Lb. rhamnosus*, *Proteus vulgaris*	Bhagat et al. (2020)
Enterococcus (E.) lactis PMD74	Turkish cheese (Ezine)	High survivability in the presence of pancreatic juice, bile salts, pepsin, and pH 3 and 41% autoaggregation	Produce enterocins	*S. aureus*, *L. sakei*, *Enterococcus (En.) faecalis*, *Listeria (L.) monocytogenes*	Uymaz Tezel (2019)
En. faecium K1	Fermented cheese (kalarei)	Thermally stable extracellular polymeric substance production and emulsification	Produce enterocins	Antioxidant, hypocholesterolemic, antibiofilm characteristics	Bhat and Bajaj (2018)
Lactiplantibacillus plantarum (DF60Mi) and *L. lactis* (DF04Mi)	Goat milk	Proteolytic activity, autoaggregation, gas generation, sensitivity to antibiotics, gelatinase synthesis, EPS production, hemolytic activity, and survival in gut	Produce Plantaricin which destroys the structure of cell membrane of bacteria	*St. mutans*, *L. lactis* subspp. *lactis*, and *L. monocytogenes*	da Silva et al. (2019)
Lactobacillus (Lb.). delbrueckii subsp. *bulgaricus* F18	Homemade butter	Cell surface hydrophobicity against *O*-xylene (33.9%), 68% autoaggregation, resistance at pH 1 for 2 h), bile resistance, and cell surface hydrophobicity	Bifidobacteria and lactobacilli have adhesiveness attributes which make it possible for them to prevent bacterial adhesion to intestinal cells	*S. aureus*, *B. subtilis*, *B. cereus*, *Enterobacter aerogenes*, *E. coli*, *Campylobacter (C.) jejuni*, *Pseudomonas (P.) aeruginosa*, *Salmonella (S.) typhimurium*, *L. monocytogenes*, *Candida (C.) albicans*	Yerlikaya et al. (2021)

Lb. rhamnosus, *Lb. helveticus*, and *En. hirae*	Fermented Xinjiang cheese	Resistance to bile, acidic pH, hydrophobic activities, autoaggregation, scavenging free radicals, and breakdown of triglycerides and cholesterol		*S. aureus*, *E. coli*, *Shigella flexneri*, *S. typhimurium*, and *L. monocytogenes*	Azat et al. (2016)
Lacticaseibacillus rhamnosus	Milk	Good adhesion to intestinal epithelium, tolerant of 0.3% bile salt and pH 3.0		*E. coli*, *S. typhimurium* LT-2, *L. monocytogenes* and *Shigella sonnei*	Aryantini et al. (2017)
Lb. fermentum SM-7	Fermented milk (koumiss)	Resistance to bile, acidic pH, and reduce cholesterol		*S. aureus* and *E. coli*	Pan et al. (2011)
Lb. casei AST18	Yogurt	Resistance to bile and acidic pH		*Penicillium chrysogenum*	Li et al. (2014)
Lb. plantarum and *Lb. helveticus*	Pakistani yogurt	Tolerance in pH 6–6.5, 0.01–1% bile salt, and 1–7% sodium chloride salt, and excellent growth in that environment		*S. aureus*, *B. cereus*, *S. typhii*, and *E. coli*	Hassan et al. (2020)
St. thermophilus	Traditional Turkish yogurts	High survivability in the presence of bile salts, acidic pH, and other stress condition in gastrointestinal tract		*S. aureus*, *B. coagulans*, *B. cereus*, *Klebsiella* (*K.*) *pneumonia*, *P. fluorescens*, *L. monocytogenes*, *E. coli*	Akpinar et al. (2011)
Lb. johnsonii 456	–	Antitoxic and anti-inflammatory effects, higher survivability in gastric juice, bile salts and at acidic condition, long time persistence in gastrointestinal tract after ingestion		Prevent attachment of pathogenic *E. coli*, *S. enterica*, *En. faecalis* with GI mucosa	Davoren et al. (2019)
Lb. reuteri	Yogurt	Higher survivability in bile salts and acidic pH		*S. aureus*, *Shigella sonnei*, *S. enterica*, *Vibrio* (*V.*) *cholera*, *L. monocytogenes*, and *E. coli*	Ortiz-Rivera et al. (2017)
Bifidobacterium lactis	Fermented milk and yogurt	Resistance toward acidic and bile conditions, adhesion properties		*S. aureus*, *B. subtilis*, *S. typhimurium*, *K. pneumonia*, *P. aeruginosa*, and *E. coli*	Tham et al. (2012)

those polymers preserve the manufacturing bacteria and enhance their lifespan (Ramchandran and Shah 2010). Different methods for regulating the post-acidification process should also be adopted (Antunes et al. 2005). The inclusion of oxygen scavengers, such as ascorbic acid, can also increase the viability of probiotics (Gaudreau et al. 2013). Gene transformation is an additional strategy for enhancing stress tolerance (Sheehan et al. 2006). To increase their survivability, probiotics are often used in combination with prebiotics. These prebiotics frequently serve as wrapping materials during alimentary passage, eventually releasing the microbes into the colon, where they subsequently serve as fermentable material (Koh et al. 2013). Probiotics have also been microencapsulated with various materials, e.g., hydrocolloids, alginate-type gels, chitosan, gelatins, supercritical carbon dioxide, and prebiotics; even 2% alginate solution can increase the viability of probiotics more than 15% (Riaz and Masud 2013).

The buffering capability of the fermented product can have a significant impact on the survivability of the probiotics. For example, increasing milk's buffering capacity would promote probiotic development and activity in fermented milk. It increases the survivability of probiotics in fermented products and the intestine by maintaining a higher pH (Mortazavian et al. 2011). The inclusion of malt, barley, and wheat extracts improves the buffering capacity and thus increases the survivability of probiotics (Charalampopoulos et al. 2003). Probiotic cells are facilitated by the solid matrix in food products, as opposed to liquid foods, by lowering their exposure to detrimental elements such as hydrogen ions (Shafiee et al. 2010).

16.5 Spoilage Microorganisms in Milk and Dairy Products and Their Significances

16.5.1 Spoilage of Milk

While milk's constituents provide a plethora of health benefits, they also create an environment conducive to microbial proliferation, resulting in deterioration and spoilage of milk. Milk spoilage is caused by a variety of factors, but the primary cause is the prolific growth of microorganisms, including bacteria and fungi, which metabolize the milk's constituents and produce undesirable by-products. A distinctive hallmark of milk spoilage is its remarkable swiftness, a direct consequence of the exponential growth of these microorganisms. The rate of spoilage is determined by several critical variables, the most important of which are the initial microflora population size and composition, as well as the storage conditions and temperatures imposed (Lu and Wang 2017).

16.5.2 Events of Spoilage

16.5.2.1 Carbohydrate Metabolism by Spoilage Microbes

Bacterial metabolism of lactose produces acids and gases, leading to undesirable attributes like sourness and gassiness in milk (De Jonghe et al. 2010). Furthermore, these bacteria can break down lactose into its constituents, releasing galactose, a monosaccharide that fuels the growth of various microorganisms. This liberation sustains and perpetuates the spoilage process, fostering a continuous cycle of bacterial proliferation.

16.5.2.2 Proteins Metabolism by Spoilage Microbes

The presence of LAB, such as thermoduric *St. salivarius*, plays a crucial role in the metabolic process. They primarily utilize lactose as a substrate for growth in both raw and pasteurized milk. This metabolic activity results in the production of lactic acid as a by-product, lowering the pH of milk from the industry standard 6.6 to a more acidic 4.5. Upon breaching this pH threshold, the acidic condition triggers a series of events, including the precipitation of casein micelles, ultimately resulting in milk curdling (O'Connor 1995).

Besides the acid-producing LAB, other microorganisms possess hydrolytic enzymes, including proteases, lipases, and phospholipases (Chen et al. 2003). Even after pasteurization eliminates the microorganisms, their enzymes remain active. This persistent enzymatic activity can cause substantial structural changes in ultra-high temperature (UHT) and sterilized milk and dairy products, leading to deformities and quality issues despite the absence of the microorganisms themselves.

16.5.2.3 Lipid Metabolism by Spoilage Microbes

Microbial-derived lipases play a pivotal role in the hydrolysis of triglycerides, the predominant lipid constituents within milk. This enzymatic action results in the breakdown of triglycerides into glycerol and free fatty acids. Lipase or phospholipase activity on milk fat can release fatty acids of different lengths, each contributing to significant changes in milk flavor.

Short-chain fatty acids, ranging from C4 to C8, are linked to the development of "rancid" flavors and accompanying odors. Conversely, free medium-chain fatty acids, within the C10–C12 range, introduce a distinctive "soapy" flavor profile to the milk (Chrisope and Marshall 1976). It is noteworthy that when the levels of free fatty acids in milk surpass a threshold of 1.5 mmol/L, the flavor modifications become significant enough to render the milk unsuitable for consumption (Deeth 2006).

16.5.3 Types of Spoilage Microorganisms in Dairy Products

16.5.3.1 Psychrotrophs

A significant proportion of the bacteria in raw milk is made up of psychrotrophic microorganisms, the most common of which are rod-shaped, gram-negative, aerobic pseudomonads. Usually, *Pseudomonas* spp. cover up 65–70% of the psychrotrophic bacterial population in raw milk (Meng et al. 2017). These bacteria can grow at 3–7 °C temperature and break down large protein and lipid, which gives them a competitive advantage over other microorganisms in cold storage. Species of *Micrococcus*, *Lactococcus*, *Bacillus*, and *Aerococcus*, as well as members of *Enterobacteriaceae* family are other significant psychrotrophs found in raw milk. Pseudomonads can lower the diacetyl level in sour cream and buttermilk, which results in an imbalance between the diacetyl and acetaldehyde ratios that provide yogurt-like flavor (Wang and Frank 1981).

16.5.3.2 Coliforms

The coliforms include *Escherichia* spp., *Klebsiella* spp., *Proteus* spp., *Enterobacter* spp., *Hafnia* spp., *Serratia* spp., and *Citrobacter* spp., originating from the digestive tract of milch animals. Their presence in raw milk indicates unsanitary production conditions. Notably, coliforms can rapidly accumulate in biofilms on milking equipment (Chambers 2002). Coliforms, like psychrotrophic microorganisms, could reduce the diacetyl content of buttermilk and sour cream (Wang and Frank 1981). The presence and metabolic activities of coliforms can be attributed to the occurrence of this flavor transformation, highlighting their influence on the flavor characteristics of such products.

16.5.3.3 Lactic Acid Bacteria (LAB)

LAB have their origins in the GI tract of milch animals shortly after birth. The dairy industry uses many LAB strains to create starter cultures used in the fermentation of dairy products such as yogurt, cheese, and butter. However, concerning spoilage, LAB that commonly cause issues belong to genera such as *Lactobacillus*, *Leuconostoc*, *Lactococcus*, *Propionibacterium*, and *Enterococcus*. These LAB are often heterofermentative, which can give rise to undesired off-flavors and textural irregularities in cheese (Bernardeau et al. 2008). In some cases, diacetyl reduction can occur in dairy products like yogurt and cheese due to the action of lactococci that thrive at lower temperatures, around 7 °C. This reduction results in the development of a yogurt-like flavor, a phenomenon directly related to the presence and activity of these lactococci in the products (Hogarty and Frank 1982).

16.5.3.4 Spore-Forming Bacteria

These microorganisms primarily belong to the genera *Bacillus*, *Geobacillus*, and *Clostridium*. They are typically gram-positive bacteria that can thrive in aerobic or facultative anaerobic conditions, except for *Clostridium* spp., which strictly require anaerobic environments. It is important to note that *Clostridium* spp. tends to proliferate more in cheese than in milk.

Among these, *Clostridium* (*C.*) *tyrobutyricum* plays a significant role in causing late-blowing defects in hard-type cheeses, characterized by high pH and low salt levels, by fermenting lactic acid to produce butyric acid, CO_2, and H_2. Other *Clostridium* spp. like *C. sporogenes* and *C. butyricum* are also implicated in causing defects in cheeses, such as putrid spots in Swiss cheese. While *C. perfringens* has not been widely associated with milk-based powdered products, its ability to survive in extreme conditions raises concerns about potential health risks associated with milk powder, underscoring the need for vigilance in quality and safety assessments (Burgess et al. 2010; Kekre et al. 2015; Hassan and Frank 2011).

16.5.3.5 Yeasts and Molds

Yeasts and molds in dairy products typically trace their origins to contamination from the surrounding environment, whether it be the dairy farm or processing plant. These are known culprits for causing defects in dairy products, particularly cheese and yogurt. The most encountered yeasts in milk include *Kluyveromyces* (*K.*) *marxianus* var. *marxianus*, *K. marxianus* var. *lactis*, *Debaryomyces hansenii*, *Saccharomyces cerevisiae*, *Candida* (*C.*) *lacticondensi*, *C. famata*, *C. lusitaniae*, *C. versatilis*, and *Yarrowia lipolytica*. On the other hand, molds found in dairy products belong to the genera *Rhizomucor*, *Aspergillus*, and *Rhizopus*. These microorganisms, when present, can lead to a wide range of quality issues and spoilage concerns in dairy products, making their control and prevention crucial in the dairy industry (Hassan and Frank 2011).

16.6 Pathogenic Microorganisms in Milk and Dairy Products and their Significance

16.6.1 Pathogenic Bacteria in Milk and Dairy Products

Pathogenic microorganisms exhibit a significant concern in milk and its derivatives, which cause foodborne illnesses and pose substantial health risks. Among these, *Escherichia* (*E.*) *coli*, *Salmonella* spp., *Staphylococcus* spp., *Bacillus* spp., *Listeria* spp. *Micrococcus* spp., *Streptococcus* spp., *Salmonella* spp., *L. monocytogenes*, *Corynebacterium* spp., *Enterobacter* spp., *Yersinia* (*Y.*) *enterocolitica*, and *C. jejuni* have the potential to contaminate both milk and dairy products (Boor et al. 2017; Bastam et al. 2021). These pathogenic and spoilage bacteria show a significant challenge to food safety leading to various illnesses of differing severity, with symptoms ranging from fever, vomiting, diarrhea, and abdominal discomfort. In severe cases, these infections can even have fatal outcomes. Many studies have documented the presence of pathogenic bacteria found in milk and dairy products. However, the variations observed in these studies can be attributed to diverse sampling strategies and methodologies, as well as factors such as geographical location, seasonality, farm size, animal density, and rearing practices. However, milk and its derivatives also contain many beneficial bacteria as well, i.e., LAB, which positively affects the dairy products quality and the human body (Shafeek et al. 2018). These microorganisms actively engage in the fermentation process, transforming milk into various products like yogurt, cheese, and kefir. A comprehensive overview of the presence of pathogenic and beneficial bacteria reported in milk and dairy products across the world is listed in Table 16.5 based on the findings of previously published studies.

16.6.2 Total and Specific Bacterial Count in Milk and Dairy Products

The microbial quality of raw milk and its derivatives can be evaluated using various parameters, such as total bacterial counts or specific counts for organisms like Enterobacteriaceae, *Staphylococcus* (*S.*) *aureus*, non-aureus staphylococci (NAS), and LABs. There is no substantial difference in the total plate counts (TPC) observed in raw milk from cattle, sheep, and goats (D'Amico and Donnelly 2010). Nonetheless, TPC in sheep and goat milk varied depending on the milking month, number of lactations, milking techniques (hand or machine), and herd size (Alexopoulos et al. 2011). Furthermore, raw milk is also tested for presence and counting of specific bacterial groups and types in cows, goats, and sheep. These include mesophilic bacteria, Enterococci, Enterobacteriaceae, *E. coli*, *S. aureus*, *Salmonella* spp., *Campylobacter* spp., and *Listeria* (*L.*) *monocytogenes*. TPC frequently ranges within the range of 2.9–8.1 log CFU/mL, with 83% of the samples testing positive for *E. coli*, and colony counts ranging from 1.0 to 4.0 CFU/mL. According to Kateřina et al. (2016), *S. aureus* was detected in 29.1% of the samples (cow: 26.9%; goat: 34.4%; sheep: 39.1%), despite the extremely low levels bacterial count (2.7 CFU/mL). The total bacterial count was between 1.1 to 2.3 log10 CFU/mL, which was considerably lower than the 5.2 log10 CFU/mL reported in raw sheep milk from Egypt (Ombarak and Elbagory 2017). In Bangladesh, buffalo raw milk contains 2.0–7.3 log CFU/mL of total bacteria count, 3.0–3.9 log CFU/mL of *S. aureus*, and 3.0–6.7 log CFU/mL of NAS, respectively (Singha et al. 2023). Raw milk with a high aerobic mesophilic bacteria count (6.6 log CFU/mL) in Bangladesh indicated potential issues related to udder infection, unhygienic milking practices, poor water quality for utensil cleaning, and milk storage system

Table 16.5 Pathogenic and beneficial bacteria found in milk and dairy products

Country	Animal	Source(s)	Types	Organisms	Year	Reference
India	–	Raw milk	Pathogenic	*Streptococcus* spp., *Enterococcus* spp., *Enterobacter* spp., *Klebsiella* spp., *Aeromonas* spp., *Pseudomonas* spp., *Porphyromonas* spp., *Fusobacteria* spp., *Acinetobacter* spp., *Micrococcus* spp., *Bacillus* spp.	2021	Mittal and Bharadwaj (2021)
			Beneficial	*Lactobacillus* spp., *Lactococcus* spp., *Leuconostoc* spp., *Propionibacterium* spp., *Bacteroides* spp., *Faecalibacterium* spp.		
Nigeria	–	Yogurt	Pathogenic	*Salmonella* spp., *Shigella* spp., *E. coli*, *Serratia* spp., *Pseudomonas* spp., *Providencia* spp., *S. aureus*, *B. subtilis*, *Enterobacter* spp.	2019	Raji and Jiya (2019)
			Beneficial	–		
New Zealand	–	Raw milk	Pathogenic	*Streptococcus* spp., *Staphylococcus* spp., *Enterococcus* spp., *E. coli*, *K. pneumonia*, *Enterobacter aerogenes*, *Acinetobacter* spp., *Aeromonas* spp., *Bacillus* spp., *Clostridium* spp., *Micrococcus* spp., *Corynebacterium* spp.	2018	Panigrahi et al. (2018)
			Beneficial	*Flavobacterium* spp., *Lactobacillus* spp., *Arthrobacter* spp., *Alcaligenes* spp.		

(continued)

Table 16.5 (continued)

Country	Animal	Source(s)	Types	Organisms	Year	Reference
Mexico	–	Cotija cheese	Pathogenic	*Salmonella* spp., *Bacillus* spp., *Staphylococcus* spp., *Enterococcus* spp., *Streptococcus* spp., *Vagococcus* spp.	2016	Escobar-Zepeda et al. (2016)
			Beneficial	*Lactobacillus* spp., *Weisella* spp., *Leuconostoc* spp.		
	Goat	Raw milk and cheese	Pathogenic	*B. cereus* 183, *L. monocytogenes*, *L. innocua*, *M. luteus*, *S. aureus*, *E. coli* (strain ATCC 25922), *P. aeruginosa*, *E. cloacae*	2016	Hernández-Saldaña et al. (2016)
			Beneficial	*Aquabacterium fontiphilum*, *Psychrobacter aquaticus*		

(continued)

Table 16.5 (continued)

Country	Animal	Source(s)	Types	Organisms	Year	Reference
New Zealand	Cow	Raw milk	Pathogenic	*S. aureus*, *E. coli*, *E. coli* O157, *E. coli* O157:H7, *Salmonella* spp., *Campylobacter* spp., *L. monocytogenes*, *M. tuberculosis*, *Brucella* spp., *Y. enterocolitica*, *Staphylococcus* spp., *Streptococcus* spp., *Salmonella* spp., *Coxiella burnetii*, *Clostridium* spp., *Aeromonas* spp., *K. pneumoniae*, *Serratia* spp., *Proteus* spp., *Hafnia alvei*, *Actinomyces* spp., *Leptospira* spp., *Enterobacter sakazakii*, *Acinetobacter* spp.	2012	Hill et al. (2012)
			Beneficial	*Lactococcus* spp., *P. fluorescens*, *P. fragi*, *P. putida*, *Psychrobacter* spp., *Flavobacterium* spp., *Alcaligenes* spp.		
	–	Raw milk	Pathogenic	*Streptococcus* spp., *Staphylococcus* spp., *Enterococcus* spp., *E. coli*, *K. pneumonia*, *Enterobacter aerogenes*, *Acinetobacter* spp., *Aeromonas* spp., *Bacillus* spp., *Clostridium* spp.	2011	Hassan and Frank (2011)
			Beneficial	*Alcaligenes* spp., *Flavobacterium* spp., *Lactobacillus* spp.		

(continued)

Table 16.5 (continued)

Country	Animal	Source(s)	Types	Organisms	Year	Reference
Canada	–	Raw milk	Pathogenic	*M. paratuberculosis*, *L. monocytogenes*, *Salmonella* spp., *Campylobacter* spp., verocytotoxin-producing *E. coli* (VTEC)	1997	Steele et al. (1997)
			Beneficial	–		
	–	Cheese	Pathogenic	*Salmonella enterica* serotype Enteritidis	2000	Ahmed et al. (2000)
			Beneficial	–		
	–	Yogurt	Pathogenic	–	2011	Savard et al. (2011)
			Beneficial	*Bifidobacterium animalis* subspp. *lactis*, *L. acidophilus*, *St. thermophilus*		
	–	Fresh cheese	Pathogenic	*Salmonella* spp., *Listeria* spp., *E. coli* O157:H7, *Staphylococcus* spp., *Brucella* (*B.*) *melitensis*, *C. botulinum*	2012	Torres-Vitela et al. (2012)
			Beneficial	–		

(continued)

Table 16.5 (continued)

Country	Animal	Source(s)	Types	Organisms	Year	Reference
US	Cattle	Raw milk	Pathogenic	*Salmonella* spp., *Campylobacter* spp., *E. coli* O157:H7, *S. typhimurium*, *Listeria* spp., *Cryptosporidium* spp., *Y. enterocolitica*, *Coxiella burnetii*, *M. avium* subspp. *paratuberculosis*	2009	Angulo et al. (2009)
			Beneficial	–		
	–	Pasteurized milk	Pathogenic	*Bacillus* spp., *Microbacterium* spp., *St. macedonicus*	2004	Fromm and Boor (2004)
			Beneficial	*Paenibacillus* spp.		
	–	Yogurts	Pathogenic	–	2011	Angelakis et al. (2011)
			Beneficial	*L. casei*, *L. lactis*, *Bifidobacterium animalis*, *L. delbrueckii*, *St. thermophilus*		
Hungary	Sheep	Raw milk	Pathogenic	*Enterobacteriaceae*, *S. aureus*, mesophilic bacteria	2020	Tonamo et al. (2020)
			Beneficial	–		
France	Cow	Cheese and yogurt	Pathogenic	*Coxiella burnetii*	2013	Eldin et al. (2013)
			Beneficial	–		
Italy	–	Yogurt	Pathogenic	*Staphylococcus* spp., *St. thermophilus*	2002	Fasoli et al. (2003)
			Beneficial	*Bifidobacterium*, *Lb. bulgaricus*, *Lb. lactis*, *Lb. acidophilus*		
	Goat	Milk	Pathogenic	*E. coli*, *Enterococci*, spores of mesophilic aerobic bacteria, *Clostridium* spp., *S. aureus*	2002	Foschino et al. (2002)
			Beneficial	*Lactococcus* spp., *Lactobacilli*		
	Sheep	Primosale cheese	Pathogenic	*S. aureus*, Enterobacteriaceae	2011	Giammanco et al. (2011)
			Beneficial	–		

(continued)

Table 16.5 (continued)

Country	Animal	Source(s)	Types	Organisms	Year	Reference
China	–	Milk powder	Pathogenic	*Bacillus licheniformis*, *Geobacillus stearothermophilus*, *Anoxybacillus flavithermus*, *Laceyella sacchari*, *Thermoactinomyces vulgaris*	2015	Sadiq et al. (2016)
			Beneficial	–		
	–	Raw milk	Pathogenic	*Bacillus* spp.	2023	Peng et al. (2023)
			Beneficial	–		
	Buffalo	Raw milk	Pathogenic	*Fungi*, *Coliforms*, *E. coli*, *S. aureus*, *Listeria* spp.	2007	Han et al. (2007)
			Beneficial	LAB		
	–	Yogurt	Pathogenic	*Enterobacter* spp., *Klebsiella* spp., *Serratia* spp.	2013	Chen et al. (2014)
			Beneficial	*Bifidobacterium* spp.		
	–	Kazak cheese	Pathogenic	*St. thermophilus* B8	2020	Li et al. (2020)
			Beneficial	*Lb. helveticus* B6, *Weissella confusa* B14, and *Lb. rhamnosus* B10		
Brazil	–	Minas cheese and raw milk	Pathogenic	Enterotoxin-producing *S. aureus*	2002	Do Carmo et al. (2002)
			Beneficial	–		
	–	Sero artisanal cheese	Pathogenic	*L. monocytogenes*, *Salmonella* spp., non aureus *Staphylococci* (NAS), diarrheagenic *E. coli*, *M. tuberculosis*, and *B abortus*	2019	Andretta et al. (2019)
			Beneficial			
	–	UHT milk	Pathogenic	*B. subtilis*, *B. amyloliquefaciens*, *B. licheniformis*, *B. pumilus* and *B. megaterium*	2017	Pinto et al. (2018)
			Beneficial	–		

(continued)

Table 16.5 (continued)

Country	Animal	Source(s)	Types	Organisms	Year	Reference
UK	–	Blue-veined cheese	Pathogenic	*S. equorum*, *Bacillus* spp., *Brevibacterium* spp., *Halomonas* spp., *Acinetobacter* spp., *Alkalibacterium* spp., *C. casei*	2018	Yunita and Dodd (2018)
			Beneficial	*L. casei*, *L. paracasei*,		
	Cow	Raw and pasteurized milk	Pathogenic	*M. paratuberculosis*	2002	Grant et al. (2002)
			Beneficial	–		
	Cow	Raw milk	Pathogenic	*E. coli* spp., *Campylobacter* spp., *Listeria* spp.	2018	Willis et al. (2018)
			Beneficial	–		
	–	Milk powder	Pathogenic	*Ent. sakazakii*, *Enterobacteriaceae*	2006	Iversen and Forsythe (2004)
			Beneficial	–		
New Zealand	–	Raw milk	Pathogenic	*Listeria* spp., *Campylobacter* spp., and *Salmonella* spp., *L. innocua*	2012	Hill et al. (2012)
			Beneficial	–		
Netherlands	Cow	Raw milk	Pathogenic	*Coxiella burnetii*	2011	Muskens et al. (2011)
			Beneficial	–		
	–	Dutch cheese	Pathogenic	–	2016	Porcellato and Skeie (2016)
			Beneficial	*Lactobacillus* spp., *Leuconostoc* spp.		
Russia	Cow	Naturally fermented milk	Pathogenic	*Streptococcus* spp., *Macrococcus* spp., *Leuconostoc* spp., *Enterococcus* spp., *Acetobacter* spp., *Citrobacter* spp., *Acinetobacter* spp., *Bacillus* spp., *Kluyvera* spp., *Enterobacter* spp.	2015	Yu et al. (2015)
			Beneficial	*Lactobacillus* spp., *Lactococcus* spp.		

(continued)

Table 16.5 (continued)

Country	Animal	Source(s)	Types	Organisms	Year	Reference
India	Cow	Milk	Pathogenic	*Coliform*, *E. coli*, *Salmonella* spp., *Shigella* spp., *S. aureus*, *L. monocytogenes*	2012	Agarwal et al. (2012)
	Cow	Raw milk	*Pathogenic*	*S. aureus*, *E. coli*, *Bacillus* spp., *P. fluorescens*, *Salmonella* spp.	2010	Mubarack et al. (2010)
			Beneficial	*Lactobacillus* spp.		
Bangladesh	–	Raw milk	Pathogenic	*E. coli*, *Klebsiella* spp., and *Staphylococcus* spp.	2011	Noor et al. (2011)
	–	Cold raw milk	Pathogenic	*Pseudomonas* spp., *Achromobacter* spp., *Aeromonas* spp., *Serratia* spp., *Alcaligenes* spp., *Chromobacterium* spp., *Flavobacterium* spp., and *Enterobacter* spp.	2014	Banik et al. (2014)
	–	Raw milk	Pathogenic	*E. coli*, *Salmonella* spp., *Shigella* spp. and *Vibrio* spp.	2015	Rahman et al. (2015)
		Pasteurized milk	Pathogenic	*E. coli* and *Shigella*-like species		
		Vendor yogurt	Pathogenic	*E. coli*, *Salmonella* spp., *Shigella* spp. and *Vibrio* spp.		
		Packed yogurt	Pathogenic	*E. coli* and *Shigella*-like species		

(Taye et al. 2021) whereas Indian milk from vendors showed varying counts of coliform, *E. coli*, fecal Streptococci, *S. aureus*, yeasts and molds, and total viable count which were found to be 0.00–4.50 (3.00 ± 1.15), 0.00–1.20 (0.70 ± 0.46), 1.00–2.80 (1.80 ± 0.64), 0.00–3.00 (2.10 ± 1.14), 0.00, 2.40 (1.00 ± 0.78), and 3.80–7.20 (5.50 ± 0.99) log10 CFU/mL, respectively (Shekhar et al. 2010). The total viable count exceeded 5 log CFU/mL in more than one-third of the milk samples. Pasteurized milk in Bangladesh showed bacterial counts ranging from 0 to 3.26 log CFU/mL with potential contamination sources being defects in pasteurization equipment, post-pasteurization handling, or poor hygiene standards among staff (Banik et al. 2014). In contrast, UHT milk exhibited no coliform count or total count (Agarwal et al. 2012).

16.7 Pathogenic Fungi, Yeasts, and Mycotoxins in Milk and Dairy Products and Their Significance

16.7.1 Contamination and Transmission of Pathogenic Fungi, Yeasts, and Mycotoxins in Milk and Dairy Products

Fungi are permitted or even necessary in some dairy products for processing from raw milk. They are widely used in various types of cheeses, including Camembert, Roquefort, and Stilton, as well as several specialty cheeses crafted from unpasteurized milk that hold Protected Designation of Origin (PDO) recognition internationally (Benkerroum 2016). Fungi play an essential role in the cheese maturation process, transitioning from local raw milk to Roquefort cheese and other types of raw milk cheeses. This maturation occurs at ambient temperature and humidity levels, contributing to the development of distinct flavors and textures in the cheese. On the other hand, certain species of fungi can be deliberately introduced into pasteurized milk as starter cultures for cheese production (Desmasures 2014). The proper use of these molds is crucial for the formation of the texture and aroma of cheese. In addition to fungi, yeasts are also important for the formation of the characteristic texture and fragrance in some fermented dairy products like cheese and koumis (Alvarez-Martin et al. 2008; Cantor et al. 2017). However, the presence of unchecked contamination and large quantities of yeasts, fungi, and their toxin metabolites in milk and dairy products can have adverse effects on health on various aspects. The contamination of yeasts and fungi in milk and dairy products can occur through multiple pathways.

1. *Inadequate hygiene during milking and processing*: Improper hygiene practices during milking, processing, and storage of milk can lead to the contamination of milk and dairy products by fungi and yeasts present in the udder or milking premises (Garnier et al. 2017).
2. *Contaminated equipment*: Insufficient cleaning and sanitization of equipment used for milking, processing, or packaging, such as milking machines, milk

cans, transport system tunnels, strainers, and bulk tanks, can create a suitable environment for the proliferation and transmission of fungi and yeasts into milk and dairy products (Schön et al. 2016; Garnier et al. 2017).

3. *Air and environment*: The air within dairy processing plants is a significant contributor to the presence of various fungi, leading to the contamination and dispersion of spores to milk and dairy products and the dairy environment (Awasti and Anand 2020).
4. *Improper storage and temperature control*: Inadequate regulation of temperature during the processing and storage of milk and dairy products has been shown to facilitate the proliferation of fungi and yeasts. The storage of yogurt at high temperature can facilitate the proliferation of 12 different species of fungi (Gougouli et al. 2011). A considerable proportion of fungi can proliferate in food items with low water content (0.7–0.8) or at a wide range of pH (3.0–8.0), even when stored under refrigeration settings (Garnier et al. 2017; Awasti and Anand 2020).
5. *Cross-contamination during processing*: Contamination of milk and dairy products may occur if they come into contact with surfaces or equipment that are contaminated with fungi or yeast during the processing or packing stages (Garnier et al. 2017; Awasti and Anand 2020).
6. *Contaminated water and feed*: Fungal species and yeasts can colonize water used in the processing and manufacturing, potentially contaminating milk and dairy products directly and indirectly. Additionally, milk and dairy products can be contaminated by fungi and yeasts present in animal feed as well (Garnier et al. 2017; Awasti and Anand 2020).
7. *Unhygienic practice during handling*: Milk and milk products may become contaminated with fungi and yeasts if they are handled unhygienically during the transportation, distribution, and retail processes (Garnier et al. 2017).

16.7.2 Pathogenic Yeasts and Fungi in Milk and Dairy Products

A variety of pathogenic fungal and yeast species have been identified from milk and dairy products in various animals, as shown in Table 16.6.

16.7.3 Mycotoxins in the Dairy Industry

Fungi contaminate milk and dairy products by producing secondary metabolites called mycotoxins, which are naturally occurring toxicants that can have severe adverse effects on health, including immunotoxic, hepatotoxic, carcinogenic, teratogenic, mutagenic, nephrotoxic, and neurotoxic effects (Prandini et al. 2009; Becker-Algeri et al. 2016; Benkerroum 2016). These mycotoxins can contaminate

Table 16.6 Pathogenic yeasts and fungi identified from milk and dairy products

Types	Species	Milk			Cheese			Yogurt	References
		Cattle	Buffalo	Goat	Cattle	Buffalo	Goat		
Yeasts	*Candida* spp.	+	+	+	+	+	+	+	Spanamberg et al. (2009), Delavenne et al. (2011), Banjara et al. (2015), Cardinali et al. (2016), Chipilev et al. (2016) and Garnier et al. (2017)
	Geotrichum spp.	+	+	–	+	+	+	–	El-Diasty et al. (2009), Delavenne et al. (2011), Marín et al. (2015) and Chipilev et al. (2016)
	Bullera spp.	+	+	+	+	+	+	+	Spanamberg et al. (2009)
	Cryptococcus spp.	+	–	+	–	–	–	+	Spanamberg et al. (2009), Delavenne et al. (2011) and Cardinali et al. (2016)
	Dekkera spp.	+	–	+	–	–	–	–	Spanamberg et al. (2009)
	Pichia spp.	+	–	+	+	+	+	+	Spanamberg et al. (2009), Delavenne et al. (2011) and Cardinali et al. (2016)
	Rhodotorula spp.	+	+	+	+	+	+	+	Spanamberg et al. (2009), Delavenne et al. (2011), Banjara et al. (2015), Cardinali et al. (2016) and Chipilev et al. (2016)
	Saccharomyces spp.	+	+	+	+	+	+	–	Banjara et al. (2015) and Chipilev et al. (2016)
	Debaryomyces spp.	+	+	+	+	+	+	+	Spanamberg et al. (2009), Delavenne et al. (2011), Banjara et al. (2015), Cardinali et al. (2016) and Chipilev et al. (2016)
	Kluyveromyces spp.	–	–	–	+	+	+	+	Spanamberg et al. (2009), Delavenne et al. (2011), Banjara et al. (2015), Cardinali et al. (2016) and Chipilev et al. (2016)
	Meyerozyma spp.	–	+	+	+	+	+	+	Cardinali et al. (2016)
	Malassezia spp.	+	+	+	+	+	+	+	Delavenne et al. (2011)

(continued)

Table 16.6 (continued)

Types	Species	Milk			Cheese			Yogurt	References
		Cattle	Buffalo	Goat	Cattle	Buffalo	Goat		
Fungi	*Penicillium* spp.	+	–	–	+	+	+	–	Pešić-Mikulec et al. (2005), El-Diasty et al. (2009), Delavenne et al. (2011), ELbagory et al. (2014), Hymery et al. (2014b), Banjara et al. (2015), Cardinali et al. (2016) and Chipilev et al. (2016)
	Cladosporium spp.	+			+	+	+	+	El-Diasty et al. (2009) and Cardinali et al. (2016)
	Aspergillus spp.	+	+	+	+	+	+	+	Pešić-Mikulec et al. (2005), Delavenne et al. (2011), ELbagory et al. (2014), Hymery et al. (2014b), Banjara et al. (2015), Marín et al. (2015) and Chipilev et al. (2016)
	Fusarium spp.	+	–	–	–	–	–	–	Pešić-Mikulec et al. (2005), Delavenne et al. (2011) and Marín et al. (2015)
	Mucor spp.	+	–	–	+	+	+	+	Pešić-Mikulec et al. (2005), El-Diasty et al. (2009) and Chipilev et al. (2016)
	Dacrymyces spp.	+	–	+	–	–	–	–	Spanamberg et al. (2009)
	Rhizopus spp.	+	–	–	+	+	+	–	Pešić-Mikulec et al. (2005), El-Diasty et al. (2009), Spanamberg et al. (2009) and Cardinali et al. (2016)
	Curvularia spp.	+	–	–	–	–	–	+	El-Diasty et al. (2009)
	Paecilomyces spp.	+	–	–	–	–	–	–	
	Absidia spp.	+	–	–	–	–	–	–	
	Alternaria spp.	+	–	–	–	–	–	–	
	Cladosporium spp.	+	–	–	–	–	–	+	Pešić-Mikulec et al. (2005), Marín et al. (2015) and Cardinali et al. (2016)
	Geotrichum spp.	+	+	+	+	+	+	+	Pešić-Mikulec et al. (2005), Marín et al. (2015), Cardinali et al. (2016) and Chipilev et al. (2016)
	Yarrowia spp.	–	–	–	+	+	+		Cardinali et al. (2016)
	Scopulariopsis spp.	+	–	–	–	–	–	–	Pešić-Mikulec et al. (2005)
	Trichoderma spp.	+	–	–	–	–	–	–	
	Trichosporon spp.	+	–	+	+	–	+	–	Spanamberg et al. (2009), Delavenne et al. (2011) and Marín et al. (2015)
	Zygomycota spp.	+	–	+	–	–	–	–	Spanamberg et al. (2009)
	Galactomyces spp.		–		+	+	+	–	Banjara et al. (2015)
	Engyodontium spp.	+	–	+	–	–	–	–	Delavenne et al. (2011)
	Torrubiella spp.	+	–	+	–	–	–	–	
	Lichtheimia spp.	+	+	+	–	–	–	–	Marín et al. (2015)
	Blastomyces spp.	–	–	–	+	+	+	–	Garnier et al. (2017)

Symbols used: "+" indicates presence of organisms; "–" indicates absence of organisms in milk and dairy products milk and dairy products both directly and indirectly. Direct mycotoxin contamination can occur in dairy products during production stages, notably evident in cheese production where fungi are utilized in the process. However, the major routes of contamination of mycotoxin-producing fungi or yeast in milk and dairy derivatives include:

1. Milk contamination by pasteurization-resistant *Byssoclamys* (*B.*) *fulva*, which can produce toxins during cheese production or storage (Prandini et al. 2009; Benkerroum 2016)
2. The use of specific wild strains of fungi to produce various PDO categorized cheeses worldwide (Pattono et al. 2013; Benkerroum 2016)
3. The utilization of different commercial strains of fungus, such as *Penicillium* (*P.*) *roqueforti*, as starter cultures for cheese maturation, leading to elevated levels of mycotoxins above EU and Codex thresholds (Fontaine et al. 2015)
4. The potential presence of certain mycotoxins, such as Zearalenone or T-2 toxin, in milk due to silage-based animal feed when the health and immune status of the animals decline (Becker-Algeri et al. 2016)

Additionally, the major significant mycotoxins found in milk and dairy products are listed in Table 16.7. Among them, aflatoxins, including types B1, B2, G1, and G2, have been identified as the most hazardous mycotoxins produced by *Aspergillus* spp. (Prandini et al. 2009; Becker-Algeri et al. 2016). In 1960, the first cases of aflatoxin M1 (AFM1) contamination in milk were identified. AFM1 is a metabolite of aflatoxin B1 produced in the animal liver and secreted into milk. Approximately 0.3–6.2% of ingested aflatoxin B1 is converted to AFM1 in the liver via cytochrome P450 enzymes (Duarte et al. 2012; Iqbal et al. 2013; Becker-Algeri et al. 2016). These AFM1 binds to milk proteins, especially casein that are present in different dairy products such as cheese and yogurt across the world. AFM1 toxin poses a wide range of health risks to humans, including immunotoxic, hepatotoxic, carcinogenic, teratogenic, and mutagenic effects (Prandini et al. 2009; Fallah 2010; Becker-Algeri et al. 2016). It remains stable throughout pasteurization, storage, and processing of milk and dairy products. Due to its significant health risks, AFM1 is considered a major concern in milk and dairy derivatives (Becker-Algeri et al. 2016).

16.8 Determination of the Microbial Composition of Raw Milk

Raw milk contains various microorganisms, classified as beneficial, spoilage, or pathogenic. While some enhance flavor and texture of dairy products, others can lead to unpleasant changes. Bacteria such as *Salmonella* spp., *E. coli*, and *Listeria* spp. in raw milk pose serious health risk to consumers (Van Kessel et al. 2011). Hence, evaluating raw milk's microbial quality is crucial to safeguard public health and ensure suitability for consumption. Field-based and enumeration tests are

Table 16.7 Significant mycotoxins identified in milk and dairy products

Mycotoxins	Producers	Main site of action	Predisposing factors	Dairy products detected in	Mode of toxicity	References
Aflatoxin M1	*Aspergillus* (*A.*) *flavus*, *A. parasiticus*, and *A. nomius*	Liver	Feed contaminated with aflatoxin	Milk, cheese, yogurt, butter, cream, and ice cream	Immunotoxic, hepatotoxic, carcinogenic, teratogenic, mutagenic	Sengun et al. (2008), Prandini et al. (2009), Fallah (2010) and Corassin et al. (2022)
Aflatoxin B1	*A. flavus*, *A. parasiticus*, *A. nomius*, *A. pseudotamarii*, *A. bombycis*, *A. ochraceoroseus*, and *A. australis*	Liver	Feed contaminated with aflatoxin	Cheese	Acute toxicity, immunosuppressive, carcinogenic, and teratogenic, mutagenic	Sengun et al. (2008)
Ochratoxin A	*A. ochraceus*, *A. melleus*, *A. sulphureus*, *A. nigri*, *A. awamori*, *A. lanosus*, *A. alliaceus*, *A. carbonarius*, *Penicillium* (*P.*) *verrucosum*, *P. chrysogenum*, and *P. nordicum*	Kidney	Feed contaminated with ochratoxin during storage	Milk and dairy products like cheese and infant formula milk	Teratogenic, immunotoxic, nephrotoxic, and retard growth rate in children	Murphy et al. (2006), Bräse et al. (2009), El Khoury et al. (2011), Duarte et al. (2012) and Pattono et al. (2013)

Citrinin	*A. carneus*, *A. niveus*, *P. citrinum*, *P. expansum*, *P. verrucosum*, and *Monascus purpureus*	Kidney, nervous system, and occasionally liver	Change of temperature, pH, and light during storage	Cheese	Nephrotoxicity, neurotoxicity, hepatotoxicity	Lund et al. (1995), Vazquez et al. (2001), Bailly et al. (2002), Sengun et al. (2008) and Kamle et al. (2022)
Zearalenone	*Fusarium* (*F.*) *graminearum*, *F. culmorum*, *F. equiseti*, *F. cerealis*, *F. verticillioides*, and *F. incarnatum*	GI tract and reproductive system	Consumption of contaminated feed (grains or crops)	Milk and dairy products	Estrogenic and impaired reproductive and carcinogenic effect	Minervini et al. (2005a, b) and Becker-Algeri et al. (2016)
Trichothecenes	*F. graminearum*, *F. sporotrichioides*, *F. poae*, and *F. equiseti*	GI tract, kidney, and skin	Consumption of contaminated feed (grains or crops)	Milk and dairy products	Acute toxicity, immunosuppressive, effect in GI tract, necrosis in skin, and nephrotoxicity	Flores-Flores et al. (2015)
Cyclopiazonic acid	*A. falvus*, *A. pseudotamarii*, *A. australis*, *P. camemberti*, and *P. verrucosum*	GI tract, *liver*, *kidney*, and *heart*	Contaminated animal feed	Different types of cheese	Cytotoxicity, immunotoxicity, and kodua poisoning	Sengun et al. (2008), Bräse et al. (2009)and Hymery et al. (2014a)

(continued)

Table 16.7 (continued)

Mycotoxins	Producers	Main site of action	Predisposing factors	Dairy products detected in	Mode of toxicity	References
Patulin	*A. clavatus*, *Byssoclamys* (*B.*) *nivea*, *B. fulva*, *P. roqueforti*, *P. expansum*, *P. clavigerum*, *P. griseofulvum*, *P. crustosum*, *P. paneum*, and *P. patulum*	Liver, spleen, and kidney	Consumption of contaminated feed (grains or crops)	Cheese	Neurotoxicity, hepatotoxicity, nephrotoxicity, genotoxicity, and teratogenicity	Erdogan and Sert (2004), Murphy et al. (2006), O'Brien et al. (2006), Bräse et al. (2009), Pattono et al. (2013) and Marín et al. (2015)
Roquefortine C	*P. roqueforti*, *P. crustosum*, and *P. paneum*	Nervous system	Consumption of contaminated silage	Cheese	Neurotoxicity	Polonsky et al. (1977), Erdogan and Sert (2004), Kokkonen et al. (2005) and Fontaine et al. (2015)
Mycophenolic acid	*B. nivea*, *P. roqueforti*, and *Neosartorya* (*N.*) *fischeri*	Liver, kidney, and GI tract	Consumption of contaminated feed	Cheese	Immunosuppressive	Kokkonen et al. (2005), Bräse et al. (2009) and Fontaine et al. (2015)

Penicillin Roquefort (PR) toxin	*P. roqueforti*, *P. expansum*, *P. verrucosum*, *P. chrysogenum*, and *P. paneum*	*Lungs*, *heart*, *liver*, and *kidneys*	Change of temperature, pH, and light during storage	Cheese	Mutagenic, carcinogenic, and acute toxicity	Erdogan and Sert (2004), O'Brien et al. (2006), Sengun et al. (2008), Bräse et al. (2009) and Hymery et al. (2014b)
Penicillic acid	*A. ochraceus*, *A. aculeatum*, *P. roqueforti*, *P. verrucosum*, *P. cyclopium*, *P. pulvivorum*, *P. polonicum*, *P. aurantiocandidum*, and *P. viridicatum*	Kidney and liver	Feed stored in lower temperature	Cheese	Carcinogenic effect, cytotoxicity, hepatotoxicity, nephrotoxicity, genotoxicity	Stoev et al. (2001), Erdogan and Sert 2004 and Bräse et al. (2009)
Iso fumigaclavines A and B	*P. roqueforti*	Liver, kidneys, and brain	Consumption of contaminated feed (maize and silage)	Cheese	Neurotoxic	Polonsky et al. (1977)
Andrastin A-D	*P. roqueforti* and *P. paneum*	Liver, kidney, and reproductive system		Cheese	Immunosuppressive and impaired reproductive effect	Nielsen et al. (2005) and McMullin et al. (2014)
T-2 toxins	*F. sporotrichioides*		Consumption of contaminated feed	Milk and dairy products	–	Flores-Flores et al. (2015)

(continued)

Table 16.7 (continued)

Mycotoxins	Producers	Main site of action	Predisposing factors	Dairy products detected in	Mode of toxicity	References
Deoxynivalenol	*F. graminearium* and *F. culmorum*	Ribosome's 60S subunit	Consumption of contaminated feed	Milk and dairy products	Immunosuppression, oxidative stress, cellular damage	Maul et al. (2012) and Shephard et al. (2013)
Fumonisins	*F. proliferatum*, and *F. verticillioides*	Lungs, kidney, and GI tract	Consumption of contaminated feed (grains or crops)	Milk and dairy products	Pulmonary edema, esophageal cancer, and leukoencephalomalacia, hepatic fibrosis	Smith et al. (2013) and Flores-Flores et al. (2015)

conducted initially for preliminary assessment, followed by molecular or immunological tests for specific pathogens. These tests are conducted at dairy processing plants, either before or after a milk tanker arrives (Clawin-Rädecker et al. 2021). Careful selection of these techniques, guided by specific attributes of interest and standard thresholds, is outlined in Table 16.8. This enables a comprehensive evaluation of the suitability of raw milk for human consumption, emphasizing the significance of microbial quality control in dairy production.

16.9 Identification of Microbiological Hazards in Milk and Dairy Products

Rapid and accurate detection of pathogenic microorganisms in dairy products is essential for outbreak source identification and food safety. Traditional culture-based methods, while sensitive and low-cost, are laborious and time-consuming, often yielding inconclusive results. Advances in microbiological diagnostics have led to novel techniques such as PCR, biosensors, and whole-genome sequencing, offering rapid, sensitive, and specific pathogen detection. These technologies, although promising, require effective sample preparation due to the complex composition of dairy products (Fig. 16.2 and Table 16.9).

16.10 Biofilm Formation in Milk and Dairy Products and Its Significance

Biofilms are complex communities of microorganisms that form three-dimensional structures by adhering to solid, liquid, or air interfaces (Sauer et al. 2022). These bacteria live in association with one another on biotic or abiotic substrates, enveloped in self-secreted extracellular polysaccharides, proteins, lipids, and DNA (Satpathy et al. 2016). It is noteworthy that up to 90% of microbial populations live in biofilms and contaminate various food sectors.

Since biofilms are accountable for approximately 60% of foodborne outbreaks, their prevalence in the food industry has been extensively studied, especially in the dairy industry. Several studies have revealed that biofilms formed by LAB have the potential to improve food quality while inhibiting the growth of foodborne pathogens (Yuan et al. 2022). Despite these advantages, microbial biofilms in dairy environments typically contain a variety of pathogenic and/or spoilage bacteria. Microbial biofilms are distinguished by their formidable resistance to control measures, including the widely used cleaning-in-place (CIP) procedures in the dairy industry (Yuan et al. 2021). This tenacity exacerbates the persistence of microbiological contamination in the dairy industry, having a discernible impact on the safety and quality of dairy products.

Table 16.8 Methods for determining microbial quality of raw milk

Methods	Principle	Target organisms	Interpretation (standard value)	Advantages	Disadvantages	References
Dye reduction test						
Methylene blue reduction test	Enzymatic reduction of methylene blue by metabolically active organisms, leading to dye decolorization in low oxygen environments	*Bacillus cereus*, *B. licheniformis*, *Bacillus subtilis*, *Citrobacter* spp., *E. coli*, *St. bovis*, *Pseudomonas* spp., and other aerobic bacteria	Decolorization in not less than 5–6 h at 36 °C	Simple, rapid, and inexpensive Dye reduction by only viable cells	Inequal reduction by different bacteria Not applicable to foods containing reducing enzymes Unsatisfactory result from well-cooled milk	Bapat et al. (2006)
Resazurin reduction test	Conversion of oxidized blue dye into pink fluorescent resorufin via aerobic respiration in metabolically active bacterial cells	*Acinetobacter baumannii*, *E. coli*, *Klebsiella* spp., *Pseudomonas aeruginosa*, and other aerobic pathogens	Decolorization in not less than 4 h at 36 °C	Relatively inexpensive Homogeneous format and more sensitive Multiplexable with other methods	Cannot estimate thermoduric or psychrophilic microorganisms Not suitable for measuring bulk milk quality	Riss et al. (2016)

Total bacterial count						
Standard plate count (SPC)	Culturing samples containing bacteria on a nutrient agar medium, where mesophilic aerobes form visible colonies	*Micrococcus* spp., *Bacillus* spp., *S. aureus*, *Acetobacter aceti*, *Pseudomonas*, *Flavobacterium*, *Saccharomyces cerevisiae*, *Candida* spp., *Penicillium* spp.	<100,000 CFU/mL	Enumeration of only viable microbes Suitable even with a low bacterial number	Long incubation time, inherent limitations, and subsequent miscalculation Both viable single cells and clumps of cells are counted as a single CFU	Parseelan et al. (2019), Gupta et al. (2020) and Elutade et al. (2023)
Plate loop count (PLC)	The combined use of a calibrated inoculating loop and a continuous pipetting syringe to facilitate the distribution of a small milk sample onto a petri dish to measure the total number of aerobic			Rapid and less requirement of dilution bottles and pipettes	Impact of particulate matter on the accuracy and reproducibility Unreliable when counts exceed 200,000 CFU/mL	Diez-Gonzalez (2014)
Coliform count	Acid and gas production from lactose in the presence of bile salts and basic dyes by coliforms	*Citrobacter* spp., *Enterobacter* spp., *E. coli*, *Klebsiella* spp., and other coliforms	<100 in raw milk	Easy and inexpensive	Cannot detect all microorganisms	Martin et al. (2016)

(continued)

Table 16.8 (continued)

Methods	Principle	Target organisms	Interpretation (standard value)	Advantages	Disadvantages	References
Somatic cell counts (SCC)	The interaction of relevant reagents with somatic cells which is an indicator of an infection that has triggered an inflammatory response.	*S. aureus*, *St. agalactiae*, *St. dysgalactiae*, *St. uberis E. coli*, *K. pneumonia*, *Serratia* spp., other mastitis pathogens	<750,000/mL	Ease of use Favorable detection time Inexpensive	Tendency to color artifacts Uncertain reads in a limited sample and samples with high SCC levels	Hossain et al. (2021)
Titratable acidity test	Measures the total acid concentration determined by exhaustive titration with a standard base	LAB	0.14–0.17% lactic acid	Runs directly in the plant in a very short time	Not a true measure	Gupta et al. (2020) and Santoso (2020)

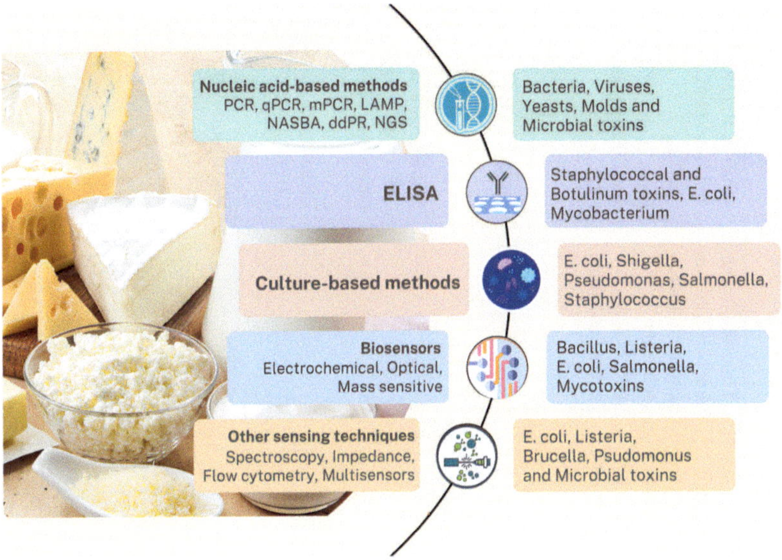

Fig. 16.2 Diagnosis techniques applied for identification of pathogenic microorganisms in milk and dairy products

16.10.1 Biofilm Distribution in Dairy Sector

Milk is widely acknowledged as the optimal medium for fostering the growth and biofilm development of microorganisms within dairy environments. The intricacies of dairy processing lines further contribute to the propitious environment for biofilm formation, owing to the intricate nature of the facilities involved. The intricate biofilm matrix, coupled with the altered physiology of biofilm cells, confers protection to microorganisms against a spectrum of disinfectants, antibiotics, pH variations, UV radiation, and desiccation strategies employed within the dairy industry (Kumari and Sarkar 2018; Wang et al. 2022a). This inherent resistance highlights the possibility of biofilms forming at various points throughout various dairy processing sites. These include raw milk tanks, spiral-wound membranes, wooden vats used in cheese production, dairy reverse osmosis and ultrafiltration membranes, and seemingly innocuous locations such as dairy farm cooling system sprinklers and the dairy wastewater system (Sun and D'Amico 2023). Previous research has shown that both Gram-positive and Gram-negative bacteria can adhere to surfaces such as metals, glass, rubber, and plastics. The formation of biofilms on these surfaces can occur in a matter of days or even hours (Woo et al. 2023). According to previous studies, dairy industry important biofilm-forming major gram-positive bacteria are *Staphylococcus* spp., *Streptococcus* spp., *Enterococcus* spp., *Micrococcus* spp., *Bacillus* spp., *Lactobacillus* spp., *Lactococcus* spp., *Listeria* spp., *Exiguobacterium* spp., *Kocuria* spp., and *Actinomyces* spp. while gram-positive bacteria are *E. coil*, *Acinetobacter* spp., *Pseudomonas* spp., *Cronobacter* spp., *Chryseobacterium* spp.,

Table 16.9 Identification methods of pathogenic microorganisms in milk and dairy products

Methods		Applications		Advantages	Disadvantages	References
Cultural-based methods	**Inhibitors**	**Target microorganism**	**Colony characteristics**			
Mannitol salt agar	10% NaCl	*Staphylococcus* spp.	Small colonies surrounded by yellow zones	Assesses living microbes and viable cells, easy quantification, high sensitivity with appropriate media	Time and resource intensive, risk of contamination, relies on phenotypic characterization	Jahan et al. (2015) and Bano et al. (2020)
Xylose lysine deoxycholate (XLD) agar	Sodium deoxycholate and selective carbohydrates	*Salmonella* spp.	A black center and a slightly red-colored translucent zone			Bano et al. (2020) and Elkenany et al. (2022)
		Shigella spp.	Red colonies			
		Pseudomonas spp.	Pink, flat, rough			
Blood agar	–	*Streptococcus* spp.	Grayish-white, smooth, glossy, and translucent			Jabber (2021)
MacConkey agar	Bile salts	*E. coli*	Flat, dry, pink, non-mucoid colonies with a surrounding darker pink area			Manishimwe et al. (2021) and Obaidat and Gharaibeh (2022)
Salmonella-Shigella agar	Bile salts, brilliant green, and sodium citrate	*Salmonella* spp.	Smooth and opaque or colorless			Elkenany et al. (2022)
		Shigella spp.	Black center			
Brilliant green agar	Brilliant green dye	*Salmonella* spp.	Pinkish-white or red colonies surrounded by a red halo in the medium			Manishimwe et al. (2021)
Sabouraud dextrose agar	–	*Aspergillus* spp.	White mycelium with black spores			Onoharigho et al. (2022)
		Candida spp.	Cream colonies			

Nucleic acid-based methods	Principles	Target microorganisms	Target genes			
Polymerase chain reaction (PCR)	In vitro primer-induced enzymatic replication of DNA segments	*L. monocytogenes*	hlyA, prfa, 16S rRNA	Rapid, sensitive, and highly specific	Expensive and pathogens identification only for included primers	Barakat et al. (2020), Salauddin et al. (2020) and Bahramian et al. (2023)
		Mycobacterium avium Paratuberculosis	IS900			
		S. aurues	23S rRNA, coa			
		E. coli	16S rRNA			
Quantitative polymerase chain reaction (qPCR)	Addition of fluorescent groups to continuously monitor the fluorescent signal	*Clostridium jejuni*	VS1	High sensitivity, rapid automated screening, and high-throughput analysis capabilities	Non-specific binding, overlap of emission spectra, and increased variation with cycle number	Farrokh et al. (2013) and Bai et al. (2022)
		M. avium Paratuberculosis	IS90			
		Shiga toxin-producing *E. coli* (STEC)	Rfb			
		L. monocytogenes	Hly, inlA			
		Salmonella spp.	invA			
		Bacillus spp.	V3–V6 16S rRNA			
		E. coli	stx1, stx2, yccT			
		Yeasts, molds	act, AccD, cox1, cytB, atp6			

(continued)

Table 16.9 (continued)

Methods		Applications		Advantages	Disadvantages	References
Multiplex polymerase chain reaction (mPCR)	Simultaneous detection of multiple targets in a single reaction well, with a different pair of primers for each target	*L. monocytogenes*	htpG, hylA, plcA, plcB, inlB, 16S rRNA	High efficiency, high throughput, and inexpensiveness	Self-inhibition among different primer sets and low amplification efficiency	Demirci et al. (2019), Barakat et al. (2020), Zhang et al. (2020), Hassani et al. (2022), Reese and Elkins (2022) and Gómez et al. (2023)
		C. jejuni	hipO, cadF, mapA			
		E. coli	bla_{SHV}, bla_{TEM} stx, stx2, uidA			
		Salmonella spp.	fbJ, fliC, fliB, and invA, ttr			
		S. aureus	bla_{Z}, hsp, mecA, nuc, lukE			
		Y. enterocolitica	ail, yst			
		E. coli O157	rfbE			
		Shigella spp.	ipaH			
		Vibrio parahaemolyticus	pR72H fragment			
		P. fluorescens	aprX,			
Loop-mediated isothermal amplification (LAMP)	Auto cycling and high DNA strand displacement activity	*L. monocytogenes*	hylA, iap	Highly selective, rapid, sensitive, conducted at a stable temperature	Heavy reliance on indirect detection methods leads to false positive results	Gadkar et al. (2018), Ledlod et al. (2020), Inaba et al. (2021) and Li et al. (2022)
		Salmonella spp.	phoP, invA			
		B. cereus	tetL			
		Shigella spp.	ipaH			
Nucleic acid sequence-based amplification (NASBA)	mRNA-based real-time enzymatic amplification without the interference of DNA	*S. aureus*	tmRNA	Selective amplification of RNA and works under isothermal conditions	Use of thermolabile enzymes, additional melting step, fragile RNA integrity	Gore et al. (2003), Gilbride (2014) and Zhai et al. (2019)
		Salmonella spp.	xcd, invA			
		Bacillus spp.	hblC			

ddPCR	Creation of micoreaction units with water-in-oil emulsion and microfluidic technology	*Salmonella* spp.	invA	Better resistance to PCR inhibitors, high-throughput capability, absolute quantification of target DNAs w/o standard curve	Narrow dynamic range and unreliable results at high bacterial concentrations (>10^6 CFU/mL)	Basanisi et al. (2022), Choi et al. (2022), Kong et al. (2022) and Mathekga et al. (2022)
		E. coli O157:H7	Z3276, rfbE			
		Coxiella burnetti	IS1111 and icd			
		L. monocytogenes	PrfA			
		B. cereus	gyrB			
Next-generation sequencing	Simultaneous, high-throughput sequencing for rapid and cost-effective generation of extensive DNA data	*Campylobacter* spp., *E coli*, *Clostridium tyrobutyricum*, *L. monocytogenes*, *M. tuberculosis*, *Shigella* spp.	Whole-genome sequencing, target sequencing, metagenomic next-generation sequencing (mNGS)	Higher specificity than RT-PCR, strain identification, new genes and pathway identification, powerful for community biodiversity	Complex in procedure and analyses, expensive	Ferrario et al. (2017) and Walsh et al. (2017)
Biosensors	**Principles**	**Target microorganisms**	**Transduction**			
Electrochemical	Redox reaction potential of associated enzymes and electrochemical signaling for the detection	*E. coli*, *Salmonella* spp., *S. aureus*, *B. cereus*, *Campylobacter* spp., *Listeria* spp. and mycotoxins	Amperometry, potentiometry	Easy miniaturization, excellent detection limits, automation, reduced costs of testing, and unaffected by sample turbidity	Lower shelf life and sensitivity to sample matrix effects	Murasova et al. (2020), Bacchu et al. (2022), Kumar and Kachhawa (2023) and Thurner and Alatraktchi (2023)

(continued)

Table 16.9 (continued)

Methods		Applications		Advantages	Disadvantages	References
Optical	Shift in phase, amplitude, frequency or polarization of light and signal emission proportionate to the conc. of an analyte	*E. coli* O157:H7, *S. enteritidis*	Absorption, fluorescence, phosphorescence, bioluminescence, reflectance, Raman scattering, surface plasmon resonance (SPR)	High specificity, sensitivity, small size, cost-effectiveness	Non-selective adsorptions and interferences due to background fluorescence	Nabok et al. (2019) and Pebdeni et al. (2022)
Mass sensitive	Responds to the mass of solute passing through it per unit time	*E. coli* O157:H7, *Salmonella* spp. and mycotoxins	Piezoelectric, surface acoustic wave, cantilever	Flexibility, real-time output, high sensitivity, simplicity and cost-effectiveness	Connection with an external electronic circuit, unsuitable for static conditions and high-temperature sensitivity	Spagnolo et al. (2022) and Chen et al. (2023)
Other sensing technologies	**Principles**	**Target microorganisms**	**Types**			
Flow cytometry	Scattering of light and emission of fluorescence upon laser beam hits the cells moving in a directed fluid stream.	*E. coli*, *Salmonella* spp., *Campylobacter* spp., *L. monocytogenes*, *S. aureus*, *Pseudomonas* spp., *L. lactis*, *E. coli* O157:H7	Traditional, acoustic focusing, imaging, mass, etc.	Speed, minimal sample requirement, and availability of stain multiplicity to examine different aspects of cell viability	Expensive equipment and results influenced by SCC exceeding 1 million/mL	Liu et al. (2021), Manohar et al. (2021), Wang et al. (2020) and Wang et al. (2023)

Impedance microbiology	Electrically measurable changes in the growth medium during microbial growth and metabolic processes	*E. coli*, *S. aureus*, *L. monocytogenes*, *Salmonella* spp., *Pseudomonas* spp., *Aeromonas* spp., Yeasts, and fungal metabolites	Electrochemical, faradic, non-faradic	Cost-effective and non-destructive method and immediate obtaining of spectra	Contamination issues, specific selective medium requirements, interference from food matrices	Bancalari et al. (2019), Wang et al. (2021), Scandurra et al. (2022) and Soares et al. (2022)
Spectroscopy	Electromagnetic radiation utilized to detect pathogens by examining their absorbed or emitted energy	*C. jejuni*, *L. monocytogenes*, *E. coli*, *Klebsiella* spp., *Brucella* spp., *S. aureus*., *Lactobacillus* spp., mycotoxins	Raman, NIR, FTIR, spectrofluorometry	Expedited identification, decreased consumable usage, automation	Not widely incorporated in clinical workflows, background interference, high technical skills	Flores-Flores et al. (2015), Shi et al. (2019), Costa et al. (2020), Shi et al. (2020), Li et al. (2022), Pandian et al. (2023) and Spagnolo et al. (2023)
Multisensory techniques	Diverse, low-selectivity sensors and advanced mathematical techniques for microbial pattern recognition and data analysis	*B. cereus*, *P. aureofaciens*, *P. fluorescens*, *P. aeruginosa*, *E. coli*, *En. faecalis*, and *S. aureus*	Electronic tongue, electronic nose	Precise, no reagent demand, easy, safe, high selectivity, and specificity	The short lifetime of sensor components due to absorption, preparation time, and intense energy consumption	Carrillo-Gómez et al. (2021) and Ghrissi (2021)
Enzyme-linked immunosorbent assay (ELISA)	Highly specific antibody-antigen interactions	*M. avium*, *Paratuberculosis*, *E. coli* and *S. aureus* toxins, AFM1, botulinum toxin serotypes A, B, E and F	Direct, indirect, sandwich, and competitive	Sensitive, rapid, inexpensive, simple, and reliable	Labor-intensive, antibody instability, unreliable in inadequate inhibition of immunogenic antigen	Sakamoto et al. (2018), Hosseiniporgham et al. (2020) and Almashhadany (2021)

Proteus mirabills, *C. jejuni*, *Salmonella* spp., *Vibrio cholerae*, *Enterobacter cloacae*, and *Burkholderia cepacian* (Woo et al. 2023).

16.10.2 Factors Influencing Biofilm Formation

Biofilm formation is a complex and dynamic process influenced by a variety of factors, such as strains, nutrient levels, temperature, pH, ionic strength, surface properties, and bacterial interactions (Wang et al. 2022b). Lactose in milk emerges as a catalyst for biofilm formation among dairy isolates, amplifying polysaccharide adhesion and activating the quorum sensing system (Duanis-Assaf et al. 2016). Bacterial adhesion, a pivotal facet of biofilm formation, is modulated by cell surface attributes (hydrophobicity, flagellation, and motility), surface properties (hydrophobicity and roughness), and environmental variables (temperature, pH, nutrient availability, and hydrodynamic conditions) (Alotaibi and Bukhari 2021).

Several variables such as feed temperature, filtered fluid, and cleaning cycles significantly influence bacterial biofilm progression on ultrafiltration membranes (Chamberland et al. 2017). Concurrently, diverse feed substrates introduce variations in biofilm compositions on reverse osmosis membranes, consequently influencing biofilm resistance to CIP procedures (Marka and Anand 2018). Moreover, extracellular appendages like fimbriae and flagella, coupled with interactions fostering cell-to-cell communication and the production of extracellular polymeric substances (EPS), confer competitive advantages within mixed microbial communities (Simões et al. 2010). The amalgamation of these factors contributes to the intricate and nuanced phenomenon of biofilm formation across diverse environmental contexts.

16.10.3 Significance of Biofilms in the Dairy Industry

Notable pathogens that exhibit a remarkable capacity to form biofilms on the surfaces of dairy plants include *Cronobacter sakazakii*, *B. cereus*, *L. monocytogenes*, *S. aureus*, and *E. coli* (Boor et al. 2017). The formation of biofilms by these pathogens is recognized as a considerable public health concern (Table 16.10), with the associated diseases imposing a substantial economic burden in recent years. It is estimated that approximately 80% of microbial infections can be attributed to the presence of microbial biofilms. The deleterious impact of biofilms extends beyond public health concerns, as they have been implicated in compromising the quality of dairy products. In two preceding studies, the proteolysis of *P. fluorescens* biofilm cells and the lipolysis of *S. aureus* biofilm cells were notably higher than those observed in their planktonic counterparts (Teh et al. 2013). The augmented heat stability of enzymes originating from biofilm cells is attributed to the protective nature of the biofilm matrix. Moreover, enzymes produced by biofilm cells have the

Table 16.10 Source and harmful effects of major pathogens forming biofilms in the dairy sector

Pathogens	Source		Harmful spoilage effects	Reference
	Dairy farm area	Processing plant		
Bacillus	Balance tank, feeding unit, tank truck, short milking tube	Welded joints of silo, inside of air separator	Diarrhea and vomiting	Grigore-Gurgu et al. (2020)
Pseudomonas	Balance tan, tank truck, short milking tube	Conveyor belt, welded joints of silo, inside of air separators	Blue discoloration on fresh cheese	Marchand et al. (2012) and González-Rivas et al. (2018)
Aeromonas	Doormat	–	Diarrhea	Papageorgiou et al. (2003)
Staphylococcus	Tank truck, short milking tube	Aging tank, welded joints of silo, inside of air separators	Methicillin resistance, diarrhea, and vomiting	Giaouris et al. (2015)
Lactic acid bacteria (LAB)	Feeding unit, doormat, tank truck, short milking tube	Conveyor belt, ultrafiltration membranes, inside of air separators	–	Marchand et al. (2012)
Enterobacteriaceae	Balance tank, floor drain, doormat, tank truck, short milking tube	Conveyor belt	Gastroenteritis, septicemia, and hemolytic uremia	Galié et al. (2018)
Listeria	Floor drain, doormat	Bulk tank outlet, rubber liners	Listeriosis in the elderly, immune-compromised patients, and pregnant women	Rothrock et al. (2017)

potential to hydrolyze amino acids in dairy products, leading to the production of BAs, which, when present in high concentrations, may pose toxicological risks to human health (Moniente et al. 2022).

Beyond compromising product quality, biofilms formed on the inner surfaces of tanks and pipes pose additional challenges, including corrosion, ultimately diminishing the heat transfer efficiency of machinery (Kimbell et al. 2020). For instance, prolonged contact with biofilms formed by *Bacillus sporothermodurans* and *Geobacillus stearothermophilus* has been linked to the development of extensive cracks on stainless steel coupons. Energy-dispersive spectroscopy analysis revealed an increased content of oxygen and sulfur, coupled with a decreased content of iron in the stainless steel (Gupta and Anand 2018).

16.10.4 Control of Biofilms in the Dairy Sector

Researchers have explored both physical and chemical methods to hinder bacterial biofilms. Previous investigations have demonstrated that mechanical treatments like CIP are unable to completely eradicate all bacterial cells, whereas chemical treatments contribute to eliminating bacterial cells and impeding biofilm formation (Yu et al. 2021). Notably, sodium hypochlorite (NaClO) has been identified as a potential disinfectant effective in inhibiting bacteria; however, there is a risk of bacterial proliferation under higher concentrations of disinfectants, creating conditions conducive to biofilm formation (Sharma et al. 2014). In addressing the threat of foodborne infections arising from biofilms, diverse strategies have emerged, encompassing the prevention of bacterial adhesion, early-stage bacterial detection, and the dissolution of biofilms, transforming sessile bacteria into susceptible planktonic cells. A substantial body of literature delves into the examination of antimicrobial agents, the inhibitory impacts of biological agents, and novel approaches to impede biofilm formation (Nahar et al. 2021). Notably, biochemical agents such as essential oils, enzymes, biosurfactants, and others are discussed as safe and environmentally friendly approaches for controlling the formation of pathogenic biofilms.

16.11 Antimicrobial Resistance in Pathogenic Microorganisms in Dairy Industry

AMR in the dairy industry is a complex and multifaceted problem that necessitates a more in-depth understanding of its causes and consequences. Bacteria adapt and become resistant to antibiotics that were once effective against them. This resistance can develop as a result of a number of factors, but one of the most significant contributors is the improper use of antibiotics. Antibiotics are commonly used in the dairy animals to treat intramammary infections (IMI) and other health issues, but they are frequently misused (Paramasivam et al. 2023). Antibiotics including beta-lactam (penicillin and cephalosporin), aminoglycosides, lincosamides, and macrolides are the most used for treating IMI worldwide (Martins et al. 2021). However, these antibiotics are losing their effectiveness against pathogens. The alarming levels of resistance observed such as 84.1% penicillin resistance and 20.5% erythromycin resistance highlight the importance of addressing AMR in the dairy industry.

Table 16.11 Prevalence of major pathogens of milk and dairy products and their antimicrobial resistance patterns in South Asia

Country	Year	Bacteria	Sample size	Positive cases	Antibiotic resistance (%)												References
					AMP	AUG	TE	P	CN	E	CIP	MET	CTX	SXT	S	NA	
Bangladesh	2019	*E. coli*	240	180	–	–	89.44	100	64.44	88.89	16.11	–	–	–	–	15.60	Sobur et al. (2019)
	2020	*S. aureus*	67	17	70.59	58.82	76.47	52.94	41.18	–	41.18	–	29.41	47.06	58.82	–	Rana et al. (2020)
	2021	*E. coli*	68	19	89.47	–	–	–	–	–	–	–	–	–	–	–	Bag et al. (2021)
	2022	*S. aureus*	219	109	92.66	–	–	–	42.20	9.17	–	–	–	–	–	–	Hoque et al. (2022)
India	2016	*E. coli*	90	73	29.31	–	34.03	–	–	–	–	–	–	–	–	–	Bhoomika et al. (2016)
	2017	*L. monocytogenes*	457	5	80	80	20	100	0	–	–	–	100	–	0	80	Sharma et al. (2017)
	2018	*E. coli*	182	22	–	45.45	40.91	–	31.82	–	–	–	–	–	–	–	Batabyal et al. (2018)
	2019	*S. aureus*	150	80	86.25	72.50	67.50	98.75	83.75	–	82.50	100	53.75	51.25	–	–	Shah et al. (2019)
	2021	*S. aureus*	200	47	–	19.15	4.26	19.15	6.38	8.51	4.26	–	–	12.77	–		Quraishi et al. (2021)
	2022	*S. aureus*	433	80	–	–	30	–	58.75	40	–	–	–	–	–	–	Brahma et al. (2022)
	2022	*S. aureus*	565	55	89.09	89.09	–	83.64	34.55	–	–	12.73	30.90	–	–	–	Neelam et al. (2022)
Nepal	2021	*S. aureus*	460	29	100	62.07	58.62	65.52	93.10	48.28	68.97	96.55	–	–	–	–	Shrestha et al. (2021)
Pakistan	2021	*E. coli*	385	30	10	30	–	–	73.33	33.33	–	–	–	–	23.33	–	Aslam et al. (2021)
	2023	*S. aureus*	150	55	–	100	72.73	100	9.09	–	100	–	–	18.18	–	–	Lubna et al. (2023)
	2023	*Klebsiella* spp.	700	180	–	–	–	–	–	–	–	–	–	–	–	–	Saddam et al. (2023)
	2023	*S. aureus*	19	14	–	–	92.86	–	42.86	57.14	71.43	–	–	–	–	–	Rasool et al. (2023)

Abbreviations: *AMP* ampicillin, *AUG* amoxicillin/clavulanate, *TE* tetracycline, *P* penicillin, *CN* gentamycin, *E* erythromycin, *CIP* ciprofloxacin, *MET* methicillin, *CTX* ceftriaxone, *STX* sulfamethoxazole-trimethoprim, *S* streptomycin, *NA* nalidixic acid

S. aureus is a major pathogen of concern in the dairy industry as it causes mastitis along with many other microbes. The ability of *S. aureus* to develop AMR and rapid transmission of antibiotic-resistant genes such as *mecA*, *mgrA*, *norA*, *arlR*, *arlS*, *DHA-1*, and *LmrS* renders conventional treatments ineffective (Tóth et al. 2020).

In addition to *S. aureus*, India has reported the presence of extended-spectrum beta-lactamases (ESBL) producing *E. coli* in milk samples (Kar et al. 2015). This finding raises concerns about AMR transmission to consumers via dairy products. ESBL-producing *K. pneumonia* has also been found in buffalo milk and many of these strains also carry AmpC type beta-lactamase and plasmid-mediated fluoroquinolone resistance genes (Bandyopadhyay et al. 2018). Furthermore, other studies in South Asian countries found that a wide range of bacteria isolated from milk samples were resistant to almost all the commonly used broad to narrow-spectrum antibiotics (Table 16.11). AMR has far-reaching implications in dairy farming that are not limited to the dairy industry. There is a substantial risk that resistant bacteria will enter the human food chain, posing a significant threat to public health. The detection of vancomycin- and methicillin-resistant *S. aureus* (MRSA) in milk, as well as multidrug-resistant *E. coli* in bovine and goat milk, exemplifies this (Bhattacharyya et al. 2016).

16.12 Conclusion

Our exploration of the dairy industry has provided valuable insights into the intricate and multifaceted interplay of microorganisms with human and livestock health, as well as its economic dimensions. In addition to producing a rich tapestry of flavors and textures, the use of microbes such as LAB in dairy products aids in the release of beneficial compounds for human health. However, the presence of spoilage and pathogenic microorganisms, as well as the formation of biofilm, poses challenges in this industry, as evidenced by extensive discussions on their prevalence and concurrent antimicrobial resistance. The rising prevalence of biological, chemical, allergenic, and physical hazards and the economic burden imposed by dairy-borne infections underscores the need for comprehensive and holistic preventive strategies encompassing vigilant monitoring and stringent quality control measures. Discussion on various methodologies and technologies to detect pathogens offers significant insights into the evolving landscape of microbiological analysis in milk and milk product samples. This industry should prioritize maximizing the benefits of probiotics for microbial enhancement of milk products while minimizing pathogenic organisms and biofilm formation. We recommend policymakers, stakeholders, and consumers channel their concerted efforts toward mitigating these risks inherent in microbial contamination within the dairy supply chain.

Author's Contributions Conceptualization and outline preparation: D.H.; data collection and curation: M.J.A., A.B.H., A.A.B; S.Z.T.B., and D.H.; figure preparation: D.H.; table preparation: M.J.A., A.B.H., A.A.B., S.Z.T.B., and D.H.; validation and supervision: D.H., M.R.K., and A.K.;

writing—original draft preparation: D.H., A.B.H., M.J.A., A.A.B., S.Z.T.B., T.S., and N.U.; writing—review and editing: M.R.K., A.K., and D.H. All authors contributed to the book chapter and approved for publication.

Conflicts of Interest Authors declare no conflicts of interest.

References

Abd El-Salam MH, El-Shafei K, Sharaf OM, Effat BA, Asem FM, El-Aasar M (2010) Screening of some potentially probiotic lactic acid bacteria for their ability to synthesis conjugated linoleic acid. Int J Dairy Technol 63:62–69. https://doi.org/10.1111/j.1471-0307.2009.00541.x

Abdul Hakim BN, Xuan NJ, Oslan SNH (2023) A comprehensive review of bioactive compounds from lactic acid bacteria: potential functions as functional food in dietetics and the food industry. Foods 12:2850. https://doi.org/10.3390/FOODS12152850

Agarwal A, Awasthi V, Dua A, Ganguly S, Garg V, Marwaha SS (2012) Microbiological profile of milk: impact of household practices. Indian J Public Health 56:88–94. https://doi.org/10.4103/0019-557X.96984

Ahmed R, Soule G, Demczuk WH, Clark C, Khakhria R, Ratnam S, Marshall S, Ng L-K, Woodward DL, Johnson WM (2000) Epidemiologic typing of Salmonella enterica serotype Enteritidis in a Canada-wide outbreak of gastroenteritis due to contaminated cheese. J Clin Microbiol 38:2403–2406. https://doi.org/10.1128/jcm.38.6.2403-2406.2000

Ahmedsham M, Amza N, Tamiru M (2018) Review on milk and milk product safety, quality assurance and control. Int J Livest Prod 9:67–78. https://doi.org/10.5897/ijlp2017.0403

Akpinar A, Yerlikaya O, Kiliç S (2011) Antimicrobial activity and antibiotic resistance of Lactobacillus delbrueckii ssp. bulgaricus and Streptococcus thermophilus strains isolated from Turkish homemade yoghurts. J Microbiol 5:675–682. https://doi.org/10.5897/AJMR10.835

Alexopoulos A, Tzatzimakis G, Bezirtzoglou E, Plessas S, Stavropoulou E, Sinapis E, Abas Z (2011) Microbiological quality and related factors of sheep milk produced in farms of NE Greece. Anaerobe 17:276–279. https://doi.org/10.1016/j.anaerobe.2011.03.011

Almashhadany DA (2021) Diagnosis of brucellosis in sheep and goats raw milk by fast and reliable techniques. Iraqi J Vet Sci 35(4):663–668. https://doi.org/10.33899/ijvs.2021.127697.1523

Alotaibi GF, Bukhari MA (2021) Factors influencing bacterial biofilm formation and development. Am J Biomed Sci Res 12:617–626. https://doi.org/10.34297/AJBSR.2021.12.001820

Alvarez-Martin P, Florez AB, Hernández-Barranco A, Mayo B (2008) Interaction between dairy yeasts and lactic acid bacteria strains during milk fermentation. Food Control 19:62–70. https://doi.org/10.1016/j.foodcont.2007.02.003

Amer MA, Lammerding AM (1983) Health maintenance benefits of cultured dairy products. Cult Dairy Prod J 18:6

Andretta M, Almeida TT, Ferreira LR, Carvalho AF, Yamatogi RS, Nero LA (2019) Microbial safety status of Serro artisanal cheese produced in Brazil. J Dairy Sci 102:10790–10798. https://doi.org/10.3168/jds.2019-16967

Angelakis E, Million M, Henry M, Raoult D (2011) Rapid and accurate bacterial identification in probiotics and yoghurts by MALDI-TOF mass spectrometry. J Food Sci 76:M568–M572. https://doi.org/10.1111/j.1750-3841.2011.02369.x

Angulo FJ, LeJeune JT, Rajala-Schultz PJ (2009) Unpasteurized milk: a continued public health threat. Clin Infect Dis 48:93–100. https://doi.org/10.1086/595007

Antunes AEC, Cazetto TF, Abolini HM (2005) Viability of probiotic micro-organisms during storage, postacidification and sensory analysis of fat-free yogurts with added whey protein concentrate. Int J Dairy Technol 58:169–173. https://doi.org/10.1111/j.1471-0307.2005.00203.x

Artursson K, Schelin J, Lambertz ST, Hansson I, Engvall EO (2018) Foodborne pathogens in unpasteurized milk in Sweden. Int J Food Microbiol 284:120–127. https://doi.org/10.1016/j.ijfoodmicro.2018.05.015

Aryantini NPD, Yamasaki E, Kurazono H, Sujaya IN, Urashima T, Fukuda K (2017) In vitro safety assessments and antimicrobial activities of lactobacillus rhamnosus strains isolated from a fermented mare's milk. Anim Sci J 88:517–525. https://doi.org/10.1111/asj.12668

Aslam N, Khan SUH, Usman T, Ali T (2021) Phylogenetic genotyping, virulence genes and antimicrobial susceptibility of Escherichia coli isolates from cases of bovine mastitis. J Dairy Res 88:78–79. https://doi.org/10.1017/S002202992100011X

Awasti N, Anand S (2020) The role of yeast and molds in dairy industry: an update. In: Dairy processing: advanced research to applications. Springer, pp 243–262. https://doi.org/10.1007/978-981-15-2608-4_12

Azat R, Liu Y, Li W, Kayir A (2016) Probiotic properties of lactic acid bacteria isolated from traditionally fermented Xinjiang cheese. J Zhejiang Univ Sci B 17:597–609. https://doi.org/10.1631/jzus.B1500250

Bacchu MS, Ali MR, Das S, Akter S, Sakamoto H, Suye SI, Rahman MM, Campbell K, Khan MZH (2022) A DNA functionalized advanced electrochemical biosensor for identification of the foodborne pathogen Salmonella enterica serovar Typhi in real samples. Anal Chim Acta 1192:339332. https://doi.org/10.1016/J.ACA.2021.339332

Bag MAS, Khan MSR, Sami MDH, Begum F, Islam MS, Rahman MM, Rahman MT, Hassan J (2021) Virulence determinants and antimicrobial resistance of E. coli isolated from bovine clinical mastitis in some selected dairy farms of Bangladesh. Saudi J Biol Sci 28:6317–6323. https://doi.org/10.1016/J.SJBS.2021.06.099

Bahramian E, Shahhosseiny MH, Shakib P, Zolfaghari MR, Ahmadi S (2023) Evaluation of hly-A and 16S rRNA genes for detection of listeria monocytogenes in raw milk sample. Chronic Dis J 11:186–190. https://doi.org/10.22122/CDJ.V11I3.835

Bai X, Wang Z, Li W, Xiao F, Huang J, Xu Q, Xu H (2022) Rapid and accurate detection for Listeria monocytogenes in milk using ampicillin-mediated magnetic separation coupled with quantitative real-time PCR. Microchem J 183:108063. https://doi.org/10.1016/J.MICROC.2022.108063

Bailly JD, Querin A, Le Bars-Bailly S, Benard G, Guerre P (2002) Citrinin production and stability in cheese. J Food Prot 65:1317–1321. https://doi.org/10.4315/0362-028X-65.8.1317

Bancalari E, D'Incecco P, Savo Sardaro ML, Neviani E, Pellegrino L, Gatti M (2019) Impedance microbiology to speed up the screening of lactic acid bacteria exopolysaccharide production. Int J Food Microbiol 306. https://doi.org/10.1016/J.IJFOODMICRO.2019.108268

Bandyopadhyay S, Banerjee J, Bhattacharyya D, Samanta I, Mahanti A, Dutta TK, Ghosh S, Nanda PK, Dandapat P, Bandyopadhyay S (2018) Genomic identity of fluoroquinolone-resistant *bla* CTX-M-*15*-type ESBL and pMAmpC β-lactamase producing *Klebsiella pneumoniae* from Buffalo Milk, India. Microb Drug Resist 24:1345–1353. https://doi.org/10.1089/mdr.2017.0368

Banik SK, Das KK, Uddin MA (2014) Microbiological quality analysis of raw, pasteurized, UHT milk samples collected from different locations in Bangladesh. Stamford J Microbiol 4:5–8

Banjara N, Suhr MJ, Hallen-Adams HE (2015) Diversity of yeast and mold species from a variety of cheese types. Curr Microbiol 70:792–800. https://doi.org/10.1007/s00284-015-0790-1

Bano SA, Hayat M, Samreen T, Asif M, Habiba U, Uzair B, Bano SA, Hayat M, Samreen T, Asif M, Habiba U, Uzair B (2020) Detection of pathogenic bacteria *Staphylococcus aureus* and salmonella sp. from raw Milk samples of different cities of Pakistan. Nat Sci (Irvine) 12:295–306. https://doi.org/10.4236/NS.2020.125026

Bapat P, Nandy SK, Wangikar P, Venkatesh KV (2006) Quantification of metabolically active biomass using Methylene Blue dye Reduction Test (MBRT): measurement of CFU in about 200 s. J Microbiol Methods 65:107–116. https://doi.org/10.1016/J.MIMET.2005.06.010

Barakat AMA, El-Razik KAA, Elfadaly HA, Rabie NS, Sadek SAS, Almuzaini AM (2020) Prevalence, molecular detection, and virulence gene profiles of Campylobacter spe-

cies in humans and foods of animal origin. Vet World 13:1430. https://doi.org/10.14202/VETWORLD.2020.1430-1438
Barman S, Ghosh R, Mandal NC (2018) Production optimization of broad spectrum bacteriocin of three strains of Lactococcus lactis isolated from homemade buttermilk. Ann Agrar Sci 16:286–296. https://doi.org/10.1016/J.AASCI.2018.05.004
Basanisi MG, La Bella G, Nobili G, Raele DA, Cafiero MA, Coppola R, Damato AM, Fraccalvieri R, Sottili R, La Salandra G (2022) Detection of Coxiella burnetii DNA in sheep and goat milk and dairy products by droplet digital PCR in south Italy. Int J Food Microbiol 366:109583. https://doi.org/10.1016/J.IJFOODMICRO.2022.109583
Bastam MM, Jalili M, Pakzad I, Maleki A, Ghafourian S (2021) Pathogenic bacteria in cheese, raw and pasteurised milk. Vet Med Sci 7:2445–2449. https://doi.org/10.1002/vms3.604
Batabyal K, Banerjee A, Pal S, Dey S, Joardar SN, Samanta I, Isore DP, Singh AD (2018) Detection, characterization, and antibiogram of extended-spectrum beta-lactamase Escherichia coli isolated from bovine milk samples in West Bengal, India. Vet World 11:1423–1427. https://doi.org/10.14202/vetworld.2018.1423-1427
Becker-Algeri TA, Castagnaro D, de Bortoli K, de Souza C, Drunkler DA, Badiale-Furlong E (2016) Mycotoxins in bovine milk and dairy products: a review. J Food Sci 81:R544–R552. https://doi.org/10.1111/1750-3841.13204
Benkerroum N (2016) Mycotoxins in dairy products: a review. Int Dairy J 62:63–75. https://doi.org/10.1016/j.idairyj.2016.07.002
Bensalah F, Delorme C, Renault P (2009) Characterisation of thermotolerant cocci from indigenous flora of 'leben'in algerian arid area and DNA identification of atypical Lactococcus lactis strains. Curr Microbiol 59:139–146. https://doi.org/10.1007/s00284-009-9411-1
Bernardeau M, Vernoux JP, Henri-Dubernet S, Guéguen M (2008) Safety assessment of dairy microorganisms: the Lactobacillus genus. Int J Food Microbiol 126:278–285. https://doi.org/10.1016/j.ijfoodmicro.2007.08.015
Bernardo D, Sánchez B, Al-Hassi HO, Mann ER, Urdaci MC, Knight SC, Margolles A (2012) Microbiota/host crosstalk biomarkers: regulatory response of human intestinal dendritic cells exposed to Lactobacillus extracellular encrypted peptide. PLoS One 7:e36262. https://doi.org/10.1371/journal.pone.0036262
Bhagat D, Raina N, Kumar A, Katoch M, Khajuria Y, Slathia PS, Sharma P (2020) Probiotic properties of a phytase producing Pediococcus acidilactici strain SMVDUDB2 isolated from traditional fermented cheese product, Kalarei. Sci Rep 10:1–11. https://doi.org/10.1038/s41598-020-58676-2
Bhat B, Bajaj BK (2018) Hypocholesterolemic and bioactive potential of exopolysaccharide from a probiotic Enterococcus faecium K1 isolated from kalarei. Bioresour Technol 254:264–267. https://doi.org/10.1016/j.biortech.2018.01.078
Bhattacharyya D, Banerjee J, Bandyopadhyay S, Mondal B, Nanda PK, Samanta I, Mahanti A, Das AK, Das G, Dandapat P, Bandyopadhyay S (2016) First report on vancomycin-resistant *Staphylococcus aureus* in bovine and caprine milk. Microb Drug Resist 22:675–681. https://doi.org/10.1089/mdr.2015.0330
Bhoomika, Shakya S, Patyal A, Gade NE (2016) Occurrence and characteristics of extended-spectrum β-lactamases producing Escherichia coli in foods of animal origin and human clinical samples in Chhattisgarh, India. Vet World 9:996–1000. https://doi.org/10.14202/vetworld.2016.996-1000
Boor KJ, Wiedmann M, Murphy S, Alcaine S (2017) A 100-year review: microbiology and safety of milk handling. J Dairy Sci 100:9933–9951. https://doi.org/10.3168/jds.2017-12969
Brahma U, Suresh A, Murthy S, Bhandari V, Sharma P (2022) Antibiotic resistance and molecular profiling of the clinical isolates of *Staphylococcus aureus* causing bovine mastitis from India. Microorganisms 10(4):833. https://doi.org/10.3390/MICROORGANISMS10040833
Bräse S, Encinas A, Keck J, Nising CF (2009) Chemistry and biology of mycotoxins and related fungal metabolites. Chem Rev 109:3903–3990. https://doi.org/10.1021/cr050001f

Burgess SA, Lindsay D, Flint SH (2010) Thermophilic bacilli and their importance in dairy processing. Int J Food Microbiol 144:215–225. https://doi.org/10.1016/j.ijfoodmicro.2010.09.027

Calahorrano-Moreno MB, Ordoñez-Bailon JJ, Baquerizo-Crespo RJ, Dueñas-Rivadeneira AA, Montenegro MCBSM, Rodríguez-Díaz JM (2022) Contaminants in the cow's milk we consume? Pasteurization and other technologies in the elimination of contaminants. F1000Res 11:91. https://doi.org/10.12688/f1000research.108779.1

Cantor MD, van den Tempel T, Hansen TK, Ardö Y (2017) Blue cheese. In: Cheese. Elsevier, pp 929–954. https://doi.org/10.1016/B978-0-12-417012-4.00037-5

Cardinali F, Taccari M, Milanović V, Osimani A, Polverigiani S, Garofalo C, Foligni R, Mozzon M, Zitti S, Raffaelli N (2016) Yeast and mould dynamics in Caciofiore della Sibilla cheese coagulated with an aqueous extract of Carlina acanthifolia All. Yeast 33:403–414. https://doi.org/10.1002/yea.3168

Carrillo-Gómez JK, Durán Acevedo CM, García-Rico RO (2021) Detection of the bacteria concentration level in pasteurized milk by using two different artificial multisensory methods. Sensing Biosensing Res 33:100428. https://doi.org/10.1016/J.SBSR.2021.100428

Chamberland J, Lessard MH, Doyen A, Labrie S, Pouliot Y (2017) Biofouling of ultrafiltration membrane by dairy fluids: characterization of pioneer colonizer bacteria using a DNA metabarcoding approach. J Dairy Sci 100:981–990. https://doi.org/10.3168/jds.2016-11829

Chambers JV (2002) The microbiology of raw milk. In: Dairy microbiology handbook. Wiley, pp 39–90. https://doi.org/10.1002/0471723959.ch2

Champagne CP, Ross RP, Saarela M, Hansen KF, Charalampopoulos D (2011) Recommendations for the viability assessment of probiotics as concentrated cultures and in food matrices. Int J Food Microbiol 149:185–193. https://doi.org/10.1016/j.ijfoodmicro.2011.07.005

Charalampopoulos D, Pandiella SS, Webb C (2003) Evaluation of the effect of malt, wheat and barley extracts on the viability of potentially probiotic lactic acid bacteria under acidic conditions. Int J Food Microbiol 82:133–141. https://doi.org/10.1016/S0168-1605(02)00248-9

Chen L, Daniel RM, Coolbear T (2003) Detection and impact of protease and lipase activities in milk and milk powders. Int Dairy J 13:255–275. https://doi.org/10.1016/S0958-6946(02)00171-1

Chen T, Wu Q, Li S, Xiong S, Jiang S, Tan Q, Zhang Z, Zhu D, Wei H (2014) Microbiological quality and characteristics of probiotic products in China. J Sci Food Agric 94:131–138. https://doi.org/10.1002/jsfa.6221

Chen Q, Meng M, Li W, Xiong Y, Fang Y, Lin Q (2023) Emerging biosensors to detect aflatoxin M1 in milk and dairy products. Food Chem 398:133848. https://doi.org/10.1016/J.FOODCHEM.2022.133848

Chinnadurai K, Kanwal HK, Tyagi AK, Stanton C, Ross P (2013) High conjugated linoleic acid enriched ghee (clarified butter) increases the antioxidant and antiatherogenic potency in female Wistar rats. Lipids Health Dis 12:1–9. https://doi.org/10.1186/1476-511X-12-121

Chipilev N, Daskalov H, Stoyanchev T (2016) Study on the prevalence of lipolytic yeasts and moulds in raw cow milk and white brined cheese. Bulg J Vet Med 19. https://doi.org/10.15547/bjvm.866

Choi CH, Kim E, Yang SM, Kim DS, Suh SM, Lee GY, Kim HY (2022) Comparison of real-time PCR and droplet digital PCR for the quantitative detection of Lactiplantibacillus plantarum subsp. plantarum. Foods 11. https://doi.org/10.3390/FOODS11091331

Chrisope GL, Marshall RT (1976) Combined action of lipase and microbial phospholipase C on a model fat globule emulsion and raw milk. J Dairy Sci 59:2024–2030. https://doi.org/10.3168/jds.S0022-0302(76)84483-9

Clawin-Rädecker I, De Block J, Egger L, Willis C, Da Silva Felicio MT, Messens W (2021) The use of alkaline phosphatase and possible alternative testing to verify pasteurisation of raw milk, colostrum, dairy and colostrum-based products. EFSA J 19:e06576. https://doi.org/10.2903/J.EFSA.2021.6576

Corassin CH, Borowsky A, Ali S, Rosim RE, de Oliveira CAF (2022) Occurrence of aflatoxin M1 in milk and dairy products traded in São Paulo, Brazil: an update. Dairy 3:842–848. https://doi.org/10.3390/dairy3040057

Costa FSL, Bezerra CCR, Neto RM, Morais CLM, Lima KMG (2020) Identification of resistance in Escherichia coli and Klebsiella pneumoniae using excitation-emission matrix fluorescence spectroscopy and multivariate analysis. Sci Rep 10:1–9. https://doi.org/10.1038/s41598-020-70033-x

D'Amico DJ, Donnelly CW (2010) Microbiological quality of raw milk used for small-scale artisan cheese production in Vermont: effect of farm characteristics and practices. J Dairy Sci 93:134–147. https://doi.org/10.3168/jds.2009-2426

da Silva LA, Lopes Neto JHP, Cardarelli HR (2019) Safety and probiotic functionality of isolated goat milk lactic acid bacteria. Ann Microbiol 69:1497–1505. https://doi.org/10.1007/s13213-019-01533-z

Davoren MJ, Liu J, Castellanos J, Rodríguez-Malavé NI, Schiestl RH (2019) A novel probiotic, Lactobacillus johnsonii 456, resists acid and can persist in the human gut beyond the initial ingestion period. Gut Microbes 10:458–480. https://doi.org/10.1080/19490976.2018.1547612

De Jonghe V, Coorevits A, De Block J, Van Coillie E, Grijspeerdt K, Herman L, De Vos P, Heyndrickx M (2010) Toxinogenic and spoilage potential of aerobic spore-formers isolated from raw milk. Int J Food Microbiol 136:318–325. https://doi.org/10.1016/j.ijfoodmicro.2009.11.007

De Vuyst L (2000) Technology aspects related to the application of functional starter cultures. Food Technol Biotechnol 38:105–112

Deeth HC (2006) Lipoprotein lipase and lipolysis in milk. Int Dairy J 16:555–562. https://doi.org/10.1016/j.idairyj.2005.08.011

Delavenne E, Mounier J, Asmani K, Jany J-L, Barbier G, Le Blay G (2011) Fungal diversity in cow, goat and ewe milk. Int J Food Microbiol 151:247–251. https://doi.org/10.1016/j.ijfoodmicro.2011.08.029

Demirci M, Yigin A, Altun S, Uysal H, Saribas S, Kocazeybek B (2019) Salmonella Spp. and Shigella Spp. detection via multiplex real-time PCR and discrimination via MALDI-TOF MS in different animal raw milk samples. Niger J Clin Pract 22:1083–1090. https://doi.org/10.4103/NJCP.NJCP_596_18

Desmasures N (2014) Mold-ripened varieties. In: Encyclopedia of food microbiology, vol 2, pp 409–416. https://doi.org/10.1016/B978-0-12-384730-0.00060-4

Desmazeaud MJ, Le Bars D (1977) Role of proteolytic enzymes of Streptococcus lactis, Penicillium roqueforti, and Penicillium caseicolum during cheese ripening. J Dairy Sci 60:1532–1538

Diez-Gonzalez F (2014) Total viable counts: specific techniques. In: Encyclopedia of food microbiology. Elsevier, pp 630–635

Do Carmo LS, Dias RS, Linardi VR, De Sena MJ, Dos Santos DA, De Faria ME, Pena EC, Jett M, Heneine LG (2002) Food poisoning due to enterotoxigenic strains of Staphylococcus present in Minas cheese and raw milk in Brazil. Food Microbiol 19:9–14. https://doi.org/10.1006/fmic.2001.0444

Duanis-Assaf D, Steinberg D, Chai Y, Shemesh M (2016) The LuxS based quorum sensing governs lactose induced biofilm formation by Bacillus subtilis. Front Microbiol 6. https://doi.org/10.3389/FMICB.2015.01517

Duarte SC, Lino CM, Pena A (2012) Food safety implications of ochratoxin A in animal-derived food products. Vet J 192:286–292. https://doi.org/10.1016/j.tvjl.2011.11.002

El Khoury A, Atoui A, Yaghi J (2011) Analysis of aflatoxin M1 in milk and yogurt and AFM1 reduction by lactic acid bacteria used in Lebanese industry. Food Control 22:1695–1699. https://doi.org/10.1016/j.foodcont.2011.04.001

ELbagory AM, Amal ME, Hammad AM, Salwa AD (2014) Prevalence of fungi in locally produced cheese and molecular characterization of isolated toxigenic molds. Benha Vet Med J 27:9–20

El-Diasty ZM, Zaied KA, Kosba ZA, Kash KS, El-Azez A (2009) Meiotic chromosomal anomalies resulting from fungal infection of maize compared with those resulting from gamma irradiation. J Agric Chem Biotechnol 34:9233–9248. https://doi.org/10.21608/jacb.2009.99293

Eldin C, Angelakis E, Renvoisé A, Raoult D (2013) Coxiella burnetii DNA, but not viable bacteria, in dairy products in France. Am J Trop Med Hyg 88:765. https://doi.org/10.4269/ajtmh.12-0212

Elkenany R, Eltaysh R, Elsayed M, Abdel-Daim M, Shata R (2022) Characterization of multi-resistant Shigella species isolated from raw cow milk and milk products. J Vet Med Sci 84:890–897. https://doi.org/10.1292/JVMS.22-0018

Elutade OO, Imohiosen JJI, Akinola OT (2023) Bacteriological quality of raw cow milk at a collection center in Iwo, Osun State, Nigeria. IOP Conf Ser Earth Environ Sci 1219:012002. https://doi.org/10.1088/1755-1315/1219/1/012002

Erdogan A, Sert S (2004) Mycotoxin-forming ability of two Penicillium roqueforti strains in blue moldy tulum cheese ripened at various temperatures. J Food Prot 67:533–535. https://doi.org/10.4315/0362-028X-67.3.533

Escobar-Zepeda A, Sanchez-Flores A, Baruch MQ (2016) Metagenomic analysis of a Mexican ripened cheese reveals a unique complex microbiota. Food Microbiol 57:116–127. https://doi.org/10.1016/j.fm.2016.02.004

Fallah AA (2010) Assessment of aflatoxin M1 contamination in pasteurized and UHT milk marketed in central part of Iran. Food Chem Toxicol 48:988–991. https://doi.org/10.1016/j.fct.2010.01.014

Farrokh C, Jordan K, Auvray F, Glass K, Oppegaard H, Raynaud S, Thevenot D, Condron R, De Reu K, Govaris A, Heggum K, Heyndrickx M, Hummerjohann J, Lindsay D, Miszczycha S, Moussiegt S, Verstraete K, Cerf O (2013) Review of Shiga-toxin-producing Escherichia coli (STEC) and their significance in dairy production. Int J Food Microbiol 162:190–212. https://doi.org/10.1016/J.IJFOODMICRO.2012.08.008

Fasoli S, Marzotto M, Rizzotti L, Rossi F, Dellaglio F, Torriani S (2003) Bacterial composition of commercial probiotic products as evaluated by PCR-DGGE analysis. Int J Food Microbiol 82:59–70. https://doi.org/10.1016/S0168-1605(02)00259-3

Ferrario C, Lugli GA, Ossiprandi MC, Turroni F, Milani C, Duranti S, Mancabelli L, Mangifesta M, Alessandri G, van Sinderen D, Ventura M (2017) Next generation sequencing-based multigene panel for high throughput detection of food-borne pathogens. Int J Food Microbiol 256:20–29. https://doi.org/10.1016/J.IJFOODMICRO.2017.05.001

Flores-Flores ME, Lizarraga E, López de Cerain A, González-Peñas E (2015) Presence of mycotoxins in animal milk: a review. Food Control 53:163–176. https://doi.org/10.1016/j.foodcont.2015.01.020

Fontaine K, Passeró E, Vallone L, Hymery N, Coton M, Jany J-L, Mounier J, Coton E (2015) Occurrence of roquefortine C, mycophenolic acid and aflatoxin M1 mycotoxins in blue-veined cheeses. Food Control 47:634–640. https://doi.org/10.1016/j.foodcont.2014.07.046

Foschino R, Invernizzi A, Barucco R, Stradiotto K (2002) Microbial composition, including the incidence of pathogens, of goat milk from the Bergamo region of Italy during a lactation year. J Dairy Res 69:213–225. https://doi.org/10.1017/S0022029902005459

Fromm HI, Boor KJ (2004) Characterization of pasteurized fluid milk shelf-life attributes. J Food Sci 69:M207–M214. https://doi.org/10.1111/j.1365-2621.2004.tb09889.x

Gadkar VJ, Goldfarb DM, Gantt S, Tilley PAG (2018) Real-time detection and monitoring of loop mediated amplification (LAMP) reaction using self-quenching and de-quenching fluorogenic probes. Sci Rep 8. https://doi.org/10.1038/S41598-018-23930-1

Galié S, García-Gutiérrez C, Miguélez EM, Villar CJ, Lombó F (2018) Biofilms in the food industry: health aspects and control methods. Front Microbiol 9:315815. https://doi.org/10.3389/FMICB.2018.00898/BIBTEX

Gao J, Li X, Zhang G, Sadiq FA, Simal-Gandara J, Xiao J, Sang Y (2021) Probiotics in the dairy industry—advances and opportunities. Compr Rev Food Sci Food Saf 20:3937–3982. https://doi.org/10.1111/1541-4337.12755

García-Burgos M, Moreno-Fernández J, Alférez MJM, Díaz-Castro J, López-Aliaga I (2020) New perspectives in fermented dairy products and their health relevance. J Funct Foods 72:104059. https://doi.org/10.1016/J.JFF.2020.104059

Garnier L, Valence F, Mounier J (2017) Diversity and control of spoilage fungi in dairy products: an update. Microorganisms 5(3):42. https://doi.org/10.3390/microorganisms5030042

Gaudreau H, Champagne CP, Remondetto GE, Bazinet L, Subirade M (2013) Effect of catechins on the growth of oxygen-sensitive probiotic bacteria. Food Res Int 53:751–757. https://doi.org/10.1016/j.foodres.2012.10.014

Ghrissi H (2021) Monitoring microorganisms' growth using multisensor electrochemical devices. Polytechnic Institute of Bragança

Giammanco GM, Pepe A, Aleo A, D'Agostino V, Milone S, Mammina C (2011) Microbiological quality of Pecorino Siciliano "primosale" cheese on retail sale in the street markets of Palermo, Italy. New Microbiol 34:179–185

Giaouris E, Heir E, Desvaux M, Hébraud M, Møretrø T, Langsrud S, Doulgeraki A, Nychas GJ, Kacániová M, Czaczyk K, Ölmez H, Simões M (2015) Intra- and inter-species interactions within biofilms of important foodborne bacterial pathogens. Front Microbiol 6:144151. https://doi.org/10.3389/FMICB.2015.00841/BIBTEX

Gilbride K (2014) Molecular methods for the detection of waterborne pathogens. In: Waterborne pathogens: detection methods and applications. Academic Press, pp 231–290

Gómez PA, Pu H, Li B, Yu M, Xu W, Chen L, Han J (2023) Comparison of PCR techniques in adulteration identification of dairy products. Agriculture 13:1450. https://doi.org/10.3390/AGRICULTURE13071450

González-Rivas F, Ripolles-Avila C, Fontecha-Umaña F, Ríos-Castillo AG, Rodríguez-Jerez JJ (2018) Biofilms in the spotlight: detection, quantification, and removal methods. Compr Rev Food Sci Food Saf 17:1261–1276. https://doi.org/10.1111/1541-4337.12378

Gore HM, Wakeman CA, Hull RM, McKillip JL (2003) Real-time molecular beacon NASBA reveals hblC expression from Bacillus spp. in milk. Biochem Biophys Res Commun 311:386–390. https://doi.org/10.1016/j.bbrc.2003.10.016

Górska-Warsewicz H, Rejman K, Laskowski W, Czeczotko M (2019) Milk and dairy products and their nutritional contribution to the average polish diet. Nutrients 11. https://doi.org/10.3390/NU11081771

Gougouli M, Kalantzi K, Beletsiotis E, Koutsoumanis KP (2011) Development and application of predictive models for fungal growth as tools to improve quality control in yogurt production. Food Microbiol 28:1453–1462. https://doi.org/10.1016/j.fm.2011.07.006

Grant IR, Ball HJ, Rowe MT (2002) Incidence of Mycobacterium paratuberculosis in bulk raw and commercially pasteurized cows' milk from approved dairy processing establishments in the United Kingdom. Appl Environ Microbiol 68:2428–2435

Grigore-Gurgu L, Ioncla Bucur F, Borda D, Alexa E-A, Neagu C, Ioana Nicolau A (2020) Biofilms formed by pathogens in food and food processing environments. In: Bacterial biofilms. IntechOpen. https://doi.org/10.5772/intechopen.90176

Gupta S, Anand S (2018) Induction of pitting corrosion on stainless steel (grades 304 and 316) used in dairy industry by biofilms of common sporeformers. Int J Dairy Technol 71:519–531. https://doi.org/10.1111/1471-0307.12444

Gupta PK, Rai DC, Paswan VK, Panta R, Yadav AK (2020) Study on physico-chemical and microbial quality of raw milk collected from different places of Assi Region in Varanasi City, Varanasi. Asian J Dairy Food Res 39:1–9. https://doi.org/10.18805/AJDFR.DR-1498

Hamann WT, Marth EH (1984) Survival of Streptococcus thermophilus and Lactobacillus bulgaricus in commercial and experimental yogurts. J Food Prot 47:781–786

Hamzehlou P, Sepahy AA, Mehrabian S, Hosseini F (2018) Production of vitamins B3, B6 and B9 by Lactobacillus isolated from traditional yogurt samples from 3 cities in Iran, winter 2016. Appl Food Biotechnol 5:105–118

Han B-Z, Meng Y, Li M, Yang Y-X, Ren F-Z, Zeng Q-K, Nout MJR (2007) A survey on the microbiological and chemical composition of buffalo milk in China. Food Control 18:742–746

Hassan AN, Frank J (2011) Microorganisms associated with milk. In: Encyclopedia of dairy sciences. Elsevier, pp 447–457. https://doi.org/10.1016/B978-0-12-374407-4.00309-5

Hassan MU, Nayab H, Shafique F, Williamson MP, Almansouri TS, Asim N, Shafi N, Attacha S, Khalid M, Ali N, Akbar N (2020) Probiotic properties of Lactobacillus helveticus and

Lactobacillus plantarum isolated from traditional Pakistani yoghurt. Biomed Res Int 2020. https://doi.org/10.1155/2020/8889198

Hassani S, Moosavy MH, Gharajalar SN, Khatibi SA, Hajibemani A, Barabadi Z (2022) High prevalence of antibiotic resistance in pathogenic foodborne bacteria isolated from bovine milk. Sci Rep 12:1–10. https://doi.org/10.1038/s41598-022-07845-6

Hernández-Saldaña OF, Valencia-Posadas M, de la Fuente-Salcido NM, Bideshi DK, Barboza-Corona JE (2016) Bacteriocinogenic bacteria isolated from raw goat milk and goat cheese produced in the center of Mexico. Indian J Microbiol 56:301–308

Hidalgo-Cantabrana C, López P, Gueimonde M, de Los Reyes-Gavilán CG, Suárez A, Margolles A, Ruas-Madiedo P (2012) Immune modulation capability of exopolysaccharides synthesised by lactic acid bacteria and bifidobacteria. Probiotics Antimicrob Proteins 4:227–237. https://doi.org/10.1007/s12602-012-9110-2

Hill B, Smythe B, Lindsay D, Shepherd J (2012) Microbiology of raw milk in New Zealand. Int J Food Microbiol 157:305–308

Hogarty SL, Frank JF (1982) Low-temperature activity of lactic streptococci isolated from cultured buttermilk. J Food Prot 45:1208–1211

Hoque MN, Talukder AK, Saha O, Hasan MM, Sultana M, Rahman AA, Das ZC (2022) Antibiogram and virulence profiling reveals multidrug resistant *Staphylococcus aureus* as the predominant aetiology of subclinical mastitis in riverine buffaloes. Vet Med Sci 8:2631–2645. https://doi.org/10.1002/VMS3.942

Hossain D, Singha S, Marie C, Thof V, Gough S, Uddin A (2021) Effect of somatic cells on milk quality and human health. Vet Sci Res Rev 7:31–34. https://doi.org/10.17582/journal.vsrr/2021/7.1.31.34

Hosseiniporgham S, Cubeddu T, Rocca S, Sechi LA (2020) Identification of Mycobacterium avium subsp. paratuberculosis (MAP) in sheep milk, a zoonotic problem. Microorganisms 8:1264. https://doi.org/10.3390/MICROORGANISMS8091264

Hymery N, Masson F, Barbier G, Coton E (2014a) Cytotoxicity and immunotoxicity of cyclopiazonic acid on human cells. Toxicol In Vitro 28:940–947

Hymery N, Vasseur V, Coton M, Mounier J, Jany J, Barbier G, Coton E (2014b) Filamentous fungi and mycotoxins in cheese: a review. Compr Rev Food Sci Food Saf 13:437–456

Inaba M, Higashimoto Y, Toyama Y, Horiguchi T, Hibino M, Iwata M, Imaizumi K, Doi Y (2021) Diagnostic accuracy of LAMP versus PCR over the course of SARS-CoV-2 infection. Int J Infect Dis 107:195–200. https://doi.org/10.1016/J.IJID.2021.04.018

Inoue K, Shirai T, Ochiai H, Kasao M, Hayakawa K, Kimura M, Sansawa H (2003) Blood-pressure-lowering effect of a novel fermented milk containing γ-aminobutyric acid (GABA) in mild hypertensives. Eur J Clin Nutr 57:490–495. https://doi.org/10.1038/sj.ejcn.1601555

Iqbal SZ, Asi MR, Jinap S (2013) Variation of aflatoxin M1 contamination in milk and milk products collected during winter and summer seasons. Food Control 34:714–718. https://doi.org/10.1016/j.foodcont.2013.06.009

Ishii S, Hosino B, Komiyama H, Uehara A, Nurtazin S (2014) Study on production and properties of Kumiss of herders in Mongolian dry steppe. J Arid Land Stud 24(1):197–197. http://purl.org/coar/version/c_970fb48d4fbd8a85

Iversen C, Forsythe S (2004) Isolation of Enterobacter sakazakii and other Enterobacteriaceae from powdered infant formula milk and related products. Food Microbiol 21:771–777. https://doi.org/10.1016/j.fm.2004.01.009

Jabber BMAN (2021) Phenotypic and molecular detection of Streptococcus uberis isolated from milk of subclinical and clinical mastitis of cow in Basrah City. Ann Rom Soc Cell Biol 25:504–522

Jahan M, Rahman M, Parvej MS, Chowdhury SMZH, Haque ME, Talukder MAK, Ahmed S (2015) Isolation and characterization of *Staphylococcus aureus* from raw cow milk in Bangladesh. J Adv Vet Anim Res 2:49–55. https://doi.org/10.5455/JAVAR.2015.B47

Jay JM (1998) Fermentation and fermented dairy products. 131–148. https://doi.org/10.1007/978-1-4615-7476-7_7

Kable ME, Srisengfa Y, Laird M, Zaragoza J, McLeod J, Heidenreich J, Marco ML (2016) The core and seasonal microbiota of raw bovine milk in tanker trucks and the impact of transfer to a milk processing facility. mBio 7. https://doi.org/10.1128/mbio.00836-16

Kamle M, Mahato DK, Gupta A, Pandhi S, Sharma N, Sharma B, Mishra S, Arora S, Selvakumar R, Saurabh V (2022) Citrinin mycotoxin contamination in food and feed: impact on agriculture, human health, and detection and management strategies. Toxins 14(2):85. https://doi.org/10.3390/toxins14020085

Kar D, Bandyopadhyay S, Bhattacharyya D, Samanta I, Mahanti A, Nanda PK, Mondal B, Dandapat P, Das AK, Dutta TK, Bandyopadhyay S, Singh RK (2015) Molecular and phylogenetic characterization of multidrug resistant extended spectrum beta-lactamase producing Escherichia coli isolated from poultry and cattle in Odisha, India. Infect Genet Evol 29:82–90. https://doi.org/10.1016/j.meegid.2014.11.003

Kateřina B, Marcela V-K, Vladimír B, Libor K, Ivana K, Renáta K (2016) Microbiological quality of raw milk in the Czech Republic. Czech J Food Sci 34:189–196

Kekre A, Bhushan A, Kumar P, Kalia VC (2015) Genome wide analysis for searching novel markers to rapidly identify *Clostridium* strains. Indian J Microbiol 55:250–257. https://doi.org/10.1007/s12088-015-0535-7

Kimbell LK, Wang Y, McNamara PJ (2020) The impact of metal pipe materials, corrosion products, and corrosion inhibitors on antibiotic resistance in drinking water distribution systems. Appl Microbiol Biotechnol 104:7673–7688. https://doi.org/10.1007/S00253-020-10777-8

Koh JH, Choi SH, Park SW, Choi N-J, Kim Y, Kim SH (2013) Synbiotic impact of tagatose on viability of Lactobacillus rhamnosus strain GG mediated by the phosphotransferase system (PTS). Food Microbiol 36:7–13. https://doi.org/10.1016/j.fm.2013.03.003

Kokkonen M, Jestoi M, Rizzo A (2005) Determination of selected mycotoxins in mould cheeses with liquid chromatography coupled to tandem with mass spectrometry. Food Addit Contam 22:449–456. https://doi.org/10.1080/02652030500089861

Kong J, Fan C, Liao X, Chen A, Yang S, Zhao L, Li H (2022) Accurate detection of Escherichia coli O157:H7 and Salmonella enterica serovar typhimurium based on the combination of next-generation sequencing and droplet digital PCR. LWT 168. https://doi.org/10.1016/J.LWT.2022.113913

Krishna M, Venkateshaiah BV, Prabha R (2019) Development of long shelf life probiotic lassi. Asian J Dairy Food Res 38(4):315–317. https://doi.org/10.18805/ajdfr.DR-1485

Kumar AM, Kachhawa K (2023) Biomedical applications of bioelectrochemical sensors. In: Multifaceted bio-sensing technology. Academic Press, pp 239–260. https://doi.org/10.1016/B978-0-323-90807-8.00014-2

Kumar SN, Sarode M, Ravi SHN (2016) Synthesis of titanium dioxide thin films and its application in reducing microbial load of milk. bioRxiv 23:045179. https://doi.org/10.1101/045179

Kumar N, Sharma G, Leahy E, Shome BR, Bandyopadhyay S, Deka RP, Shome R, Dey TK, Lindahl JF (2021) Understanding antibiotic usage on small-scale dairy farms in the Indian States of Assam and Haryana using a mixed-methods approach-outcomes and challenges. Antibiotics 10. https://doi.org/10.3390/ANTIBIOTICS10091124

Kumari S, Sarkar PK (2018) Optimisation of Bacillus cereus biofilm removal in the dairy industry using an in vitro model of cleaning-in-place incorporating serine protease. Int J Dairy Technol 71:512–518. https://doi.org/10.1111/1471-0307.12454

Lavoie K, Touchette M, St-Gelais D, Labrie S (2012) Characterization of the fungal microflora in raw milk and specialty cheeses of the province of Quebec. Dairy Sci Technol 92:455–468. https://doi.org/10.1007/s13594-011-0051-4

Ledenbach LH, Marshall RT (2009) Microbiological spoilage of dairy products. In: Compendium of the microbiological spoilage of foods and beverages. Springer, pp 41–67. https://doi.org/10.1007/978-1-4419-0826-1_2

Ledlod S, Bunroddith K, Areekit S, Santiwatanakul S, Chansiri K (2020) Development of a duplex lateral flow dipstick test for the detection and differentiation of Listeria spp. and Listeria monocytogenes in meat products based on loop-mediated isothermal amplification. J Chromatogr B Anal Technol Biomed Life Sci 1139. https://doi.org/10.1016/J.JCHROMB.2019.121834

Li H, Zhang S, Lu J, Liu L, Uluko H, Pang X, Sun Y, Xue H, Zhao L, Kong F (2014) Antifungal activities and effect of Lactobacillus casei AST18 on the mycelia morphology and ultrastructure of Penicillium chrysogenum. Food Control 43:57–64. https://doi.org/10.1016/j.foodcont.2014.02.045

Li J, Huang Q, Zheng X, Ge Z, Lin K, Zhang D, Chen Y, Wang B, Shi X (2020) Investigation of the lactic acid bacteria in kazak cheese and their contributions to cheese fermentation. Front Microbiol 11:228. https://doi.org/10.3389/fmicb.2020.00228

Li X, Wang S, Zhai Z, Wang W, Hao Y, Lin J (2022) Slipchip-based immunomagnetic separation combined with loop-mediated isothermal amplification for rapid detection of Bacillus cereus with tetracycline resistance gene tetL in pasteurized milk. Food Control 140:109122. https://doi.org/10.1016/J.FOODCONT.2022.109122

Liu S, Wang B, Sui Z, Wang Z, Li L, Zhen X, Zhao W, Zhou G (2021) Faster detection of *Staphylococcus aureus* in milk and milk powder by flow cytometry. Foodborne Pathog Dis 18:346–353. https://doi.org/10.1089/FPD.2020.2894

Lu M, Wang NS (2017) Spoilage of milk and dairy products. Elsevier Ltd.

Lubna, Hussain T, Shami A, Rafiq N, Khan S, Kabir M, Khan NU, Khattak I, Kamal M, Usman T (2023) Antimicrobial usage and detection of multidrug-resistant *Staphylococcus aureus*: methicillin- and tetracycline-resistant strains in raw milk of lactating dairy cattle. Antibiotics 12. https://doi.org/10.3390/ANTIBIOTICS12040673

Lund F, Filtenborg O, Frisvad JC (1995) Associated mycoflora of cheese. Food Microbiol 12:173–180

Manishimwe R, Moncada PM, Bugarel M, Scott HM, Loneragan GH (2021) Antibiotic resistance among Escherichia coli and Salmonella isolated from dairy cattle feces in Texas. PLoS One 16. https://doi.org/10.1371/JOURNAL.PONE.0242390

Manohar SM, Shah P, Nair A (2021) Flow cytometry: principles, applications, and recent advances. Bioanalysis 13:185–202. https://doi.org/10.4155/BIO-2020-0267

Marchand S, De Block J, De Jonghe V, Coorevits A, Heyndrickx M, Herman L (2012) Biofilm formation in milk production and processing environments; influence on milk quality and safety. Compr Rev Food Sci Food Saf 11:133–147. https://doi.org/10.1111/J.1541-4337.2011.00183.X

Marín P, Palmero D, Jurado M (2015) Occurrence of moulds associated with ovine raw milk and cheeses of the Spanish region of Castilla La Mancha. Int J Dairy Technol 68:565–572. https://doi.org/10.1111/1471-0307.12208

Marka S, Anand S (2018) Feed substrates influence biofilm formation on reverse osmosis membranes and their cleaning efficiency. J Dairy Sci 101:84–95. https://doi.org/10.3168/jds.2017-13249

Martin NH, Trmcic A, Hsieh TH, Boor KJ, Wiedmann M (2016) The evolving role of coliforms as indicators of unhygienic processing conditions in dairy foods. Front Microbiol 7:220072. https://doi.org/10.3389/FMICB.2016.01549/BIBTEX

Martins L, Gonçalves JL, Leite RF, Tomazi T, Rall VLM, Santos MV (2021) Association between antimicrobial use and antimicrobial resistance of Streptococcus uberis causing clinical mastitis. J Dairy Sci 104:12030–12041. https://doi.org/10.3168/jds.2021-20177

Mathekga BSP, Nxumalo Z, Thimiri Govinda Raj DB (2022) Micro and nanofluidics for high throughput drug screening. In: Progress in molecular biology and translational science. Academic Press, pp 93–120

Maul R, Müller C, Rieß S, Koch M, Methner F-J, Irene N (2012) Germination induces the glucosylation of the Fusarium mycotoxin deoxynivalenol in various grains. Food Chem 131:274–279. https://doi.org/10.1016/j.foodchem.2011.08.077

McMullin DR, Nsiama TK, Miller JD (2014) Isochromans and α-pyrones from Penicillium corylophilum. J Nat Prod 77:206–212. https://doi.org/10.1021/np4005486

Meng L, Zhang Y, Liu H, Zhao S, Wang J, Zheng N (2017) Characterization of Pseudomonas spp. and associated proteolytic properties in raw milk stored at low temperatures. Front Microbiol 8:287552. https://doi.org/10.3389/FMICB.2017.02158/BIBTEX

Minervini F, Fornelli F, Lucivero G, Romano C, Visconti A (2005a) T-2 toxin immunotoxicity on human B and T lymphoid cell lines. Toxicol 210:81–91. https://doi.org/10.1016/j.tox.2005.01.007

Minervini F, Giannoccaro A, Cavallini A, Visconti A (2005b) Investigations on cellular proliferation induced by zearalenone and its derivatives in relation to the estrogenic parameters. Toxicol Lett 159:272–283

Mittal N, Bharadwaj A (2021) Surveillance of pathogenic bacteria from milk samples. In: 2021 Fifth international conference on information systems and computer networks (ISCON). IEEE, pp 1–5

Moniente M, Botello-Morte L, García-Gonzalo D, Pagán R, Ontañón I (2022) Analytical strategies for the determination of biogenic amines in dairy products. Compr Rev Food Sci Food Saf 21:3612–3646. https://doi.org/10.1111/1541-4337.12980

Mortazavian AM, Ghorbanipour S, Mohammadifar MA, Mohammadi M (2011) Biochemical properties and viable probiotic population of yogurt at different bacterial inoculation rates and incubation temperatures. Philipp Agric Sci 94:155–160

Mubarack HM, Doss A, Dhanabalan R, Balachander S (2010) Microbial quality of raw milk samples collected from different villages of Coimbatore District, Tamil Nadu, South India. Indian J Sci Technol 3:61

Muehlhoff E, Bennett A, McMahon D (2013) Milk and dairy products in human nutrition. FAO, pp 75–110

Murasova P, Kovarova A, Kasparova J, Brozkova I, Hamiot A, Pekarkova J, Dupuy B, Drbohlavova J, Bilkova Z, Korecka L (2020) Direct culture-free electrochemical detection of Salmonella cells in milk based on quantum dots-modified nanostructured dendrons. J Electroanal Chem 863. https://doi.org/10.1016/J.JELECHEM.2020.114051

Murphy PA, Hendrich S, Landgren C, Bryant CM (2006) Food mycotoxins: an update. J Food Sci 71:R51

Muskens J, Van Engelen E, Van Maanen C, Bartels C, Lam T (2011) Prevalence of Coxiella burnetii infection in Dutch dairy herds based on testing bulk tank milk and individual samples by PCR and ELISA. Vet Rec 168:79. https://doi.org/10.1136/vr.c6106

Nabok A, Al-Rubaye AG, Al-Jawdah AM, Tsargorodska A, Marty JL, Catanante G, Szekacs A, Takacs E (2019) Novel optical biosensing technologies for detection of mycotoxins. Opt Laser Technol 109:212–221. https://doi.org/10.1016/J.OPTLASTEC.2018.07.076

Nagaraja Roopashri A, Savitha J, Divyashree MS, Mamatha BS, Usha Rani KA, Kumar A (2023) Indian traditional fermented foods: the role of lactic acid bacteria. In: Lactobacillus—a multifunctional genus. IntechOpen

Nahar S, Jeong HL, Kim Y, Ha AJ-W, Roy PK, Park SH, Ashrafudoulla M, Mizan MFR, Ha S-D (2021) Inhibitory effects of flavourzyme on biofilm formation, quorum sensing, and virulence genes of foodborne pathogens Salmonella typhimurium and Escherichia coli. Food Res Int 147. https://doi.org/10.1016/j.foodres.2021.110461

Neelam, Jain VK, Singh M, Joshi VG, Chhabra R, Singh K, Rana YS (2022) Virulence and antimicrobial resistance gene profiles of *Staphylococcus aureus* associated with clinical mastitis in cattle. PLoS One 17. https://doi.org/10.1371/JOURNAL.PONE.0264762

Nielsen KF, Dalsgaard PW, Smedsgaard J, Larsen TO (2005) Andrastins A–D, penicillium roqueforti metabolites consistently produced in blue-mold-ripened cheese. J Agric Food Chem 53:2908–2913. https://doi.org/10.1021/jf047983u

Noor R, Uddin MA, Motazzim-Ul-Haque HM (2011) Isolation and identification of pathogenic Escherichia coli, Klebsiella spp. and Staphylococcus spp. in raw milk samples collected from different areas of Dhaka City, Bangladesh. Stamford J Microbiol 1:19–23. https://doi.org/10.3329/sjm.v1i1.9098

O'Brien M, Nielsen KF, O'Kiely P, Forristal PD, Fuller HT, Frisvad JC (2006) Mycotoxins and other secondary metabolites produced in vitro by Penicillium paneum Frisvad and Penicillium roqueforti Thom isolated from baled grass silage in Ireland. J Agric Food Chem 54:9268–9276. https://doi.org/10.1021/jf0621018

O'Connor C (1995) Rural dairy technology. ILRI (aka ILCA and ILRAD)

Obaidat MM, Gharaibeh WA (2022) Sheep and goat milk in Jordan is a reservoir of multidrug resistant extended spectrum and AmpC beta-lactamases Escherichia coli. Int J Food Microbiol 377. https://doi.org/10.1016/J.IJFOODMICRO.2022.109834

Ombarak RA, Elbagory AM (2017) Bacteriological quality and occurrence of some microbial pathogens in goat's and ewe's milk in Egypt. Int Food Res J 24:847

Onoharigho FO, Ighede PA, Edo GI, Akpoghelie PO, Akpoghelie EO (2022) Isolation and identification of bacterial and fungal spoilage organisms in branded and unbranded milk; consumer perception of safety hazard for milk. Appl Microbiol Theory Technol:31–48. https://doi.org/10.37256/AMTT.3220221766

Ortiz-Rivera Y, Sánchez-Vega R, Gutiérrez-Méndez N, León-Félix J, Acosta-Muñiz C, Sepulveda DR (2017) Production of reuterin in a fermented milk product by Lactobacillus reuteri: inhibition of pathogens, spoilage microorganisms, and lactic acid bacteria. J Dairy Sci 100:4258–4268. https://doi.org/10.3168/jds.2016-11534

Owusu-Kwarteng J, Akabanda F, Agyei D, Jespersen L (2020) Microbial safety of milk production and fermented dairy products in Africa. Microorganisms 8(5):752. https://doi.org/10.3390/microorganisms8050752

Padilla B, Ruiz-Matute AI, Belloch C, Cardelle-Cobas A, Corzo N, Manzanares P (2012) Evaluation of oligosaccharide synthesis from lactose and lactulose using β-galactosidases from Kluyveromyces isolated from artisanal cheeses. J Agric Food Chem 60:5134–5141. https://doi.org/10.1021/jf300852s

Pan DD, Zeng XQ, Yan YT (2011) Characterisation of Lactobacillus fermentum SM-7 isolated from koumiss, a potential probiotic bacterium with cholesterol-lowering effects. J Sci Food Agric 91:512–518. https://doi.org/10.1002/jsfa.4214

Pandian S, Lakshmi SA, Priya A, Balasubramaniam B, Zaukuu JLZ, Durgadevi R, Abe-Inge V, Sohn SI (2023) Spectroscopic methods for the detection of microbial pathogens and diagnostics of infectious diseases—an updated overview. Processes 11:1191. https://doi.org/10.3390/PR11041191

Panigrahi S, Devi B, Swain K, Priyadarshini P (2018) Microbiology of milk: public health aspect. Pharma Innov J 7:260–264

Papageorgiou DK, Melas DS, Abrahim A, Koutsoumanis K (2003) Growth and survival of Aeromonas hydrophila in rice pudding (milk rice) during its storage at 4°C and 12°C. Food Microbiol 20:385–390. https://doi.org/10.1016/S0740-0020(03)00022-4

Paramasivam R, Gopal DR, Dhandapani R, Subbarayalu R, Elangovan MP, Prabhu B, Veerappan V, Nandheeswaran A, Paramasivam S, Muthupandian S (2023) Is AMR in dairy products a threat to human health? An updated review on the origin, prevention, treatment, and economic impacts of subclinical mastitis. Infect Drug Resist 16:155–178. https://doi.org/10.2147/IDR.S384776

Parseelan A, Muthu S, Kannan P, Ayyasamy E, Narayanan R (2019) Aerobic plate count of milk and dairy products marketed in different zones of Chennai. Int J Livest Res. https://doi.org/10.5455/IJLR.20180711104721

Parvez S, Malik KA, Ah Kang S, Kim H (2006) Probiotics and their fermented food products are beneficial for health. J Appl Microbiol 100(6):1171–1185. https://doi.org/10.1111/j.1365-2672.2006.02963.x

Pattono D, Grosso A, Stocco PP, Pazzi M, Zeppa G (2013) Survey of the presence of patulin and ochratoxin A in traditional semi-hard cheeses. Food Control 33:54–57. https://doi.org/10.1016/j.foodcont.2013.02.019

Pebdeni AB, Roshani A, Mirsadoughi E, Behzadifar S, Hosseini M (2022) Recent advances in optical biosensors for specific detection of E. coli bacteria in food and water. Food Control 135:108822. https://doi.org/10.1016/J.FOODCONT.2022.108822

Penedo LA, Nunes JC, Gama MAS, Leite PEC, Quirico-Santos TF, Torres AG (2013) Intake of butter naturally enriched with cis9, trans11 conjugated linoleic acid reduces systemic inflammatory mediators in healthy young adults. J Nutr Biochem 24:2144–2151. https://doi.org/10.1016/j.jnutbio.2013.08.006

Peng Z, Li Y, Yan L, Yang S, Yang D (2023) Correlation analysis of microbial contamination and alkaline phosphatase activity in raw milk and dairy products. Int J Environ Res Public Health 20(3):1825. https://doi.org/10.3390/ijerph20031825

Pešić-Mikulec D, Stojanović L, Jovanović L (2005) Moulds associated with milk depending on macroclimate and geographical location. Appl Ecol Environ Res 3:61–65

Pierezan MD, Maran BM, Maran EM, Verruck S, Pimentel TC, da Cruz AG (2022) Relevant safety aspects of raw milk for dairy foods processing. In: Advances in food and nutrition research. Elsevier, pp 211–264. https://doi.org/10.1016/bs.afnr.2022.01.001

Pinto CLO, Souza LV, Meloni VAS, Batista CS, Silva R, Martins EMF, Cruz AG, Martins ML (2018) Microbiological quality of Brazilian UHT milk: identification and spoilage potential of spore-forming bacteria. Int J Dairy Technol 71:20–26. https://doi.org/10.1111/1471-0307.12339

Polonsky J, Merrien MA, Scott PM (1977) Roquefortine and isofumigaclavine A, alkaloids from "*Penicillium Roqueforti*". In: Annales de la nutrition et de l'alimentation. JSTOR, pp 693–698

Porcellato D, Skeie SB (2016) Bacterial dynamics and functional analysis of microbial metagenomes during ripening of Dutch-type cheese. Int Dairy J 61:182–188. https://doi.org/10.1016/j.idairyj.2016.05.005

Prandini A, Tansini G, Sigolo S, Filippi L, Laporta M, Piva G (2009) On the occurrence of aflatoxin M1 in milk and dairy products. Food Chem Toxicol 47:984–991. https://doi.org/10.1016/j.fct.2007.10.005

Quigley L, O'Sullivan O, Beresford TP, Ross RP, Fitzgerald GF, Cotter PD (2011) Molecular approaches to analysing the microbial composition of raw milk and raw milk cheese. Int J Food Microbiol 150:81–94. https://doi.org/10.1016/j.ijfoodmicro.2011.08.001

Quraishi A, Kaur P, Singh Sharma N, Arora AK (2021) Antibiotic sensitivity patterns in Staphylococcus spp. isolated from goat milk in association with molecular detection of antibiotic resistance genes. Iran J Vet Res 22:239–243. https://doi.org/10.22099/IJVR.2021.38465.5599

Rahman T, Akon T, Sheuli IN, Hoque N (2015) Microbiological analysis of raw milk, pasteurized milk and yogurt samples collected from different areas of Dhaka city, Bangladesh. J Bangladesh Acad Sci 39:31–36. https://doi.org/10.3329/jbas.v39i1.23655

Rahman MS, Hassan MM, Chowdhury S (2021) Determination of antibiotic residues in milk and assessment of human health risk in Bangladesh. Heliyon 7. https://doi.org/10.1016/J.HELIYON.2021.E07739

Raji MIO, Jiya MH (2019) Evaluation of pathogenic bacteria in packaged milk products sold in sokoto metropolis, Nigeria. Asian J Appl Sci 12:85–90

Ramchandran L, Shah NP (2010) Characterization of functional, biochemical and textural properties of synbiotic low-fat yogurts during refrigerated storage. LWT Food Sci and Technol 43:819–827. https://doi.org/10.1016/j.lwt.2010.01.012

Rana EA, Das T, Dutta A, Rahman M, Bostami MB, Akter N, Barua H (2020) Coagulase-positive methicillin-resistant *Staphylococcus aureus* circulating in clinical mastitic goats in Bangladesh. Vet World 13:1303–1310. https://doi.org/10.14202/VETWORLD.2020.1303-1310

Rasool Z, Noreen H, Anjum A, Rizvi A, Rabaan AA, Halwani MA, Sabour AA, Aljeldah M, Shammari BRA, Alhajri SM, Alshubaith IH, Garout M, Firyal S, Ahmed N (2023) Genotypic and phenotypic characterization of erythromycin-resistant *Staphylococcus aureus* isolated from bovine mastitis and humans in close contact. Trop Med Infect Dis 8. https://doi.org/10.3390/TROPICALMED8010026

Reese KR, Elkins KM (2022) Simultaneous detection of foodborne pathogens using a real-time PCR triplex high-resolution melt assay. Appl Microbiol 2:453–459. https://doi.org/10.3390/APPLMICROBIOL2030034

Riaz QUA, Masud T (2013) Recent trends and applications of encapsulating materials for probiotic stability. Crit Rev Food Sci Nutr 53:231–244. https://doi.org/10.1080/10408398.2010.524953

Riss TL, Moravec RA, Niles AL, Duellman S, Benink HA, Worzella TJ, Minor L (2016) Cell viability assays. Assay guidance manual

Rothrock MJ, Davis ML, Locatelli A, Bodie A, McIntosh TG, Donaldson JR, Ricke SC (2017) Listeria occurrence in poultry flocks: detection and potential implications. Front Vet Sci 4:269657. https://doi.org/10.3389/FVETS.2017.00125/BIBTEX

Saddam S, Khan M, Jamal M, Rehman SU, Slama P, Horky P (2023) Multidrug resistant *Klebsiella Pneumoniae* reservoir and their capsular resistance genes in cow farms of district Peshawar, Pakistan. PLoS One 18. https://doi.org/10.1371/JOURNAL.PONE.0282245

Sadiq FA, Li Y, Liu T, Flint S, Zhang G, He G (2016) A RAPD based study revealing a previously unreported wide range of mesophilic and thermophilic spore formers associated with milk powders in China. Int J Food Microbiol 217:200–208. https://doi.org/10.1016/j.ijfoodmicro.2015.10.030

Sakamoto S, Putalun W, Vimolmangkang S, Phoolcharoen W, Shoyama Y, Tanaka H, Morimoto S (2018) Enzyme-linked immunosorbent assay for the quantitative/qualitative analysis of plant secondary metabolites. J Nat Med 72:32. https://doi.org/10.1007/S11418-017-1144-Z

Salauddin M, Akter MR, Hossain MK, Nazir KHMNH, Noreddin A, El Zowalaty ME (2020) Molecular detection of multidrug resistant *Staphylococcus aureus* isolated from bovine mastitis milk in Bangladesh. Vet Sci 7. https://doi.org/10.3390/VETSCI7020036

Sanders ME, Merenstein DJ, Reid G, Gibson GR, Rastall RA (2019) Probiotics and prebiotics in intestinal health and disease: from biology to the clinic. Nat Rev Gastroenterol Hepatol 16:605–616. https://doi.org/10.1038/s41575-019-0173-3

Santoso KA (2020) The effects of milk age on the titratable acidity of raw milk. Int J Sci Res 9:1041–1049. https://doi.org/10.21275/SR20709091753

Satpathy S, Sen SK, Pattanaik S, Raut S (2016) Review on bacterial biofilm: a universal cause of contamination. Biocatal Agric Biotechnol 7:56–66. https://doi.org/10.1016/J.BCAB.2016.05.002

Sauer K, Stoodley P, Goeres DM, Hall-Stoodley L, Burmølle M, Stewart PS, Bjarnsholt T (2022) The biofilm life cycle: expanding the conceptual model of biofilm formation. Nat Rev Microbiol 20:608–620. https://doi.org/10.1038/S41579-022-00767-0

Savard P, Lamarche B, Paradis M-E, Thiboutot H, Laurin É, Roy D (2011) Impact of Bifidobacterium animalis subsp. lactis BB-12 and, Lactobacillus acidophilus LA-5-containing yoghurt, on fecal bacterial counts of healthy adults. Int J Food Microbiol 149:50–57. https://doi.org/10.1016/j.ijfoodmicro.2010.12.026

Scandurra G, Cardillo E, Ciofi C, Ferro L (2022) UHT milk characterization by electrical impedance spectroscopy. Appl Sci (12):7559. https://doi.org/10.3390/APP12157559

Schön K, Schornsteiner E, Dzieciol M, Wagner M, Müller M, Schmitz-Esser S (2016) Microbial communities in dairy processing environment floor-drains are dominated by product-associated bacteria and yeasts. Food Control 70:210–215. https://doi.org/10.1016/j.foodcont.2016.05.057

Sengun I, Yaman DB, Gonul S (2008) Mycotoxins and mould contamination in cheese: a review. World Mycotoxin J 1:291–298

Shafeek M, El-Malt L, Abdel Hameed K, El-Zamkan M (2018) Some virulence genes of pathogenic enterococci isolated from raw milk and some milk products. SVU Int J Vet Sci 1:102–113

Shafiee G, Mortazavian AM, Mohammadifar MA, Koushki MR, Mohammadi A, Mohammadi R (2010) Combined effects of dry matter content, incubation temperature and final pH of fermentation on biochemical and microbiological characteristics of probiotic fermented milk. Afr J Microbiol Res 4:1265–1274

Shah MS, Qureshi S, Kashoo Z, Farooq S, Wani SA, Hussain MI, Banday MS, Khan AA, Gull B, Habib A, Khan SM, Dar BA (2019) Methicillin resistance genes and in vitro biofilm formation among *Staphylococcus aureus* isolates from bovine mastitis in India. Comp Immunol Microbiol Infect Dis 64:117–124. https://doi.org/10.1016/J.CIMID.2019.02.009

Sharma IM, Petchiappan A, Chatterji D (2014) Quorum sensing and biofilm formation in mycobacteria: role of c-di-GMP and methods to study this second messenger. IUBMB Life 66:823–834. https://doi.org/10.1002/IUB.1339

Sharma S, Sharma V, Dahiya DK, Khan A, Mathur M, Sharma A (2017) Prevalence, virulence potential, and antibiotic susceptibility profile of Listeria monocytogenes isolated from bovine raw milk samples obtained from Rajasthan, India. Foodborne Pathog Dis 14:132–140. https://doi.org/10.1089/FPD.2016.2118

Sheehan VM, Sleator RD, Fitzgerald GF, Hill C (2006) Heterologous expression of BetL, a betaine uptake system, enhances the stress tolerance of Lactobacillus salivarius UCC118. Appl Environ Microbiol 72:2170–2177. https://doi.org/10.1128/AEM.72.3.2170-2177.2006

Shekhar C, Motina E, Kumar S (2010) Microbiological quality of raw milk and its public health significance. J Dairy Foods Home Sci 29:15–18

Shephard GS, Burger H-M, Gambacorta L, Gong YY, Krska R, Rheeder JP, Solfrizzo M, Srey C, Sulyok M, Visconti A (2013) Multiple mycotoxin exposure determined by urinary biomarkers in rural subsistence farmers in the former Transkei, South Africa. Food Chem Toxicol 62:217–225. https://doi.org/10.1016/j.fct.2013.08.040

Shi J, Hu X, Zou X, Guo Z, Holmes M, Tahir HE, Huang X, Li Z (2019) Rapid identification of Lactobacillus species using near infrared spectral features of bacterial colonies. J Near Infrared Spec 27:302–313. https://doi.org/10.1177/0967033519852012

Shi L, Xu L, Xiao R, Zhou Z, Wang C, Wang S, Gu B (2020) Rapid, quantitative, high-sensitive detection of Escherichia coli O157:H7 by gold-shell silica-core nanospheres-based surface-enhanced Raman scattering lateral flow immunoassay. Front Microbiol 11:596005. https://doi.org/10.3389/FMICB.2020.596005/BIBTEX

Shrestha A, Bhattarai RK, Luitel H, Karki S, Basnet HB (2021) Prevalence of methicillin-resistant *Staphylococcus aureus* and pattern of antimicrobial resistance in mastitis milk of cattle in Chitwan, Nepal. BMC Vet Res 17. https://doi.org/10.1186/S12917-021-02942-6

Simões M, Simões LC, Vieira MJ (2010) A review of current and emergent biofilm control strategies. LWT Food Sci Technol 43:573–583. https://doi.org/10.1016/J.LWT.2009.12.008

Singha S, Ceciliani F, Rahman MM, Mannan MA, Chowdhury S, Nath SC, Paul OB, Persson Y, Boqvist S (2023) Factors influencing somatic cell counts and bacterial contamination in unpasteurized milk obtained from water buffalo in Bangladesh. Trop Anim Health Prod 55:242. https://doi.org/10.1007/s11250-023-03644-x

Smith RM, Nguyen TA, Ha HTT, Thang PH, Thuy C, Xuan Lien T, Bui HT, Le TH, Struminger B, McConnell MS (2013) Prevalence of cryptococcal antigenemia and cost-effectiveness of a cryptococcal antigen screening program–Vietnam. PLoS One 8:e62213. https://doi.org/10.1371/journal.pone.0062213

Soares JC, Soares AC, Popolin-Neto M, Paulovich FV, Oliveira ON, Mattoso LHC (2022) Detection of *Staphylococcus aureus* in milk samples using impedance spectroscopy and data processing with information visualization techniques and multidimensional calibration space. Sens Actuators Rep 4:100083. https://doi.org/10.1016/J.SNR.2022.100083

Sobur MA, Sabuj AAM, Sarker R, Rahman AMMT, Kabir SML, Rahman MT (2019) Antibiotic-resistant Escherichia coli and Salmonella spp. sssociated with dairy cattle and farm environment having public health significance. Vet World 12:984–993. https://doi.org/10.14202/VETWORLD.2019.984-993

Spagnolo S, De La Franier B, Davoudian K, Hianik T, Thompson M (2022) Detection of E. coli bacteria in milk by an acoustic wave aptasensor with an anti-fouling coating. Sensors 22:1853. https://doi.org/10.3390/S22051853

Spagnolo S, Davoudian K, De La Franier B, Hianik T, Thompson M (2023) *Staphylococcus aureus* detection in milk using a thickness shear mode acoustic aptasensor with an antifouling probe linker. Biosensors 13:614. https://doi.org/10.3390/BIOS13060614/S1

Spanamberg A, Ramos JP, Leoncini O, Alves SH, Valente P (2009) High frequency of potentially pathogenic yeast species in goat's raw milk and creamed cheese in Southern Brazil. Acta Sci Vet 37:133–141

Steele ML, McNAB WB, Poppe C, Griffiths MW, Chen S, Degrandis SA, Fruhner LC, Larkin CA, Lynch JA, Odumeru JA (1997) Survey of Ontario bulk tank raw milk for food-borne pathogens. J Food Prot 60:1341–1346

Stoev SD, Vitanov S, Anguelov G, Petkova-Bocharova T, Creppy EE (2001) Experimental mycotoxic nephropathy in pigs provoked by a diet containing ochratoxin A and penicillic acid. Vet Res Commun 25:205–223. https://doi.org/10.1023/A:1006433709685

Sun L, D'Amico DJ (2023) Characterization of microbial community assembly on new wooden vats for use in cheese production. Food Microbiol 109. https://doi.org/10.1016/J.FM.2022.104154

Tadesse S, Berhanu T, Woldegiorgis AZ (2020) Aflatoxin M1 in milk and milk products marketed by local and industrial producers in Bishoftu town of Ethiopia. Food Control 118:107386. https://doi.org/10.1016/j.foodcont.2020.107386

Tamime AY, Thomas LV, Wiley J (2005) Probiotic dairy products. Wiley Online Library. https://doi.org/10.1002/9781119214137

Taye Y, Degu T, Fesseha H, Mathewos M (2021) Isolation and identification of lactic acid bacteria from cow milk and milk products. Sci World J. https://doi.org/10.1155/2021/4697445

Tegegne B, Tesfaye S (2017) Bacteriological milk quality: possible hygienic factors and the role of *Staphylococcus aureus* in raw bovine milk in and around Gondar, Ethiopia. Int J Food Contam 4:1–9. https://doi.org/10.1186/S40550-016-0046-2/TABLES/4

Teh KH, Lindsay D, Palmer J, Andrewes P, Bremer P, Flint S (2013) Lipolysis within single culture and co-culture biofilms of dairy origin. Int J Food Microbiol 163:129–135. https://doi.org/10.1016/J.IJFOODMICRO.2013.02.009

Tham CS-C, Peh K-K, Bhat R, Liong M-T (2012) Probiotic properties of bifidobacteria and lactobacilli isolated from local dairy products. Ann Microbiol 62:1079–1087. https://doi.org/10.1007/s13213-011-0349-8

Thurner F, Alatraktchi FAZ (2023) Recent advances in electrochemical biosensing of aflatoxin M1 in milk—a mini review. Microchem J 190:108594. https://doi.org/10.1016/J.MICROC.2023.108594

Tonamo A, Komlósi I, Varga L, Czeglédi L, Peles F (2020) Bacteriological quality of raw ovine milk from different sheep farms. Animals 10:1163. https://doi.org/10.3390/ani10071163

Torres-Vitela MR, Mendoza-Bernardo M, Castro-Rosas J, Gomez-Aldapa CA, Garay-Martinez LE, Navarro-Hidalgo V, Villarruel-López A (2012) Incidence of Salmonella, Listeria monocytogenes, Escherichia coli O157: H7, and staphylococcal enterotoxin in two types of Mexican fresh cheeses. J Food Prot 75:79–84. https://doi.org/10.4315/0362-028x_jfp-11-258

Tóth AG, Csabai I, Krikó E, Tőzsér D, Maróti G, Patai ÁV, Makrai L, Szita G, Solymosi N (2020) Antimicrobial resistance genes in raw milk for human consumption. Sci Rep 10:7464. https://doi.org/10.1038/s41598-020-63675-4

Uymaz Tezel B (2019) Preliminary in vitro evaluation of the probiotic potential of the bacteriocinogenic strain Enterococcus lactis PMD74 isolated from ezine cheese. J Food Qual. https://doi.org/10.1155/2019/4693513

Van Asselt ED, Marvin HJP, Boon PE, Swanenburg M, Zeilmaker M, Mengelers MJB, Van Der Fels-Klerx HJ (2016) Chemical and physical hazards in the dairy chain. WUR. https://library.wur.nl/WebQuery/wurpubs/fulltext/447318

Van Kessel JAS, Karns JS, Lombard JE, Kopral CA (2011) Prevalence of Salmonella enterica, Listeria monocytogenes, and Escherichia coli virulence factors in bulk tank milk and in-line filters from U.S. dairies. J Food Prot 74:759–768. https://doi.org/10.4315/0362-028X.JFP-10-423

Vazquez BI, Fente C, Franco CM, Vazquez MJ, Cepeda A (2001) Inhibitory effects of eugenol and thymol on Penicillium citrinum strains in culture media and cheese. Int J Food Microbiol 67:157–163. https://doi.org/10.1016/S0168-1605(01)00429-9

Velázquez-Ordoñez V, Valladares-Carranza B, Tenorio-Borroto E, Talavera-Rojas M, Varela-Guerrero JA, Acosta-Dibarrat J, Puigvert F, Grille L, Revello ÁG, Pareja L (2019) Microbial contamination in milk quality and health risk of the consumers of raw milk and dairy products. In: Nutrition in health and disease-our challenges: now and forthcoming time. https://doi.org/10.5772/intechopen.86182

Walsh AM, Crispie F, Daari K, O'Sullivan O, Martin JC, Arthur CT, Claesson MJ, Scott KP, Cotter PD (2017) Strain-level metagenomic analysis of the fermented dairy beverage nunu highlights potential food safety risks. Appl Environ Microbiol:83. https://doi.org/10.1128/AEM.01144-17

Wang JJ, Frank JF (1981) Characterization of psychrotrophic bacterial contamination in commercial buttermilk. J Dairy Sci 64:2154–2160

Wang B, Liu S, Sui Z, Wang J, Wang Y, Gu S (2020) Rapid flow cytometric detection of single viable salmonella cells in milk powder. Foodborne Pathog Dis 17:447–458. https://doi.org/10.1089/FPD.2019.2748

Wang S, Zhu X, Meng Q, Zheng P, Zhang J, He Z, Jiang H (2021) Gold interdigitated microimmunosensor based on Mn-MOF-74 for the detection of Listeria monocytogens. Biosensors Bioelectron 183:113186. https://doi.org/10.1016/J.BIOS.2021.113186

Wang N, Jin Y, He G, Yuan L (2022a) Dynamic tracing of bacterial community distribution and biofilm control of dominant species in milk powder processing. LWT 154. https://doi.org/10.1016/J.LWT.2021.112855

Wang N, Jin Y, He G, Yuan L (2022b) Intraspecific and interspecific extracellular metabolites remodel biofilms formed by thermophilic spoilage bacteria. J Appl Microbiol 133:2096–2106. https://doi.org/10.1111/JAM.15338

Wang M, Bai Z, Liu S, Liu Y, Wang Z, Zhou G, Gong X, Jiang Y, Sui Z (2023) Accurate quantification of total bacteria in raw milk by flow cytometry using membrane potential as a key viability parameter. LWT 173. https://doi.org/10.1016/J.LWT.2022.114315

Weber M, Liedtke J, Plattes S, Lipski A (2019) Bacterial community composition of biofilms in milking machines of two dairy farms assessed by a combination of culture-dependent and -independent methods. PLoS One 14(9):e0222238. https://doi.org/10.1371/journal.pone.0222238

Willis C, Jørgensen F, Aird H, Elviss N, Fox A, Jenkins C, Fenelon D, Sadler-Reeves L, McLauchlin J (2018) An assessment of the microbiological quality and safety of raw drinking milk on retail sale in England. J Appl Microbiol 124:535–546. https://doi.org/10.1111/jam.13660

Woo JH, Guk JH, Yi S, Lee J, Song H, Kim WH, Cho S (2023) Effect of biofilm formation by antimicrobial-resistant gram-negative bacteria in cold storage on survival in dairy processing lines. Int J Food Microbiol 386. https://doi.org/10.1016/J.IJFOODMICRO.2022.110019

Yan M, Niu C, Li X, Wang F, Jiang S, Li K, Yao Z (2022) Heavy metal levels in milk and dairy products and health risk assessment: a systematic review of studies in China. Sci Total Environ 851:158161. https://doi.org/10.1016/j.scitotenv.2022.158161

Yerlıkaya O, Saygili D, Akpinar A (2021) Evaluation of antimicrobial activity and antibiotic susceptibility profiles of Lactobacillus delbrueckii subsp. Bulgaricus and streptococcus thermophilus strains isolated from commercial yoghurt starter cultures. Food Sci Technol (Brazil) 41:418–425. https://doi.org/10.1590/fst.03920

Yu J, Wang HM, Zha MS, Qing YT, Bai N, Ren Y, Xi XX, Liu WJ, Menghe BLG, Zhang HP (2015) Molecular identification and quantification of lactic acid bacteria in traditional fermented dairy foods of Russia. J Dairy Sci 98:5143–5154. https://doi.org/10.3168/jds.2015-9460

Yu T, Ma M, Sun Y, Xu X, Qiu S, Yin J, Chen L (2021) The effect of sublethal concentrations of benzalkonium chloride on the LuxS/AI-2 quorum sensing system, biofilm formation and motility of Escherichia coli. Int J Food Microbiol 353. https://doi.org/10.1016/j.ijfoodmicro.2021.109313

Yuan L, Sadiq FA, Wang N, Yang Z, He G (2021) Recent advances in understanding the control of disinfectant-resistant biofilms by hurdle technology in the food industry. Crit Rev Food Sci Nutr 61:3876–3891. https://doi.org/10.1080/10408398.2020.1809345

Yuan L, Fan L, Hou J, Luo R, Wang S, Zhou W, Yang Z (2022) Metagenomic analysis reveals microbial community and functional capacity in Kombucha. Qual Assur Saf Crops Foods 14:1–8. https://doi.org/10.15586/QAS.V14I3.1102

Yunita D, Dodd CER (2018) Microbial community dynamics of a blue-veined raw milk cheese from the United Kingdom. J Dairy Sci 101:4923–4935. https://doi.org/10.3168/jds.2017-14104
Zeng X, Jia H, Shi Y, Chen K, Wang Z, Gao Z, Yuan Y, Yue T (2022) Lactobacillus kefiranofaciens JKSP109 and Saccharomyces cerevisiae JKSP39 isolated from Tibetan kefir grain co-alleviated AOM/DSS induced inflammation and colorectal carcinogenesis. Food Funct 13:6947–6961. https://doi.org/10.1039/D1FO02939H
Zhai L, Liu H, Chen Q, Lu Z, Zhang C, Lv F, Bie X (2019) Development of a real-time nucleic acid sequence–based amplification assay for the rapid detection of Salmonella spp. from food. Braz J Microbiol 50:255. https://doi.org/10.1007/S42770-018-0002-9
Zhang Y, Hu X, Wang Q (2020) Sensitive and specific detection of E. coli, Listeria monocytogenes, and Salmonella enterica serovar Typhimurium in milk by microchip electrophoresis combined with multiplex PCR amplification. Microchem J 157:104876. https://doi.org/10.1016/J.MICROC.2020.104876

MIX
Papier aus verantwortungsvollen Quellen
Paper from responsible sources
FSC® C105338

If you have any concerns about our products,
you can contact us on
ProductSafety@springernature.com

In case Publisher is established outside the EU,
the EU authorized representative is:
Springer Nature Customer Service Center GmbH
Europaplatz 3, 69115 Heidelberg, Germany

Printed by Libri Plureos GmbH
in Hamburg, Germany